Introduction to Flat Panel Displays

Wiley – SID Series in Display Technology

Display Systems: Design and Applications
Lindsay W. MacDonald and Anthony C. Lowe (Eds.)

Reflective Liquid Crystal Displays
Shin-Tson Wu and Deng-Ke Yang

Colour Engineering: Achieving Device Independent Colour
Phil Green and Lindsay MacDonald (Eds.)

Display Interfaces: Fundamentals and Standards
Robert L. Myers

Digital Image Display: Algorithms and Implementation
Gheorghe Berbecel

Flexible Flat Panel Displays
Gregory Crawford (Ed.)

Polarization Engineering for LCD Projection
Michael G. Robinson, Jianmin Chen, and Gary D. Sharp

Fundamentals of Liquid Crystal Devices
Deng-Ke Yang and Shin-Tson Wu

Introduction to Microdisplays
David Armitage, Ian Underwood, and Shin-Tson Wu

Mobile Displays: Technology and Applications
Achintya K. Bhowmik, Zili Li, and Philip Bos (Eds.)

Photoalignment of Liquid Crystalline Materials: Physics and Applications
Vladimir G. Chigrinov, Vladimir M. Kozenkov, and Hoi-Sing Kwok

Projection Displays, Second Edition
Mathew S. Brennesholtz and Edward H. Stupp

Introduction to Flat Panel Displays
Jiun-Haw Lee, David N. Liu, and Shin-Tson Wu

LCD Backlights
Shunsuke Kobayashi, Shigeo Mikoshiba, and Sungkyoo Lim (Eds.)

Liquid Crystal Displays: Addressing Schemes and Electro - Optical Effects, Second Edition
Ernst Lueder

Transflective Liquid Crystal Displays
Zhibing Ge and Shin-Tson Wu

Liquid Crystal Displays: Fundamental Physics and Technology
Robert H. Chen

OLED Displays: Fundamentals and Applications
Takatoshi Tsujimura

Interactive Displays
Achintya K. Bhowmik

Illumination, Color and Imaging: Evaluation and Optimization of Visual Displays
P. Bodrogi, T. Q. Khan

3D Displays
Ernst Lueder

Addressing Techniques of Liquid Crystal Displays
Temkar N. Ruckmongathan

Flat Panel Display Manufacturing
Jun Souk, Shinji Morozumi, Fang-Chen Luo, and Ion Bita

Modeling and Optimization of LCD Optical Performance
Dmitry A. Yakovlev, Vladimir G. Chigrinov, and Hoi-Sing Kwok

Physics and Technology of Crystalline Oxide Semiconductor CAAC-IGZO: Fundamentals
Noboru Kimizuka, Shunpei Yamazaki

Physics and Technology of Crystalline Oxide Semiconductor CAAC-IGZO: Application to LSI
Shunpei Yamazaki, Masahiro Fujita

Physics and Technology of Crystalline Oxide Semiconductor CAAC-IGZO: Application to Displays
Shunpei Yamazaki, Tetsuo Tsutsui

Introduction to Flat Panel Displays

Jiun-Haw Lee
National Taiwan University
Taipei City, Taiwan

I-Chun Cheng
National Taiwan University
Taipei City, Taiwan

Hong Hua
University of Arizona
Arizona, USA

Shin-Tson Wu
University of Central Florida
Florida, USA

Second Edition

This edition first published 2020

Edition History:
1e Wiley, 2008

Registered Offices
John Wiley & Sons, Inc., 111 River Street, Hoboken, NJ 07030, USA
John Wiley & Sons Ltd, The Atrium, Southern Gate, Chichester, West Sussex, PO19 8SQ, UK

Editorial Office
The Atrium, Southern Gate, Chichester, West Sussex, PO19 8SQ, UK

For details of our global editorial offices, customer services, and more information about Wiley products visit us at www.wiley.com.

Wiley also publishes its books in a variety of electronic formats and by print-on-demand. Some content that appears in standard print versions of this book may not be available in other formats.

Library of Congress Cataloging-in-Publication Data

Names: Lee, Jiun-Haw, author. | Cheng, I-Chun, 1974- author. | Hua, Hong, 1973- author. | Wu, Shin-Tson, author.
Title: Introduction to flat panel displays / Jiun-Haw Lee, National Taiwan University, Taipei City, Taiwan, I-Chun Cheng, National Taiwan University, Taipei City, Taiwan, Hong Hua, University of Arizona, Arizona, USA, Shin-Tson Wu, University of Central Florida, Florida, USA.
Description: Second edition. | Hoboken, NJ : John Wiley & Sons, Inc., [2020] | Series: Wiley-SID series in display technology | Includes bibliographical references and index.
Identifiers: LCCN 2020004365 (print) | LCCN 2020004366 (ebook) | ISBN 9781119282273 (cloth) | ISBN 9781119282198 (adobe pdf) | ISBN 9781119282228 (epub)
Subjects: LCSH: Flat panel displays.
Classification: LCC TK7882.I6 L436 2020 (print) | LCC TK7882.I6 (ebook) | DDC 621.3815/422—dc23
LC record available at https://lccn.loc.gov/2020004365
LC ebook record available at https://lccn.loc.gov/2020004366

Cover Design: Wiley
Cover Image: Yuichiro Chino/Getty Images

Set in 10/12pt WarnockPro by SPi Global, Chennai, India

10 9 8 7 6 5 4 3 2 1

Contents

Series Editor's Foreword

The first edition of *Introduction to Flat Panel Displays* has proved to be a popular and valued resource, which has been widely used both as a textbook and for reference. However, it was published over a decade ago in 2008, and established readers of the SID book series will not need reminding how fundamentally the subject matter has changed in that time. It is worth recalling that 2008 is also reported to be the first year in which worldwide sales of LCD televisions exceeded those of CRT sets.

Continuing demand for the first edition demonstrates that there is still a need for a broad-based introductory but authoritative account of flat panel displays, and it followed that the editors of the present book should consider writing a new and revised edition. It soon became clear though, that a simple revision would not be sufficient, and the volume you are holding represents a comprehensively updated and rewritten book which reflects the present state and latest developments in flat panel display technologies and applications. In order to provide the reader with a book which is a reasonable size and properly focused on contemporary topics, chapters in the first edition which described display technologies of lesser current importance – plasma and field emission devices – have been dropped. Important new chapters have been added on topics which are now central to flat panel applications: near-eye displays, reflective/e-paper displays and touch panel devices. The chapters describing the well-established, dominant display technologies such as LCDs, OLEDs, and LEDs have been comprehensively revised and updated to reflect the full range of technologies used in commercial displays and to describe the most recent important advances in these devices. Chapters describing AM backplane devices and structures, and the key principles of vision and color science have likewise been thoroughly updated to reflect their evolution and importance. Each chapter has been authored by an expert in display science, and the enthusiasm of the writers for their subjects is evident in their work. The authors' work in preparing this new edition has been virtually the same as writing a new book from the beginning, and I am grateful to all of them for their persistence and dedication to the task.

Flat panel display technology has revolutionized the ways in which we interact with electronic systems and through that, shapes the way we lead our lives. The pace of innovation shows no signs of slowing and the new cohorts of scientists and engineers who take the subject forward will need a range of training and reference works. Providing these resources is a key objective of the SID book series, and I believe that the present volume will make an important contribution to this aim.

Ian Sage
Great Malvern

1

Flat Panel Displays

1.1 INTRODUCTION

Displays provide a man–machine interface through which information can be passed to the human visual system. The information may include pictures, animations, and movies, as well as text. One can say that the most basic functions of a display are to produce, or re-produce, colors and images. The use of ink to write, draw, or print on a paper as in a painting or a book might be regarded as the longest established display medium. However, the content of such a traditional medium is static and is typically difficult or impossible to modify or update. Also, a natural or artificial source of light, is needed for reading a book or viewing a picture. In contrast, there are now many electronic display technologies, which use an electronic signal to create images on a panel and stimulate the eyes. In this chapter, we first introduce flat panel display (FPD) classifications in terms of emissive and non-emissive displays, where non-emissive displays include both transmissive and reflective displays. Then, specifications of FPDs will be outlined. Finally, the FPD technologies described in the later chapters of this book will be briefly introduced.

Displays can be subdivided into emissive and non-emissive technologies. Emissive displays emit light from each pixel which forms a part of the image on the panel. On the other hand, non-emissive displays modulate light by means of absorption, reflection, refraction, and scattering, to display colors and images. For a non-emissive display, a light source is needed. Such non-emissive displays can then be further classified into transmissive and reflective types. In historical terms, one of the most successful technologies for home entertainment has been the cathode ray tube (CRT), which enabled the widespread adoption of television (TV). It exhibits the advantages of being self-emissive and offering wide viewing angle, fast response, good color saturation, long lifetime, and good image quality. However, one of its major disadvantages is its size and bulk. The depth of a CRT is roughly equal to the length or width of the panel. For example, for a 19 in. (38.6 cm × 30.0 cm) CRT with aspect ratio of 4 : 3 the depth of a monitor is about 40 cm. Hence, it is hardly portable; its bulky size and heavy weight limit its applications.

In this book, we introduce various types of FPDs. As the name implies, these displays have a relatively thin profile, several centimeters or less, which is largely independent of the screen diagonal. Specifying a display or the design and optimization of a display-based product require selection of an appropriate technology, and are strongly dependent on the application and intended conditions of use. These issues, together with the intense pace of FPD development, which has made available many options and variations of the different display types, have made a thorough understanding of displays essential for product engineers. The options can be illustrated by some typical examples. For instance, the liquid crystal display (LCD) is presently the dominant FPD technology and is available with diagonal sizes ranging from less than 1 in. (microdisplay) to over 100 in. Such a display is usually driven by thin-film-transistors (TFTs). The liquid crystal cell acts as a light modulator which does not itself emit light. Hence, a backlight module is usually used behind a transmissive LCD panel to form a complete display module. In most LCDs, two crossed polarizers are employed which can provide a high contrast ratio. However, the use of polarizers limits the maximum optical transmittance to about 35–40%, unless a polarization conversion scheme is implemented. Moreover, at oblique angles the

Introduction to Flat Panel Display, Second Edition. Jiun-Haw Lee, I-Chun Cheng, Hong Hua, and Shin-Tson Wu.

optical performance of the assembly is degraded by two important effects. Firstly the projections of optic axes of two crossed polarizers onto the E vector of the light are no longer perpendicular to each other when light is incident at an oblique angle, so it is difficult to maintain a good dark state in the display over a wide viewing cone. Secondly, the liquid crystal (LC) is a birefringent medium, which means that electro-optic effects based on switching an LC are dependent on the relative directions of the incident light and the LC alignment in the cell. Hence, achieving a wide viewing angle and uniform color rendering in an LCD requires special care. To achieve wide-view, multi-domain architectures and phase compensation films (either uniaxial or biaxial) are commonly used; one for compensating the light leakage of crossed polarizer at large angles and another for compensating the birefringent LC layer. Using this phase compensation technique, transmissive multi-domain LCDs exhibit a high contrast ratio, high resolution, crisp image, vivid colors (when using quantum dots or narrow-band light emitting diodes), and a wide viewing angle. It is still possible for the displayed images to be washed out under direct sunlight. For example, if we use a smartphone or notebook computer in the high ambient light conditions found outdoors in clear weather, the images may not be readable. This is because the reflected sunlight from the LCD surface is much brighter than that transmitted from the backlight, so the ambient contrast ratio is greatly reduced. A broadband anti-reflection coating and adaptive brightness control help improve the sunlight readability.

Another approach to improve sunlight readability is to use reflective LCDs [1]. A reflective LCD uses ambient light to illuminate the displayed images. It does not need a backlight, so its weight, thickness, and power consumption are reduced. A wrist watch is such an example. Most reflective LCDs have inferior performance compared to transmissive ones in terms of contrast ratio, color saturation, and viewing angle. Moreover, in fully dark conditions a reflective LCD is not readable at all. As a result, its application is rather limited.

To overcome the sunlight readability issue while maintaining high image quality, a hybrid display termed a transflective liquid crystal display (TR-LCD) has been developed [2]. In a TR-LCD, each pixel is subdivided into two sub-pixels which provide, respectively, transmissive (T) and reflective (R) functions. The area ratio between T and R can be adjusted depending on the applications. For example, if the display is mostly used out of doors, then a design which has 80% reflective area and 20% transmissive area might be used. In contrast, if the display is mostly used indoors, then we can use 80% transmissive area and 20% reflective area. Within this TR-LCD family, there are various designs: double cell gap versus single cell gap, and double TFTs versus single TFT. These approaches attempt to solve the optical path-length disparity between the T and R sub-pixels. In the transmissive mode, the light from the backlight unit passes through the LC layer once, but in the reflective mode the ambient light traverses the LC medium twice. To balance the optical path-length, we can make the cell gap of the T sub-pixels twice as thick as that of the R sub-pixels. This is the dual cell gap approach. The single cell gap approach, however, has a uniform cell gap throughout the T and R regions. To balance the different optical path-lengths, several approaches have been developed, e.g. dual TFTs, dual fields (providing a stronger field for the T region and a weaker field in the R region), and dual alignments. Although TR-LCDs can improve sunlight readability, the fabrication process is much more complicated and the performance inferior to transmissive devices. Therefore, TR-LCD has not been widely adopted in products.

Light-emitting diodes (LEDs) consist of a semiconductor p–n junction, fabricated on a crystalline substrate. Under a forward bias, electrons and holes are injected into the device where they recombine and emit light. The emission wavelength of the LED is determined by the bandgap of the semiconductor. For longer wavelength (such as red and yellow) emission, an AlGaInP-based semiconductor is needed. Three group III (Al, Ga, and In) and one group IV (P) atoms are needed to allow tuning of the emission wavelength and lattice-matching to the substrate (e.g. GaAs). However, for shorter wavelength (green and blue) emission, it was not easy to find a lattice-matched substrate. Besides, there were other technological difficulties in fabricating nitride-based LEDs such as p-type doping and InGaN growth. In recognition of their successful demonstration of the InGaN-based blue LED, Professor Isamu Akasaki, Professor Hiroshi Amano, and Professor Shuji Nakamura were awarded the Nobel Prize in Physics in 2014. By combining the blue LED with phosphors, white emission is possible from a single chip. LEDs have been used for many display and lighting applications, such as traffic lights, very large diagonal (over 100 in.) signage, backlights of LCD, and general lighting, due to their long lifetime and high efficiency. A detailed description of LEDs from the viewpoints of materials, devices, fabrication, and applications will be presented in Chapter 5.

In Chapter 6, organic light-emitting devices (OLEDs) will be introduced. The operating principle of OLEDs is quite similar to that of the LED. It is also an electroluminescence (EL) device, but fabricated from organic materials rather than a semiconductor. In contrast to LEDs, it is not necessary to fabricate OLEDs on a crystalline substrate. From the manufacturing viewpoint, the OLED is similar to an LCD because it can be fabricated on a very large glass substrate. Apart from the usual glass substrate, OLEDs can be also fabricated on a flexible substrate if suitable processes are used. The device structure of the OLED is quite simple, comprising a stack of thin organic layers (~200 nm) sandwiched by anode and cathode electrodes. When transparent conductors are used for both the anode and cathode, a transparent display can be fabricated, while a metallic cathode layer can provide a mirror-like appearance. When the OLED is not activated the panel appears highly reflective, while information displayed on the OLED is superimposed on the mirror-like background. In addition to displays, OLEDs can provide a flat, large-area, and diffuse light source for general illumination. This is quite different from LED lighting which provides a point source and highly directional emission of light.

In Chapter 7, the basic working principles of several reflective display technologies, including electrophoretic displays, reflective liquid crystal displays, interferometric modulator displays and electrowetting displays, will be reviewed. These reflective displays do not require an internal light source. They possess some attractive features, providing low eye strain, low power consumption, and excellent optical contrast under high ambient light levels, and are favored for portable reading applications and for outdoor use. Some reflective displays require the image being displayed to be constantly refreshed, while some are bistable and retain the image without power. In bistable displays, energy is only consumed during switching operations. In addition, some have a video-rate switching capability, while others are more suitable for displaying still images. Today most monochrome reflective display technologies match the typical contrast ratio standard of 10 : 1 for printed images on paper, but the reflectance of their bright states are still less than the typical value of 80% for white paper. Many color reflective displays rely on color filters or side-by-side pixel subdivision. However, to achieve color images with good brightness and saturation, multiple colors within the same pixel area is desirable.

By fabricating a display on a flexible substrate rather than rigid glass, flexible displays (using technologies including LCD, OLED, and electrophoretic effects) can be fabricated with the advantages of being thin, robust, and lightweight.

Most FPDs have been developed to provide a format for direct-view applications, such as TVs, computer monitors, laptop screens, tablets, and smart phones. However, several FPD technologies including LCDs and OLEDs, can readily be made into microdisplays with panel sizes less than 1 in. and pixel sizes of tens of microns or less. Such microdisplays are not suitable for direct-viewing, but they have found applications in an emerging class of head-mounted displays (HMDs) which are key enablers for virtual reality and augmented reality systems. In Chapter 8, the working principles and recent development of head-mounted displays will be reviewed. Unlike a direct-view display, an HMD system requires an optical system to collect light from a microdisplay source and couple it into the viewer's eye. The system may use a single microdisplay and optical system to display a two-dimensional image to one eye, yielding a monocular information display. Alternatively, it may be configured with a microdisplay and viewing optics for each eye, yielding a binocular system with the capability of rendering stereoscopic views. In some of the most advanced HMD systems, each set of optics may be capable of rendering light fields which replicate the configuration of light rays originating from a real scene, enabling a true 3D viewing experience. The proximity of an HMD system to the eye allows it to be configured into one of two different types – either an immersive or a see-through display. An immersive HMD blocks a user's view of the real world and places the user in a purely computer-rendered virtual environment, creating the immersive visual experience known as virtual reality. A see-through HMD, on the other hand, blends views of the real world and a computer-rendered digital environment, creating an experience variously known as augmented reality, mixed reality or increasingly as spatial computing. Chapter 8 will start with a brief introduction to the optical principles of HMD systems and an overview of historical developments, then follow with a brief review of the human visual system parameters critical to the design of an HMD system. It will then review paraxial optical specifications, common miniature display sources, optical principles and architectures, summarize optical design methods and optical performance specifications critical to HMD system design, and the chapter concludes with a review of several emerging HMD technologies with advanced capabilities, such as eyetracking, addressable focus cues, occlusion capability, high dynamic range, and light field rendering.

A touch panel (TP) is not a "flat panel display." However, it provides an intuitive interface which provides input to the machine, and provides an enhancement to many displays which is critical to their application. In some cases, a single touch sensing function is enough, such as in an automatic teller machine (ATM). On the other hand, a multi-touch function is needed for controlling many mobile devices (such as mobile phones and tablet computers). Usually, electrical parameters (such as resistance or capacitance values) of the TP are changed by touch and the x–y positions at which these changes occur provide the input function. So, a TP must be transparent to allow mounting on top of the display, and a separate TP increases the thickness of the display module. Integration of the TP and the display can reduce the module thickness. TP technologies will be introduced in Chapter 9.

1.2 EMISSIVE AND NON-EMISSIVE DISPLAYS

Both emissive and non-emissive FPDs have been developed. In emissive displays, each pixel emits light with a different intensity and color which stimulate the human eyes directly. CRTs, LED panels, and OLEDs are emissive displays. When the luminance of the panel viewed from different directions is constant, the device is called a Lambertian emitter and this represents an ideal performance for an emissive display because it results in a wide viewing angle performance. Due to the self-emissive characteristics, it can be used in conditions of very low ambient light. When such displays are turned off, they are completely dark (ignoring any ambient reflections). Hence, the display contrast ratio (see also Section 1.3.3) under low ambient light can be very high. On the other hand, displays which do not emit light themselves are called non-emissive displays. An LCD is a non-emissive display in which the liquid crystal molecules in each pixel work as a light switch, independently of the other pixels. An external voltage reorients the LC director which controls an optical phase retardation. As a result, light incident from the backlight unit or from the ambient is modulated. Most high-contrast LCDs use two crossed polarizers. The applied voltage controls the transmittance of the light through the polarizers. If the light source is behind the display panel, the display is termed a *transmissive* display. On the other hand, it is also possible to use the ambient light as the illumination source, imitating the principle of traditional media, such as reading a book, and the device is then called a *reflective* display. Different technologies for reflective displays such as electrophoretic, interferometric modulators, and electrowetting displays as well as LCDs will be introduced in Chapter 7. Since no extra light source is needed in a reflective display, its power consumption is relatively low. Under high ambient light conditions, images on emissive displays and transmissive LCDs can be washed out. In contrast, reflective displays exhibit a higher luminance as the ambient light increases. However, they cannot be used in dark conditions. Hence, transflective LCDs have been developed, which will be described in Chapter 4.

1.3 DISPLAY SPECIFICATIONS

In this section, we introduce some specifications which are generally used to describe and evaluate FPDs in terms of their mechanical, electrical, and optical characteristics. FPDs can be smaller than 1 in. for projection displays, 2–6 in. for cell phones, 7–9 in. for car navigation, ~8–20 in. for tablets and notebooks, ~10–25 in. for desktop computers, and ~30–110 in. for direct-view TVs. For different FPDs, their requirements for pixel resolution also differ. Luminance and color are two important characteristics which directly affect the display performances. Dependencies of these two parameters on viewing angle as well as image uniformity, device lifetime, and response time should be addressed when describing the performances of a FPD. Contrast ratio is another important parameter, which strongly depends on the ambient environment.

1.3.1 Physical Parameters

The basic physical parameters of a FPD include the display size, aspect ratio, resolution, and pixel format. The size of a display is typically specified by the diagonal length, in units of inches. For example, a 15 in. display indicates that the diagonal of the viewable area is 38.1 cm. Display formats, include landscape, square, and portrait types, corresponding to display widths larger than, equal to, and smaller than the height, respectively. Most monitors and TVs use a landscape format with the width-to-height ratio, also called the "aspect ratio," of 4 : 3, 16 : 9, or 16 : 10, typically.

FPDs typically provide a rectangular "dot matrix" of addressable pixels which can display images and characters. To increase image quality, one may use more pixels in a display. Table 1.1 lists some standard resolutions of FPDs. For example, video graphics array (VGA) indicates a display 640 pixels in width and 480 in height. The aspect ratio is 4 : 3. Higher resolution typically (but not necessarily) provides better image quality. The HD series includes several wide screen standards with an aspect ratio of 16 : 9. FHD has a resolution of 1920 × 1080, which may be abbreviated as 2K1K. Doubling the pixels count in columns and rows results in 4× the resolution, which is termed UHD, 4K2K, or 4K. Similarly, an 8K standard is proposed with still higher resolution. Once the resolution, display size, and aspect ratio are known, one may obtain the pitch of pixels. For example, a 5.5 in. display with aspect ratio of 16 : 9 and FHD resolution has a pitch of ~63 μm. Or, we can use pixel per inch (ppi) to describe the pixel density of the display. The above example corresponds to ~401 ppi.

In the case of an HMD system for VR or AR applications, a microdisplay source is used. Although the pixel resolution of the microdisplay is a critical contributor to the system performance, the image resolution perceived by the viewer also depends on the optical magnification of the viewing optics. For instance in an HMD system, a VGA resolution microdisplay can produce an image with an apparent angular resolution equivalent to or better than an image provided via a FHD microdisplay if the optical magnification to the VGA panel is substantially lower than that to the FHD panel, this angular resolution being traded off against the field of view of the image. More detailed discussion on the resolution metrics of HMD systems can be found in Chapter 8.

Note that not all of the panel area contributes to the displayed image; the active area of each pixel is normally surrounded by a small inactive area occupied by inter-electrode gaps and possibly other structures such as stray light barriers. One can define the "fill factor" or "aperture ratio" as the ratio of the active display area in a pixel over the whole pixel size, with its maximum value of 100%. Also, for a full-color display, at least three primary colors are needed to compose a color pixel. Hence, each color pixel is divided into three subpixels, red, green, and blue (RGB) which share the total pixel area. For example, if we assume that a color pixel has

Table 1.1 Resolution of FPDs.

Abbreviation	Full name	Resolution
VGA	Video graphics array	640 × 480
SVGA	Super video graphics array	800 × 600
XGA	Extended graphics array	1024 × 768
HD	High definition	1280 × 720
FHD	Full high definition	1920 × 1080
UHD (4K)	Ultra-high definition	3840 × 2160
8K		7680 × 4320

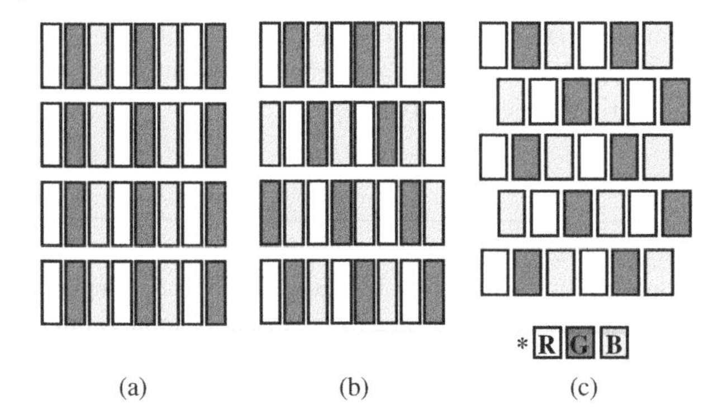

Figure 1.1 Subpixel layout of a FPD: (a) stripe, (b) mosaic, and (c) delta configurations.

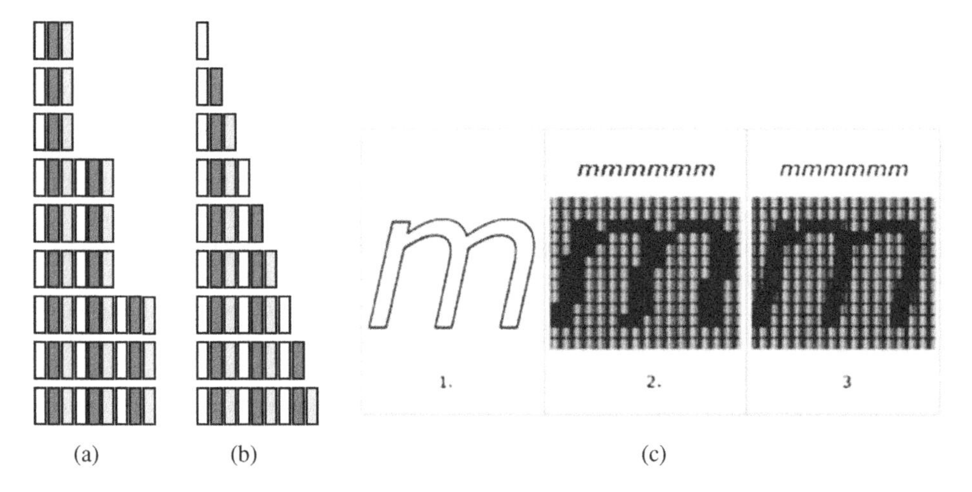

Figure 1.2 (a) White (red + green + blue) pixels lit-on at the edge of the slope, and (b) with subpixel rendering in a stripe configuration. (c) "*m*" in italic (2) without and (3) with subpixel rendering on a display [3].

a size of 63 μm × 63 μm, then the dimension of each subpixel will be 21 μm × 63 μm. If the area of each active, switchable sub-pixel which contributes to light emission or transmission is 18 μm × 60 μm, then the fill factor will be ~82%.

Different layouts of RGB subpixels are possible, as shown in Figure 1.1. A stripe configuration, is straightforward and makes fabrication and driving circuit design relatively easy. However, for a given display area and resolution it provides a poor color mixing performance. Both mosaic and delta configurations make the fabrication process and/or the driving circuits more complicated, but the resulting image quality is higher because of their better color mixing capabilities.

When displaying an oblique black-on-white pattern on a display with a stripe subpixel configuration as shown in Figure 1.2, a clear sawtooth can be seen at the edge. However, because each pixel is formed of three subpixels, these can be switched on in a controlled sequence from the top to the bottom such that edge of the pattern appears smoother – a technique called "subpixel rendering." [3] Obviously, the colors at the edges of some rows are no longer white. For the first and the fourth rows, the red subpixel is switched on at the edge while for the second and the fifth rows, red and green emission results in a yellow color at the edge. This is called a "color fringing artifact." Figure 1.3 shows the letter "*m*" in italic, without and with subpixel rendering. A smoother edge can be clearly seen when subpixel rendering is used. Advanced sub-pixel rendering algorithms not only switch different sub-pixels on or off at an oblique edge, but also adjust their luminance values to optimize the visual quality of the image.

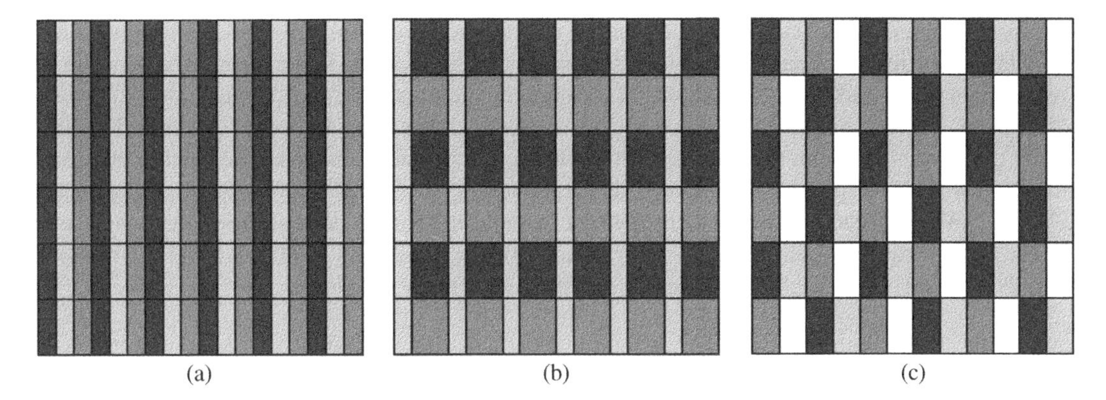

Figure 1.3 (a) stripe, (b) PenTile™ RGRB, and (c) PenTile RGBW configurations [3].

There are three kinds of photoreceptor cells in human eyes, which respond to long, medium, and short wavelength regions of the visible spectrum. That is the major reason we use red, green, and blue as three primaries for the display and will be discussed further in Chapter 2. The arrangement of the photoreceptor cells does not correspond to a stripe configuration. Besides this, the numbers for different types of cell are not the same. The PenTile™ configuration has been proposed to mimic the layout of different photoreceptors in the eye to achieve better color mixing [4]. Here, "Pen" is a contraction of the Greek prefix "penta" meaning five and indicates that five subpixels form a pixel. There are many possible formats; Figure 1.3b shows one of them which is called the RGBG format. Combined with the subpixel rendering technique, a high display resolution can be obtained with a larger subpixel size. Figure 1.3a,b show the stripe and PenTile layouts with the same resolution. One can see that PenTile configuration allows larger red and blue subpixels. This is important for some displays, such as OLEDs, because it is not easy to reduce the subpixel size during fabrication. Hence, the PenTile arrangement can relax the design rules required in manufacture of displays for a given resolution. Also for OLED displays, the lifetime of the blue sub-pixel is an issue and by enlarging the blue emitting region its current density can be reduced and its lifetime extended. Another type of PenTile pattern is the RGBW arrangement, shown in Figure 1.3c. Here, a white subpixel is added alongside the three primary colors. In some displays such as LCD, where different colors are obtained by filtering unwanted colors from a white backlight, the blocked light is responsible for a major loss in efficiency. With the introduction of the white subpixel, the efficiency can be raised.

A device in which the pixel density is increased to the point where the human eye cannot resolve the individual pixels, is called a "retina display." This implies a very high resolution so that when projected onto the retina of the eye, the pixel density is higher than that of the photoreceptors in the retina. Evidently, a higher ppi is required for displays such as phones which are used closer to the eye, in order to satisfy the requirement for a retina display. Typically, ~300 ppi is required for a phone with a typical viewing distance of ~30 cm. With larger viewing distances such as those normal when watching a TV, a larger pixel is acceptable in a retina display. A detailed illustration of retina displays will be presented in Chapter 2.

1.3.2 Brightness and Color

Luminance and color gamut are two of the most important optical characteristics of a FPD. A display with high luminance looks dazzling under dark conditions. On the other hand, a display with insufficient brightness appears washed out under high ambient light levels. Typically, the luminance of a FPD should be as bright as (or a little brighter than) real objects under the ambient light in which the display is used. In an ordinary indoor environment, a monitor has a luminance of 200–300 cd m^{-2}. For a large screen TV, a higher luminance (500–1000 cd m^{-2}) may be needed. A FPD is used to produce or reproduce colors, hence, the number of colors a FPD can display, and how closely the color displayed on a FPD matches that of the real object (color fidelity) are two important characteristics to judge display performance. Since the color on a FPD is produced by mixing together (at least) three primary colors, i.e. RGB, more "pure" (narrow-band) primaries result in a broader range of colors which can be displayed, which is called the "color gamut" (see Section 2.4.5). As well as the usual three primaries (red, green, and blue), more primaries (such as yellow and cyan) can be added into the subpixels and can further broaden the color gamut. Besides, with suitable design of the driving method, the power consumption of the display can be reduced simultaneously [5, 6]. The perceived range of brightness from dark to bright can be divided into equal steps defined by numbers with 2, 4, 8, or more bits, which are called "gray levels" or collectively a "gray scale" (see Section 2.4.3). For example, a FPD can display 16 million colors ($2^8 \times 2^8 \times 2^8 \sim 16.8\,\mathrm{M}$) when each RGB subpixel can show 256 (8 bit) gray levels.

1.3.3 Contrast Ratio

The device contrast ratio (CR) of a FPD is defined as:

$$CR = L_\mathrm{w}/L_\mathrm{b} CR = L_w/L_b, \tag{1.1}$$

where L_w and L_b are the luminances of the white and black states, respectively. A higher CR requires a higher on/off ratio and hence potentially better image quality and higher color saturation. When CR is equal to or below 1, the information content of a FPD is lost or inverted. For most emissive displays, the off-state luminance is zero. Hence, the contrast ratio is infinite under perfectly dark viewing conditions. However, under ambient light conditions, surface reflections from the display mean that Eq. (1.1) should be modified to:

$$\mathrm{CR_A} = (L_\mathrm{w} + L_\mathrm{ar})/(L_\mathrm{b} + L_\mathrm{ar}), \tag{1.2}$$

where $\mathrm{CR_A}$ stands for the ambient contrast ratio, and L_ar is the luminance from ambient reflection. From Eq. (1.2), as the ambient reflection increases, $\mathrm{CR_A}$ decreases sharply. To maintain a high ambient contrast under increasing ambient light levels, one can: (i) increase the on-state luminance, and (ii) reduce the reflectance of the display surface. However, under a very high ambient such as outdoor sunlight, luminance from the direct sun is 4 orders of magnitude higher than a FPD, which severely washes out the information content of any emissive or transmissive FPD. Sunlight readability is an important issue especially for mobile displays. On the other hand, an adequate ambient light is required for viewing conventional media such as a book or newspaper. A similar consideration applies to reflective displays, such as reflective LCDs.

1.3.4 Spatial and Temporal Characteristics

Uniformity of a FPD refers to any unwanted change in the luminance and color over a display area. Human eyes are sensitive to luminance and color differences. For example, a 5% luminance difference is noticeable between two adjacent pixels. In the case of a gradual change, human eyes can tolerate up to 20% luminance change over the whole display.

Optical characteristics (luminance and color) may also change at different viewing angles. For Lambertian emitters, such as CRTs, the viewing angle performance is quite good. The emission profile of LEDs and OLEDs can be engineered by packaging and optimizing their layer structure, respectively. However, the viewing angle of LCDs requires careful attention because LC materials are birefringent and crossed polarizers are no longer crossed when viewed at oblique angles. There are several ways to quantify the viewing angle of a FPD. For example, one may masure the viewing cone with: (i) a luminance threshold, (ii) a minimum contrast ratio, say 10 : 1, or (iii) a specified maximum value of color shift. In some cases the contrast ratio viewed at oblique angle can be smaller than 1, resulting in "gray level inversion."

Response time is another important metric. If a FPD has a slow response time, one may see blurred images of fast moving objects. By switching the pixel from "off" to "on" and from "on" to "off," one can obtain rise and fall times, respectively which are typically specified between 10% and 90% luminance levels,. One may also define the response time from one to another gray level – the so-called "gray-to-gray" (GTG) response time. Most displayed scenes contain extensive areas of different luminance pixels, so the GTG response time is more meaningful. For LCDs, this GTG response time may be much longer than the black-white rise and fall time [7]. The TFT matrix used to address many FPDs provides a voltage set-and-hold function. This is quite different from the CRT's impulse type response, and requires a different metric to characterize it. Therefore, a motion picture response time (MPRT) [8] is commonly used to define the response time of a TFT LCD, which will be further discussed in Chapters 2 and 4.

After a long period of operation, the luminance of a FPD (especially for emissive display) may decay. If a fixed pattern is displayed on an emissive panel for a long period of time, then all the pixels are turned on to display a blank white screen, one can see a "ghost image" of the fixed pattern displayed with a lower brightness, which is called a "residual image" or "burn-in." As mentioned before, human eye can detect less than 5% nonuniformity between two adjacent pixels. Hence the lifetime of a FPD is crucial for static images. An alternative solution is to use only moving pictures, rather than static images for information display. Then the luminances of all the pixels decay uniformly, since the average on time for all pixels is the same.

1.3.5 Efficiency and Power Consumption

Power consumption is a key parameter, especially for mobile displays, as it affects the battery life. For displays with wall-plug electrical input, lower power consumption implies a lower heat generation, which means heat dissipation is easier and "green" environment targets can be met more easily. Typically, one uses the unit lm/W to describe power efficacy of a FPD (see Section 2.2). A portable display with lower power consumption leads to a longer battery life. For notebooks and TVs, a high optical efficiency also translates into less heat dissipation and a lower electricity bill. Thermal management in a small chassis device is an important issue. Energy Star is a program which defines the "power consumption" for electronic products, such as displays (https://www.energystar.gov). For example, in Energy Star Display Specification 7.1 (released in April 2017), the maximum power consumption of the display under on-state operation is defined, which is related to the screen area and the maximum luminance of the display. For example, the maximum power consumption of a 60 in. TV with aspect ratio 4 : 3 and maximum luminance 500 cd/m^2 should be less than 144 W.

1.3.6 Flexible Displays

A FPD is usually fabricated on thin glass plates which provide a rather rigid substrate. On the other hand, conventional media are printed on paper, which is flexible. An important current research and development theme is fabrication of FPDs on a flexible substrate, to provide a conformable or "paper-like" display [9]. Compared to glass-based FPDs, flexible displays are thin and lightweight. In addition, flexible displays can potentially be fabricated by a roll-to-roll process at low cost. Potential substrates for flexible FPDs include ultra-thin glass, plastic, and stainless steel. A bendable ultra-thin glass substrate is possible, but the cost is high. Plastic substrates are suitable for flexible displays, but the highest temperature which can be tolerated in the manufacturing process is typically lower than 200 °C. A stainless steel substrate is bendable, and resistant to high temperatures, however, it is opaque and therefore not suitable for transmissive displays. There are many technical bottlenecks for flexible FPDs, such as material selection, fabrication processes, device configurations, display packaging and measurement.

1.4 APPLICATIONS OF FLAT PANEL DISPLAYS

The following sections briefly outline the applications of each technology. Details will be described in the related chapters.

1.4.1 Liquid Crystal Displays

Although LC materials were discovered more than a century ago [10, 11], their useful electro-optic effects and stable materials are developed only in late 1960s and 1970s. In the early stage, passive matrix LCDs were adopted in electronic calculators and wrist watches [12]. With the advance of thin film transistors [13], color filters [14], and low voltage LC effects [15], active matrix LCDs gradually penetrated into the market of notebook computers, desktop monitors, and televisions. Today, LCDs have found widespread uses in our everyday life, including smartphones, tablets, virtual reality and augmented reality displays, automotive displays, navigation systems, notebook computers, desktop monitors, and large screen TVs [16].

To satisfy this wide range of applications, three types of LCDs have been developed: transmissive, reflective, and transflective. Transmissive LCDs can be further separated into projection and direct-view devices. In a high-resolution smartphone display, the pixel size is around 30–40 μm. Thus, the TFT aperture ratio becomes particularly important because it limits the light throughput [17]. To enlarge the aperture ratio, poly-silicon (p-Si) TFTs are commonly used because their electron mobility is about two orders of magnitude higher than that of amorphous (a-Si) silicon. High mobility allows a smaller TFT to be used which, in turn, enlarges the aperture ratio. For a detailed structure of a TFT LCD, please see Figure 4.1.

For a large-sized LCD TV, say 65 in. diagonal, 16 : 9 aspect ratio, and 3840 × 2160 resolution, the pixel size is about 350 μm by 350 μm, which is much larger than that of a microdisplay. Thus, a-Si silicon is adequate although its electron mobility is relatively low. Amorphous silicon is easy to fabricate and has good uniformity. Thus, a-Si TFT dominates large screen LCD panel market.

Similarly, reflective LCDs can also be divided into projection and direct-view displays. In projection displays using Liquid-Crystal-on-Silicon (LCoS) [18], the pixel size can be as small as 4 μm because of the high electron mobility of crystalline silicon (c-Si). In an LCoS panel, the electronic driving circuits are hidden beneath the metallic (aluminum) reflector. Therefore, the aperture ratio can reach >90% and the displayed picture is quite smooth. On the other hand, most reflective direct-view LCDs use a-Si TFTs and a circular polarizer. Its sunlight readability is excellent, but it is not readable in a dark ambient. Therefore, a thin front light is needed for reflective direct-view LCDs.

To obtain a high quality transmissive display and good sunlight readability, a hybrid TR-LCD has been developed. In a TR-LCD, each pixel is divided into two sub-pixels: one for transmissive and another for reflective display [19]. In a dark to normal ambient, the backlight is turned on and the TR-LCD works as a transmissive display. Under direct sunlight, TR-LCD works in the reflective mode. Therefore, its dynamic range is wide and its functionality does not depend on the ambient lighting condition. TR-LCD can overcome sunlight readability issues, but its fabrication is much more complicated and the cost is higher than its transmissive counterpart. As a result, its application is limited. For a detailed discussion of TR-LCDs, please refer to Chapter 4.

1.4.2 Light-Emitting Diodes

The LED is an electroluminescent (EL) device based on crystalline semiconductors [20]. To convert electrical to optical power, one has to inject carriers into the LED through electrodes, which then recombine to give light. The emission wavelength is mainly determined by the semiconductor material, and can be fine-tuned by device design.

Since it is difficult to grow large size single crystals, the wafer diameter of LEDs is limited to about 8 in. After device processing, LEDs are diced from the wafer followed by a packaging process. The size of a single packaged LED is typically several millimeters, which means that the pixel size of an LED panel is large, and suitable for use in huge area displays. Due to their self-emissive characteristic, LEDs are commonly used for large displays, such as outdoor signage (mono color, multi-color, and full color), traffic signals, and general lighting to replace light bulbs. Compared to conventional devices using light bulbs, LED displays exhibit the advantages of lower power consumption, greater robustness, longer lifetime, and lower driving voltage (so safer). There are also many outdoor screens with diagonals over 100 in. which consist of millions of LED pixels.

Rather than providing a display itself, an LED can be also used as the light source, such as in a backlight module for an LCD, and for general lighting. Compared to the conventional cold cathode fluorescent lamp (CCFL), which resembles a thin fluorescent tube, LED exhibits a better color performance, longer lifetime, and faster response. Another important motivation to use LEDs in LCD backlighting is that the mercury in CCFLs is harmful to the environment. When using LEDs for general lighting applications, a broad emission spectrum is preferred to simulate natural light, such as sunlight, and obtain a faithful color rendering of the reflective objects (Section 2.4.6). This is quite different from the requirements for LED displays and LCD backlights, which usually need a narrow spectrum.

1.4.3 Organic Light-Emitting Devices

The OLED is also an EL device, like the LED, except its materials are organic thin films with amorphous structures [21]. Amorphous organic materials have a much lower carrier mobility (typically five order of magnitude lower) than crystalline semiconductors, which results in a higher driving voltage for OLEDs. Besides, the operational lifetime of OLEDs is one order of magnitude shorter than semiconductor LEDs. However, due to its amorphous characteristics, fabrication of large size panels (e.g. 55 in.) is possible.

Since the conductivity of amorphous organic materials is very low, very thin organic films (100–200 nm in total) are required to reduce the driving voltage to a reasonable value (i.e. <10 V). It is quite a challenge in thin film formation, especially for large size substrates. Several fabrication technologies have been proposed, such as physical vapor deposition, spin-coating, ink-jet printing, and laser-assisted patterning. OLEDs are widely used in display applications (such as TVs and mobile phones). Besides, OLEDs can be used in lighting applications [22, 23]. Two advantages of OLEDs are: (i) low process temperature, and (ii) compatibility with different substrate materials, which makes them suitable for flexible displays. One of the strategies for OLED development is to improve the device performance (especially driving voltage and lifetime) to match (or at least approach) those of LEDs. Due to their large size fabrication capability, the potential manufacturing cost of OLEDs is lower than LEDs. Because OLEDs have some advantages in performance and fabrication cost over LEDs, it has a chance to replace LEDs in some applications since they are both EL devices with similar operational principles.

1.4.4 Reflective Displays

A wide variety of reflective display technologies are available today. They are quite different in working principles and performance. Some of them, such as interferometric modulator displays, electrowetting displays and guest-host polymer dispersed liquid crystal displays exhibit fast response and are capable of video frame rate operation. However, most of them are still some way from commercial success because of their poor color gamut and relatively high power consumption for video rate operation. On the other hand, bistable reflective display technologies with sufficiently good reflectivity and contrast ratio, such as electrophoretic displays and cholesteric liquid crystal displays, are attractive for displaying (quasi)static images where a low switching speed is not a major concern. With the advantages of low power consumption and good outdoor readability, these reflective displays are suitable for portable reading devices, wearable or mobile devices and signage applications. For instance, electrophoretic technologies have been adopted in many e-book readers and electronic paper displays. Because many of the reflective display technologies can be made thin and flexible, they are suited to billboards, signage and shelf-edge labels. In the application of wearable or mobile devices, these low-power paper-like reflective display technologies have been incorporated into an electronic paper watch, electronic wristband, and similar devices. For detailed discussions of the above-mentioned reflective displays, please refer to Chapter 7.

1.4.5 Head-Mounted Displays

Head-mounted displays (HMDs), also known as head-worn or near-eye displays, are typically attached in close proximity to a user's eye and require an optical system to couple the light from a microdisplay source into the user's eye. The basic principles of an HMD system can be dated back to 1830s when Sir Charles Wheatstone proposed the concept of the stereoscope for viewing a pair of static photographs with slight disparities. Through over a century of technical development, the stereoscope has evolved into a new class of display technology enabling a new paradigm of applications. Instead of static photographs and simple mirrors, modern HMD systems enjoy a wide range of choices of electronic displays as the image sources, a wide range of advanced optics technologies in the optical viewer, and a wide range of sensing, computing, and communication capabilities.

A modern HMD system can be as simple as a microdisplay source plus a single magnifier-like eyepiece, providing a monocular display for information access and navigation. It can also be configured into a very sophisticated system integrating not only advanced microdisplays and optics but also a suite of advanced sensors and computing hardware and software, yielding a computing platform for advanced missions and visual experiences. Some advanced HMD systems go beyond the traditional route of displaying a 2D image or rendering a 3D perception of depth via binocular viewing, and create a true 3D viewing experience via light field rendering.

Over many decades of development, HMD technology has become a key enabler for virtual reality and augmented reality applications. To satisfy the needs of VR and AR applications, two types of HMDs have been developed: immersive and see-through. Immersive HMDs, primarily used for VR systems, block a user's view of the real world and immerse him or her in a purely computer-rendered virtual environment. See-through HMDs, mainly used for AR systems, blend views of the real world and a computer-rendered digital world digitally or optically. Both types of HMD technology share most of the same fundamental optical principles and requirements, but see-through HMDs, especially those providing optical see-through, confront many unique optical challenges. For instance, an optical combiner which combines the light paths of the real-world and virtual world views, plays a critical role in the architecture of optical see-through HMDs. It can be as simple as a beamsplitter or as sophisticated as a holographic waveguide.

The rapidly growing interest in VR and AR applications, the ever-increased bandwidth and accessibility of wireless networks, the miniaturization of electronics, and the ever-growing power of computers have collectively boosted the rapid development of HMD technologies in recent years. Please refer to Chapter 8 for a detailed discussion of the historic development, basic working principles, optical design fundamentals, and recent advances in head-mounted displays.

1.4.6 Touch Panel Technologies

When a TP is touched, an electrical, optical, or magnetic parameter is changed and the point of touch can be identified. As electrical signals, typically we can use changes of resistance or capacitance. A resistive TP consists of two substrates. The inner sides of the substrates are coated with transparent resistive layers, separated by an air gap. The outer substrate is deformable. When the resistive TP is touched, contact is made between the upper and lower conductive layers. The contact position affects the resistance value read out from the driving circuit. However, the air gap between the two substrates results in a low optical transmittance, which reduces the luminance from the display panel. An important use case is when the object which touches the TP is a finger. This can be regarded as equivalent to a capacitor connected to the ground which therefore changes the capacitance measured at the TP. That is the basic idea of the capacitive TP. With suitable design of vertical and horizontal electrodes on the substrate, self- and mutual-capacitance TPs can be obtained, respectively. Note that a conductive object such as a finger is needed to activate the touch function on a capacitive TP. When a TP is physically stacked on top of the display, it is called an "out-cell" configuration. To reduce the thickness of the TP-display module and simplify the fabrication process, on-cell and in-cell TPs have been introduced. Taking the LCD as an example, it consists of two glass substrates. By fabricating the TP onto the outer substrate, an on-cell configuration is created. Note that there is a dense array of TFTs and conductors on the bottom substrate of the TFT panel. With a suitable layout and driving scheme, a TP can be integrated inside the display, which is called the "in-cell" configuration.

References

1 Wu, S.T. and Yang, D.K. (2001). *Reflective Liquid Crystal Displays*. Wiley.
2 M. Okamoto, H. Hiraki, and S. Mitsui, U.S. Patent 6,281,952, Aug. 28 (2001).
3 Fang, L., Au, O.C., Tang, K., and Wen, X. (2013). Increasing image resolution on portable displays by subpixel rendering – a systematic overview. *APSIPA Trans. Signal Inform. Process.* 1: 1.
4 Brown Elliott, C.H., Credelle, T.L., Han, S. et al. (2003). Development of the PenTile Matrix™ color AMLCD subpixel architecture and rendering algorithms. *J. SID* 11 (/1): 89.
5 Cheng, H.C., Ben-David, I., and Wu, S.T. (2010). Five-primary-color LCDs. *J. Disp. Technol.* 6: 3.
6 Luo, Z. and Wu, S.T. (2014). A spatiotemporal four-primary color LCD with quantum dots. *J. Disp. Technol.* 10: 367.
7 Wang, H., Wu, T.X., Zhu, X., and Wu, S.T. (2004). Correlations between liquid crystal director reorientation and optical response time of a homeotropic cell. *J. Appl. Phys.* 95: 5502.
8 Song, W., Li, X., Zhang, Y. et al. (2008). Motion-blur characterization on liquid-crystal displays. *J. SID* 16: 587.
9 Crawford, G.P. (2005). *Flexible Flat Panel Displays*. Wiley.
10 Reinitzer, F. (1888). Beiträge zur kenntniss des cholesterins. *Monatsh. Chem.* 9: 421.
11 Lehmann, O. (1889). Über fliessende Krystalle. *Z. Phys. Chem.* 4: 462.
12 Ishii, Y. (2007). The world of liquid crystal display TVs-Past, Present and Future. *J. Disp. Technol.* 3: 351.
13 Lechner, B.J., Marlowe, F.J., Nester, E.O., and Tults, J. (1971). Liquid crystal matrix displays. *Prof. IEEE* 59: 1566.
14 Fischer, A.G., Brody, T.P., and Escott, W.S. (1972). Design of a liquid crystal color TV panel. In: *Proc. IEEE Conf. on Display Devices*, New York, NY, 64.
15 Schadt, M. and Helfrich, W. (1971). Voltage-dependent optical activity of a twisted nematic liquid crystal. *Appl. Phys. Lett.* 18: 127.
16 Liu, C.T. (2007). Revolution of the TFT LCD technology. *J. Disp. Technol.* 3: 342.
17 Stupp, E.H. and Brennesholtz, M. (1998). *Projection Displays*. New York: Wiley.
18 Armitage, D., Underwood, I., and Wu, S.T. (2006). *Introduction to Microdisplays*. Wiley.
19 Zhu, X., Ge, Z., Wu, T.X., and Wu, S.T. (2005). *J. Disp. Technol.* 1: 15.
20 Round, H.J. (1907). A note on carborundum. *Electr. World* 19: 309.
21 Tang, C.W. and Vanslyke, S.A. (1987). Organic electroluminescent diodes. *Appl. Phys. Lett.* 51: 913.
22 Iino, S. and Miyashita, S. (2006). Printable OLEDs promise for future TV market. *SID Symp. Dig.* 37: 1463.
23 Hirano, T., Matsuo, K., Kohinata, K. et al. (2007). Novel laser transfer technology for manufacturing large-sized OLED displays. *SID Symp. Dig.* 38: 1592.

2

Color Science and Engineering

2.1 INTRODUCTION

Display systems are used to produce and reproduce color images which makes the topic "color science and engineering" very important for evaluating their performance. Typically, the perception of colors can be treated as a four-stage process: (i) existence of a light source – either man made or natural, (ii) light-object interaction – such as reflection, absorption, and transmission, (iii) stimulation of the eyes, and (iv) recognition by the brain. Figure 2.1a, illustrates the human eye seeing the color of an object under sunlight, which is a "white" light source because its spectral bandwidth covers the entire visible range. If there was no light source, there would be no photons to stimulate the human eye and, therefore, no color could be formed. Under illumination, the object (e.g., paper in Figure 2.1a) absorbs a portion of the incident photons and reflects the rest. As shown in Figure 2.1b, there are yellow and green inks on the white paper. When incident white light illuminates the yellow ink, the "blue" component of the white light is most strongly absorbed. The reflected light contains a higher proportion of red and green wavelengths, resulting in a perception of yellow. Similarly, the green ink absorbs "red" and "blue" light. Where there is no ink, the white paper reflects all components of the white light almost equally, so it appears white. It follows from the above discussion, that the color of an object is also dependent on the spectral content of the incident light. For example, if the light source is red, then the yellow ink will have the same appearance as the white paper. After the light–object interaction, reflected photons are received by the detector; here it is a human eye. To properly describe a light wave, there are four basic parameters: intensity, wavelength, phase, and polarization. Photons with different emission wavelengths in the visible region (~380–780 nm) stimulate the photosensitive cells (cone and rod cells, as discussed later) of the eye which generate the perception of different colors, such as violet, blue, green, yellow, orange, and red. The light intensity gives the observer a perception of bright and dark. However, the human eye cannot resolve the polarization state and phase of the light.

In human eyes, in individuals with "normal"color vision, there are three different types of cone cells with different spectral sensitivities. This makes it possible to use three primary colors (red, green, and blue) to generate different (but not all) the colors and to describe the colors quantitatively [1]: this is called "trichromatic space." In 1931, the Commission Internationale de l'Eclairage (CIE) suggested the (X, Y, Z) colorimetric system, which can specify all the colors by their distinct coordinates, and also indicates the brightness of the target object [2]. It is a convenient system for describing colors. However, the CIE 1931 system is not suitable for discussing the magnitude of the perceived difference between two colors. Besides, the 1931 system is set up to quantify the colors of self-luminous objects without any ambient reference – which is not unrealistic for some display applications. To solve this problem, the uniform color spaces are proposed (e.g., CIE 1976 (L*u*v*)- and (L*a*b*)-spaces). In these systems, a numerical "color difference" can be specified for two colors. In different areas of the CIE 1976 color diagrams (for example, for two similar greenish colors or for two reddish colors), just-distinguishable color difference is nearly identical in magnitude. Since the trichromatic space can be quantitatively described by CIE colorimetric systems, different colors can be produced or reproduced in a display device by mixing three primary emitters. Although the reflection spectrum of a "real" object is

Introduction to Flat Panel Display, Second Edition. Jiun-Haw Lee, I-Chun Cheng, Hong Hua, and Shin-Tson Wu.

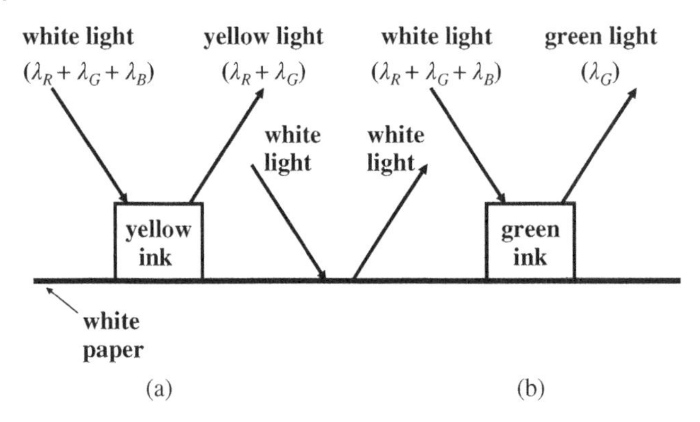

Figure 2.1 Formation of colors.

different from the one appearing on the display, they appear the same color to the human visual system. This ability of different spectral power distributions to produce identical perceived colors, is called "metamerism."

In this chapter, we first describe photometry, then the structure of the human eye and its functionalities, followed by the formulation of colorimetry which includes the CIE standards, light sources, and finally metamerism.

2.2 PHOTOMETRY

Due to the spectral sensitivity of the human eye, we perceive brighter or dimmer illumination from light sources with the same optical output power (in terms of Watts) emitting at different wavelengths. Here, the photometric unit, lumen (lm), is defined as: the luminous flux (F) from a monochromatic light at 555 nm emitting the optical power of 1/683 W. The spectral sensitivity of the human eye can be represented as $V(\lambda)$ under the photopic region and reaches its highest sensitivity at 555 nm, which will be illustrated in Section 2.3. For example, $V(\lambda)$ is 0.1 at 650 nm, which means the sensitivity is 10× less than at 555 nm. So, 1/68.3 W is needed for monochromatic light at 650 nm to obtain 1 lm. Actually, the primary photometric unit is not "lm," but candela (cd), which is defined as one lumen per unit solid angle (lm/sr) and is called luminous intensity (I). The initial definition of 1 cd was the luminous intensity of a standardized plumber's candle. As shown in Figure 2.2, the candle emits light in all directions, hence we use "lm" to describe the radiant flux. When

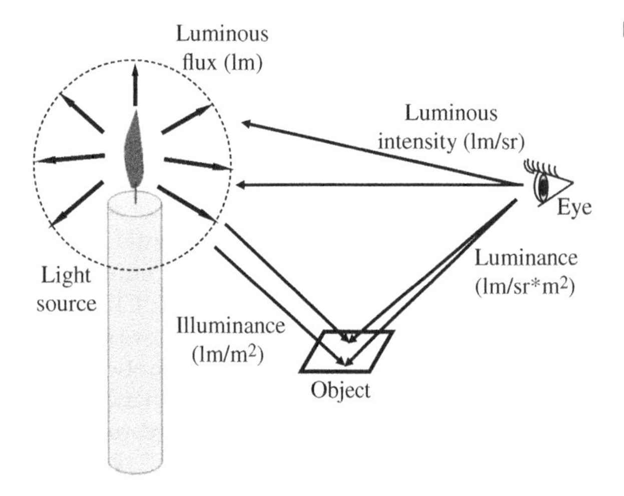

Figure 2.2 Illustrations of photometric units.

Table 2.1 Definitions of photometric units.

Photometric terms	Symbol	Units	Definition
Luminous flux	F	lm	lm
Luminous intensity	I	cd	lm/sr
Illuminance	M	lux	lm/m^2
Luminous exitance	E	lux	lm/m^2
Luminance	L	nit	cd/m^2

human eyes view the candle, they only admit light within a limited solid angle, so we receive the luminous intensity in terms of "cd." The candle can be used as a light source to illuminate an object. Then, we can define the "illuminance" (E) of the light source in units of lux, or lm/m^2. After light–object interaction, the light is modulated (reflected, transmitted, scattered, or absorbed) by the object and can be regarded as being re-emitted from the object, where apparent emission is referred to as the luminous exitance (M) which again has units of lux. When people see the object illuminated by the light source, the human eyes receive light only within a certain angular range, so the luminance (L) of the object can be defined as cd/m^2, or nits. Definitions of photometric units are also shown in Table 2.1.

Example 2.1 A perfect diffuse surface means its luminance observed from different viewing angles is constant, which is also called a "Lambertian surface." For example, rough paper approximates a Lambertian surface. For a Lambertian surface (with a size A) illuminated by a light source with illuminance E, what is the luminance (L) of this surface? Assume this surface can perfectly reflect all the light, i.e. luminous flux of the incident beam on the surface is equal to that of the exiting light.

Answer

From Table 2.1, luminance (L, in terms of cd/m^2) can be also regarded as the luminous intensity (I; in terms of lm/sr) per unit area (A):

$$L = \mathrm{d}I/\mathrm{d}A \tag{2.1}$$

when viewing from a larger angle, the area looks smaller as compared to that at normal direction with a cosθ relation. θ is the angle between the viewing direction and the surface normal. That is

$$\mathrm{d}A = \mathrm{d}A_0 \cos\theta \tag{2.2}$$

where A_0 is the area viewing from normal direction. Because luminance of a Lambertian is the same for any viewing direction, one can obtain the luminous intensity as:

$$I = I_0 \cos\theta \tag{2.3}$$

where I_0 is the luminous intensity at normal direction of the surface. Incident flux to the Lambertian surface can be represented as:

$$F_{\mathrm{in}} = EA \tag{2.4}$$

The total luminous flux which radiates from the surface is:

$$\begin{aligned} \mathrm{F}_{\mathrm{out}} &= \int I\mathrm{d}\,\omega = \int I(\theta)\mathrm{d}\,\omega = \int\int I(\theta)\mathrm{d}\,\phi \sin\,\theta\mathrm{d}\,\theta \\ &= I_0 \int_0^{2\pi}\int_0^{\pi/2} \cos\,\theta\,\sin\,\theta\mathrm{d}\theta\mathrm{d}\phi = 2\pi I_0 \int_0^{\pi/2} \cos\,\theta\,\sin\,\theta\mathrm{d}\theta = \pi I_0 = \pi LA \end{aligned} \tag{2.5}$$

Since the luminous flux of the incident beam on the surface is equal to that of the exiting light ($F_{in} = F_{out}$), one can obtain:

$$E = \pi L \tag{2.6}$$

Typically, power efficiency (in terms of lm/W) is used to describe the efficiency of a display system. For example, if the total input electrical power (wall-plug power) of the display is 10 W and the total radiated flux is 20 lm, the power efficiency of the display is 2 lm/W. The power efficiency describes how much optical power emitted from a display (lm) which is produced by an electrical power input (W). For electroluminescence (EL) devices such as LED, current efficiency is also defined in terms of cd/A. The denominator is current, which quantifies the number of electron–hole pairs provided to the display in unit time. The electron–hole pairs recombine and generate photons which are received by human eyes (cd). For example, consider a Lambertian emitting EL display which, as above, emits a total luminous flux of 20 lm with the current = 300 mA. Then the current efficiency of the display is 21.22 cd/A.

2.3 THE EYE

Figure 2.3a shows a schematic diagram of a human eye [3]. The incoming light passing through the cornea, the aqueous humor, eye lens, and vitreous body, is received by the retina. Primary refraction and approximate focusing of light is achieved at the air/cornea interface. The eye lens, with a higher refractive index ($n = 1.42$) than the cornea, the aqueous humor, and vitreous body ($n = 1.33$–1.37), functions to focus a clear image to the retina, as shown in Figure 2.3b,c [4]. The shape of the eye lens can be adjusted by the ciliary muscle around it. Such a system can be approximately described by the Gaussian Lens formula [5]:

$$\frac{1}{d_1} + \frac{1}{d_2} = \frac{1}{f}, \tag{2.7}$$

where d_1 is the distance from the object to the eye lens, d_2 is the distance from the eye lens to the retina (which is 17 mm typically), and f is the focal length. The image on the retina is totally reversed (upside-down and right–left). However, after interpretation by the brain we can recognize the images in their normal orientation in real space. When the object viewed is more distant, the eye lens becomes flatter, as shown in Figure 2.3b. On the other hand, the ciliary muscle will contract the eye lens to increase its curvature in order to focus nearby subjects, Figure 2.3c.

The retina receives the incoming photons and transfers them into bio-potential signals. After some processing within the eye, those signals are then transmitted through the optic nerve to the brain and interpreted

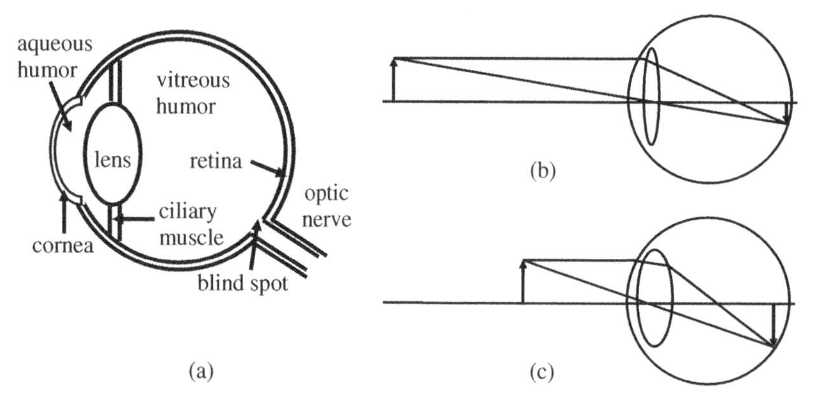

Figure 2.3 (a) Cross section of the eye; (b) and (c): formation of image in the human eye.

Table 2.2 The dimensions and quantities of the cone and rod cells.

	Diameter	Length	Quantity
Rod	2 μm	40–60 μm	100 000 000
Cone	2.5–7.5 μm	28–58 μm	6 500 000

by the viewer. The retina is a multilayer structure which can be divided into three parts: (i) photoreceptor, (ii) connecting nerve tissue (including outer and inner plexiform and nuclear layers, and ganglion cell layer), and (iii) optic nerve [6]. There are two kinds of photoreceptor cells in the eye which are most important in vision, and which are called "rod" and "cone" cells, named according to their shapes [7]. The dimensions and quantities of the cone and rod cells are listed in Table 2.2. Rods are more sensitive than cones and, thus, are more easily saturated when the ambient illumination is high (e.g., sunlight). Rods are sensitive to the light intensity but are inert to colors. On the other hand, the cone cells function well under brighter ambient conditions and are color sensitive. This explains why people can only see grayscales rather than colors under dark ambient conditions (e.g., moonless night).

Figure 2.4a,b show the spatial distribution of the photoreceptors. Solid and dashed lines show the mean values and one-standard deviation away from the mean value of different specimens studied [8]. The symbols show the results obtained from previous reports [9]. We can see that the cones have a maximum distribution density near the visual axis of the eye. Besides, there are almost no rods on the visual axis, because it is occupied by the cones. The number of the rod cells increases and reaches a maximum away from the visual axis. As shown in Figure 2.4a, there is a blind spot in each human eye where the optic nerve passes through it and

Figure 2.4 Spatial distributions of (a) cone and (b) rod cells.

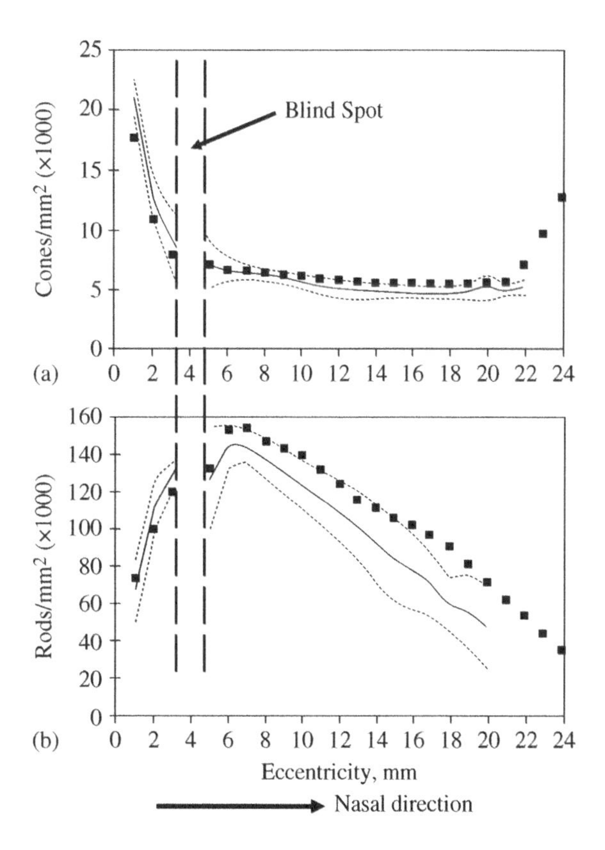

hence there are no cone or rod cells. Typically, in bright ambient conditions the eyesight is most sensitive within the 10° viewing cone. Outside this region, colors are almost indistinguishable. In addition, we do not receive any optical signal from the blind spot. However, in actual life, we "feel" that we can see quite a large viewing angle and there is no blind spot. This is because our eyeballs move and rotate, and construction of a complete visual scene is accomplished by the interpretation of the brain.

To investigate the temporal response of the human eye, a flicker experiment was conducted. That is, the human eye was stimulated with flashing light with different modulation frequencies [10]. When the frequency is high enough, human eyes cannot distinguish the flickers and regard the light as continuous, due to the persistence of vision. The response time of the photoreceptor, cones and rods, is ~15 and ~100 ms, respectively, and hence it is widely accepted that the critical flicker fusion rate is ~50–90 Hz, although some reports claim that the human eye can perceive flicker artifacts up to 500 Hz under certain viewing conditions [11]. The "frame rate" of a display is typically set at this range (50–90 Hz). For example, a 60 Hz frame rate means there are 60 frames in one second, and this is basically enough for human eyes to see continuous images without flickers.

However, when considering moving pictures, the situation becomes more complex [9]. First of all, for some displays with a slower response time than cone cells (such as some liquid crystal displays [LCDs]), image blurring happens when displaying moving pictures. In addition, human eyes tend to track the moving object, resulting in the image blurring even though the response time is high enough (e.g., organic light-emitting device [OLED] with ~μs response time). As shown in Figure 2.5, for example, considering a pixel moving from the left to the right side in different frames. For the emission from pixel 1 (spatial domain) during frame 1 (temporal domain), the signal persists at pixel 1 position during the beginning of the frame 2, resulting in the image blurring at the edge of the moving object, even though the response time of the display is much shorter than that of the eye. Considering the color display, not only will luminance blur, but the color may also change at the edge of the moving object on the display, which is typically observed in the LCD with field sequential color operation, as discussed in Chapters 4 and 5. One method to reduce the blurring of the image of the moving object is to insert the black frame between two display frames. Or, in other words, reduce the emission time in one frame. For example, emission of the 4.2 ms for one frame with 60 Hz frame rate, which means only ~25% of frame time is turned-on. Obviously, the luminance decreases with reducing the duty cycle, and a higher peak luminance is required to keep the same average luminance. The definition to quantitatively describe the response time of a moving object in a display, moving picture response time (MPRT) is [12]:

$$\text{MPRT (ms)} = \text{BEW (pixel)}/v\,\text{(pixel/frame)} \times T_{\text{f}}\,\text{(ms/frame)} \tag{2.8}$$

BEW is the perceived blurred edge width, which is defined as the average pixel numbers for the luminance value from 10 to 90% and from 90 to 10%. v is the velocity of the moving object, in terms of pixels per frame. T_{f} is the inverse of the frame rate. Obviously, to achieve reasonable low MPRT, BEW should be as low as possible. Besides, higher frame rate also helps to reduce the MPRT, which will be further discussed in Chapter 4.

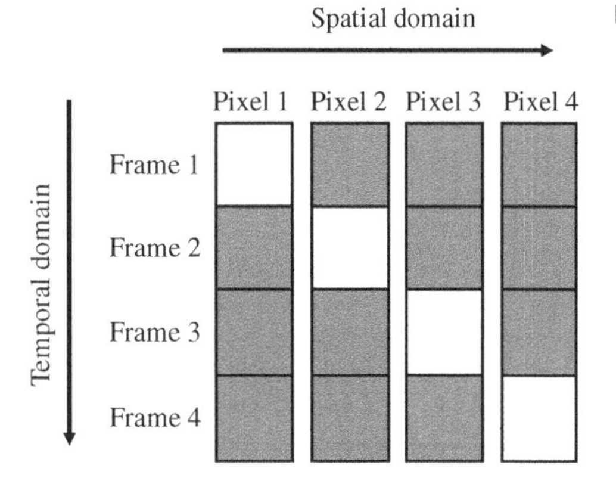

Figure 2.5 Illustration of a moving object in a display.

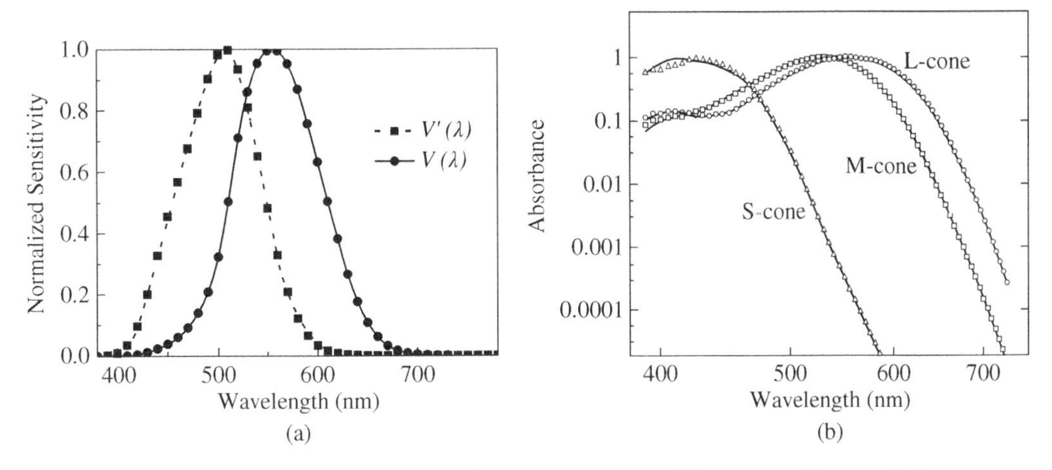

Figure 2.6 Spectral responses of (a) photopic and scotopic regions, and (b) L-, M-, and S- cones [13].

Figure 2.6a shows the normalized spectral sensitivity of human eyes at "scotopic" and "photopic" regions (which means the low- and high-level ambient conditions), in terms of $V'(\lambda)$ and $V(\lambda)$, respectively. It is not completely true to regard the scotopic and photopic vision regimes as exclusive contributions from the rod and cone cells, respectively. Actually, there is an overlap of the intensity responsivity between rod and cone cells. When the ambient light level is higher than that of a full moon and lower than ordinary indoor lighting, both cones and rods can sense light. Higher intensity saturates the rods, but a lower one cannot stimulate the cone cells. At low and high ambient light levels, the eye is most sensitive at $\lambda = 507$ nm and 555 nm, respectively. Note that there are three different kinds of cone cells which have different spectral responses, as shown in Figure 2.6b. The S-, M-, and L-cones are sensitive to the short, medium, and long wavelength, respectively.

To measure the sharpness of human vision, typically we use the pattern "C" and "E" as shown in Figure 2.7a with different sizes and directions. A standard distance to view the pattern is 6 m (corresponding to ~20 ft). And the standard "visual acuity" is defined that the visual angle of the whole character is 5' (5 minutes, 5/60°). And the gap of the "C" and "E" is 1' (1/60°), which is 1.75 mm (i.e. d = 1.75 mm in Figure 2.6a) when viewing at 6 m [14]. In this case, the visual acuity is called 6/6, 20/20, or 1.0. The "6" and "20" in the numerator mean the distance when the gap to the visual angle is just at 1'. When people cannot see the 1.75 mm gap, but they can just distinguish the larger gap, say 3.50 mm, they have the visual acuity of 6/12, 20/40, or 0.5. The "12" and "40" in the denominator mean the viewing distance when the gap to the visual angle is just at 1'. When the gap is 3.50 mm, this distance is 12 m or 40 ft. "0.5" is straightforward because 6/12 = 0.5, and 20/40 = 0.5. When people can see a smaller gap, say 0.88 mm, the visual acuity is 6/3, 20/10, or 2.0. Considering a sphere shaped eyeball with a diameter ~1.7 cm, 1' corresponds to ~5 μm on the retina, about the diameter of the cone, as shown in Figure 2.7b. In Section 1.3.1, we discussed the resolution of a display. With the pixel shrinks to a certain small size, the human eye cannot distinguish the individual pixels. Consider a standard observer with visual acuity of 1.0, when the pixel size is less than the 1' visual angle, it is called the "retina display." Obviously, this critical pixel size is dependent on the viewing distance. For mobile phone applications, the viewing distance is ~25 cm. And hence the pixel size should be smaller than 72.7 μm, or pixel density should be higher than 349 ppi; note that this is just a rough approximation. The calculation is based on the "standard" observer with visual acuity of 1.0. Besides, the arrangement of the display pixel is not the same as the that of human cone cells.

Of the colors of visible light, blue and violet have the shortest wavelengths with the highest photon energy and may possibly damage the retina cells (rods and cones) with high intensity or long exposure. In addition, human eye exposure to blue light at night is believed to affect the circadian clock by suppressing melatonin production. The sensitivity function for melatonin suppression is centered at ~460 nm with a

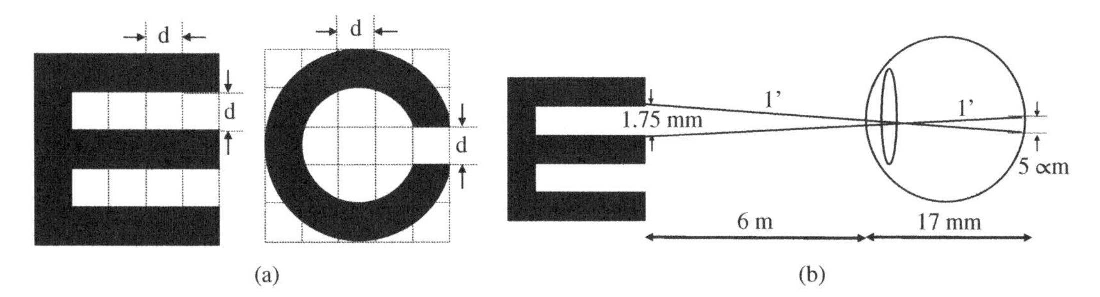

Figure 2.7 (a) Two test patterns for vision acuity, and (b) schematic diagram of the test pattern image into the human eye (not in scale).

full-width-at-half-maximum (FWHM) of 100 nm. The lack of melatonin directly results in sleeping disorders. These effects are sometimes collectively referred to as "blue hazard" to highlight the suggested health issues caused to humans by blue light exposure [15–17]. However, in a display, three primary colors (red, green, blue) are required, and hence the proper usage of displays at night is a very important topic for health. Regarding the melatonin suppression issue, note that we are concerned about light quality during the night only. During the daytime, melatonin should be suppressed to keep attentive and awake, and this is achieved by various light sources such as sunlight. Home lighting is typically yellowish, and therefore includes a lower proportion of blue wavelengths, which is better for relaxation (and sleep). On the other hand, in the workplace, concentration is needed and lighting is typically white or even bluish-white. Further discussion about lighting is included in Chapter 5 LED and Chapter 6 OLED.

Humans have two eyes. The images formed in each eye are different, which allows us to determine the distance of objects in the scene. By supplying different, tailored images to our right and left eyes, it is possible to create a compelling illusion of a 3D scene in the human brain, and that is the basic idea of the 3D display. However, note that as shown in Figure 2.3, the lens in a human eye has a different focal length when we see objects at different distances, this adjustment being automatically triggered by the human brain. However, when viewing a 3D display (for example a head mounted display for virtual reality), the images are on a flat screen at a fixed distance. This conflict between lens adaptation and apparent distance for different images and objects, is called "binocular visual fatigue" and can cause discomfort when viewing a 3D display for a period. This issue will be further discussed in Chapter 8.

2.4 COLORIMETRY

2.4.1 Trichromatic Space

In human eyes, the recognition of a color comes from the combination of the stimulus of different cones. Hence, it may be possible to match an arbitrary color by mixing red, green, and blue colors. (This statement is not completely correct which will be shown later.) Figure 2.8 shows the setup of the color-matching experiments. An arbitrary light with a target color illuminates the lower half of the white screen which produces the stimulus to the human eye through a viewing hole in the black barrier. The hole in the barrier limits the viewing angle of light passing through and only allows stimulation of a certain region of the photoreceptors of the subject's eyes. Red, green, and blue lights illuminate the upper half of the white screen. The intensities of red, green, and blue lights are adjusted to match the color of the lower half of the screen. The idea of the color-matching experiment is that the stimuli from the upper and the lower half screen to L-, M-, and S-cones of the human eyes are exactly the same, and hence the color is matched, although their spectra are different.

According to the color-matching experiment, because there are three kinds of cones, any stimulus can be represented by a vector in a three-dimensional vector space, which is called a "trichromatic space," as shown in

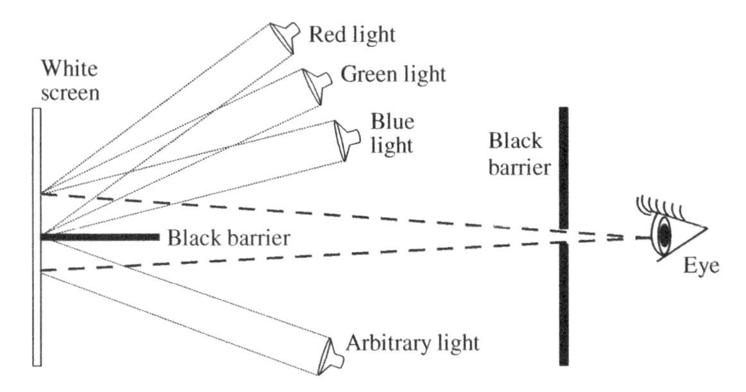

Figure 2.8 Experimental setup of color-matching experiments.

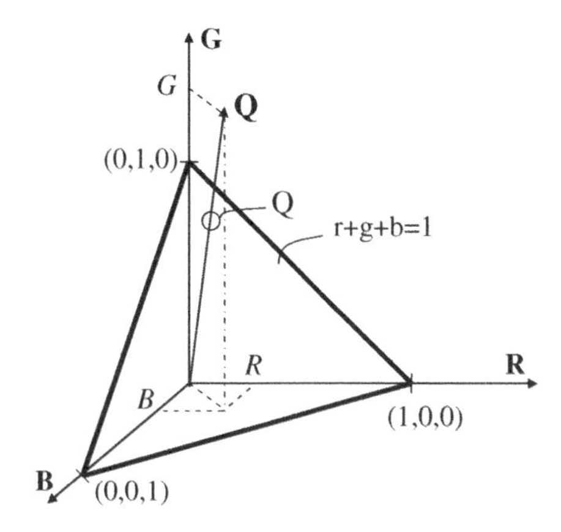

Figure 2.9 Tristimulus space of (R, G, B) primary colors (redraw from Ref. 1).

Figure 2.9. Here, **R**, **G**, and **B** represent the unit vectors of the stimulus of the red, green, and blue, respectively. Or we can think it is the red, green, and blue lights, respectively, in Figure 2.8. The unit vector is different for different wavelengths. For example: light emissions with wavelengths of 630 and 650 nm are both red, which corresponds to different vectors in trichromatic space. For an arbitrary light **Q** (in Figure 2.8), one can find unique tristimulus values R_Q, G_Q, and B_Q to match the color and obtain the following equation:

$$\mathbf{Q} = R_Q\mathbf{R} + G_Q\mathbf{G} + B_Q\mathbf{B} \tag{2.9}$$

These R_Q, G_Q and B_Q values are the intensities of the red, green, and blue lights. For example, as shown in Figure 2.9, the G is larger than its unit vector while R and B are smaller. That means the stimulus **Q** includes a lager component of green than red and blue. The length of the stimulus **Q** represents the intensity. The longer the length, the higher the intensity. The vector **Q** intersects the plane formed by three points (R, G, B) = (1, 0, 0), (0, 1, 0), and (0, 0, 1), which represents pure red, green, and blue, respectively, as shown in Figure 2.9. We can also observe that this point is closer to (R, G, B) = (0, 1, 0), which also echoes the color of **Q** containing more "green" component than "red" and "blue". On this plane, we define:

$$r = \frac{R}{R+G+B} \tag{2.10}$$

$$g = \frac{G}{R+G+B} \tag{2.11}$$

$$b = \frac{B}{R+G+B} \tag{2.12}$$

$$r+g+b=1 \tag{2.13}$$

Each point in the triangle represents a distinct color. The line which links the blue and red points, represents the color mixing between these two colors. The colors inside the triangle correspond to the mixing of red, green, and blue.

However, in some cases, one cannot obtain a matched color no matter how the (R, G, B) intensities are adjusted. However, it is possible to move one of the primaries, for example the "red" one, from the opposite to the same side of the barrier, as the color being matched. So, in this case, Eq. (2.9) can be rewriten as:

$$\mathbf{Q} + R_Q\mathbf{R} = G_Q\mathbf{G} + B_Q\mathbf{B} \tag{2.14}$$

Or, in another form:

$$\mathbf{Q} = -R_Q\mathbf{R} + G_Q\mathbf{G} + B_Q\mathbf{B} \tag{2.15}$$

That means the tristimulus value can be negative ($-R_Q$ in Eq. (2.15)). The basic idea of the color-matching experiments is to produce the stimuli to L-, M-, and S-cones by red, green, and blue lights, respectively, and it can be matched to arbitrary light by varying the intensities. However, as shown in Figure 2.6b, the responsivities of the L-, M-, and S-cones are not independent. Although it is possible to excite the L-cone with a sufficiently long wavelength (say 700 nm), however, there is no wavelength to excite the M-cone only without the L- and S-cones. For the wavelength ~450–500 nm, L-, M-, and S-cone are all excited simultaneously. Hence, a negative tristimulus value is needed to match the color, which will be further discussed in the next section. This concept is important and is related to the color gamut of a display, which will be discussed in Section 2.4.5. In addition, two or more stimuli can be added linearly to form a new stimulus. For example:

$$\mathbf{Q}_1 = R_1\mathbf{R} + G_1\mathbf{G} + B_1\mathbf{B} \tag{2.16}$$

$$\mathbf{Q}_2 = R_2\mathbf{R} + G_2\mathbf{G} + B_2\mathbf{B} \tag{2.17}$$

Then,

$$\mathbf{Q} = \mathbf{Q}_1 + \mathbf{Q}_2 = (R_1 + R_2)\mathbf{R} + (G_1 + G_2)\mathbf{G} + (B_1 + B_2)\mathbf{B} \tag{2.18}$$

2.4.2 CIE 1931 Colormetric Observer

In 1931, the CIE used three primary colors with the wavelengths at 700, 546.1, and 435.8 nm to match all the visible monochromatic lights, which is called the CIE 1931 (R, G, B) system Mathematically, a spectrum can be expressed by superposition of monochromatic units. Hence, a broadband light can be viewed as a mixing of many monochromatic components, as shown in Eqs. (2.3)–(2.5). To obtain the coordinates of each color, the first step is to obtain the RGB-stimulus of the monochromatic light. Here, we introduce the equal energy spectrum (E), with constant optical power (in terms of Watts) for all the wavelengths

$$E = \int E(\lambda)\mathrm{d}\lambda \tag{2.19}$$

where $E(\lambda)$ is the monochromatic component of E. $E(\lambda)$ is used as the matched monochromatic light for the color-matching experiment and hence its intensity is kept constant in terms of the radiometry unit (Watt). That means:

$$E(\lambda) = \bar{r}(\lambda)\mathrm{R} + \bar{g}(\lambda)G + \bar{b}(\lambda)B \tag{2.20}$$

where $\bar{r}(\lambda)$, $\bar{g}(\lambda)$, and $\bar{b}(\lambda)$ are the tristimulus values with stimulus $E(\lambda)$ at wavelength λ, as shown in Figure 2.10a. By normalizing the $\bar{r}(\lambda) + \bar{g}(\lambda) + \bar{b}(\lambda) = 1$ at different wavelengths, we can obtain the CIE 1931 (R, G, B) chromaticity diagram, shown in Figure 2.10b. Due to the linear summation of different color stimuli, a spectrum with many different wavelengths can be divided into the monochromatic lights and be viewed as the color mixing of each monochromatic component. Hence, any color can be represented within the horseshoe shaped region, formed by the monochromatic-light locus and the connection between the 380

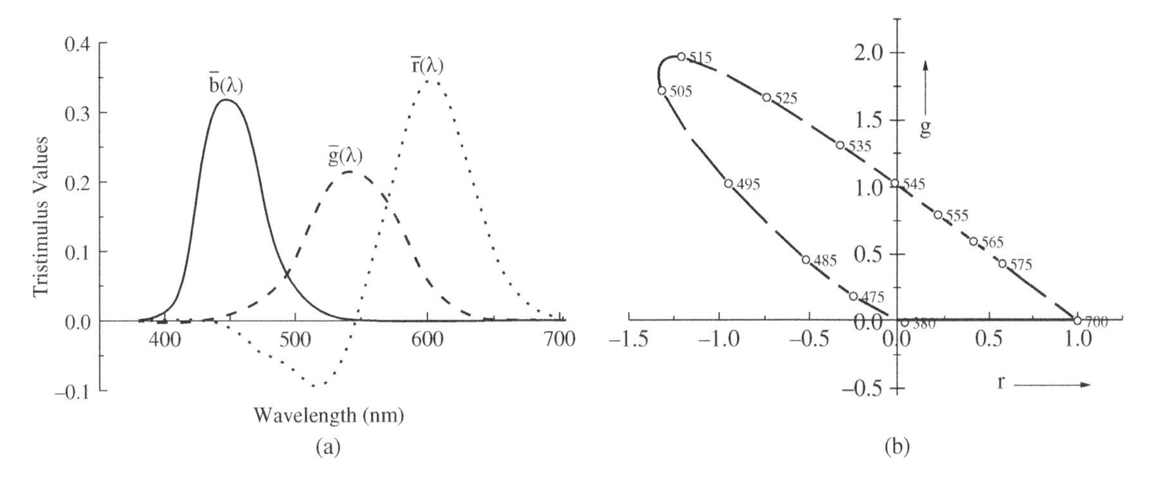

Figure 2.10 (a) Tristimulus value for different wavelengths and (b) CIE 1931 (R, G, B) chromaticity diagram [1].

and 780 nm. Besides, we can note that in this color system, the $\overline{r}(\lambda)$ values are negative when the matched wavelength lies between 435.8 and 546.1 nm, as shown in Figure 2.10a, as indicated in Section 2.4.1.

So, in Figure 2.10b, we have to use the negative r values when describing some colors which is not desirable. To improve this situation, the CIE 1931 (X, Y, Z) system was proposed. The X, Y, Z coordinates are obtained by linear transformations of the CIE 1931 (R, G, B) system according to the following equations:

$$x = \frac{0.49000r + 0.31000g + 0.20000b}{0.66697r + 1.13240g + 1.20063b} \tag{2.21}$$

$$y = \frac{0.17697r + 0.81240g + 0.01063b}{0.66697r + 1.13240g + 1.20063b} \tag{2.22}$$

$$z = \frac{0.00000r + 0.01000g + 0.99000b}{0.66697r + 1.13240g + 1.20063b} \tag{2.23}$$

Then, we can obtain the CIE 1931 (X, Y, Z) chromaticity diagram as shown in Figure 2.11. In this system, the horseshoe lies in the first quadrant which means all the colors can be described by positive values of the coordinates. Another important feature of the CIE 1931 (X, Y, Z) color system is that the Y value is set equal to the luminance of the stimulus, in units of (cd/m^2), using the relations:

$$X = \frac{x}{y}V, \qquad Y = V, \qquad Z = \frac{z}{y} \quad V \tag{2.24}$$

where V is the luminance of the stimulus. Then, it is straightforward to understand:

$$V(\lambda) \equiv \overline{y}(\lambda) \tag{2.25}$$

To obtain the CIE 1931 (X, Y, Z) color coordinates from a spectrum, we can use the following equations, relying on the linear summation of tristimulus values for each wavelengths:

$$X = k\int_{\lambda} P(\lambda)\overline{x}(\lambda)\mathrm{d}\lambda \tag{2.26}$$

$$Y = k\int_{\lambda} P(\lambda)\overline{y}(\lambda)\mathrm{d}\lambda \tag{2.27}$$

$$Z = k\int_{\lambda} P(\lambda)\overline{z}(\lambda)\mathrm{d}\lambda \tag{2.28}$$

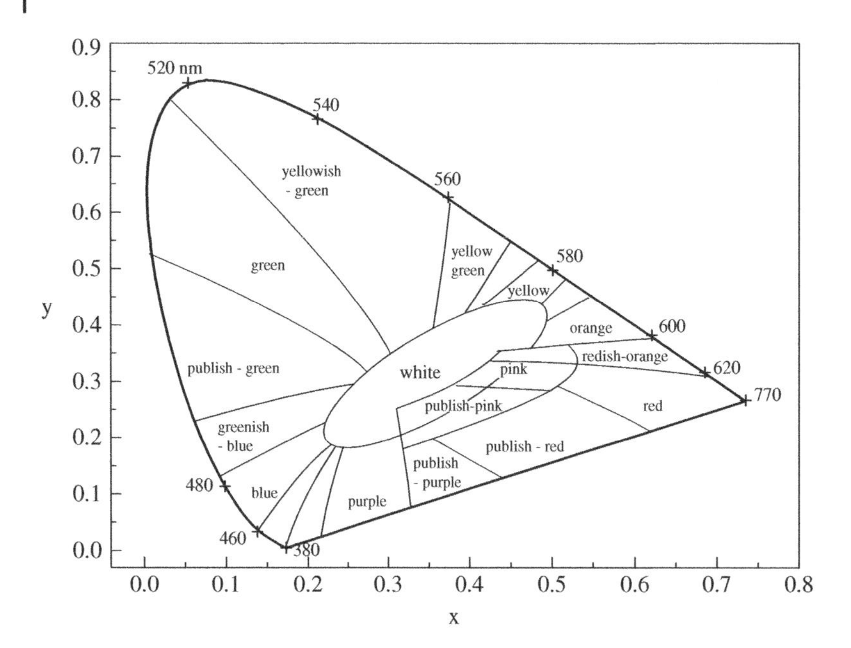

Figure 2.11 CIE 1931 (X, Y, Z) chromaticity diagram.

$$x = \frac{X}{X + Y + Z} \tag{2.29}$$

$$y = \frac{Y}{X + Y + Z} \tag{2.30}$$

where $k = 683\,\mathrm{lm/W}$ which represents the unit transformation from radiometry (Watt) to photometry (lm), and $P(\lambda)$ is the spectral distribution of the stimulus in terms of W/sr^*m^2.

Example 2.2 Find the luminance and color coordinates of the mixture of the two color stimuli listed below:

	(x, y)	Luminance (nits)
Stimulus 1	(0.6, 0.3)	30
Stimulus 2	(0.2, 0.7)	21

Answer

From Eqs. (2.24) and (2.25),

$$X/x = Y/y = Z/z = X + Y + Z$$

$$X_1/0.6 = 30/0.3 = Z_1/0.1$$

$$X_2/0.2 = 21/0.7 = Z_2/0.1$$

$$X_1 = 60, Z_1 = 10$$

$$X_2 = 6, Z_2 = 3$$

$$X = X_1 + X_2 = 66$$

$$Y = Y_1 + Y_2 = 51$$

$$Z = Z_1 + Z_2 = 13$$

$$x = X/(X + Y + Z) = 66/130 = 0.51$$

$$y = 51/130 = 0.39$$

$$\text{Luminance} = Y = 51 \text{ (nits)}$$

2.4.3 CIE 1976 Uniform Color System

Although CIE 1931 (X, Y, Z) color system can describe the color exactly, there is a problem when dealing with the color difference and tolerance. Figure 2.12 shows the famous MacAdam ellipses in CIE 1931 (X, Y, Z) chromaticity diagram [18]. Color differences within the area of each ellipse in this figure cannot be distinguished by people with normal color vision. Note that the ellipses are magnified by 10 times for clarity so the ellipses are really small. Hence, we can see that ellipses in the blue color region are much smaller than those in the green and red areas which means that a small change in color coordinates in the CIE 1931 (X, Y, Z) color system in the blue region results in an observable difference as perceived by human eyes while the same magnitude of change in other parts of the CIE 1931 (X, Y, Z) diagram results in a change which is invisible. In other words, when the difference of the color coordinate, i.e. $(\Delta x, \Delta y)$, between two color stimuli is kept constant, the perceived differences are largest in the blue region. To better quantify the "color difference" between two stimuli, for example: a real object and its image on a display, it is necessary to have a "uniform color system."

Figure 2.13 illustrates a more uniform color system, denoted the CIE 1976 $(L^*u^*v^*)$ color system. Although the sizes of the MacAdam ellipses in CIE 1976 $(L^*u^*v^*)$ color system are still not equally sized, the size difference is smaller compared to CIE 1931 (X, Y, Z) color system. It is also possible to create a color space with MacAdam "circles" with the same size, which needs non-linear transformation of CIE 1931 (X, Y, Z) color system [19]. Hence, the color mixing as shown in Example 2.2 is not valid for such a system. There is another common uniform color system, CIE 1976 $(L^*a^*b^*)$ color system, which is also non-linearly transformed from CIE 1931 (X, Y, Z) color system. The coordinate transform from CIE 1931 (X, Y, Z) to 1976 $(L^*u^*v^*)$ color

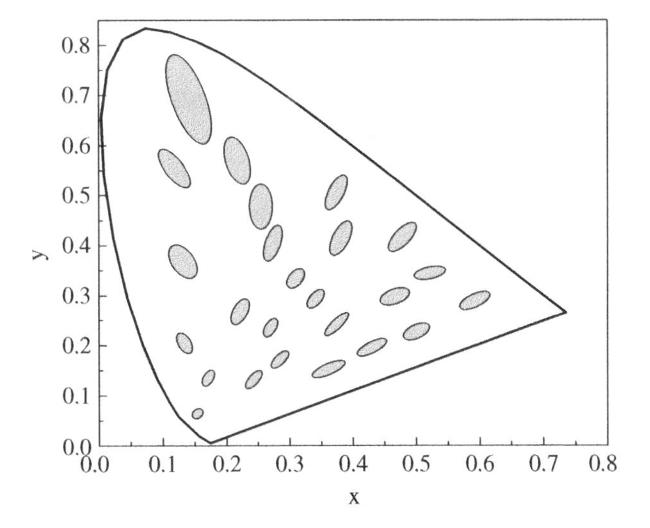

Figure 2.12 MacAdam ellipses of just-visible color differences in the CIE 1931 (X, Y, Z) chromaticity diagram [14].

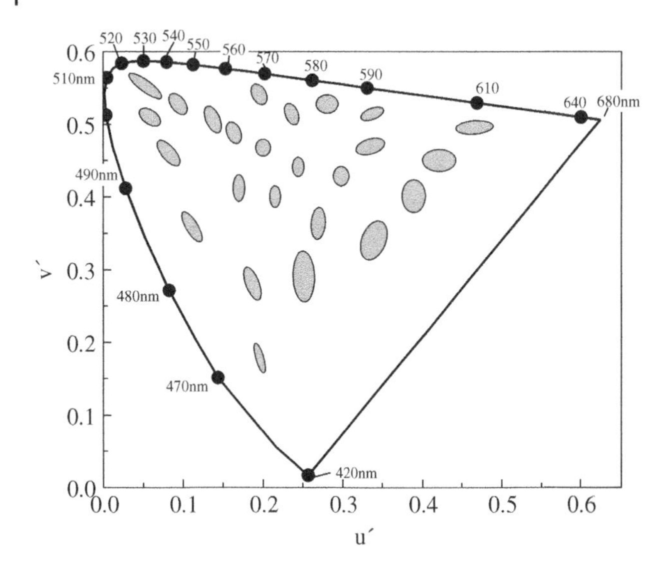

Figure 2.13 CIE 1976 *(u′, v′)* chromaticity diagram and MacAdam ellipses.

system is described in Eqs. (2.31)–(2.38). Since the transformation from (x, y) to (u^*, v^*) is linear, equations for color mixing are still valid in this color space with some modifications. Hence, 1976 $(L^*u^*v^*)$ uniform color system is typically used for display application.

$$L = 116\left(\frac{Y}{Y_n}\right)^{13} - 16 \qquad \text{For } YY_n > 0.01 \tag{2.31}$$

$$u = 13L(u' - u'_n) \tag{2.32}$$

$$v = 13L(v' - v'_n) \tag{2.33}$$

$$L_m = 903.3\frac{Y}{Y_n} \qquad \text{for} \quad \frac{Y}{Y_n} \leq 0.008856 \tag{2.34}$$

where u', v' and u_n', v_n' are calculated from:

$$u' = \frac{4X}{X + 15Y + 3Z} \tag{2.35}$$

$$v' = \frac{9Y}{X + 15Y + 3Z} \tag{2.36}$$

$$u'_n = \frac{4X_n}{X_n + 15Y_n + 3Z_n} \tag{2.37}$$

$$v'_n = \frac{9Y_n}{X_n + 15Y_n + 3Z_n} \tag{2.38}$$

where X_n, Y_n, Z_n are tristimulus values of the reference light source. The reference light is typically a white point, such as daylight or black body radiation, which is used for describing the color performances of an object under light source illumination. When considering an emissive or transmissive display, X_n, Y_n, Z_n values are set to zero, and (u^*, v^*) converges to (u', v') as shown in Figure 2.13.

Compared to the CIE 1931 (X, Y, Z) color system, the CIE 1976 $(L^*u^*v^*)$ color space not only improves the uniformity of color differences, but also the "brightness" difference is made more uniform by a non-linear transformation between the Y and L^* values in CIE 1931 and 1976 color systems, respectively. The luminance

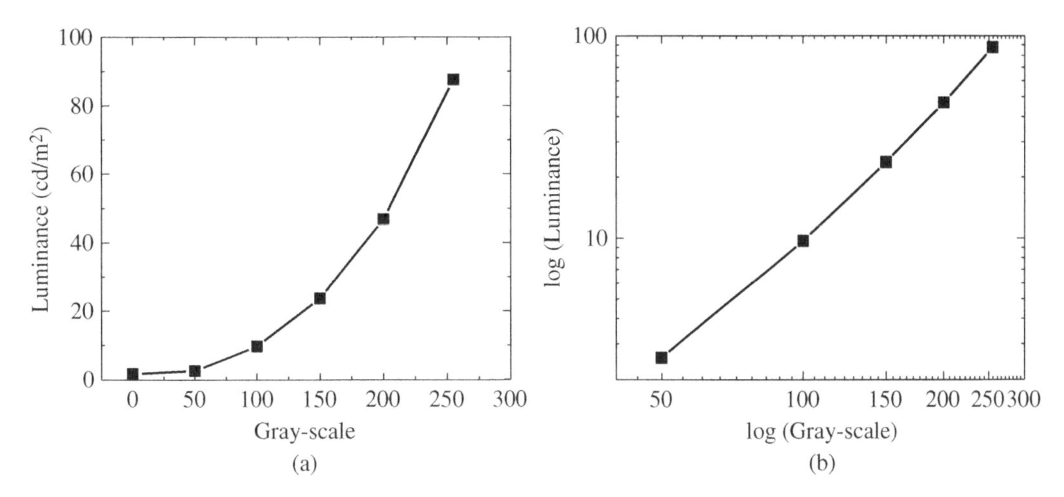

Figure 2.14 Luminance versus gray-level with (a) linear and (b) log scales.

(Y value) is proportional to the "physical" optical intensity, in Watts. However, due to the non-linear response of human eyes, it is not appropriate to use the Y value to describe the "intensity difference" between two stimuli with the same color but different brightness. The L value is more visually "uniform" in the intensity axis than the Y value. As shown in Eq. (2.25), L is proportional to Y to the power of 1/3, which means that human eyes can distinguish a smaller change in luminance, when the luminance value is low. That means when the stimulus is dim, only a small difference in luminance value (Y) can result in a large change in visual perception (L^*). However, when the stimulus is bright, it requires a larger difference in luminance to obtain the same change in apparent brightness. In a display, we use the term "gray scale" to determine the "brightness difference." As discussed in Section 1.3.2, brightness of a display can be digitally-defined by different gray levels from a graphics driver. For example, the darkest and brightest gray scale of one primary color is 0 and 255, respectively. So, 16 million colors can be shown in this display. The apparent differences in brightness between any two adjacent gray levels should be the same. Figure 2.14a,b show the gray levels versus the luminance with linear and log scales, respectively, measured from a LCD monitor, lit at the white points set at different gray scales (0, 50, 100, 150, 200, and 255). The slope of the line in Figure 2.14b is called the γ-value and is 2.157 in this case, which is evidently larger than 1 and slightly smaller than 3, close to the reverse in Eq. (2.30). So, for example: a display has gray scales from 0 (black) to 255 (white). Gray scale 1 to 254 represents different "grays" from dark to light. Then, the luminance difference between gray scale 0 and 1 is smaller than that between 254 and 255 since the γ-value is not equal to 1.

Considering the luminance of the real world, starlight in the night sky without the moon can be as low as 10^{-3} cd/m^2, and the daylight can be as high as 10^5 cd/m^2. Or, more practically, for an observer in a room during daytime without indoor lighting, the sunlight from the window may be as high ~10 000 cd/m^2. The objects in the room may be ~100 cd/m^2, and the dark region in the room may be ~1 cd/m^2 [20, 21]. To describe such a scene, contrast ratio (CR) (or dynamic range) should be higher than ~10 000. Consider the display shown in Figure 2.13, the maximum and minimum luminance is 87.6 cd/m^2 and 1.72 cd/m^2 for gray scale 255 and 0, respectively. Obviously, it cannot reproduce the luminance range in the real world. Then, the contrast ratio is only 50.9, far below 10 000. Besides, the 8-bit gray scale is not enough to describe the luminance detail for the real world. Hence, the Radiocommunication sector of the International Telecommunication Union (ITU-R) recommends a standard (BT 2020, where "BT" means broadcasting service) for high dynamic range (HDR) with 10-bit or 12-bit, corresponding to 1024 or 4096 luminance steps, respectively [22, 23] (https://www.itu.int/rec/R-REC-BT.2020/en). Not only HDR, BT 2020 also defines the color coordinates of three primaries, which will be discussed in Section 2.4.5. To display HDR content, typically a display should exhibit the maximum and minimum luminance at 1000 and 0.01 cd/m^2, to achieve the contrast ratio of 100 000. Note that we only consider the contrast ratio of the display in this stage. When considering the ambient effect, the situation is more complex. For example, for a display under indoor lighting (1000 lx) where the surface reflection of the display is ~1%, reflection of the ambient is ~3.2 cd/m^2, which is much higher than the minimum luminance of a HDR display. Further description of HDR for different display technologies will be given in Chapter 4.

2.4.4 CIECAM 02 Color Appearance Model

The CIE 1931 (X,Y,Z) and CIE 1976 UCS systems are successfully used to describe a "single" color patch and the difference between two colors. However, when viewing an image from a display, there are many colors and brightness levels on the screen. In addition, the "color appearance" is affected by the background and surroundings of the object in question. As an example, Figure 2.15a, shows a round cylinder on a checkerboard (http://persci.mit.edu/_media/gallery/checkershadow_double_med.jpg). The lightsource coming from the right sight casts a shadow on the checkerboard. Visually, the black square at region "A" gives a compelling appearance of being darker than the white square at "B." However, as shown in Figure 2.15b, a straight gray line with uniform color, intersecting through regions "A" and "B" demonstrates that in fact the squares have exactly equal brightness and color. In other words, squares "A" and "B" in Figure 2.15a, appear different solely due to their surroundings. Such an effect cannot be accounted for by the CIE 1931 (*X,Y,Z*) and CIE 1976 UCS systems. Hence, a new model describing the color "appearance" is required. One of the most successful models is CIECAM 02 [24] (www.cie.co.at/index.php/index.php?i_ca_id=435).

To construct the color appearance model, first we need to define some terms. "Brightness" is an absolute value which can be quantified (such as by the CIE 1931 (*X,Y,Z*) system *Y* value), and "lightness" is the relative brightness, depending on different surroundings. For example, when we read a book under dim and bright ambient conditions, the white region exhibits lower and higher "brightness," due to different illumination conditions. However, the "lightness" of the white region remains the same, because the reflectivity of the white paper is the same, regardless of the luminance level. "Hue" describes the visually perceived color, such as red, yellow, green, and blue. "Colorfulness" describes the color difference between the test color and the achromatic color (such as white point). For example, the color "deep red" means the hue of this color is red and colorfulness is high. The achromatic color (white, gray, and black) has no hue. "Chroma" can be regarded as the relative colorfulness, with the same relation to "colorfulness" as between lightness and brightness. Color saturation was used to describe the "colorfulness" of a stimulus. Color saturation increases toward the boundary of the chromaticity diagram. The monochromatic stimulus exhibits highest color saturation.

To include the effects from the surroundings, a model as shown in Figure 2.16 is employed. Here, color appearance of the center region ("stimulus") is affected by the "background" and "surrounding field." The stimulus and background show uniform color areas with angular sizes of 2° and 10° to the eye. The input

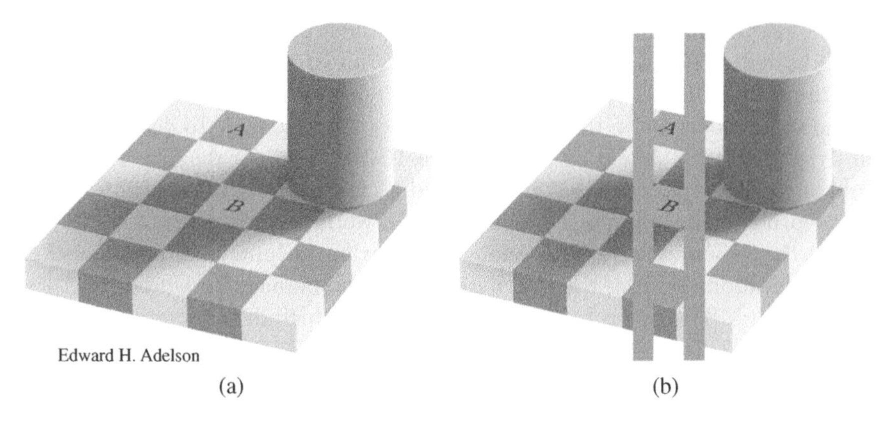

(a) (b)

Figure 2.15 Illusion of checkerboard with the shadow.

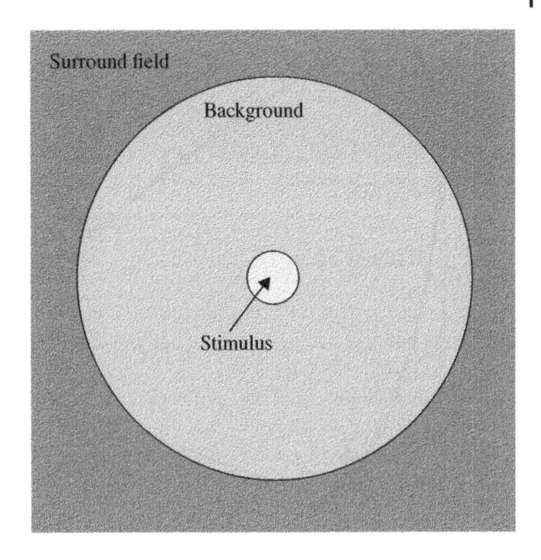

Figure 2.16 Schematic diagram of the observing model.

parameters for the calculation include 1931 (X,Y,Z) values of the stimulus and the white reference, together with the luminance of the background and the surrounding field. Also, the viewing conditions, (i.e. average, dim, and dark surroundings), should be taken into consideration. Then, a full description of the color appearance can be described by obtaining the values of all the parameters described above: brightness, lightness, hue, colorfulness, chroma, and saturation.

Although there are still some constraints (such as treating only static images without considering dynamic effects), CIECAM 02 is one of the most successful color appearance models describing the surrounding effect of the image. So, let us go back and qualitatively explain the illusion in Figure 2.15a. Although the "brightness" of the shadow region is gradually reduced, this change is typically disregarded because human eyes tend to focus on sharp edges, such as those defining the checks. Then, region "B" is "brighter" than the surrounding checks, which provides a higher "lightness," and appears "brighter" than the region "A," although the absolute physical brightness of regions "A" and "B" is exactly the same.

2.4.5 Color Gamut

Let us go back to CIE 1931 (X,Y,Z) system, as shown in Figure 2.11. The boundary of the horseshoe shape of the chromaticity diagram is formed by the monochromatic line and the purple line, which connects the shortest and longest wavelengths. Any spectrum can be divided into monochromatic components. It is straightforward to understand that the color coordinate of a stimulus with a broader bandwidth lies further from the boundary of the chromaticity diagram. The center of the 1931 (X,Y,Z) chromaticity diagram, i.e. (0.33, 0.33), corresponds to white light with a wavelength-independent energy spectrum (and also to an indefinite number of different energy distributions which provide a metameric color match).

In a display, three primary colors are used to generate all the colors in images, based on the trichromatic color space theory. By mixing the three primaries, the colors within the triangle, as shown in Figure 2.17, can be obtained. When the triangle is larger, corresponding to the three primaries being closer to monochromatic light, such a display can demonstrate more colors. Such a triangle is called the "color gamut" of the display. Figure 2.17 shows the triangles defined by two standards. The smaller triangle is based on the definitions of the National Television System Committee (NTSC). The color coordinates of the red, green, and blue stimuli are (0.67, 0.33), (0.21, 0.71), and (0.14, 0.08), respectively. The larger one is from ITU-R, which recommends BT 2020 with red, green, and blue at (0.708, 0.292), (0.170, 0.797), and (0.131, 0.046), respectively. Actually, as shown in Figure 2.17, the three primaries are on the monochromatic locus with the wavelength of 630, 532, and 467 nm, respectively.

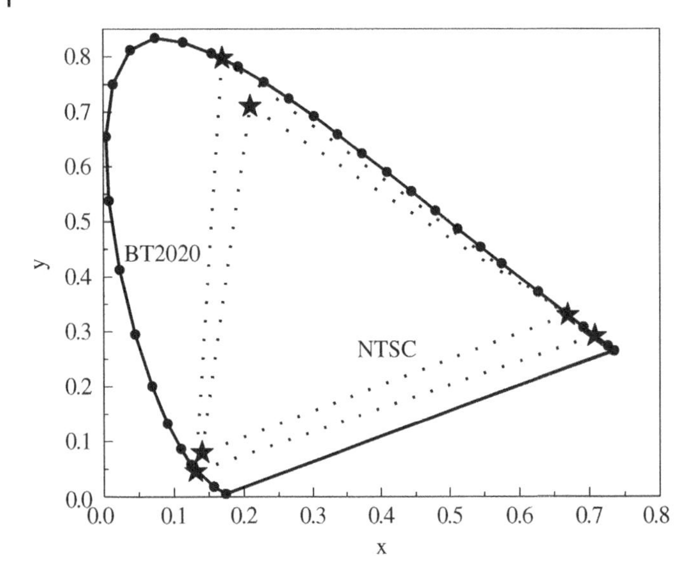

Figure 2.17 Color gamut of NTSC and BT 2020 standards.

2.4.6 Light Sources

As mentioned above, a light source is essential for generating color stimuli. The same object may appear different colors under different ambient lights. For example, peoples' skin can appear pale under street lights at night. Light-sources are very important for reflective displays since they show different colors under different lights. One of the purposes of displays is to reproduce realistic color images, which are illuminated by the natural or artificial "white lights." Hence, there are some "white standards" included in displays which allow the display to present the same image stimuli as the real object. For example, a display typically exhibits the function to switch among different white standards depending on the user's choice. Two common white standards are D65 and D93, which means the "white" looks like daylight with color temperatures of 6500 K and 9300 K (c.f. Section 2.4.6.1), respectively. Besides, for non-emissive and transmissive displays, a backlight module is needed since such a display acts as a light valve rather than an emitter, and these components will also be introduced in this section.

2.4.6.1 Sunlight and Blackbody Radiators

One of the most important natural lights is "sunlight," which is typical of the light from a "black body radiator," which provides a spectral power distribution that depends on the temperature of the emitter alone. The spectral density of radiant power (with the unit of W/m^3) of blackbody radiation can be described by the equation: [25]

$$M_e \frac{c}{4} u_{e\lambda} = c_1 \lambda^{-5} (e^{c_2 T \lambda} - 1)^{-1} \tag{2.39}$$

where, c, h, and k are light velocity, Planck's constant, and Boltzman's constant, respectively. c_1 and c_2 are constants with values:

$$c = 2.99792458 \times 10^8 (\text{m} \cdot \text{s}^{-1}) \tag{2.40}$$

$$h = 6.626176 \times 10^{-34} (\text{J} \cdot \text{s}) \tag{2.41}$$

$$k = 1.380662 \times 10^{-23} (\text{J} \cdot \text{K}^{-1}) \tag{2.42}$$

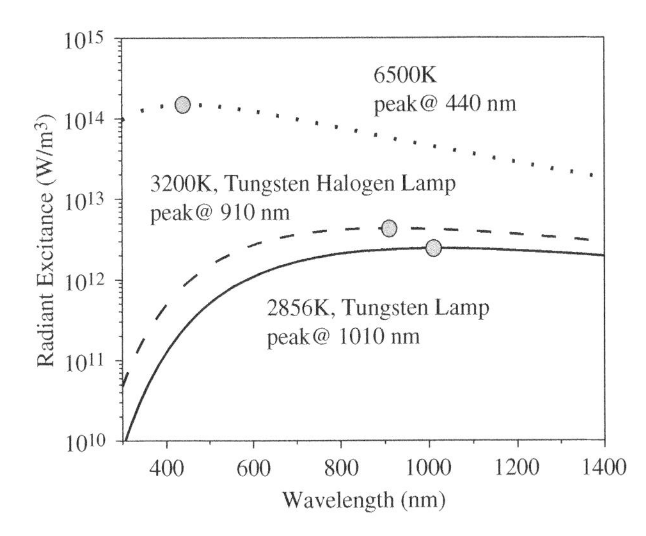

Figure 2.18 Power intensity spectrum of a blackbody radiator at different temperatures.

$$c_1 = \frac{c}{4} 8\pi hc = 2\pi hc^2 = 3.741832 \times\ 10^{-16} (\mathrm{W} \cdot \mathrm{m}^2) \tag{2.43}$$

$$c_2 = \frac{hc}{k} = 1.438786 \times\ 10^{-2} (\mathrm{m} \cdot \mathrm{K}) \tag{2.44}$$

With increasing temperature of the blackbody radiator, the radiant power increases and its spectral peak is shifted toward the blue, as shown in Figure 2.18. A "blackbody locus" can be defined on the CIE 1931-(X, Y, Z) chromaticity diagram. When the chromaticity coordinates of a stimulus lie on the blackbody locus, we can define the "color temperature" of the color stimulus which is the temperature of the blackbody radiator at these chromaticity coordinates. When the chromaticity coordinates of a stimulus do not lie on the blackbody radiator locus, a correlated color temperature (CCT) of the lightsource can often be defined by constructing a line in the CIE 1976 ($L^*u^*v^*$) chromaticity diagram, through the coordinates of the lightsource and perpendicular to a tangent to the blackbody locus. The point on the blackbody locus at which this tangent lies, gives the CCT. A tungsten lamp is a typical blackbody radiator with the color temperature at 2856 K, which looks yellowish white and is denoted "illuminant A" by the CIE. The CCT of sunlight at the surface of the earth ranges from 4000 to 25 000 K, depending on the observation time and place, due to the absorption of the atmosphere. The "air mass" (m) is defined as the ratio of the path length within the atmosphere for sunlight to the observation point, to the thickness of the atmosphere. For example, sunlight observed from the equator with a solar elevation of 30° has an air mass = 2. Since the atmosphere absorbs light with short wavelengths more than long ones, the CCTs is lower with increasing air mass. This explains the observation that sunlight is "redder" during sunrise and sunset, but looks "white" around noon. There is a widely used white standard for displays, D65, which means the sunlight with a CCT of 6500 K. Consider a display with white point at 6500 K used in different lighting condition from 2800 to 9300 K, the "color appearance" of the white point of the display changes from bluish and reddish, according to the CIECAM 02 model.

2.4.6.2 Light Sources for Transmissive, Reflective, and Projection Displays

Unlike the continuous blackbody radiators, one of the criteria of a light source for transmissive displays is to provide sharp peaks for RGB primaries, with high efficiency to decrease the power consumption. The light source of the transmissive display is behind the display, and is called a "backlight." Cold cathode fluorescent lamps (CCFLs) previously provided the main backlight technology for LCDs [26]. The operating principle of CCFL is based on the excitation of mercury vapor and rare gas under low pressure by a high voltage provided at two end electrodes inside the tube, causing a "glow discharge." In contrast to an incandescent lamp, CCFL provides a "cold" light without a hot filament. Typically, CCFL can provide a stable white backlight with the power efficiency over 60 lm/W. However, mercury is harmful to the environment. Also, a robust solid-state device is preferred to the fragile CCEL tube. Hence, LEDs have now replaced CCFL as the backlight for LCDs. The backlight configuration and operating principle of LEDs will be discussed in Chapters 4 and 5.

An extra light source is needed for a reflective display under dim ambient conditions. Such a light source is placed on the top of the display, and is called a front light. The basic criteria for front light systems include transparency and thinness. The emitter (typically CCFL or LEDs) is hidden at the side of the display, and emits light into a transparent light guide. Light propagates inside the light guide, and structures on the light guide redirect the light to the reflective display, which modulates the light and reflects back the images to the viewer. Note that the front light should direct its light toward the display and emit minimal light toward the outer viewer's side in order to optimize the contrast ratio of the reflective display.

For the light sources of projectors, the most important criteria are high brightness and high efficiency, due to the large projected image size. Typically, gas discharge lamps are used. A laser source can also be used as the light source for the projector. With the narrow emission spectra of the laser system, the color gamut can be greatly increased. The power consumption can be reduced due to the high efficiency of the lasers and the full usage of the photons, compared to the broad emission spectra from the lamps. One important issue for the laser projector comes from the "speckle" effect, due to the highly coherent nature of the laser. Due to the interference of the laser light, it shows speckle on the screen and reduces the image quality. Such an effect can be ameliorated by destroying the laser coherence.

2.4.6.3 Color Rendering Index

For a reflective display, an external light source is needed in order to readout the displayed images. Hence, the color and the brightness of a reflective display strongly depend on the optical characteristics (such as spectrum and intensity) of the light source. For example, a green apple under sunlight looks yellow when viewed under yellow street light (which is a so-called "high pressure sodium light"). To quantitatively describe the color difference of a reflective object between a light-source under test and a blackbody radiator (such as sunlight), the color rendering index (CRI) is defined for evaluating a light source. The maximum value of CRI is 100, which means the color is exactly the same when viewed from a light source under test and the blackbody radiator. On the other hand, a low CRI value means the "color difference" of the same reflective object under the two different light sources is large. Here, color difference (ΔE^*) is defined as the absolute distance between two points in the UCS coordinates, such as 1976 (L^*, u^*, v^*) color system:

$$\Delta E = [(\Delta L^*)^2 + (\Delta u^*)^2 + (\Delta v^*)^2]^{0.5} \tag{2.45}$$

where ΔL^*, Δu^*, and Δv^* represent the coordinate differences between the two colors. By choosing eight different objects which appear (i) light grayish red, (ii) dark grayish yellow, (iii) strong yellow-green, (iv) moderate yellowish green, (v) light bluish green, (vi) light blue, (vii) light violet, and (viii) light reddish purple, under daylight, CRI can be defined as:

$$\mathrm{CRI} = 100 - \Sigma_{\mathrm{i}=1\ldots8}\, \Delta E\mathrm{i}^* \tag{2.46}$$

where i indexes the eight different objects. To achieve a high CRI, the spectrum of the light source must be broad enough to reflect the light from different wavelengths. Light sources with a continuous spectrum, such as an incandescent bulb, exhibit a high CRI value. On the other hand, for a fluorescent lamp with discrete emission wavelengths, the CRI value is low (~50). For a monochromatic light, the CRI value can be a negative number.

2.5 PRODUCTION AND REPRODUCTION OF COLORS

Displays are used to produce and reproduce color images. Typically, three primary colors: red, green, and blue are needed to generate all the colors due to the trichromatic color space generated by the human eye. In other

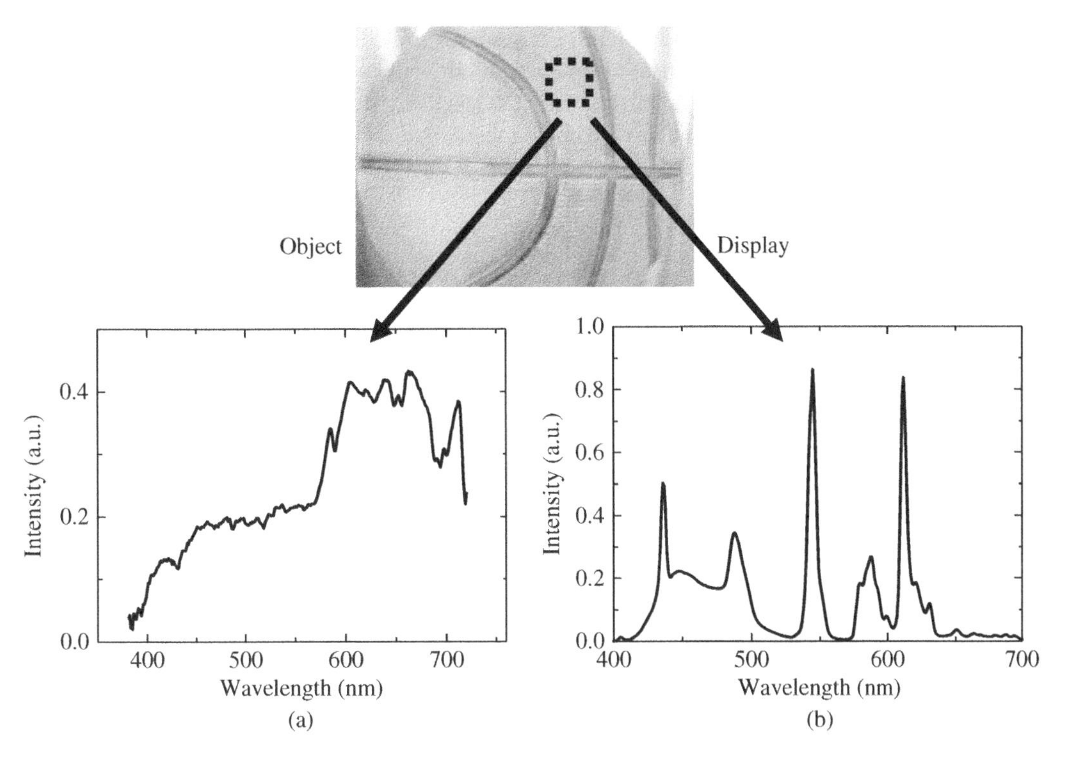

Figure 2.19 Spectra of a basketball from (a) the real object and (b) the display.

words, three primary colors were used to generate the same stimulation of the L-, M-, and S-cones as the real object, although the spectra from the real object and the display are quite different. Figure 2.19a shows the measured spectra of a basketball under sunlight, which shows a continuous curve. We can also take a picture of the basketball, display it on a LCD monitor, and measure the spectrum again. As shown in Figure 2.19b, the basketball color from the display consists of red, green, and blue emissions, which is quite different from the real object. However, the two colors (from the real object and the display) should appear the same. Using different spectra to obtain the same perceived color is possible due to "metamerism" which was described previously.

2.6 DISPLAY MEASUREMENTS

Standard measurements of displays are very important, from scientific, technological, and industrial viewpoints. The Information Display Measurements Standards (IDMSs) compiled by the International Committee for Display Metrology (ICDM) in the Society for Information Display (SID) provides a platform for characterizing electronic displays (http://www.icdm-sid.org/downloads/idms1.html). Reporting is important from an industrial viewpoint. A chapter in IDMS illustrates several templates for the reporting of display measurements from different display types (such as emissive display and 3D display). For example, when measuring the uniformity of a panel, luminance (L, in terms of cd/m^2) and color (CIE 1931 (X, Y, Z)) from five or nine points should be obtained. The positions of the points are identified. Not only the bright image characteristics, but also the "black" state should be measured, for calculating the contrast ratio (CR) values. From these measurement data, other characteristics (such as CCT, $\Delta u'v'$, and uniformity) can be obtained. The measurement conditions should be controlled. For example, the luminance of the display should be measured in a dark room, and hence darkroom conditions should be defined. Then, optical performance is measured using different test patterns (such as checkerboard and different gray scales and colors). Luminance, chromaticity, uniformity, viewing angles, and temporal responses are recorded. After the optical measurements, physical, mechanical, and electrical characterization is performed (such as display strength and power consumption). Some special characterization of 3D display (such as optimized viewing distance and cross talk) is discussed. Characterization of touch panel is also introduced.

Homework Problems

2.1 What are the rod and cone cells in a human eye? What are the differences between them?

2.2 Is color printing a kind of additive or subtractive mixing? Why?

2.3 The CIE 1931 coordinate of a white light source E is (0.33, 0.33). Please find the R:G:B luminance ratio (in terms of photometric unit) for such a light source. (Hint: The wavelengths for the R, G, and B primary lights are 700 nm, 546.1 nm and 435.8 nm)

	R	G	B
λ	700 nm	546.1 nm	435.8 nm
x	0.735	0.273	0.166
y	0.265	0.718	0.008
z	0	0.01	0.826
$V(\lambda)$	0.0041	0.9841	0.018

2.4 Find the R:G:B radiance ratio to obtain illuminant E?

References

1 Wyszecki, G. and Stiles, W.S. (2000). *Color Science—Concepts and Methods, Quantitative Data and Formulae*, 2e. New York: Wiley.
2 CIE No. 15:2004 (1995). *Colorimetry*, 2e. Vienna, Australia: Commission Internationale de l'Éclairage.
3 Dowling, J.E. (1987). *The Retina—An Approachable Part of the Brain*. the Belknap Press of Harvard University Press.
4 Smith, G. and Atchison, D.A. (1997). *The Eye and Visual Optical Instruments*. Cambridge University Press.
5 Hecht, E. (2002). *Optics*, 4e. Baker & Taylor Books.
6 Pocock, G. and Richards, C.D. (2006). *Human Physiology - The Basis of Medicine*, 3e. Oxford University Press.
7 Wandell, B.A. (1995). *Foundation of Vision*. Sinauer Associates.
8 Curcio, C., Sloan, K.R., Kalina, R.E., and Hendrickson, A.E. (1990). Human photoreceptor topography. *J. Comp. Neurol.* 292: 497.
9 Osterberg, G.A. (1935). Topography of the layer of rods and cones in the human retina. *Acta Ophthalmol.* 13 (Suppl. 6): 1.
10 A. Wilkins, J. Veitch, and B. Lehman, "LED Lighting Flicker and Potential Health Concerns: IEEE Standard PAR1789 Update," 2010 IEEE Conversion Congress and Exposition.
11 Davis, J., Hsieh, Y.H., and Lee, H.C. (2015). Humans perceive flicker artifacts at 500 Hz. *Sci. Rep.* 5 (7861).

12 Igarashi, Y., Yamamoto, T., Tanaka, Y. et al. (2003). Proposal of the perceptive parameter motion picture response time (MPRT). *SID 03 Digest*: 1039.
13 Nathans, J. (1999). The evolution and physiology of human review color vision: insights from molecular genetic studies of visual pigments. *Neuron* 24: 299.
14 Consilium Ophthalmologicum Universale (1988). Visual functions committee: visual acuity measurement standard. *Ital. J. Ophthalmol* 1: 15.
15 Lockley, S.W., Brainard, G.C., and Czeisler, C.A. (2003). High sensitivity of the human circadian melatonin rhythm to resetting by short wavelength light. *J. Clin. Endocrinol. Metab.* 88: 4502.
16 Pauley, S.M. (2004). Lighting for the human circadian clock: recent research indicates that lighting has become a public health issue. *Med. Hypotheses* 63: 588.
17 Lewy, A.J., Wehr, T.A., Goodwin, F.K. et al. (1980). Light suppresses melatonin secretion in humans. *Science* 210: 1267.
18 MacAdam, D.L. (1942). Visual sensitivities to color differences in daylight. *J. Opt. Soc. Am.* 32: 247.
19 D. Farnsworth, "A temporal factor in colour discrimination," *Visual problems of colour*, National Physical Laboratory Symposium No. 8, London: HMSO (1958).
20 François, E., Fogg, C., He, Y. et al. (2016). High dynamic range and wide color gamut video coding in HEVC: status and potential future enhancements. *IEEE Trans. Circuits Syst. Video Technol.* 26: 63.
21 Xiao, F., DiCarlo, J.M., Catrysse, P.B., and Wandell, B.A. (2002). High dynamic range imaging of natural scenes. In: *Proc. 10th Color Imag. Conf*., Scottsdale, AZ, USA, 337–342.
22 Seetzen, H., Heidrich, W., Stuerzlinger, W. et al. (2004). High dynamic range display systems. *ACM Trans. Graph.* 23: 760.
23 Zhu, R., Chen, H., and Wu, S.T. (2017). Achieving 12-bit perceptual quantizer curve with liquid crystal display. *Opt. Express* 25: 10939.
24 Fairchild, M.D. (2013). *Color Appearance Models*, 3e. Wiley.
25 Beiser, A. (2003). *Concepts of Modern Physics*, 6e. MacGraw-Hill Companies Inc.
26 T. Nishihara and Y. Takeda, "Improvement of lumen maintenance in cold cathode fluorescent lamp," *IDW'00*, pp 379.

3

Thin Film Transistors

3.1 INTRODUCTION

Thin film transistors (TFTs) are widely used as electronic switches to turn on/off the pixels of an active-matrix (AM) liquid crystal display (LCD) or organic light-emitting device (OLED) [1, 2]. Most commonly, the TFTs used in active matrix displays are made from amorphous or poly-crystalline silicon, rather than from single-crystalline Si wafer. The reasons to use these disordered forms of Si include: First, the diagonal size of a crystalline Si wafer is limited to 12 in. which makes large size displays on crystalline Si impossible. Second, the Si-substrate is opaque to visible light, making it unsuitable as the substrate of transmissive LCDs. Finally, it is technically difficult to grow epitaxial single crystal Si on a glass substrate, since the softening temperature of the glass is below 600 °C which is far below that of Si (1200 °C). On the other hand, amorphous Si can be grown as a uniform thin film over a large substrate (up to 2160 mm × 2460 mm is possible) by low temperature plasma-enhanced chemical vapor deposition (PECVD). It is compatible with LCD fabrication processes and meets the basic requirements for LCD driving. The poly-crystalline Si (poly-Si), made by excimer laser melting and recrystallization of an amorphous Si thin film, has an increased grain size and hence a higher carrier mobility. In a display backplane this can bring the advantages of: (i) an increased aperture ratio, as the higher mobility permits physically smaller TFTs, (ii) a capability for system on panel (SOP) fabrication due to the higher performance of the transistors [3].

However, the non-uniformity of poly-Si which results from the recrystallization process, and the higher leakage current resulting from the increased surface roughness are two major technological issues for poly-Si. Therefore, most TFTs for large-size LCD applications are based on amorphous Si and its related fabrication processes. In addition to silicon-based TFTs, significant progress has been made in the development of oxide semiconductor based TFTs in the past two decades. In particular, indium–gallium–zinc oxide (IGZO) TFTs have today been adopted in the industrial production of some high-end displays.

In this chapter, we will first introduce some basic semiconductor concepts, which are essential background knowledge not only for TFTs, but also for light-emitting diodes (LEDs, Chapter 5). Next, we will discuss the electronic properties of amorphous- and polycrystalline silicon, followed by the electrical characteristics of various types of TFTs, including those based on amorphous silicon, polycrystalline silicon, organic semiconductors, and metal oxides. Techniques to make the TFTs flexible will be introduced later. Finally, methods to drive a display will be illustrated.

3.2 BASIC CONCEPTS OF CRYSTALLINE SEMICONDUCTOR MATERIALS

Before describing the characteristics of amorphous semiconductors with a disordered structure, which are typically used in a conventional AM TFT, we first introduce some basic properties of semiconductors with a single crystal structure in this section. Semiconductors are a class of solid-state materials which have an electrical conductivity intermediate between those of insulators (e.g. glass, and quartz) and conductors (e.g.

Introduction to Flat Panel Display, Second Edition. Jiun-Haw Lee, I-Chun Cheng, Hong Hua, and Shin-Tson Wu.

Ag, Al, and Au) [4]. Typical semiconductor conductivity ranges from 10^{-8} to 10^{3} S/cm. For an insulator, the conductivity is too low for useful carrier transport. In a conductor, on the other hand, carriers (electrons and holes) can move with little obstruction. The conductivity of a semiconductor can be tuned by means of electric fields and impurity concentrations, which opens the possibility for various electronic applications. For example, the conductivity of a TFT is modulated by an electric field and hence it can be used as an electronic "switch." When doped with different impurity atoms, a semiconductor can be modified to conduct either holes or electrons, and is then called a p- or n-type semiconductor material, respectively. When p- and n-type semiconductors are brought in contact with each other, a p–n junction is formed that allows unidirectional carrier transport. This is the basic structure of an LED as described in Chapter 5.

A semiconductor consisting of only one kind of atom is called an "elementary semiconductor." Examples include silicon (Si) and germanium (Ge), which are located in group IV of the periodic table. In this chapter, we mainly focus on Si which is widely used in TFT switches due to its advantages of high performance, mature fabrication techniques, and comparatively low cost. For LED applications, however, due to its indirect bandgap characteristics (shown in Figure 3.4 of Section 3.2.1), single crystal Si with a bulk structure cannot emit light efficiently. Hence, compound semiconductors that combine two or more elements (e.g. gallium arsenide (GaAs) which combines gallium (Ga) from group III and arsenic (As) from group V) are employed as is discussed in Chapter 5.

3.2.1 Band Structure of Crystalline Semiconductors

A semiconductor with a single crystal structure has its atoms arranged in a 3D periodic array. A Group IV atom has four "valence electrons" (which will be defined later) in its outer shell and has a strong tendency to form four covalent bonds. Hence, each atom in a Si-crystal is connected by covalent bonds to four neighboring atoms, each at equal distance but in different directions so that they lie at the vertices of an equilateral tetrahedron. Each atom shares four electrons with its neighboring atoms, and all atoms in the lattice thereby gain a highly stable closed electron shell configuration. Figure 3.1a shows the 3D structure of single crystal Si. Each sphere represents a Si atom that is connected to four other atoms by a covalent bond. For a compound semiconductor such as GaAs, the bonding configuration is similar to that of Si, as shown in Figure 3.1b. Larger black and smaller white spheres represent Ga and As atoms, respectively. However, since different atoms exhibit different attractive forces to the valence electrons, the bonding contains some ionic character, rather than being purely covalent. The periodic structure of the crystalline semiconductor is usually called a "lattice." Also, as shown from Figure 3.1a,b, one can define a "unit cell" of the lattice, which is the smallest repeat unit able to generate the whole lattice. The "lattice constant" corresponds to the edge length of the unit cell and is critical for the epitaxial growth of semiconductor layers used in LED fabrication, as will be discussed in Chapter 5.

Considering now a single isolated atom, the nucleus at the center attracts the electrons, creating a Coulomb potential that is proportional to the inverse of the electron to nucleus distance. Solutions to the quantum mechanical Schrödinger equation for the atom, define many discrete energy levels in the potential well, as shown in Figure 3.2a. Each discrete energy state can be filled by two electrons according to Pauli's exclusion principle, and in the ground state of the atom these are occupied from lower to higher energy states in order. Lower states are filled first with the inner shell electrons. The outermost electrons, which partly or completely fill the highest occupied energy levels, are called valence electrons, as mentioned above. In single atoms and in the more complex systems described next, electrons in low-lying energy levels are also spatially confined to smaller volumes of space near the atomic nuclei. When two atoms approach one another and form a bond, each energy level splits into two discrete states, as shown in Figure 3.2b. Electrons from the two atoms again successively occupy the energy levels in this new system in order from low to high energy. In a crystal there are many atoms, typically with a concentration of about 10^{23} cm^{-3}. Hence, each energy level of an isolated atom is split into innumerable closely packed levels, which collectively are called an "energy band," as shown in Figure 3.2c [5, 6]. In this system too, electrons fill the bands from low to high energy states in order. Finally, at a

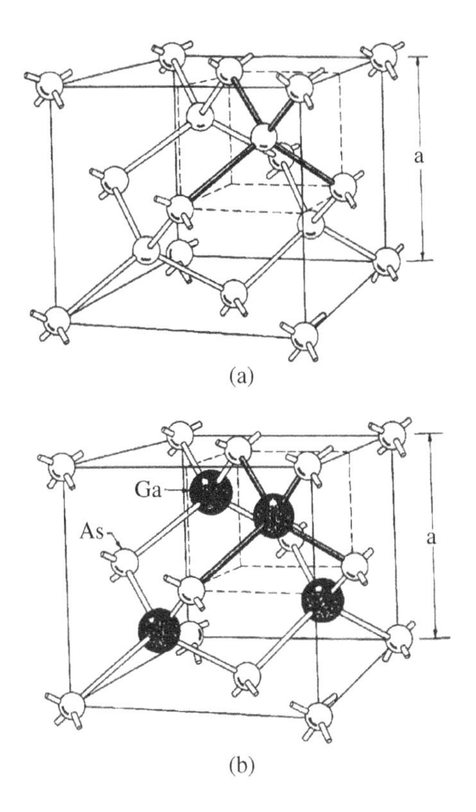

Figure 3.1 3D crystalline structures of (a) Si and (b) GaAs [4].

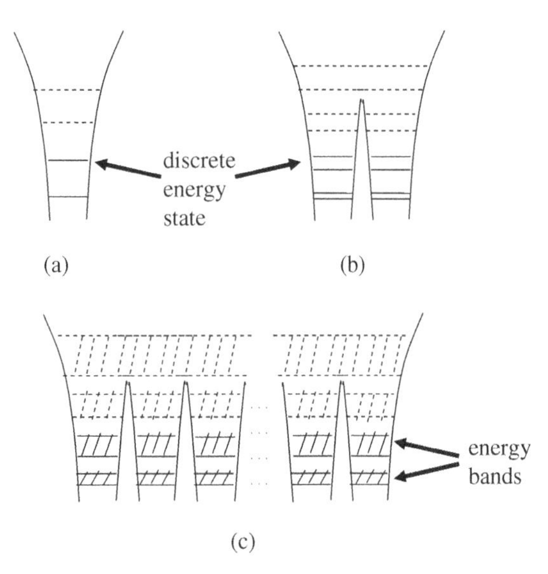

Figure 3.2 Energy levels in (a) an isolated atom, (b) two atoms, and (c) a crystal structure. Solid and dashed lines depict the occupied and unoccupied energy states (bands), respectively.

temperature of 0 K, valence electrons completely fill an energy band termed the "valence band" that represents the highest energy states occupied by electrons.

Sometimes, a valence electron is also called a "bonding electron" since it is shared by the neighboring atoms. Because the valence electrons are the outermost electrons of an atom, they experience a much smaller influence from the nuclei due to the shielding effect by the inner shell electrons. Those electrons can gain thermal

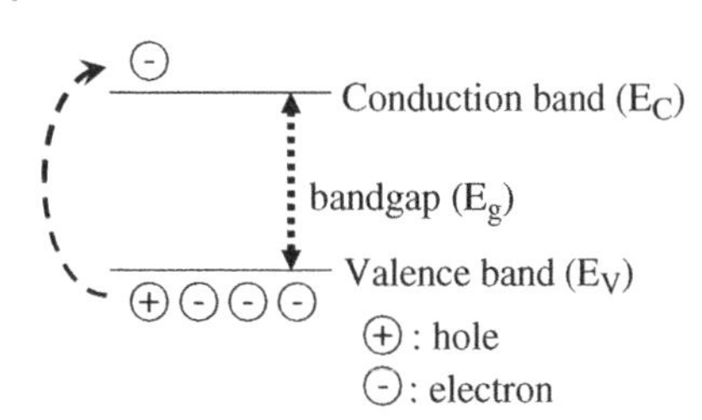

Figure 3.3 Energy band structure.

energy and escape out of the covalent bond, and can then be treated as "free electrons" with suitable modifications. In the energy band diagram described above, this means that a valence electron jumps to the next higher energy state, which is in the so-called "conduction band," as shown in Figure 3.3. The top of the valence band and the bottom of the conduction band are denoted as "E_v" and "E_c," respectively. The energy difference between the conduction and valence bands is called the "bandgap (E_g)," which reflects the fact that no energy level exists between these two bands. A higher temperature results in more electrons being excited to the conduction band. At the same time, each excited electron leaves a "hole" with positive charge in the original covalent bond. If a valence band electron moves its position and neutralizes a hole, its movement creates a hole in another position, and it appears that the hole is transported through the crystal, in a direction opposite to the electron movement.

Due to the wave properties of a particle, an electron exhibits a de Broglie wavelength (λ) of

$$\lambda = h/\mathbf{p} \tag{3.1}$$

or

$$\lambda = 2\pi/\mathbf{k} \tag{3.2}$$

where $h = 6.626 \times 10^{-34}$ (J s) is the Planck constant and $\mathbf{p}$ and $\mathbf{k}$ are the momentum and wavenumber of the electron "wave," respectively. For a free electron, the kinetic energy (E) can be represented by:

$$E = \mathbf{p}^2/2\mathrm{m} \tag{3.3}$$

or

$$E = (h\mathbf{k}/2\pi)^2/2m \tag{3.4}$$

or

$$\mathrm{d}^2E/\mathrm{d}\mathbf{k}^2 = (h/2\pi)^2/m \tag{3.5}$$

where m is the mass of the electron. The curvature of the dispersion curve, $\mathrm{d}^2E/\mathrm{d}\mathbf{k}^2$, is inversely proportional to the mass. However, for an electron moving in the conduction band of a semiconductor, the De Broglie wavelength is close to the lattice constant of the crystal structure, so the electron wave will experience "diffraction" by the lattice. This diffraction effect becomes more serious when the wavelength of the electron is smaller, i.e. the momentum becomes higher. Then, the E–$\mathbf{k}$ relationship is no longer a parabolic function as shown in Eq. (3.3). The detailed E–$\mathbf{k}$ band structure can be calculated from quantum mechanical theory. Figure 3.4 shows the E–$\mathbf{k}$ band structures of Si and GaAs. $E = 0$ corresponds to the maximum energy of the valence band. One can see that there are bandgaps between the maximum of the valence band and the minimum of the conduction band. Excited electrons accumulate near the energetically favorable minimum of the conduction band. On the other hand, the holes left by the excitation of electrons are located near the maximum of the valence band for the same reason. The vertical lines show different $\mathbf{k}$ vectors corresponding to different values of the electron momentum. Since the atoms are arranged in an ordered array as shown in Figure 3.1, electrons moving in different directions experience different periods and diffraction effects from the nuclei. In Si, the top of the valence band and the bottom of the conduction band occur at different values of momentum, at the Γ and X points on the $\mathbf{k}$-axis, respectively. Because of this mis-match, silicon is referred to as an

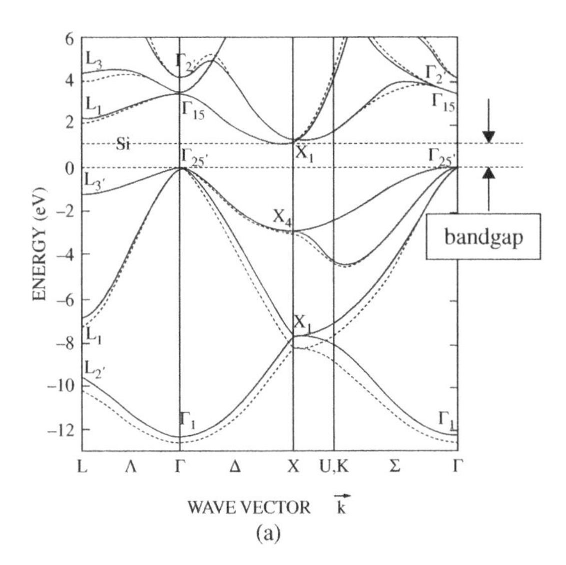

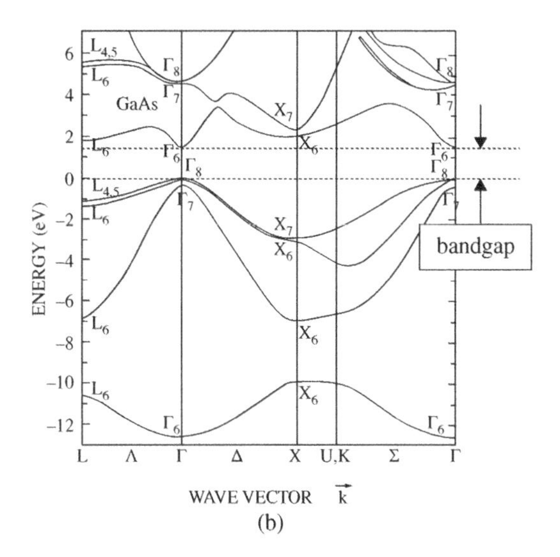

Figure 3.4 Calculated E–$\mathbf{k}$ band structures of (a)Si and (b)GaAs [7]. Solid and dashed lines in Si band structure show results by two different kinds of calculations.

"indirect bandgap" semiconductor. On the other hand, in GaAs, the two extremes coincide at the Γ-point and this is called a "direct bandgap" semiconductor. The different band structures of direct and indirect bandgap semiconductors result in different optical characteristics, which will be discussed further in Chapter 5. We can note that near the minimum of the conduction band and the maximum of the valence band where most carriers accumulate, the E–$\mathbf{k}$ curve shows a nearly parabolic shape. Hence, one can modify Eqs. (3.4) and (3.5) to describe the motion of an electron in the conduction band of a semiconductor by:

$$E = (h\mathbf{k}/2\pi)^2/2m^* \tag{3.6}$$

and

$$d^2E/d\mathbf{k}^2 = (h/2\pi)^2/m^* \tag{3.7}$$

where m^* is the effective mass, which can be obtained provided that the E–$\mathbf{k}$ relationship is known.

3.2.2 Intrinsic and Extrinsic Semiconductors

A material which behaves as a semiconductor without the presence of any impurity atoms is called an "intrinsic semiconductor." In such a material, electrons and holes in the conduction and valence bands (free electrons and holes) are simultaneously generated by thermal excitation of a carrier. Hence, the concentrations of the free electrons and holes can be described as:

$$n = p = n_i \tag{3.8}$$

where n and p are electron and hole concentrations, respectively. n_i is the intrinsic carrier density dependent on the temperature and energy bandgap. Obviously, with a higher temperature or a smaller bandgap, it is easier to generate more electron–hole pairs and a correspondingly larger n_i value.

On the other hand, one may intentionally add impurities into a semiconductor to vary the conductivity, in which case the material is called an "extrinsic semiconductor." As shown in Figure 3.5, group V atoms like As, or group III atoms like B can substitute for Si atoms in the crystal. Each As or B atom respectively, generates one more electron or one more hole in the semiconductor, either of which effectively increases the carrier concentration. Addition of these impurities is called n-type and p-type doping since the added carriers are

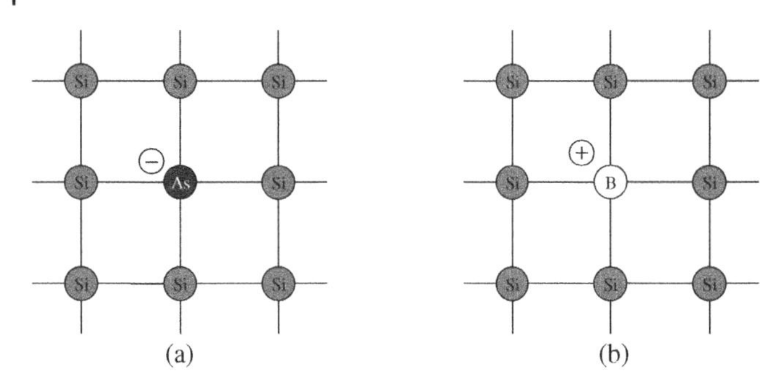

Figure 3.5 Schematic diagram of bonding of Si with (a) As and (b) B impurities.

(negatively charged) electrons and (positively charged) holes, respectively. Group V and III dopants are also called "donors" and "acceptors," since in comparison to silicon, they donate or accept an electron to or from the band structure, respectively.

Typically, in many crystalline semiconductors, this extra electron or hole can be treated as nearly free, which is called the "complete ionization" condition. However, this is not the case for some III–V semiconductors, especially for the wide bandgap materials to be described in Chapter 6. To describe the probability (F) of finding an electron at energy level (E), it is usual to use the concept of the Fermi energy (E_F) since electrons in a solid obey Fermi–Dirac statistics.

$$F(E) = \frac{1}{1 + exp[(E - E_F)/(kT/q)]} \tag{3.9}$$

where E_F is the Fermi level in electron volts (eV), k the Boltzmann constant (1.38×10^{-23} J/K), T the absolute temperature in K, and q the charge (1.6×10^{-19} C). From the above equation, one can note that when $E = E_F$, the $F(E) = 0.5$, which means this energy level has a probability of 50% of being occupied by an electron. $F(E)$ curves versus energy at different temperatures are shown in Figure 3.6. At T close to 0 K, $F(E)$ is a step function, i.e. $F(E)$ is 1 below E_F, and 0 above E_F, respectively. As the temperature is increased, the $F(E)$ curves tend to adopt a less steep slope in the vicinity of E_F which means the probability of the electron occupying higher energy levels increases. This mechanism explains the fact that carrier concentration increases at higher temperature.

In an n-type semiconductor, since free electrons are released from the donors, the concentration of electrons is higher than that of holes. That means it is easier to find a free electron in the conduction band than in the

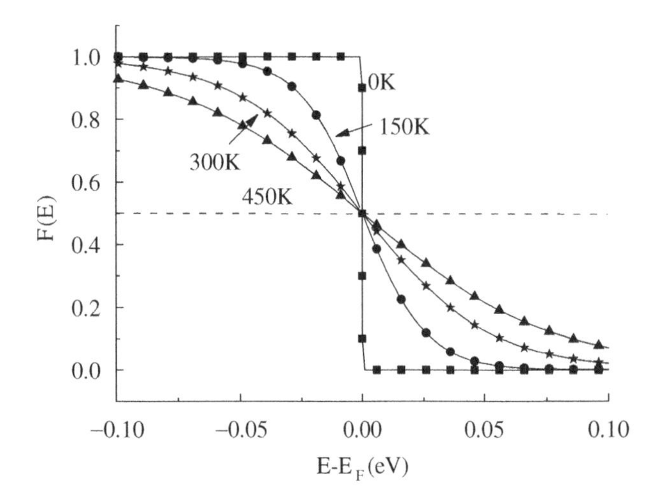

Figure 3.6 $F(E)$ curves at different temperatures.

intrinsic case, and the E_F level shifts toward the conduction band level E_c. On the other hand, for a *p*-type semiconductor, E_F moves downward toward the valence band level E_v. For an intrinsic semiconductor, E_F is typically very close to the midpoint between E_c and E_v. Hence, one can determine whether a semiconductor is *n*- or *p*-type by observing whether E_F lies in the upper or lower half of the bandgap, respectively. Theoretically, the electron concentration (n) can be obtained from:

$$n = \int N(E)F(E)\mathrm{d}E \tag{3.10}$$

where $N(E)$ is the "density of states," which quantifies how many carriers can be accommodated at any given energy. In the bulk of a crystalline material, $N(E)$ is proportional to $E^{0.5}$, and the value is 0 at $E < E_c$. That means there is no state available within the bandgap. When the energy is increased above E_c, there are more and more available states. On the other hand, the probability of a state being occupied decays exponentially with increasing energy, as shown in Eq. (3.9). Typically, carrier concentrations (intrinsic or extrinsic) can be described by:

$$p = n_i \exp\left(\frac{E_i - E_F}{kT/q}\right) = N_v \exp\left(-\frac{E_F - E_v}{kT/q}\right) \tag{3.11}$$

$$n = n_i \exp\left(\frac{E_F - E_i}{kT/q}\right) = N_c \exp\left(-\frac{E_c - E_F}{kT/q}\right) \tag{3.12}$$

$$pn = n_i^2 = N_c N_v \exp\left(-\frac{E_g}{kT/q}\right) \tag{3.13}$$

where E_i is the middle of the bandgap in units of eV, and N_c and N_v are the effective density of states of the conduction band and valence band, respectively. One can note that when the Fermi level is shifted from E_i toward the valence band, the hole concentration increases. On the other hand, when E_F moves up from E_i toward E_c, the electron concentration increases. The product of Eqs. (3.11) and (3.12) yields (3.13), which shows that the product of electron and hole concentrations is a constant that is the same for intrinsic and extrinsic semiconductors. That means, for example, it is possible to dope the semiconductor with donors in order to increase the electron concentration. However, at the same time, the hole concentration decreases. Corresponding (opposite) changes occur in the case of acceptor doping. For an extrinsic semiconductor under the complete ionization condition, one can also write:

$$p = N_A \tag{3.14}$$

$$n = N_D \tag{3.15}$$

where N_A and N_D are acceptor and donor concentrations, respectively.

Example 3.1 Find the carrier concentrations in crystalline Si doped with boron at a concentration of 10^{16} atoms/cm^3 at 300 K. Also, find the energy difference (in terms of eV) between the Fermi level and the top of the valence band. ($n_i = 1.45 \times 10^{10}$ cm^{-3} and $E_g = 1.12$ eV at this temperature)

Answer

Assuming complete ionization, from Eqs. (3.14) and (3.13),

$$p = N_A = 10^{16}(\mathrm{cm}^{-3})$$

$$n = n_i^2/p = (1.45 \times 10^{10})^2/10^{16} = 2.1 \times 10^4(\mathrm{cm}^{-3})$$

From Eq. (3.11),

$$E_i - E_F = \frac{kT}{q}\ln(p/n_i) = \frac{1.38\times10^{-23}\times300}{1.6\times10^{-19}}\ln(10^{16}/1.45\times10^{10}) = 0.35(\text{eV})$$

$$E_F - E_v = 1.12/2 - 0.35 = 0.21(\text{eV})$$

3.3 CLASSIFICATION OF SILICON MATERIALS

Silicon is the most commonly used semiconductor material in our daily life. We can classify Si in terms of its crystallinity (grain size), as shown in Figure 3.7 [8]. In amorphous silicon (a-Si), which is the least ordered form of the material, the mobility is as low as 1 and 0.01 cm^2/Vs for electrons and holes, respectively. When Si forms a crystalline structure with domains in the order of several to several tens of nm, it is called nanocrystalline Si (nc-Si), and has higher mobility values. As the grain size is increased into the range of 100s of nm and microns, the material is referred to as microcrystalline (μc-Si) or polycrystalline Si (poly-Si). Increase of grain size from a-Si to nc- and even μc-Si can be achieved by optimization of the fabrication parameters during thin film deposition. However, to obtain poly-Si, a process of melting and re-crystallization of an a-Si layer is typically needed which requires a temperature (1200 °C) that is much higher than the softening temperature of a glass substrate (600 °C). But local heating by laser irradiation makes it possible to form high quality poly-Si thin films on glass or even plastic substrates. In the following sections, we will describe the material characteristics, preparation, and carrier transport of a-Si and poly-Si, which are the two most commonly used active materials in display back-plane technologies today.

3.4 HYDROGENATED AMORPHOUS SILICON (A-SI:H)

Amorphous silicon (a-Si) consists of Si atoms without a crystalline structure (long-range order). However, it still retains short-range order; each Si atom forms covalent bonds with its four neighboring atoms but with some variations in bonding length and angle [9]. Due to the non-periodic structure, it is not possible to draw

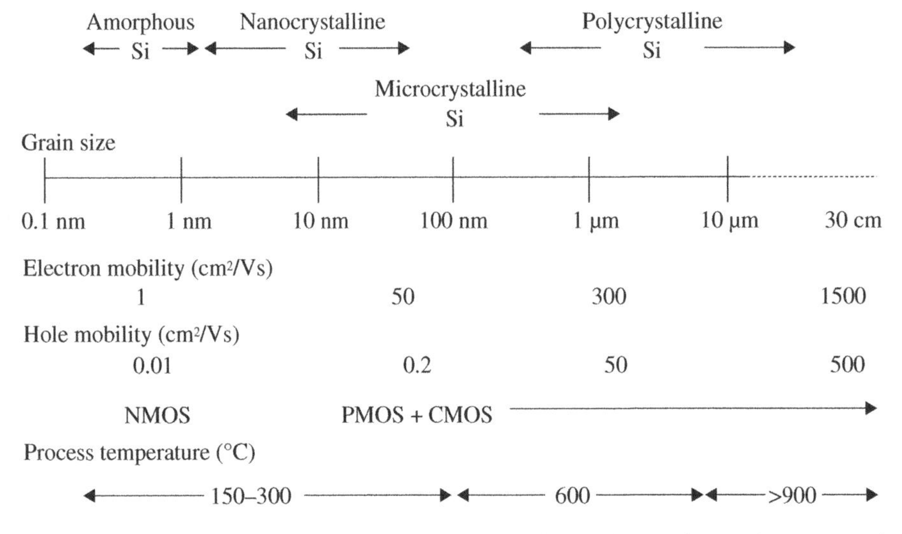

Figure 3.7 Classification of Si by grain size together with their mobility values and process conditions [8].

E-**k** band diagrams. Instead, the concept of "density of states" is used to describe the electrical and optical characteristics of a-Si. Since atoms in the a-Si network exhibit statistical differences in their bonding, there are some localized states which extend from the conduction and valence bands into the bandgap, and which are called "tail states." In contrast to the case of a crystalline material, where any out-of-place atom is regarded as a defect, the elementary defect in a-Si is the "coordination defect." Dangling bonds, corresponding to unpaired valence electrons, are a type of defect which commonly exists in a-Si. They can be described as "deep states" with their energy inside the bandgap. Carriers with lower energy will be trapped in these states and become immobile. Carriers with sufficiently high energy can move by hopping between these localized states, but their resulting mobility is low. To improve the electronic properties of a-Si, H atoms are introduced to passivate these dangling bonds. Therefore, electronic grade a-Si is often called hydrogenated a-Si, denoted as a-Si:H.

3.4.1 Electronic Structure of a:Si-H

Figure 3.8a,b show a schematic structure and density of states of a-Si:H. One can see that a-Si:H has a random structure without long-range order [10]. Hence, it is not possible to obtain an E-**k** band diagram, like Figure 3.4. On the other hand, as described above one can describe the carrier concentrations at different energies using the concept of "density of states," i.e. how many states are allowed at certain energy level, and then determine the "electrical bandgap" and "optical bandgap" of a-Si:H.

a-Si:H can be viewed as a random network connected by covalent bonds with most of the Si atoms being fourfold coordinated, as shown in Figure 3.8a. Owing to the preservation of this short-range order, the overall electronic structure of a-Si:H is similar to that of crystalline silicon. However, deviations of bonding length and angle from the ideal values result in the extension of the density of states into the forbidden energy gap. These are the tail states in Figure 3.8b. The larger and smaller spheres in Figure 3.8a represent silicon and hydrogen atoms, respectively. H atoms are introduced during thin film formation by PECVD, which will be described later. H atoms effectively reduce the dangling bonds in a-Si by forming Si–H bonds. Although hydrogen atoms can passivate a large proportion of the dangling bonds in a-Si:H networks, there are still many bonding defects, which are the origin of the "deep states" in Figure 3.8b within the bandgap. The solid lines indicate the density of states and the shaded regions indicate the states which are occupied by the electrons.

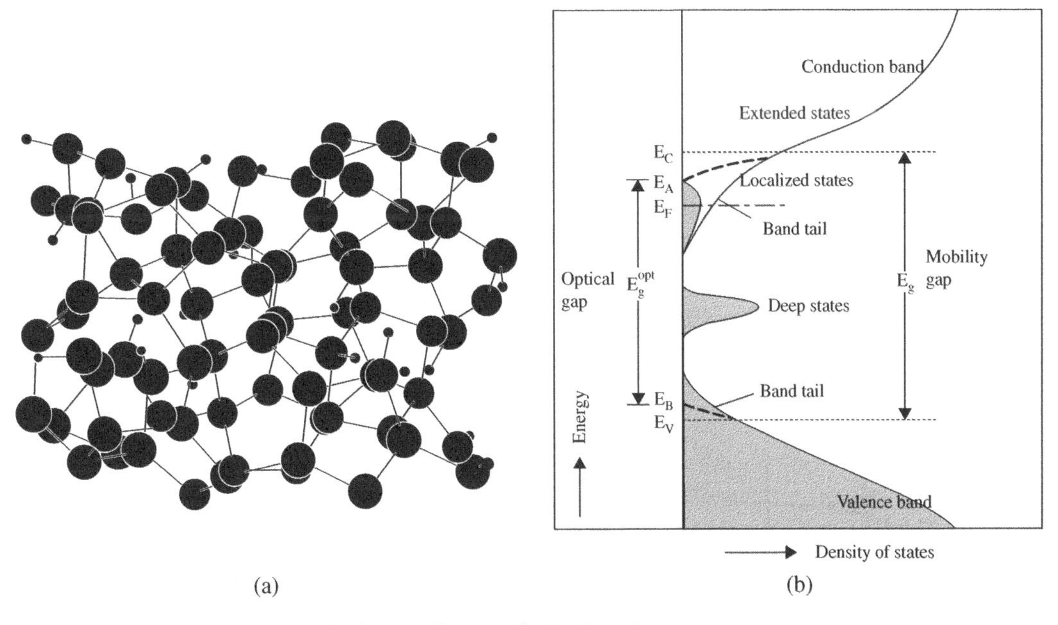

Figure 3.8 (a) 3-D structures, and (b) density of states of a-Si:H [9, 10].

3.4.2 Carrier Transport in a-Si:H

In crystalline Si, electrons move within the ordered lattice; this is analogous to the motion of free electrons and can be described by use of the effective mass concept. On the other hand, the disordered potential which exists in a-Si:H can cause strong scattering and localization of electrons. The "mobility edge," E_c or E_v in Figure 3.8b, is introduced to distinguish the extended and localized states. Due to their frequent scattering, the mobility of electrons even above the mobility edge is significantly reduced to 2–5 cm^2/V s, which is far lower than that in crystalline silicon. Electrons which are confined in the "localized states" can also contribute to charge conduction via hopping to neighboring localized states at elevated temperatures. These localized states include both the band tail localized states and the deep defect states.

When electrical charge is injected into a-Si:H, the excess carriers fill the deep levels first, and then the localized tail states. Thermal activation of electrons from the Fermi level to states above the mobility edge gives conduction in extended states. As noted, carrier mobility is reduced by scattering and by trapping and escape processes, which in turn limits the operational speed of a-Si:H based devices. The low mobility value results in high resistivity according to the equation:

$$\rho = (nq\mu)^{-1} \tag{3.16}$$

where ρ is the resistivity in units of Ohm/cm, n the carrier density in cm^{-3}, q the electron charge (1.6×10^{-19} C), and μ the carrier mobility in units of cm^2/V s. Typically, for undoped or n-type a-Si, conduction from holes can be ignored since the hole mobility is much lower than that of electrons. The "mobility gap" (electrical bandgap) of a-Si:H has a value of about 1.85 eV, which is larger than the bandgap of crystalline Si at 1.1 eV. As mentioned above, it is not possible to obtain an E–**k** band diagram in a-Si due to its random structure. Hence, the selection rule for optical transitions which distinguishes direct from indirect bandgaps is relaxed in this material. Due to the complex structure of a-Si:H, there is no sharp absorption edge. One can define an "optical bandgap" whose value is typically about 1.7 eV, and which represents the energy of onset of optical absorption. Because the absorption coefficient of a-Si:H for photons is larger than that of crystalline Si between ~1.7 eV, the optical bandgap of a-Si:H, and ~3.5 eV, the direct bandgap of crystalline Si, one important application for a-Si:H is in photovoltaic devices.

3.4.3 Fabrication of a-Si:H

PECVD is the usual technique used to fabricate large-area a-Si:H thin films with high quality at low temperature [8]. Figure 3.9 shows a schematic diagram of a PECVD system. Here, SiH_4 is used as the source for silicon deposition. H_2 gas is added to further passivate dangling bonds. Plasma generated by radio frequency (rf) power decomposes the SiH_4 into SiH_x and H radicals, which then deposit on the substrates. The standard substrate temperature is about 250 °C, which allows the radical species to migrate across the surface until they meet and chemically bond to one another and ensures sufficient H incorporation in the resulting a-Si:H. To

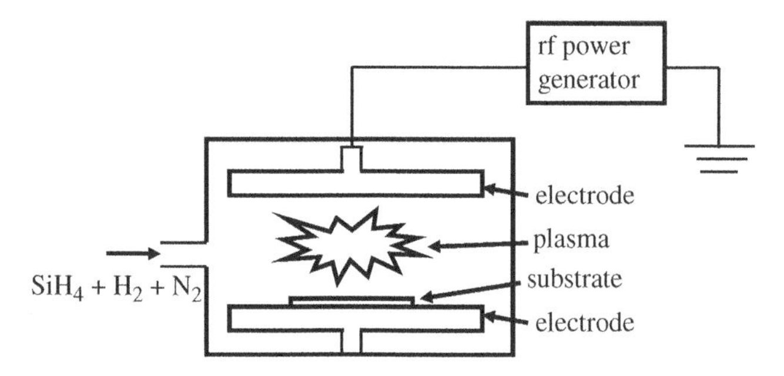

Figure 3.9 Schematic diagram of PECVD system.

obtain high quality a-Si:H with high throughput under the temperature limits imposed by the glass (<600 °C) or flexible plastic (<200 °C) substrate, process parameters such as the feed gas flow rate, rf power, chamber pressure, substrate temperature, and SiH_4/H_2 ratio must be controlled and optimized. Under some process conditions, such as at low SiH_4/H_2 ratios, high plasma power density, and thicker film deposition, there is a tendency to form nanocrystalline or microcrystalline silicon, rather than a-Si:H.

3.5 POLYCRYSTALLINE SILICON

Polycrystalline Si (poly-Si) exhibits a larger grain size and hence better electrical properties than a-Si:H, as shown in Figure 3.7. As mentioned above, TFTs are opaque and sensitive to visible light, so shielding is needed to block light from the device, which means that in LCD and other non-emissive devices the TFT area cannot contribute to information display. Better TFT electrical properties give the possibility to shrink the TFT dimension which can in consequence: (i) increase the fractional area of each pixel which modulates light (referred to as the "aperture ratio"), and in turn to achieve a higher display luminance from a given backlight power, and (ii) increase display resolution. In addition, the increased hole mobility in poly-Si makes it possible to fabricate not only *n*-type, but also *p*-type TFTs. The performance (especially the mobility and speed) of both *p*- and *n*-type poly-Si TFTs is good enough to be applied in the peripheral (e.g. row and column driver) circuits on the glass substrate as well as for the switch elements in the pixels of a poly-Si TFT based display, which can greatly reduce the cost of integrated display driver circuits (IC) and the manufacturing complexity of the panel.

3.5.1 Carrier Transport in Polycrystalline Silicon

Poly-Si consists of grains joined together at grain boundaries. Within the grain, the atoms are arranged in a periodic manner. At the grain boundaries, there are many defects, such as strained Si–Si bonds and dangling bonds. The defects associated with the grain boundaries create trapping states within the bandgap, which degrade the TFT performance. For instance, the mobility and subthreshold slope are reduced, the threshold voltage and dark leakage current are increased, and the electrical bias stress stability is worse.

A simplified model which includes carrier trapping at the grain boundary has been proposed by Seto and can be used to understand the electrical transport properties of poly-Si [11]. In this model, poly-Si is assumed to consist of identical crystallites having a grain size of L. Carriers are introduced into the poly-Si by doping, but these free carriers are trapped by defect states associated with the grain boundary and depleted from the grains. Due to this trapping of charge, potential barriers are formed at the grain boundaries. Under the condition that the crystallite is completely depleted of carriers and the traps are partially filled, the potential barrier height at the grain boundary, V_B, is given by

$$V_B = qLN_d/8\varepsilon \tag{3.17}$$

where q, L, N_d, and ε are the electron charge in Coulombs (C), grain size in cm, doping concentration in units of cm^{-3}, and dielectric permittivity in units of Farad/cm (F/cm), respectively. The potential barrier height increases linearly with the number of charges localized there, and therefore with the doping concentration. When all the traps are filled, any additional carriers introduced by doping can screen the trapped charges at the grain boundary, resulting a potential barrier height of

$$V_B = qQ_t^2/8\varepsilon N_d \tag{3.18}$$

where Q_t is the number of traps at the grain boundary in units of cm^{-2}. Under this condition, V_B is inversely proportional to the carrier concentration. Figure 3.10 illustrates the potential barrier height as a function of doping concentration. One can see that the potential barrier at the grain boundary reaches a maximum of $V_B = qQ_t/8\varepsilon$ at $N_d = Q_t/L$, where all the traps are filled and the grain is completely depleted. Based on the

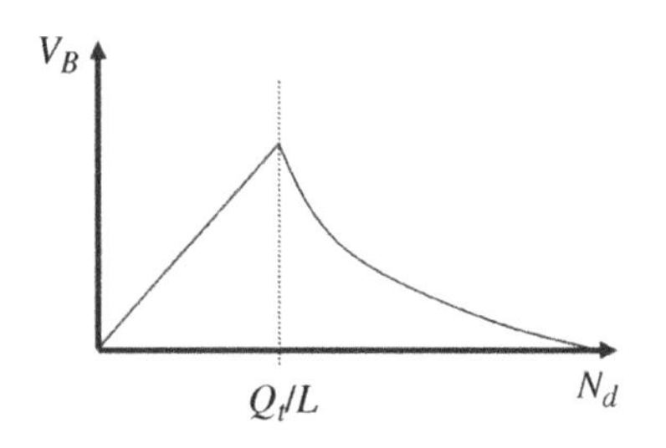

Figure 3.10 Potential barrier height at the grain boundary as a function of doping concentration in poly-Si.

analysis above, it is evident the electrical properties of poly-Si improve as the grain size increases and the defect density reduces.

3.5.2 Fabrication of Polycrystalline-Silicon

To obtain a high quality poly-Si layer, a feasible way is to melt a-Si with no (or little) heating of the substrate. During the cooling process of Si from the liquid to solid state, it crystallizes from some Si seeds, and forms a larger grain size. Typically, a laser with a suitable wavelength, which can be effectively absorbed by the a-Si but to which the glass substrate is transparent, is selected for melting the a-Si and transforming it to poly-Si [12]. As mentioned above, a-Si contains a substantial proportion of hydrogen after thin film growth by PECVD. Under high power laser irradiation, the Si–H bonds break quickly to release H_2 gas which, in escaping out of the poly-Si, can lead to film ablation. Hence, a dehydrogenation process is needed before crystallization which involves heating the substrate or irradiating the layer with a laser at a lower power density. However, after forming a large grain size poly-Si layer, there are still many traps formed by dangling bonds at the grain boundaries which require another hydrogenation process to passivate them. Typically, the defect density decreases by one order of magnitude after this passivation process, compared to the as-grown poly-Si.

When choosing a suitable laser to melt Si, a shorter wavelength is preferred since the associated absorption coefficient is larger. However, the laser wavelength cannot be too short or it will be absorbed in the glass substrate. Typically, an excimer laser such as XeCl with its emission wavelength at 308 nm is used for Si recrystallization, which is therefore termed excimer laser anneal (ELA). Although Xenon (Xe) is an inert gas, in the excited state created by an electric discharge it reacts with Cl and forms XeCl, also in an excited state. It is this species which emits light as it transitions to the ground state and then dissociates into the constituent atoms once more. XeCl is a pulsed laser with a pulse duration of several tens of nanosecond (ns). Within a short period, the Si film will be heated above its melting point. Then it cools down and crystallizes into larger grains. Due to the short pulse width, the heat delivered to the layer is not too much, hence it can dissipate with little heating effect on the substrate. Glass exhibits a low thermal conductivity and the heat may melt the surface of the glass substrate beneath the Si, but it can quickly dissipate through the bulk of the glass to the air at the other side.

The laser crystallization process converts optical energy into thermal energy which melts the Si. Since the light propagates from the surface of the thin-film toward the glass substrate, the Si on the top side melts first. As the laser fluence is increased, more Si becomes liquid and finally the film is melted completely. Figure 3.11 shows the mobility and grain size resulting from different laser fluences. It shows a clear dependence between the two parameters. Micrographs of the poly-Si thin film are shown in the insets. One can see that grain size and TFT mobility at first increase and then decrease as the laser fluence increases. Under exposure to the low energy laser, the film retains its uniform a-Si structure without being transformed into poly-Si, so the grain size and TFT mobility remain low. When the pulse energy increases, a-Si starts to melt and forms poly-Si at the top of the thin-film. At this stage, the melt depth is less than the film thickness, and poly-Si solidifies from the seeds provided by the underlying a-Si, This is called the "partial-melting" region. As the laser fluence further increases, the thin film is melted through. However, some Si still remains in the solid phase at the Si–glass interface. It persists as discrete islands, rather than a continuous film, but still seeds the

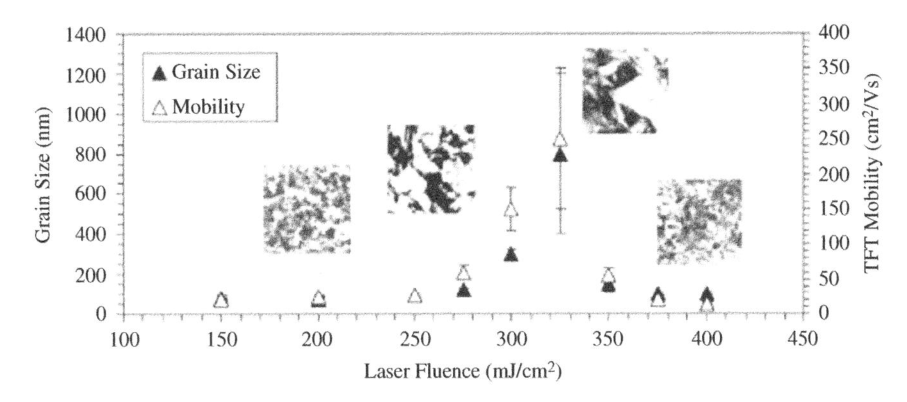

Figure 3.11 Grain size, TFT mobility and photographs of thin-film under different laser fluence irradiation [13].

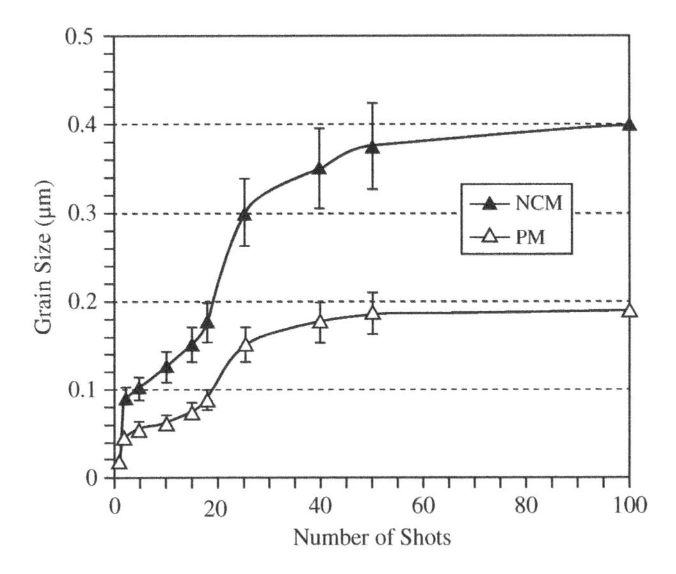

Figure 3.12 Grain size with different shot numbers. NCM and PM: near-complete-melting and partial-melting [13].

recrystallization process. If the pulse energy is further increased, the number of nucleation sites decreases, resulting in an increase in the grain size. The crystallite growth direction is both vertical and lateral. During this region, grain size and TFT mobility increase with increasing laser fluence until it reaches a maximum at about 325 mJ/cm^2, as shown in Figure 3.11. This is called the "near-complete-melting" region. When the laser energy is very high, the whole thin film becomes molten and no nucleation sites remain. Crystallization nucleates spontaneously somewhere in the liquid silicon, and hence the crystallization is nearly homogeneous without any preferred orientation, and the grain size decreases again. This is called the "complete-melting" region.

One can see that, although it is possible to obtain a large grain-size and high mobility poly-Si, the process window is quite small (<50 mJ/cm^2) which results in difficulty in achieving large area, stable TFT manufacture. Besides, since the irradiation area of the excimer laser is typically much smaller than the glass substrate area, a scanning stage is used in the ELA process which necessitates a uniform and stable laser beam. To increase the process tolerance and further improve the device performance, one can use a multi-shot line beam scanning technique. As shown in Figure 3.12, with increasing the shot numbers, the grain size shows a significant increase and the effect of shot-to-shot variation decreases.

3.6 THIN-FILM TRANSISTORS

3.6.1 Fundamentals of TFTs

A TFT is a device with three terminals. One is called the "gate (G)" which on application of a voltage can switch open or closed, a conduction channel between the other two terminals, called "source (S)" and "drain (D)," as shown in Figure 3.13. In an *n*-channel TFT, the conducting carriers are electrons. Hence, electrons drift from the source electrode to the drain in the electric field applied between those terminals, which results in current flow. The switching structure is based on a metal–insulator–semiconductor (MIS) structure, which is also used widely in the integrated circuit (IC) industry with crystalline Si. In such a structure, a potential applied to the (gate) metal can attract or repel carriers at the semiconductor side of the insulator, thereby changing the carrier concentration in the semiconductor channel. Effectively this modulates the channel between high and low conductance states, and hence can be used to switch the device on and off.

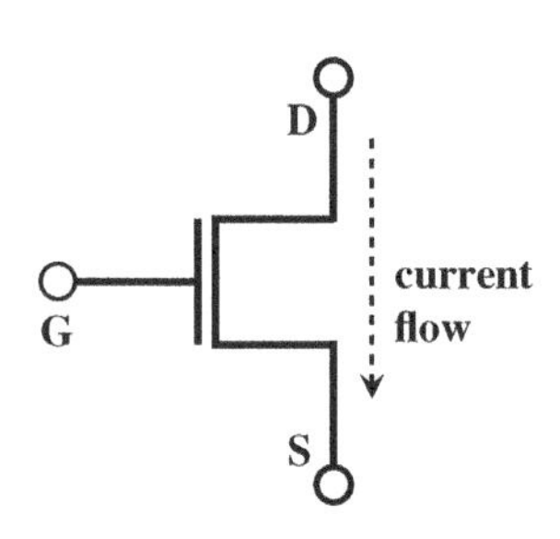

Figure 3.13 Schematic diagram and notation of a TFT.

Figure 3.14a shows the energy band diagram of an ideal MIS structure when $V_G = 0$. Here, we first consider the case when the semiconductor is crystalline Si with a perfect lattice and no traps. Also in this figure, we suppose that the work function of the metal (Φ_m, which is the energy needed to promote an electron from the metal to the vacuum) is aligned with the Fermi level of the semiconductor, which is not always the case when using different metals and semiconductor materials. The insulator has a very large bandgap and there are no carriers inside this layer. When a positive voltage is applied to the metal, it attracts negative charges toward the insulator interface side of the semiconductor, as shown in Figure 3.14b. Those carriers accumulate at the interface which increases the electron density at this boundary. From Eq. (3.12), we can see that this increase in electron density implies that the E_F and E_c come closer to the same value, which results in band bending of both conduction and valence bands, as shown in Figure 3.14b. In a uniform semiconductor layer (having the same material and doping concentrations throughout), the Fermi level must remain flat. If that were not the case, since the Fermi level describes a constant probability of an electron occupying a state at that energy, electrons would flow from the region of high E_F to one of lower E_F until it becomes flat. In Figure 3.14a, since all the bands are flat, this case is usually called the "flat band" condition which is however not necessarily realized at $V_G = 0$ for non-ideal cases.

TFT performance can be modeled and evaluated in a similar way to a conventional metal-oxide semiconductor field-effect transistor (MOSFET). Gate voltage affects the channel conductance and controls the switching

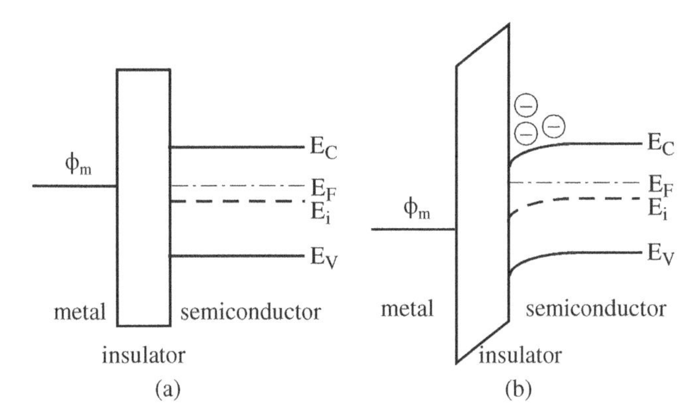

Figure 3.14 Band diagram of an ideal MOS structures at (a) $V_G = 0$, and (b) $V_G > 0$.

on and off of this conductance. In an n-channel TFT, (conventional) current flowing from drain to source (I_D) is modulated by the gate, and also depends on the drain voltage (V_D). Typically, the source is grounded ($V_S = 0$). The I–V characteristics of an n-channel TFT can be described by the following equations:

$$I_D = \frac{1}{2}(\mu_n C_i)\left(\frac{W}{L}\right)[2(V_G - V_T)V_D - V_D{}^2], 0 \leq V_D \leq V_G - V_T \tag{3.19}$$

$$I_D = \frac{1}{2}(\mu_n C_i)\left(\frac{W}{L}\right)(V_G - V_T)^2, V_D > V_G - V_T \tag{3.20}$$

where μ_n is electron mobility (cm^2/V·s), C_i is the capacitance per unit area of the insulator layer (F/cm^2), W is the channel width, L is the channel length (the distance between the source and drain), and V_T is the threshold voltage, the minimum gate voltage needed to create a conducting path between source and drain.

Figures 3.15a–c show typical electrical characteristics of an a-Si TFT. The I_D–V_D curves with different V_G values are called output characteristics, as shown in Figure 3.15a. For a positive and sufficiently large V_G, there are mobile electrons in the channel, which make it conductive. Then, with an increase of V_D, the I_D continues increasing until $V_D = V_G - V_T$, which is called the "linear region." Since the channel region resembles a resistance, it is reasonable that higher V_D results in higher current. With a further increase in the

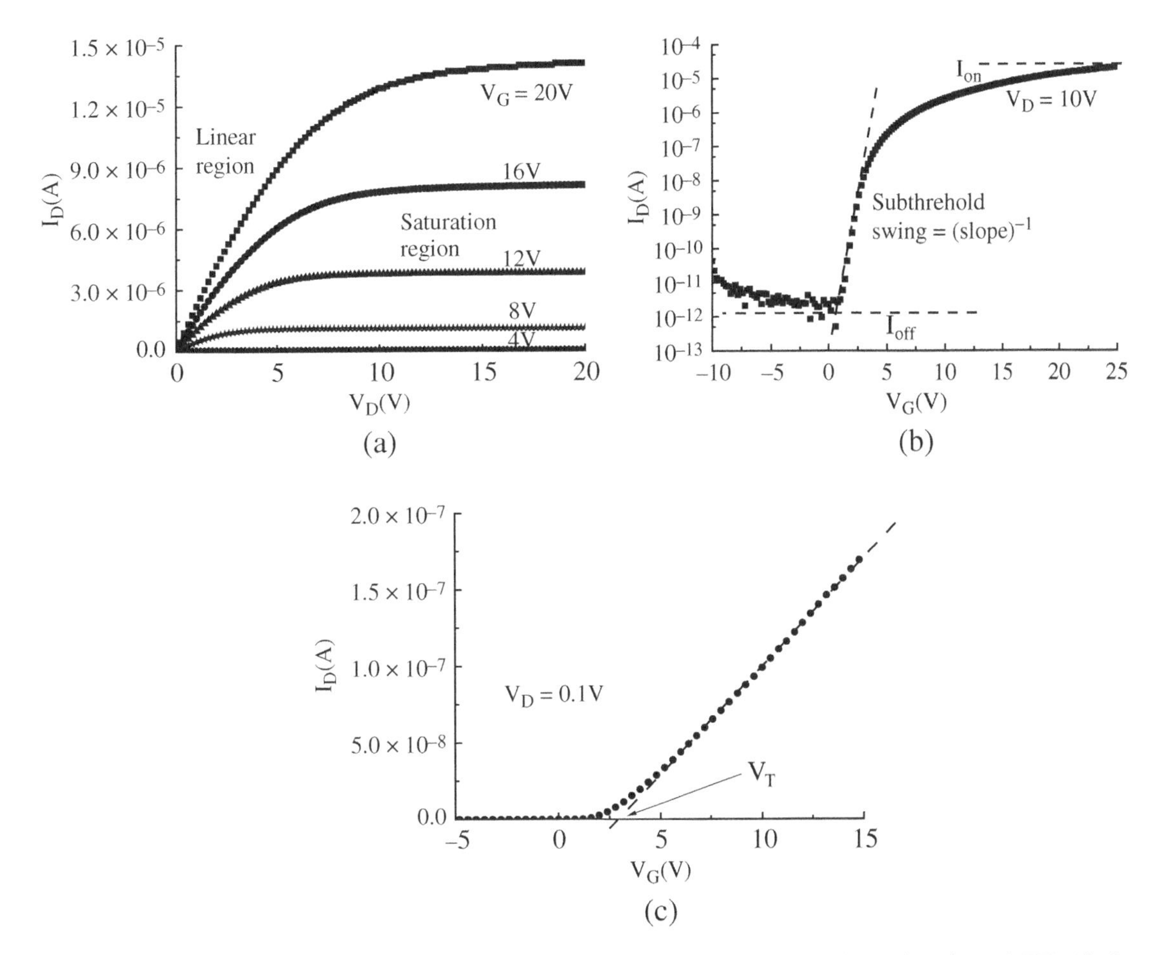

Figure 3.15 (a) Output characteristic, and (b) transfer characteristic, and (c) I_D-V_G at small V_D value of an a-Si TFT with channel width and length of 200 μm and 27 μm, respectively.

voltage, the channel cannot support a higher current density, which leads to a "saturation region." The saturation current is greater at higher values of gate voltage since this attracts more electrons to accumulate at the insulator-semiconductor interface which can then contribute to current conduction.

Figure 3.15b shows the log(I_D)-V_G curves at V_D = 0.1 and 10 V; these are called "transfer characteristic curves." When V_D = 10 V, one can see that with increasing V_G, the I_D is very low (at about 10^{-12} A) for V_G less than about −1 V, sharply increases at V_G from 0 to 5 V (from 10^{-12} to 10^{-6} A), and reaches an ON-state current of $\sim 10^{-5}$ A. From the transfer characteristic, one can obtain the "ON/OFF ratio" of a TFT which is the ratio of the ON current to the minimum measured current. Higher ON/OFF ratios are preferred since a higher ON current can provide better driving of a pixel, and lower OFF current corresponds directly to lower current leakage. As shown in Eqs. (3.19) and (3.20), increasing the mobility value effectively helps to improve the ON current. To quantitatively determine the V_T value, it is easier to observe the I_D–V_G curve with a linear scale. Figure 3.15c shows the I_D–V_G curve at a small V_D = 0.1 V in the low I_D region. Here, the TFT is operated in the linear region due to the small V_D value. Hence, Eq. (3.19) can be approximately described as:

$$I_D \sim (\mu_n C_i)\left(\frac{W}{L}\right)(V_G - V_T)V_D \tag{3.21}$$

since the ${V_D}^2$ term is very small and can be neglected. The V_T is the voltage required to switch on this TFT, and its value can be measured from the x-intercept since $I_D = 0$ at $V_G = V_T$. Once V_T is obtained, one can determine the mobility value in such a device from a fit of data to Eqs. (3.19) and (3.20). Above V_T, there would be a significant current passing through the channel. On the other hand, one can define a "subthreshold swing" by reference to Figure 3.15b. It is the inverse of the slope of log(I_D) versus V_G, in terms of (V/decade). This parameter quantifies how much voltage change is needed to switch on and off the TFT. A smaller subthreshold swing is preferred since it results in higher speed and lower power consumption.

Typical TFTs can be categorized as normal or inverted structures, depending on whether their gate is on the top or bottom, respectively. Another classification depends on whether the drain/source and gate electrodes are on the same or opposite sides of the channel, termed coplanar or staggered structures, respectively. Hence, there are basically four different kinds of TFT structures, which are shown in Figure 3.16. Different structures require different fabrication procedures and result in different device performances. The staggered bottom-gate structure is commonly used in a-Si:H TFTs, and a coplanar top-gate structure is widely adopted in poly-Si TFTs. The majority of organic TFTs are built with a bottom-gate configuration, either staggered or coplanar. For oxide semiconductor TFTs, all of these four configurations have been employed.

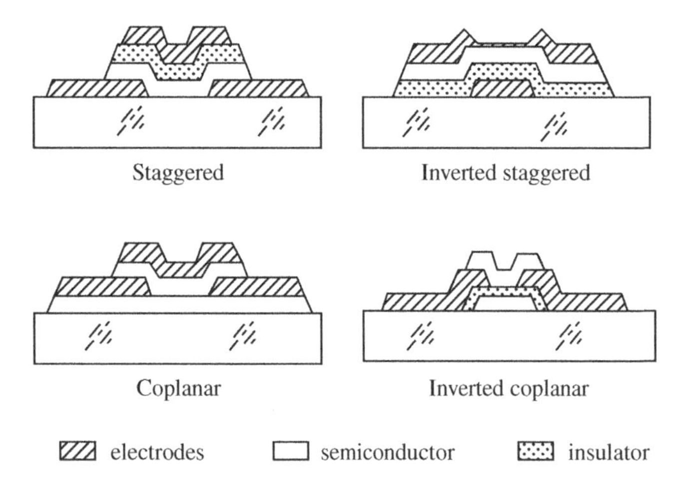

Figure 3.16 TFTs with different configurations [14].

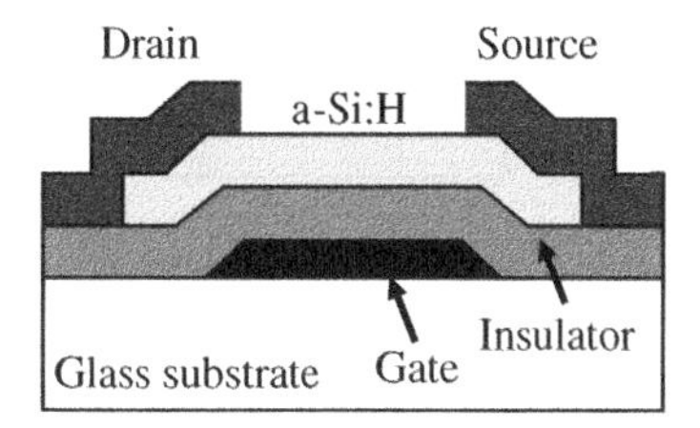

Figure 3.17 Cross section of an a-Si:H TFT.

3.6.2 a-Si:H TFTs

a-Si:H TFTs are widely used as switching devices in active matrix arrays. Figure 3.17 shows a typical cross section of an a-Si TFT fabricated on a glass substrate. An inverted staggered structure with a SiN_x gate dielectric offers advantages of easy fabrication and better electrical performance compared to other structures. In addition, the a-Si:H channel is sensitive to light. With this device structure the bottom metal gate can effectively shield the channel from light illuminating the bottom side of an AMLCD display system to avoid light-induced degradation of the a-Si:H.

When using a-Si:H as the semiconductor material, the band bending behavior is similar to that of crystalline Si, shown in Figure 3.14, with some small modifications. The flat band condition at $V_G = 0$ is typically true for an a-Si:H TFT [14]. However, when a positive voltage is applied to the metal gate electrode (small V_G), the trap states in a-Si become important: electrons accumulated at the oxide/semiconductor interface first fill deep states and tail states which have zero mobility. Those carriers are localized and immobile. When V_G is further increased above the threshold voltage (V_T), it forms a thin layer with mobile carriers at the interface between SiN_x gate insulator and a-Si:H channel layer. Hence, one can use the V_G to create and control an electron path at this interface. The typical field-effect mobility, threshold voltage and ON/OFF ratio of a-Si:H TFTs are 1 to 1.5 cm^2/V s, 1 to 3 V and 10^6 to 10^8, respectively.

The fabrication process for an a-Si:H TFT is as follows. The gate electrode, which is typically Cr, Ta, or Al, is deposited first. Then, a uniform insulating layer is formed on top of the gate electrode, usually of silicon nitride (SiN_x) deposited by PECVD. Next, the a-Si:H is deposited on the top of the insulator to form a MIS structure. On the right and left sides of the a-Si:H layer, drain and source electrodes are formed, which typically consist of multiple layers, i.e. heavily doped *n*-type a-Si:H (n^+ a-Si:H) followed by Cr/Al. The n^+ a-Si:H forms an ohmic contact with the channel layer (a-Si:H) and provides better electron injection.

3.6.3 Poly-Si TFTs

Field-effect mobilities in poly-Si are considerably higher than in a-Si:H. Depending on the preparation method and quality of the poly-Si, the field-effect mobility ranges from 10 to 500 cm^2/V s for electrons and 10 to 200 cm^2/V s for holes. The high mobility allows use of poly-Si TFTs as both driving and switching devices. In addition, compared to a-Si:H TFTs, narrower channels can be used in switching devices. This allows a higher aperture ratio and reduces the parasitic gate capacitance. Furthermore the ability to fabricate both *n*- and *p*-channel TFTs in poly-Si, makes complementary metal-oxide-semiconductor (CMOS) circuitry possible which promises fully-integrated circuits with low-power consumption.

Due to the limitations of the laser recrystallization process used to convert a-Si to poly-Si, the starting point for poly-Si TFT fabrication is usually a flat and continuous silicon thin film without other layers beneath. Therefore, they are usually fabricated with a coplanar top-gate structure. As with crystalline silicon MOSFETs, SiO_2 gate insulators offer a better quality interface with poly-Si channels. Figure 3.18 shows the typical process flow for a poly-Si TFT. A dielectric buffer layer is first deposited on the substrate, followed by the deposition of an a-Si layer. The film is then crystallized by excimer laser annealing as described in Section 3.5.2. After crystallization, a SiO_2 layer is deposited by chemical vapor deposition as the gate insulator, followed by the a-Si gate deposition. After the gate definition, ion implantation is performed to make the source and drain regions and to dope the a-Si gate, and a furnace anneal step is carried out to activate the implants. Crossover

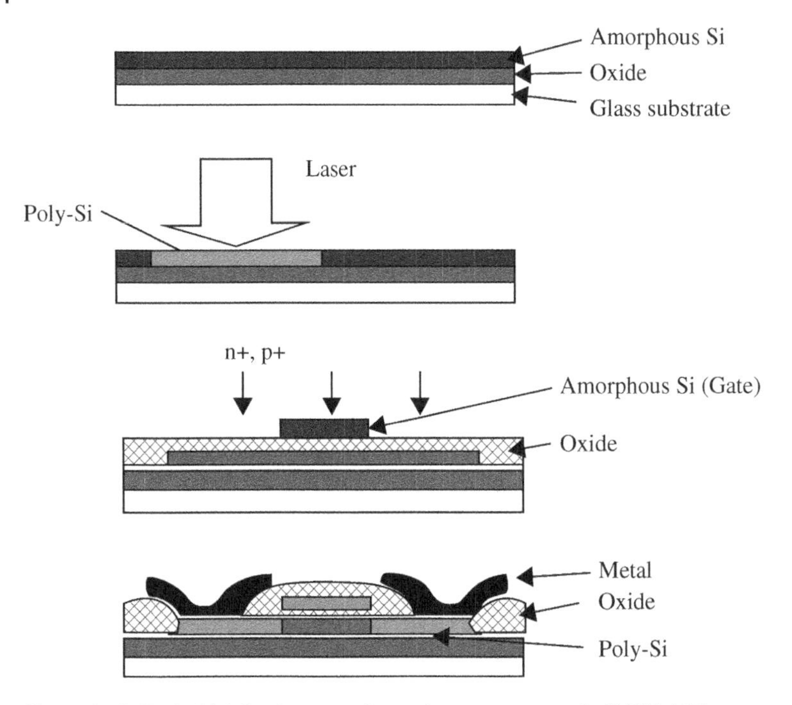

Figure 3.18 Typical fabrication steps for coplanar top-gate poly-Si TFTs [15].

isolation oxide is then deposited, and contact holes are etched. Finally, a metal layer is deposited and patterned to form the source, drain and gate contacts.

3.6.4 Organic TFTs

One of the most important advantages for organic thin film transistors (OTFTs) is that the fabrication process can be conducted at low temperatures, compatible with flexible substrates [16]. There are several methods to form the organic thin film. Organic semiconductors with a low molecular weight, often have sufficient vapor pressure that films can be deposited by thermal evaporation. Most OTFTs are based on p-type channels. The charge transport in organic semiconductors will be discussed in Section 6.4.2. Pentacene is a common choice of organic material with low molecular weight which can form the basis of a p-channel TFT. The molecular structure of pentacene is shown in Figure 3.19a. The mobility value achieved in an OTFT is dependent on many parameters, such as the configuration of molecular stacking in crystallites, the surface morphology, and the grain size. By controlling the fabrication parameters such as deposition rate and substrate temperature, one may increase the grain size of thin film pentacene to tens of nanometers, and improve the mobility to $\sim$1 cm^2/V s [17]. By growing a single crystal organic thin film from a compound such as rubrene (molecular structure as shown in Figure 3.19), OTFTs can be made with a record mobility value as high as $\sim$20 cm^2/V s [18] or more.

Many organic semiconductors are sufficiently soluble in well-chosen solvents, to permit solution processing. Then techniques such as spin-coating, ink-jet printing, stamping, or imprinting the solution onto the substrate can be used [19]. After evaporation of the solvent, the organic material is left as a thin film structure. Details of the fabrication of organic thin films will be described in Section 6.5.3.

Many OTFTs are sensitive to environmental species, such as water and oxygen, which makes them suitable for sensor applications [20]. However, that also means it is difficult to obtain a stable device for display applications. During device fabrication, formation of the organic thin film is typically directly followed by a passivation process to avoid any attack by environmental species. Figure 3.20 shows two common device

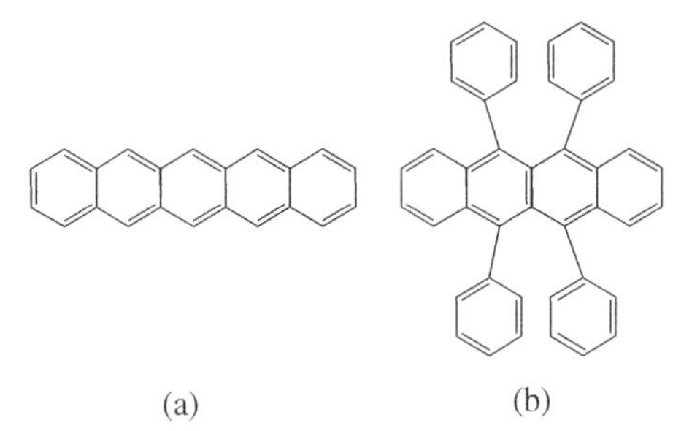

Figure 3.19 Molecule structure of (a) pentacene, and (b) rubrene.

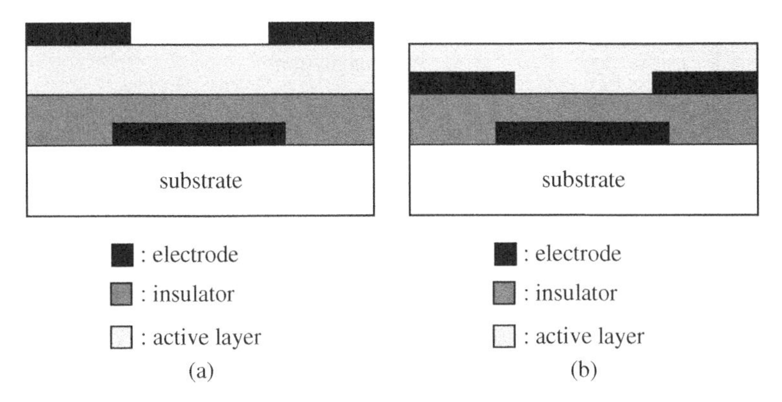

Figure 3.20 Device configurations of OTFT: (a) top and (b) bottom contact.

configurations [21]. Both are inverted structures since the gate insulator must be formed before the sensitive organic thin film. For a bottom contact device, the organic layer is deposited on the top of the structure. The drain, source, gate, and insulator layers can be defined by conventional photolithography which exhibits a high resolution (less than 1 μm). On the other hand, for a top contact device, the organic material is deposited followed by evaporation of the drain and source electrodes through a shadow mask. Resolution is limited to several tens of μm in this configuration. Typically OTFTs with a top contact configuration exhibit performance superior to bottom contact devices due to their larger contact area and lower contact resistance. Molecular thin-films usually exhibit a much lower mobility value than semiconductor materials with covalent bonds, which means that the electrical current provided by an OTFT is typically smaller than a-Si:H and poly-Si TFTs. It is also difficult to change the carrier concentrations of the molecules by intentional doping, which limits the flexibility for device design. The long term stability under operation and storage conditions is also still an issue for OTFTs [22].

3.6.5 Oxide Semiconductor TFTs

Due to their light sensitive characteristics, an opaque shield is needed for the active region of most TFTs which limits the aperture ratio of a display. By using wide bandgap semiconductors, such as zinc oxide (ZnO), there is little absorption within the visible wavelength range. Therefore, it is possible to fabricate a light insensitive, transparent thin film transistor (TTFT), which can greatly increase the aperture ratio [23]. Figure 3.21 shows the device structure and transmission spectrum of a ZnO TFT. Here, all the electrodes (gate, drain, and source) are made of transparent indium tin oxide (ITO). The insulator is made of aluminum–titanium oxide (ATO), which is also transparent. As shown in Figure 3.21b, the average transmission over the visible wavelength range through the channel and source/drain region is about 75%.

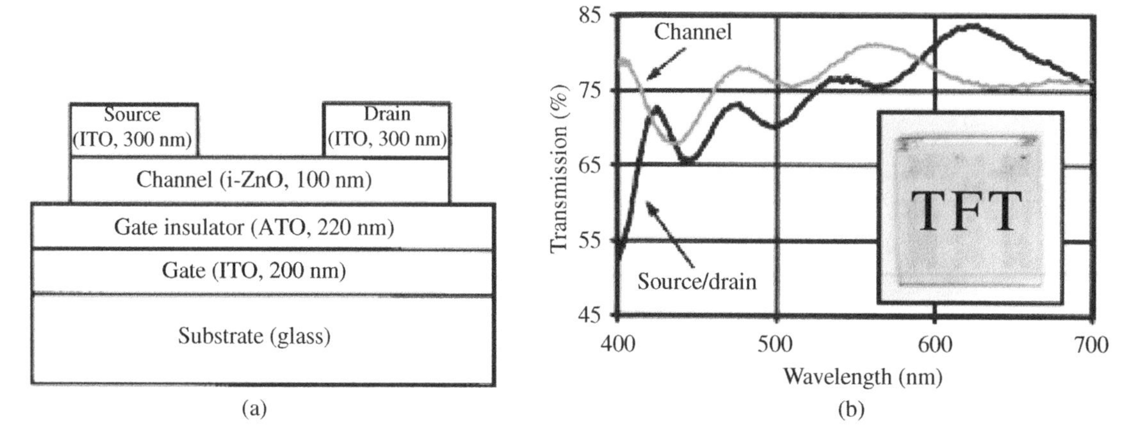

Figure 3.21 (a) Device structure and (b) transmission spectra of TTFT [23].

Another advantage of using oxide semiconductors as TFT active layers comes from their high mobility (typically 10s of $cm^2/V\ s$) in comparison to a-Si:H. Oxide thin films can be formed by physical vapor deposition (PVD), such as sputtering or pulsed laser deposition [24]. Unlike silicon-based TFTs, an amorphous phase is preferred for oxide semiconductor TFTs since crystallization results in a high surface roughness and large leakage current, which reduces the on/off ratio of TFTs. It is interesting that the mobility value does not differ a great deal for oxide based materials in their crystalline and amorphous phases. In such materials, electrons mainly propagate along networks of large metal atoms, which have good wavefunction overlap both in the crystalline and amorphous phases, as shown in Figure 3.22. There are some requirements on the material selected: (i) a tendency to form an amorphous phase during PVD, (ii) a high carrier mobility, and (iii) a

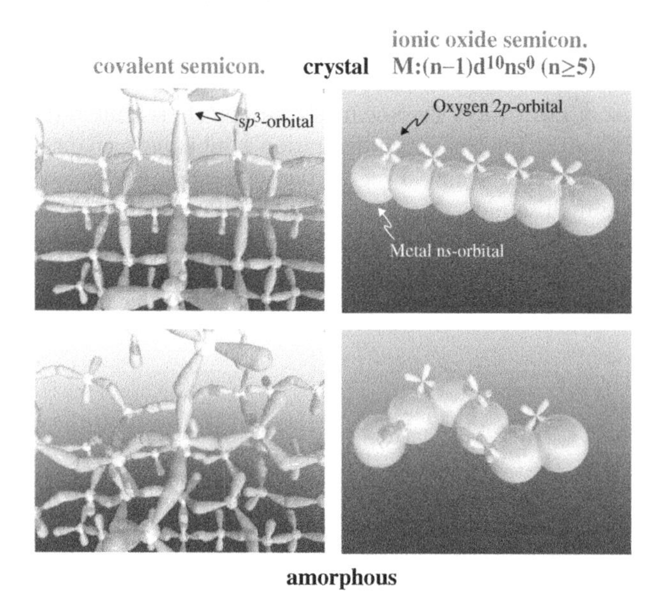

Figure 3.22 Schematic orbital drawing of electron pathway in conventional compound semiconductor (left) and ionic oxide semiconductors (right). Oxide semiconductors composed of post-transition-metal cations. Spheres denote metal s orbitals. The contribution of oxygen 2p orbitals is small. Direct overlap between neighboring metal s orbitals is rather large, and is not significantly affected even in an amorphous structure. [25]

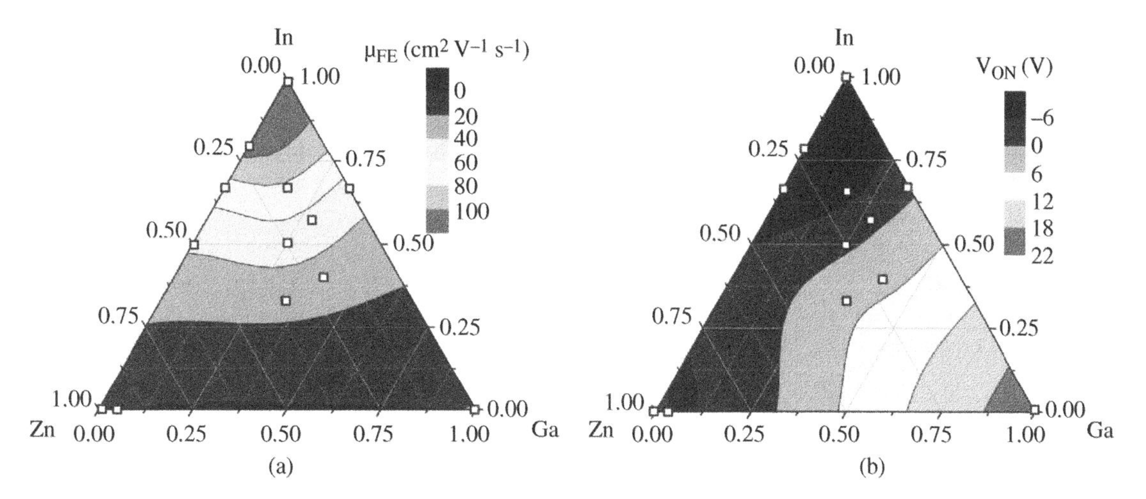

Figure 3.23 (a) Field-effect mobility and (b) threshold voltage obtained for TFTs with different oxide semiconductor compositions, in the IGZO system [27].

low carrier concentration in order to achieve a low off current. By reducing the process temperature, oxide semiconductor TFTs can be also fabricated on flexible substrates.

IGZO TFTs have shown the best performance among various types of oxide transistors [26]. In the IGZO material system, the performance of TFTs is influenced by the composition, as shown in Figure 3.23. Raising the indium content leads to an increase in both field-effect mobility and background carrier concentration. Because gallium bonds strongly with oxygen, the addition of gallium can suppress the background carrier concentration, resulting in a considerable shift of threshold voltage toward more positive values. Today IGZO TFTs have been adopted in the industrial production of some high-end displays.

3.6.6 Flexible TFT Technology

Flexible displays exhibit the advantages of being rugged, thin, and lightweight. Potentially, displays on flexible substrates may be fabricated by a roll-to-roll process, and therefore achieve a very low manufacturing cost. To fabricate a flexible TFT backplane, three types of substrates are commonly used: flexible glass substrates, metal foil substrates and polymer film (plastic) substrates. A comparison of various types of flexible substrates is presented in Table 3.1. Because flexible glass is inherently fragile and difficult to handle, metal foil and polymer substrates are better choices. Metal foil substrates are optically opaque and are only suitable for reflective or top emissive displays. Their major advantages include: good durability, compatibility with high process temperature, high dimensional stability, and excellent permeation barrier properties against moisture and oxygen. Moreover, they can serve as heat sinks and provide electromagnetic shielding. However, most metal foil substrates come with rolling marks and inclusions, which must be smoothed by surface polishing and/or a planarization layer. In addition, to reduce the substrate leakage current and parasitic capacitance a thick dielectric layer is often deposited prior to device fabrication. Plastic substrates are highly flexible, compatible with the roll-to-roll process and can be optically transparent. However, to fabricate electronic devices on plastic substrates, several important issues need to be taken into consideration, such as a low maximum process temperature, the poor dimensional stability of the substrate, and a mismatch of the thermal expansion coefficient between device layers and the substrate. To mitigate these issues, plastic substrates with a high glass transition temperature and low coefficient of thermal expansion are preferred. Because most polymer film substrates are highly permeable to gas and moisture, it is crucial to coat them with thin-film permeation barrier layers to improve the environmental stability of organic-based electronics implemented on top. Thin-film barrier coating technology can also be applied to encapsulate the organic devices.

Table 3.1 A comparison of various types of flexible substrates.

Property	Glass	Stainless steel	Polyethylene naphthalate (PEN)	Polyethylene terephthalate (PET)	Polyimide (PI)
Thickness (μm)	100	100	100	100	100
400–700 nm Optical Transmission (%)	> 92	No	87	89	Yellow
Young's Modulus (GPa)	70–80	200	6.1	5.3	2.5
Max process temperature (°C)	600	1000	~180 (T_g = 121 °C)	~150 (T_g = 78 °C)	300–400 (T_g = 360–410 °C)
Coefficient of Thermal Expansion (ppm/K)	3–5	10	13	15	16–17
Moisture Absorption (%)	None	None	0.14	0.14	1.8
Permeable to O_2 and H_2O	No	No	Yes	Yes	Yes
Electrical Conductivity	None	High	None	None	None

The TFT fabrication process involves accurate registration between device layers. The "transfer technique" and "direct fabrication method" are two major approaches for producing TFTs on compliant substrates. In the transfer technique, the device is fabricated by a standard process on a conventional rigid substrate, and then it is transferred onto a flexible substrate. An example is the surface-free technology by laser annealing (SUFLTA) method developed by Seiko Epson Corporation [28]. Although this approach can provide flexible TFTs with optimal performance, the cost is high compared to a direct fabrication method. Many efforts have been made to develop direct fabrication methods. In one method, the compliant substrate is laminated onto a rigid carrier with or without a temporary adhesive applied in between [29, 30]. If an adhesive is used, the highest processing temperature may be limited by the adhesive. Another method, coat-release, is more often used in industry. In this method, a sacrificial layer or a release agent is applied to a rigid carrier, prior to solution coating of the polymer film which will form the flexible substrate. After the device fabrication process is completed, the devices along with the flexible substrate can be released from the carrier substrate by laser irradiation, in processes such as electronics on plastic by laser release (EPLaR™) developed by Philip [31], or by mechanical force, as in the flexible universal plane (FlexUP) process introduced by Industrial Technology Research Institute [32]. To further reduce the fabrication cost and improve throughput, solution-based and printing processes can be employed for TFT fabrication [33–35].

To make a flexible TFT backplane, interconnections and contacts must be flexible. Percolating networks of metals, conductive carbon-based nano-materials and their combinations have been studied extensively for flexible transparent conductor applications [36, 37]. For instance, nanoimprinted metal grids [38] or metal nanomeshes obtained by grain boundary lithography [39] and solution processed metallic nanowires and carbon nanotubes have shown promising electrical conductance [36, 37]. Graphene and graphene derivatives have also been investigated as the flexible transparent conductive materials [40]. It is worth noting that a compromise must be made between the optical transparency and electrical conductance in all the above mentioned flexible conductors.

In addition to flexible conductors, flexible encapsulation is another essential component in a flexible display, especially when organic functional materials are used. In principle, a thin layer of an inorganic material such as SiO_2, SiN_x or Al_2O_3 can serve as a permeation barrier against oxygen and moisture. However, most single-layered inorganic encapsulants have their barrier properties seriously degraded by diffusion paths caused by microscopic defects, in particular granular film growth, cracks above occluded dust particles, and pinholes. To overcome these difficulties, multilayered structures are adopted instead to provide long and tortuous pathways for permeating species. For instance, the Barix™ layer developed by Vitex Systems consists of alternating layers of thin ceramic inorganic layers and thick organic polymer layers. The inorganic sublayers serve as the permeation barrier to oxygen and moisture, while the organic sublayers mechanically decouple the inorganic sublayers from one another to suppress crack or defect propagation and provide sufficient mechanical flexibility. Based on this concept, research efforts have been directed towards reducing the number of sublayers, decreasing the thickness by adopting atomic layer and molecular layer deposited sublayers [41], and developing organic–inorganic hybrid materials as single-layered structures [42]. These encapsulation techniques can be applied to encapsulate both flexible polymeric substrates and organic devices, such as OLED, in flexible displays.

The influence of mechanical stress and strain on the electrical performance of flexible a-Si:H TFTs has been studied extensively. In general, the normalized field-effect mobility of a-Si:H is linearly proportional to the applied mechanical strain. Reduction and enhancement in the field-effect mobility are observed when respectively compressive and tensile mechanical strain is applied to the TFTs [43, 44]. When the strain becomes large, failure of a-Si:H TFTs occurs due to fracture (or buckling) of the device layers.

Regarding the effect of mechanical strain on the electrical performance of poly-Si TFTs made by excimer laser irradiation, the results are quite diverse. For instance, Kuo et al. found that the electron mobility increases and the hole mobility decreases when the poly-Si TFTs are subjected to a mechanical tensile strain, and the change in mobility saturates as the strain increases above 0.6% [45]. Peng et al. observed an increase and decrease of hole mobility when a compressive and tensile strain are applied to a poly-Si TFT, respectively [46]. Kim et al. showed unchanged performance of poly-Si TFTs under both tensile and compressive strains [47].

The changes in electrical performance of oxide semiconductor TFTs under mechanical strain reported from different research groups are more consistent. It has been shown that the mobility of n-channel oxide semiconductor TFTs, such as IGZO and ZnO TFTs, is enhanced and decreased when the TFT is subjected to tensile and compressive strains within the elastic region [48–50]. The threshold voltage of the oxide TFTs exhibits negative and positive shifts under tension and compression, respectively. The detailed mechanisms responsible for the change in the electrical performance of these flexible TFTs under the influence of mechanical strain remains a wide open field.

3.7 PM AND AM DRIVING SCHEMES

To drive a LCD or an OLED display, both passive matrix (PM) and active matrix (AM) driving techniques are commonly used [51, 52]. For a PM-LCD, due to the capacitor-like characteristics of the LC layer, voltage applied to a selected pixel inevitably affects the neighboring pixels and degrades contrast of the display, an effect which is called "crosstalk." [53] Moreover, the RMS voltage selection ratio in a PM-LCD with n addressed lines, is limited to a maximum value of $\sqrt{\frac{\sqrt{n}+1}{\sqrt{n}-1}}$ [54]. At large n, this ratio rapidly approaches unity, and an LCD pixel is unable to provide a high contrast which is independent of the angle of view in response to this small voltage change. Furthermore the small voltage discrimination ratio means that both on and off pixels experience a voltage almost equal to the threshold, so there is a critical slowdown in the switching speed. The contrast in complex PM displays is further degraded by distortion of the addressing waveforms within the RC network formed by the pixels and conductive tracks. On the other hand, by using TFTs in LCD and OLED pixels, the turn-on/off and different gray levels required on each pixel can be controlled independently by active matrix (AM) driving, and therefore achieve much higher image quality.

Figure 3.24 shows the operating principle of a PM-LCD. Electrodes in stripe configurations are placed on the two glass substrates. Stripes on the bottom and top glass substrates are aligned perpendicularly to work as column and row electrodes of a PM-LCD, and each row/column intersection area forms a pixel. The LC is filled between the two glass substrates, and hence they together can be regarded as a capacitor. To display a "T" in a 4×4 LCD display, the first row is first selected with a pulse voltage V_s during the first time slot, while a select data voltage of $-V_d$ is applied to the column lines 1 to 4. Each of the pixels in row 1 therefore has a

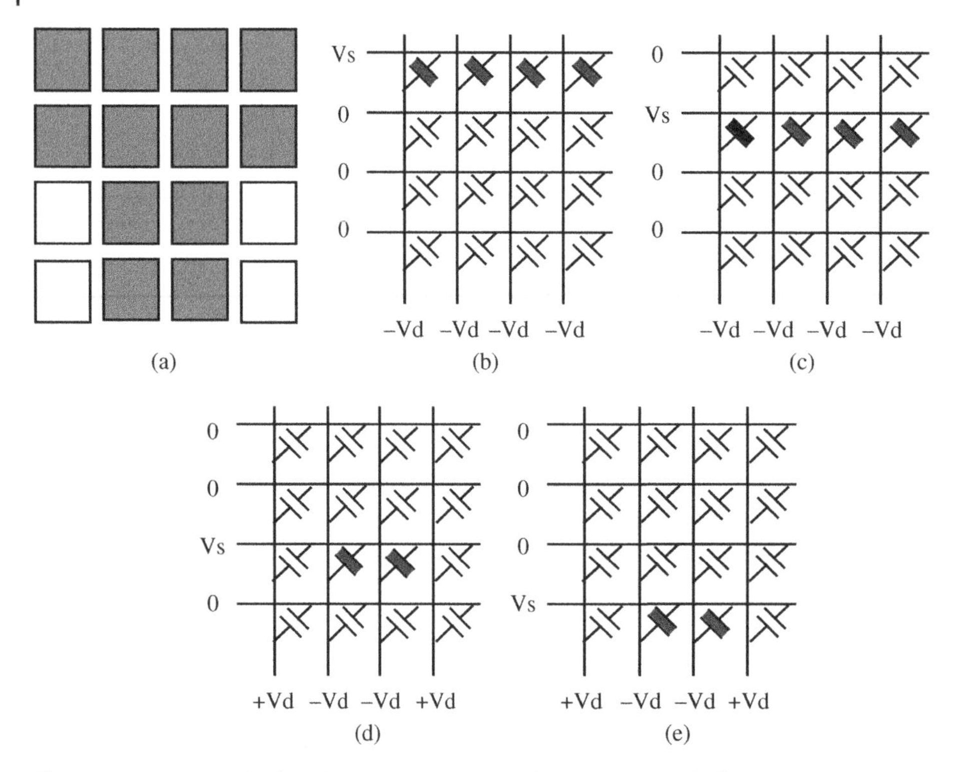

Figure 3.24 (a) image displayed on PM-LCD, and the voltage during the (b) first, (c) second, (d) third, and (e) fourth time slots on each row and column. Solid black capacitors show where the pixel is being switched to the "on" state during the row select time period.

voltage difference of $V_s + V_d$ applied during the first timeslot. The remaining row electrodes 2–4 are held at 0 V during this timeslot, so the voltage difference across all the pixels in these rows during the first time slot is V_d. Then in the second time slot, the voltage on row 1 electrode is set back to 0 V, and the scanning voltage V_s is applied to row 2. The driving voltages for columns 1 to 4 are set to $-V_d$, hence only the pixels in row 2 have a voltage difference of $V_s + V_d$, and all other pixels experience a voltage difference of V_d. During the third time slot, the scanning voltage V_s is applied to row 3, and the driving voltages applied to columns 2 and 3 where pixels are to be turned on are $-V_d$, but for columns 1 and 4 where pixels should remain off a driving voltage of $+V_d$ is used. Thus the voltage difference on the selected pixels (3, 2) and (3, 3) is $V_s + V_d$, while the smaller potential difference $V_s - V_d$ is applied to the non-selected pixels (3, 1) and (3, 4). During this time slot, all other pixels have a voltage applied of either $-V_d$ or $+V_d$. Similar driving pulses are then applied, to selectively address the pixels in row 4 during the next time slot. The LC will respond to the RMS value of the voltages they experience, averaged over all the time slots. In a practical display, slightly more elaborate waveforms are used so that DC balance and optimization of the use of the supply voltage by level shifting are assured during each addressing cycle.

For a PM-LCD with N row lines, only one row is selected at any time. A voltage of V_s is applied to that row to select all the pixels there, and either V_d or $-V_d$ is simultaneously applied from the column line to turn each pixel off or on. Neighboring pixels in the same column will also be affected by the voltage from the same column line. As a result, over a whole frame scan period, each on-pixel has its voltage set to $V_s + V_d$ only for $1/N$ of the frame time, and has either V_d or $-V_d$ applied for the remaining $(N-1)/N$ of the frame time. Similarly, each off-pixel experiences a voltage of $V_s - V_d$ for $1/N$ of the frame time, and V_d or $-V_d$ for the remaining $(N-1)/N$ of the frame time. Since LCs respond to the RMS value of the applied voltage, the RMS voltage applied to on and off-pixels from the above schemes can be calculated:

For "on" pixels: $V_{on}^2 = \frac{1}{N}(V_s + V_d)^2 + \frac{N-1}{N}(V_d)^2$

For "off" pixels: $V_{off}^2 = \frac{1}{N}(V_s - V_d)^2 + \frac{N-1}{N}(V_d)^2$

Each pixel of a PM display can therefore be independently driven to its on or off state, by setting appropriate V_{column} values (of either $\pm V_d$) during the time slot when the corresponding row is selected by addressing it with V_s. The difference between V_{on} and V_{off} in this display depends on the number of rows N, and on the relative magnitudes of V_s and V_d. However, whatever the choice of voltages, the difference between V_{on} and V_{off} approaches zero as N becomes larger. In order to achieve satisfactory contrast in a display, the LCD pixel must be fully switched by V_{on} but unswitched by V_{off}. Practical LC effects always require a finite voltage difference to switch, so the maximum number of rows which can be addressed in a PM display is limited.

Figure 3.25 shows the operational principle of PM-OLED. Here, we take a diode to indicate each pixel of the OLED display. As in the LCD case, to display a "T" in a 4×4 OLED display, the first row is selected with a low voltage. Four OLEDs at the first row are biased at high voltage from the column lines to be turned on. The pixels in row 1 are then switched to an emissive state, during this time slot only. Then, during the next time slot, the second row is biased with a low voltage, and so on through the following time slots and rows. The advantages of the PM-OLED include easy-fabrication and low-cost. However, since each row is selected only during a short time slot, the average luminance is the peak luminance over the number of rows. When the number of rows is large, the peak luminance needs to be very high and this shortens the device lifetime [55].

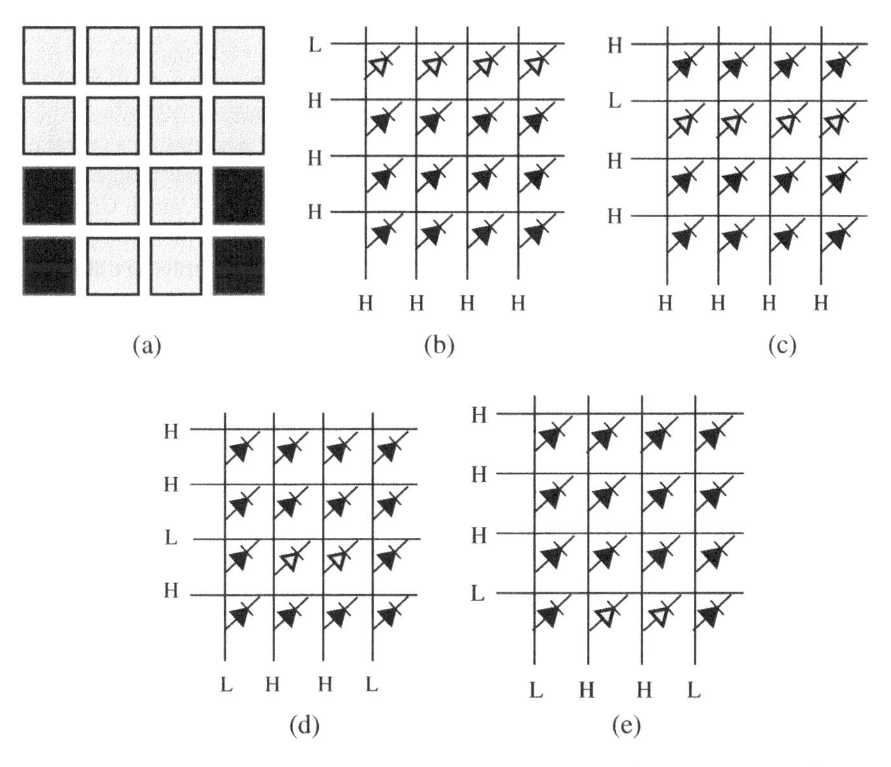

Figure 3.25 (a) image displayed on PM-OLED, and their voltage at (b) first, (c) second, (d) third, and (e) fourth time slots at each row and column.

Also, the row-by-row addressing of the display results in a short burst of light emission during the frame period. The whole addressing cycle must be completed at a rate faster than the critical fusion frequency of the eye (above about 60 Hz), in order for it to be perceived as a constant light level.

There are some differences between OLEDs and LCDs under PM driving. For OLEDs, each individual pixel has a fast turn-on time that is close to or faster than each slot time, thus the luminance for each pixel reaches a peak during this turn-on slot, and drops to near zero at all other times. The apparent luminance is the average over the whole panel scan period. On the other hand, for LCDs, the LC response time is much slower than the turn-on slot time, thus during the single scanning slot when a high voltage is applied to the pixel, it does not fully respond to the instantaneous signal. Rather, LCs respond to the RMS voltage applied over the whole addressing cycle and change their transmittance according to this RMS value.

According to the addressing schemes described up to this point, each pixel in a PM-LCD or PM-OLED, can be switched to ON or OFF states, providing only two gray levels, i.e. 0 and 1. To achieve more gray levels in a PM display, pulse width modulation (PWM) is usually applied as part of the common driving schemes. That is, the pulse width on the pixel is changed to achieve more gray levels. Due to the persistence of vision, human eyes perceive the time-average brightness. In the scheme shown in Figure 3.26, the line addressing pulse is divided into three time-intervals which are applied sequentially to provide four gray levels. When the three time-intervals are all dark, the resulting screen is also dark (gray level 0). However, if one of the three is bright, then a dark gray level is displayed (gray level 1). By increasing the number of bright time-intervals, the displayed shade becomes light gray (gray level 2) and then white (gray level 3). Hence, there are now four gray levels, determined by controlling the pulse width. For PM-LCD, due to the slow response of LC, image blur may occur when displaying moving pictures.

Modern high resolution and large-area displays are invariably addressed by an AM driving technique using TFTs. Figure 3.27 shows the equivalent circuit of an AM-LCD pixel. Only one scan line is selected at a time. When a scan line is selected, the TFTs in this row are all switched on to a conducting state. For each pixel which is intended to be turned on to a bright state, a voltage is fed from the data line to charge both the LC pixel capacitance and the storage capacitor. The scan line is then de-selected and TFT is switched off for the remainder of the frame address period. The LC pixel maintains a constant optical transmission throughout this period, because its state is maintained by the voltage held on the storage capacitor. This is sometimes called a "hold type" display. The extra, storage capacitor connected in parallel with the LC pixel improves the voltage retention characteristics in the presence of finite leakage current through the TFT and the LC pixel. Unlike a PM-LCD, where the driving voltage is dependent on the row number, AM-LCDs are hold-type displays and cross-talk is much weaker. The gray levels can be controlled independently for each pixel by applying different values of V_D from the data lines directly.

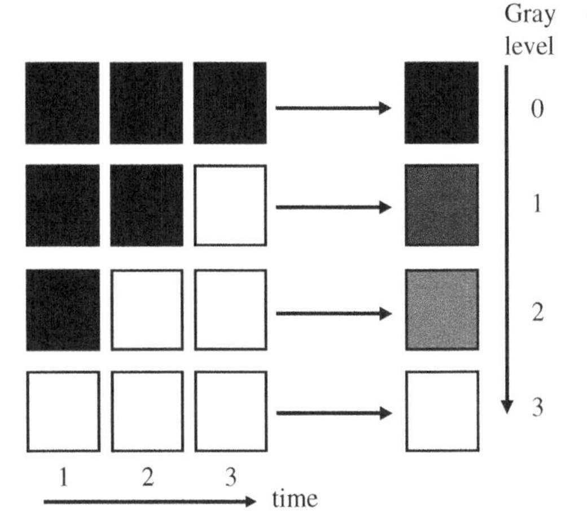

Figure 3.26 Four gray levels achieved by PWM.

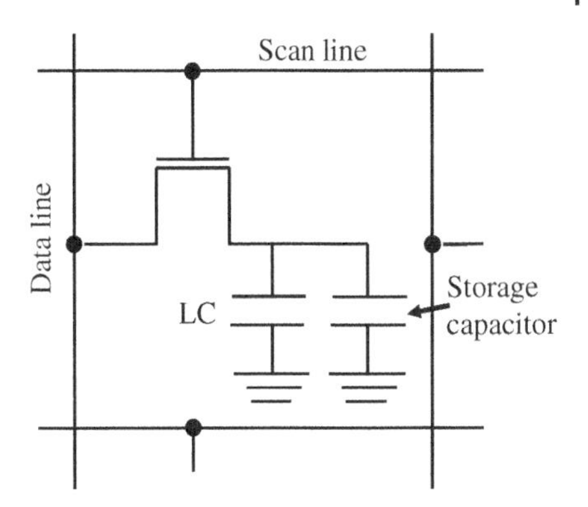

Figure 3.27 Equivalent circuit of AM LCD.

Example 3.2 For the equivalent circuit shown in Figure 3.27, find the charging time of the LC, i.e. V_S, from 0 to 4.5 V with $V_D = 5$ V and 9 V, respectively. V_D is always kept at 5 or 9 V, while V_G is a step function which jumps from 0 to 20 V at $t = 0$. There is no charge stored on the LC ($V_S = 0$ V) for $t < 0$. Timing diagrams of V_G and V_D are shown in Figure 3.28. When $V_G = 0$ V, the leakage current can be neglected (open circuit, $R_{off} = \infty$). When $V_G = 20$ V, the TFT can be replaced by an equivalent resistor with $R_{on} = 5$ MΩ. The LC layer (including the storage capacitor) exhibits $C_{LC} = 3$ pF.

Answer

Figure 3.29a,b show the equivalent circuits corresponding to $V_G = 0$ and $V_G = 20$ V, respectively. When the gate is open at $t = 0$:

$$I = \frac{V_D - V_S}{R_{on}} \tag{3.22}$$

$$I = C_{LC}\frac{dV_S}{dt} \tag{3.23}$$

Equating Eqs. (3.22) and (3.23) yields:

$$\frac{V_D - V_S}{R_{on}} = C_{LC}\frac{dV_S}{dt} \tag{3.24}$$

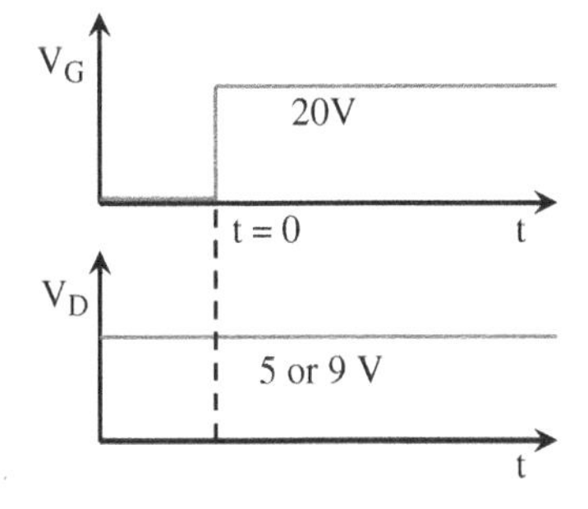

Figure 3.28 Timing diagram of V_G and V_D of Example 3.2.

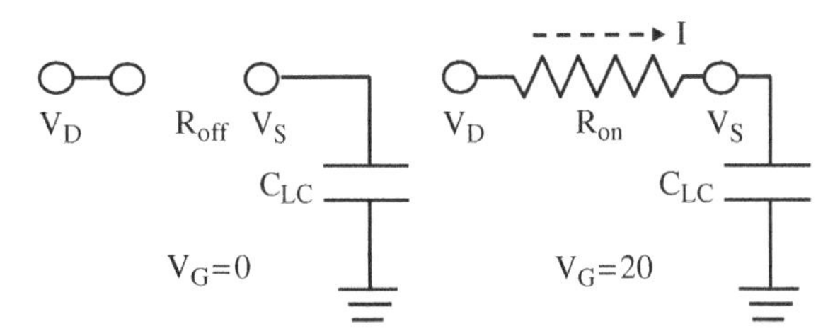

Figure 3.29 Equivalent circuit for (a) $V_G = 0$, and (b) $V_G = 20$ V.

or

$$R_{\text{on}}C_{\text{LC}}\frac{\mathrm{d}V_{\text{S}}}{\mathrm{d}t} + V_{\text{S}} = V_{\text{D}} \tag{3.25}$$

Solving this differential equation, we can obtain

$$V_{\text{S}} = V_{\text{D}}[1 - \exp(-t/\tau)] \tag{3.26}$$

where

$$\tau = R_{\text{on}} \times C_{\text{LC}} \tag{3.27}$$

Hence, for V_{S}, from 0 to 4.5 V with $V_{\text{D}} = 5$ V and 9 V, the charging times are 34.5 and 10.4 μs, respectively. Actually, the gate signal is only applied during a very short time in a TFT-LCD. For example, in a panel with a resolution of 1024 × 768 at a frame rate of 60 Hz, the V_{G} duration is at most, $1/(60 \times 768) = 21.7$ μs. Hence, a suitable V_{D} value is needed in order to achieve the desired V_{S} in such a short time slot. As shown in Example 3.2, if $V_{\text{D}} = 5$ V, the charging time from 0 to 4.5 V is longer than the V_{G} duration, which would result in insufficient charging of the LC cell.

For AM driving of an OLED, a row-at-a-time addressing technique similar to the driving method of an LCD is adopted. However, since the OLED is a current-driven device, rather than a capacitor as is the case for an LCD, a more complicated circuit is needed in an AM-OLED display. Figure 3.30 shows the equivalent circuit of an AM-OLED pixel. At least two TFTs are required, which are marked as the address and drive TFTs, respectively. When a row select pulse is applied to a particular row, the address TFT is switched on. If the pixel is intended to be turned on, the data line is also selected, and current provided by the power line passes through the drive TFT to the OLED. Note that there is a capacitor, which can hold the voltage when the scan line is de-selected, which means the OLED keeps on emitting light during the remainder of the frame address period when the row is not selected. Hence, the peak luminance of the AM-addressed OLEDs can be set much lower than in the PM-OLED case, while achieving the same average brightness, which also prolongs the operational lifetime, and makes larger size displays possible. This pixel design is usually called the 2-transistor and 1-capacitor (2T1C) configuration. The requirements for the address and drive TFTs are quite different. The address TFT is used as a switch, and hence the on/off ratio is important. The drive TFT is used to provide current to the OLED, and hence a high current density capability is required. To provide an adequate current from the drive TFT, a high mobility is typically required. From this viewpoint, LTPS backplanes are more suitable for driving OLEDs. However, the uniformity of LTPS is not very good due to the laser annealing process. Since the J–V characteristics of OLEDs follow steep curves described by space charge limited current (SCLC) or trap-charge limited current (TCLC) equations, which will be described in Chapter 6,

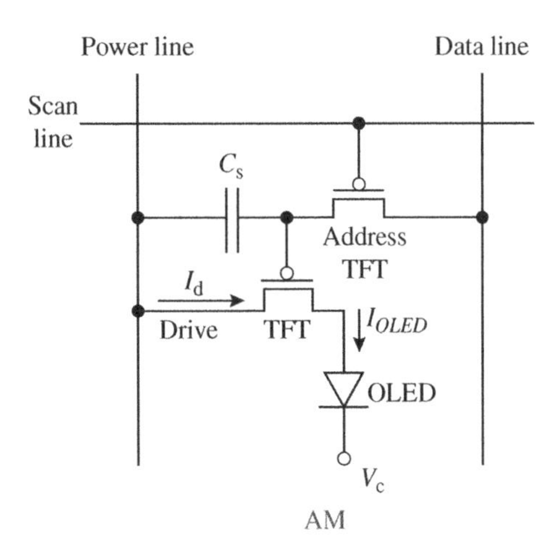

Figure 3.30 Equivalent circuit of AM OLED with two-transistor one-capacitor (2T1C) configuration [52].

the current density varies a lot even with a small difference in driving voltage. To overcome this problem, more than two TFTs are often used, which allow compensation of the non-uniformity of the LTPS-TFTs to assure a consistent current density. In order to achieve different gray levels in the AM-OLED, both voltage modulation from the data line and PWM can be applied. The voltage modulation scheme used in AM-OLEDs is similar to that applied to AM-LCDs. However, due to the steep luminance–voltage curve of OLEDs, it is not easy to control the gray levels precisely and uniformly. For PWM driving, because of the higher frame rate, the power consumption is higher and the control system is more complicated.

Homework Problems

3.1 A 4 in. Si wafer is doped with 10^{17} cm^{-3} phosphorus. The wafer thickness is 350 μm. Assume the shape of the Si-wafer is a complete circle. Also, assume that the electron current is much higher than the hole current with electron mobility $\mu_n = 700$ cm^2/V s. Find the current when 10 V is applied between the top and bottom faces of this wafer.

3.2 GaAs and Si exhibit effective electron masses of 0.07 and 0.19m_0, respectively at the bottom of the conduction band. What is the physical meaning of the "effective" mass of an electron in a semiconductor being less than the "real" mass of a free electron? Which material exhibits a steeper E–$\mathbf{k}$ curve at the bottom of the conduction band? Why?

3.3 The band diagram of a MIS device is shown below. Is the flat band voltage (V_{FB}) positive, negative, or zero? Draw the band diagram under short circuit conditions ($V = 0$).

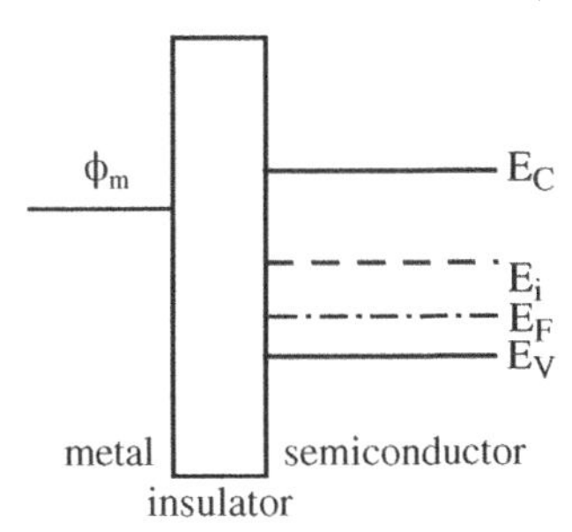

3.4 For a TFT with $V_T = 1$ V, $W = 50$ μm, and $L = 10$ μm, $\mu_n = 0.5$ cm^2/V s, and $C_i = 15$ nF/cm^2. Calculate the drain current at $V_G = 10$ V and $V_D = 0.1$ V. Repeat the calculation for $V_G = 20$ V and $V_D = 20$ V.

References

1 Le Comber, P.G., Spear, W.E., and Ghaith, A. (1979). Amorphous-silicon field-effect device and possible application. *Electron. Lett.* 15: 179.

2 Kimura, M., Yudasaka, I., Kanbe, S. et al. (1999). Low temperature polysilicon thin film transistor driving with integrated driver for high resolution light emitting polymer display. *IEEE Trans. Electron Devices* 46: 2282.

3 Mimura, A., Konishi, N., Ono, K. et al. (1989). High performance low-temperature poly-Si n-channel TFT's for LCD. *IEEE Trans. Electron Devices* 36: 351.

4 Sze, S.M. (2001). *Semiconductor Devices—Physics and Technology*, 2nde. Wiley.

5 Streetman, B.G. and Banerjee, S.K. (2005). *Solid State Electronics Devices*, 6e. Prentice-Hall.

6 Peyghambarian, N., Koch, S.W., and Mysyrowicz, A. (1993). *Introduction to Semiconductor Optics*. Prentice-Hall.
7 Chelikowsky, J.R. and Cohen, M.L. (1976). Nonlocal pseudopotential calculations for the electronic structure of eleven diamond and zinc-blende semiconductors. *Phys. Rev. B* 14: 556.
8 Wagner, S., Gleskova, H., Cheng, I.C., and Wu, M. (2003). Silicon for thin-film transistors. *Thin Solid Films* 430: 15.
9 Tsukada, T. (2003). *TFT/LCD: Liquid-Crystal Displays Addressed by Thin-Film Transistors*. Taylor & Francis.
10 Street, R.A. (ed.) (1999). *Technology and Applications of Amorphous Silicon*. Berlin Heidelburg: Springer-Verlag.
11 Seto, J.Y.W. (1975). The electrical properties of polycrystalline silicon films. *J. Appl. Phys.* 46: 5247.
12 Shih, A., Meng, C.Y., Lee, S.C., and Chern, M.Y. (2000). Mechanism for pillar-shaped surface morphology of polysilicon prepared by excimer laser annealing. *J. Appl. Phys.* 88: 3725.
13 Voutsas, A.T. (2003). A new era of crystallization: advances in polysilicon crystallization and crystal engineering. *Appl. Surf. Sci.* 208–209: 250.
14 Powell, M.J. (1989). The physics of amorphous-silicon thin-film transistors. *IEEE Trans. Electron Devices* 36: 2753.
15 Boyce, J.B., Mei, P., Fulks, R.t., and Ho, J. (1998). Laser processing of polysilicon thin-film transistors: grain growth and device fabrication. *Phys. Status Solidi A* 166: 729.
16 Rogers, J.A. and Bao, Z. (2002). Printed plastic electronics and paperlike displays. *J. Polym. Sci. Pol. Chem.* 40: 3227.
17 Lin, Y.Y., Gundlach, D.J., Nelson, S.F., and Jackson, T.N. (1997). Pentacene-based organic thin-film transistors. *IEEE Trans. Electron Devices* 44: 1325.
18 Briseno, A.L., Tseng, R.J., Ling, M.M. et al. (2006). High-performance organic single-crystal transistors on flexible substrates. *Adv. Mater.* 18: 2320.
19 Ling, M.M. and Bao, Z. (2004). Thin film deposition, patterning, and printing in organic thin film transistors. *Chem. Mater.* 16: 4824.
20 Zhu, Z.T., Mason, J.T., Dieckmann, R., and Malliaras, G.G. (2002). Humidity sensors based on pentacene thin-film transistors. *Appl. Phys. Lett.* 81: 4643.
21 Dimitrakopoulos, C.D. and Mascaro, D.J. (2001). Organic thin- film transistors: a review of recent advances. *IBM J. Res. Dev.* 45: 11.
22 Benor, A., Hoppe, A., Wagner, V., and Knipp, D. (2007). Electrical stability of pentacene thin film transistors. *Org. Electron.* 8: 749.
23 Hoffman, R.L., Norris, B.J., and Wager, J.F. (2003). ZnO-based transparent thin-film transistors. *Appl. Phys. Lett.* 82: 733.
24 Nomura, K., Ohta, H., Takagi, A. et al. (2004). Room-temperature fabrication of transparent flexible thin-film transistors using amorphous oxide semiconductors. *Nature* 432: 488.
25 Hosono, H. (2006). Ionic amorphous oxide semiconductors: Material design, carrier transport, and device application. *J. Non-Cryst. Solids* 352: 851.
26 Nomura, K., Takagi, A., Kamiya, T. et al. (2006). Amorphous oxide semiconductors for high-performance flexible thin-film transistors. *Jpn. J. Appl. Phys.* 45: 4303.
27 Fortunato, E., Barquinha, P., and Martins, R. (2012). Oxide semiconductor thin-film transistors: a review of recent advances. *Adv. Mater.* 24: 2945.
28 Inoue, S., Utsunomiya, S., and Shimoda, T. (2003). Transfer mechanism in surface free technology by laser annealing / ablation (SUFTLA®). *SID 03 Digest* 34: 984–987.
29 Chen, J. and Liu, C.T. (2013). Technology advances in flexible displays and substrates. *IEEE Access* 1: 150–158.
30 Kaltenbrunner, M., White, M.S., Głowacki, E.D. et al. (2012). Ultrathin and lightweight organic solar cells with high flexibility. *Nat. Commun.* 3: 770.

31 French, I., George, D., Kretz, T. et al. (2007). Flexible displays and electronics made in AM-LCD facilities by the EPLaRTM process. *SID 07 Digests* 38: 1680–1683.
32 Lee, C.-C., Chang, Y.-Y., Cheng, H.-C. et al. (2010). A novel approach to make flexible active matrix displays. *SID 10 Digest*: 810–813.
33 Sirringhaus, H. (2009). Materials and applications for solution-processed organic field-effect transistors. *Proc. of the IEEE* 97 (9): 1570–1579.
34 Heo, S.J., Yoon, D.H., Jung, T.S., and Kim, H.J. (2013). Recent advances in low-temperature solution-processed oxide backplanes. *J. Inform. Disp.* 14 (2): 79–87.
35 Choi, C.-H., Lin, L.-Y., Cheng, C.-C., and Chang, C.-h. (2015). Pinted oxide thin film transistors: a mini review. *ECS J. Solid State Sci. Technol.* 4 (4): P3044–P3051.
36 Cuo, C.F. and Ren, Z. (2015). Flexible transparent conductors based on metal nanowire networks. *Mater. Today* 18 (3): 143–154.
37 López-Naranjo, E.J., González-Ortiz, L.J., Apátiga, L.M. et al. (2016). Transparent electrodes: a review of the use of carbon-based nanomaterials. *J. Nanomater.* 2016, 4928365-1-12.
38 Kang, M.-G. and Guo, L.J. (2007). *Adv. Mater.* 19 (10): 1391–1396.
39 Guo, C.F., Sun, T., Wang, Y. et al. (2013). Conductive black silicon surface made by silver nanonetwork assisted etching. *Small* 9 (14): 2415–2419.
40 Eda, G., Fanchini, G., and Chhowalla, M. (2008). Large-area ultrathin films of reduced graphene oxide as a transparent and flexible electronic material. *Nat. Nanotechnol.* 3: 270–274.
41 Park, J.-S., Chae, H., Chung, H.K., and Lee, S.I. (2011). Thin film encapsulaton for flexible AM-OLED: a review. *Semicond. Sci. Technol.* 26, 034001-1-8.
42 Mandlik, P., Gartside, J., Han, L. et al. (2008). A single-layer permeation barrier for organic light-emitting displays. *Appl. Phys. Lett.* 92: 103309.
43 Gleskova, H., Hsu, P.I., Xi, Z. et al. (2004). Field-effect mobility of amorphous silicon thin-film transistors under strain. *J. Non-Cryst. Solids* 338–340: 732–735.
44 Won, S.H., Chung, J.K., Lee, C.B. et al. (2004). Effect of mechanical and electrical stresses on the performance of an a-Si:H TFT on plastic substrate. *J. Electrochem. Soc.* 151: G167–G170.
45 Kuo, P.-C., Jamshidi-Roudbari, A., and Hatalis, M. (2007). Effect of mechanical strain on mobility of polycrystalline silicon thin-film transistors fabricated on stainless steel foil. *Appl. Phys. Lett.* 91, 243507-1-3.
46 Peng, I.-H., Liu, P.-T., and Wu, T.-B. (2009). Effect of bias stress on mechanically strained low temperature polycrystalline silicon thin film transistor on stainless steel substrate. *Appl. Phys. Lett.* 95, 041909-1-3.
47 Kim, M., Cheon, J., Lee, J. et al. (2011). World-best performance LTPS TFTs with robust bending properties on AMOLED displays. *SID 11 Digest*: 194–197.
48 Kim, J.-M., Nam, T., Lim, S.J. et al. (2011). Atomic layer deposition ZnO:N flexible thin film transistors and the effects of bending on device properties. *Appl. Phys. Lett.* 98, 142113 -1-3.
49 Münzenrieder, N., Cherenack, K.H., and Tröster, G. (2011). The effects of mechanical bending and illumination on the performance of flexible IGZO TFTs. *IEEE Trans. Electron Devices* 58: 2041–2048.
50 Lin, C.-Y., Chien, C.-W., Wu, C.-C. et al. (2012). Effects of mechanical strains on the characteristics of top-gate staggered a-IGZO thin-film transistors fabricated on polyimide-based nanocomposite substrates. *IEEE Trans. Electron Devices* 59: 1956–1962.
51 den Boer, W. (2005). *Active Matrix Liquid Crystal Displays: Fundamentals and Applications*. Newnes.
52 Meng, Z. and Wong, M. (2002). Active-matrix organic light emitting diode displays realized using metal-induced unilaterally crystallized polycrystalline silicon thin-film transistors. *IEEE Trans. Electron Devices* 49: 991.
53 Yeh, P. and Gu, C. (1999). *Optics of Liquid Crystal Display*, 248. Wiley.
54 Alt, P.M. and Pleshko, P. (1974). Scanning limitations of liquid-crystal displays. *IEEE Trans. Electron Devices* 21: 146–155.
55 Kijima, Y., Asai, N., Kishii, N., and Tamura, S. (1997). RGB luminescence from passive-matrix organic LED's. *IEEE Trans. Electron Devices* 44: 1222.

4

Liquid Crystal Displays

4.1 INTRODUCTION

Three types of liquid crystal displays (LCDs) have been developed: (i) transmissive, (ii) reflective, and (iii) transflective. A transmissive LCD uses an efficient backlight to illuminate the display panel, resulting in high luminance, high dynamic range (HDR), and wide color gamut. By implementing multi-domain structures and including phase compensation films, direct-view transmissive LCDs can exhibit a wide viewing angle and have been used extensively for smartphones, tablets, notebook computers, desktop monitors, and televisions. The pixel size of these direct-view LCDs is usually in the 50–300 μm range, depending on the required resolution density. On the other hand, transmissive microdisplays are commonly used in projection displays, such as data projectors [1]. There, a high power arc lamp or high luminance light-emitting diode (LED) array is used as a light source. Using a projection lens, the displayed image can be magnified by more than 50×. To reduce the size and cost of optics, the LCD panel is usually made small (less than 25 mm in diagonal) and the size of each pixel is ~20–40 μm. Thus, a poly-silicon based thin-film-transistor (TFT) LCD is the common choice.

Similarly, reflective LCDs can be subdivided into direct-view and projection displays. A direct-view reflective LCD, e.g. a simple twisted (TN) nematic display, cholesteric liquid crystal (CLC) display [2] or bistable nematic LCD [3], uses ambient light to allow readout of the displayed images. A Ch-LCD has a helical structure which reflects colored light so that the display does not require color filters or polarizer. Thus, the reflectance within a given color band – which is determined by the pitch length and refractive index of the liquid crystal (LC) – is relatively high (in theory ~50%). Moreover, it does not require a backlight so it is light in weight and the whole device can be thin (<200 μm). Therefore, it is a strong contender for flexible displays. CLC is a bistable device so its power consumption is low, provided that the display is not refreshed too frequently. A major drawback of reflective direct-view LCDs is their poor readability under low ambient light.

Another type of reflective LCD is designed for projection displays, such as augmented reality (AR), using liquid-crystal-on-silicon (LCoS) microdisplay panels [4]. Unlike a transmissive microdisplay, LCoS is a reflective device. Here the reflector employed is an aluminum metallic mirror. Because the active matrix components are hidden beneath the reflector, LCoS can tolerate a high flux from the light source and deliver high luminance images. This feature is particularly attractive for see-through AR displays used under strong ambient light. Crystalline silicon has high mobility so the pixel size can be made small (<10 μm) and the aperture ratio is >90%. Therefore, the image not only has high resolution but also appears seamless. In comparison, the aperture ratio of a transmissive microdisplay is normally less than 50%. The light blocked by the black matrix shows up on the screen as a dark pattern (also known as screen door effect). The viewing angle of a LCD is less critical in projection than for direct-view displays because in a projection display the polarizing beam splitter has a narrow acceptance angle and a wide viewing cone is provided by scattering at the screen.

For outdoor applications, the displayed images of a transmissive LCD can be washed out by sunlight. A reflective LCD would be a better choice. However, such a reflective display is unreadable in a dark ambient. Therefore, the transflective LCD which combines the features of a transmissive display and a reflective display

Introduction to Flat Panel Display, Second Edition. Jiun-Haw Lee, I-Chun Cheng, Hong Hua, and Shin-Tson Wu.

seems an ideal choice. In a dark ambient the backlight is switched on and the display works in transmissive mode. Under bright ambient light, the backlight is turned off and the reflective mode is operational.

Several monographs have been dedicated to advanced projection displays [5], reflective displays [6], transflective LCD [7], and flexible displays [8]. Therefore, in this chapter we will focus on the TFT-addressed wide-viewing angle transmissive LCDs which form the mainstream of LCD production and development. We will start by introducing the twisted nematic (TN), in-plane switching (IPS), and fringe field switching (FFS) displays, and then the multi-domain vertical alignment (MVA) device mode. Phase compensation methods for achieving wide viewing angle will also be addressed.

4.2 TRANSMISSIVE LCDS

Figure 4.1 shows the device structure of a transmissive TFT addressed LCD. An LCD is a non-emissive display, i.e. it does not emit light. Instead, it functions as a two-dimensional spatial light modulator. Thus, a backlight is needed. Two types of backlight configurations have been widely employed: edge-lit (Figure 4.1a) and direct-lit (Figure 4.1b). In an edge-lit LCD, the LED array is at the edge of a display panel. The emitted light passes through a light guide plate (LGP) and is steered toward the TFT LCD panel. In a direct-lit LCD, the backlight unit can be divided into thousands or more local dimming zones using mini-LEDs each with a chip size of about 100 μm, and each zone may contain several mini-LEDs depending on the required luminance. Each zone can be controlled independently. Such a mini-LED based LCD can achieve $10^6:1$ contrast ratio (CR), which is comparable to an organic LED [9]. As depicted in Figure 4.1, a diffuser is used to homogenize the backlight in order to avoid illumination hot spots. A series of tailored optical films may be stacked on top of the backlight to steer its Lambertian emission into a central ±40° cone to improve the display brightness. Since most LCDs require linearly polarized light in order to obtain a high contrast ratio, two sheets of stretched dichroic polarizers are used for large size direct-view displays. The first glass substrate carries a TFT array, which allows independent switching of the LCD pixels. Each display sub-pixel is controlled by a TFT. For a 4K2K resolution (3840 × 2160 × RGB) LCD, there are ~24 million sub-pixels. Since TFTs are light sensitive and must be shielded from backlight illumination by an opaque mask, the aperture ratio (the proportion of transparent ITO electrode area) is reduced to ~80%, depending on the pixel density. As the pixel density increases, the aperture ratio decreases. The LC layer is sandwiched between two ITO substrates each of whose inner surfaces are coated with a thin (80 nm) polyimide layer or 30 nm photo-alignment layer. Some LCDs (twisted-nematic, IPS and FFS) require surface alignment treatment (mechanical buffing or photoalignment) while others (MVA and patterned vertical alignment [PVA]) do not. The cell gap is usually controlled at ~3–4 μm for a transmissive LCD. The performance of the display such as light throughput, response time, and viewing angle are all influenced by the LC configuration employed.

For direct-view LCDs, compact size, light weight, and low power consumption are equally as important as viewing angle, color gamut, and contrast ratio. For direct-view LCDs, color filters are embedded in the inner side of the top (second) substrate. Three sub-pixels (red, green, and blue) form a color pixel. The size of each sub-pixel in a standard resolution display is ~80 μm × 240 μm. Each sub-pixel transmits only one color; the other colors are absorbed. Figure 4.2 depicts the emission spectra of a backlight (white light-emitting diode, WLED) and blue-LED pumped green and red quantum dots (QDs), and the transmission spectra of RGB color filters.

As shown in Figure 4.2, the transmission spectra of RGB color filters are relatively broad. The advantage of this is that more light is transmitted, however the color purity is degraded. The peak transmission of the RGB color filters is ~95%, 85%, and 80%, respectively. However, roughly speaking, each color filter only transmits ~25% of incident white light. The other ~75% is absorbed by the color pigments. Moreover, a WLED emits a relatively broad spectrum. Blue-green light will be transmitted through both the blue and green color filters. Similarly, the orange light will be transmitted through both green and red filters. This leaked light will degrade

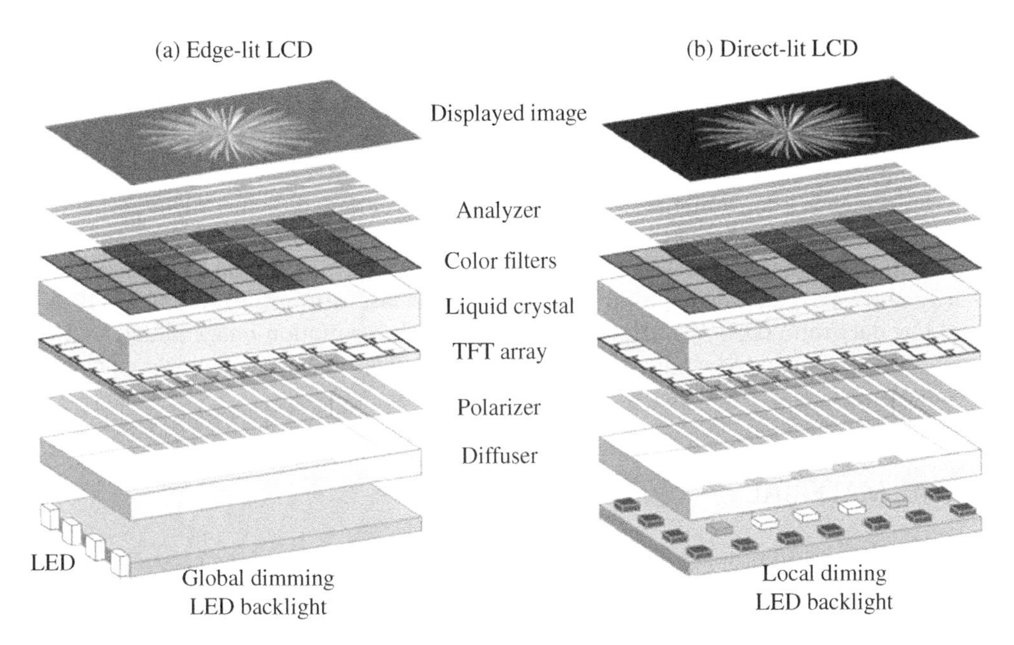

Figure 4.1 Device structure of (a) edge-lit and (b) direct-lit transmissive TFT LCDs.

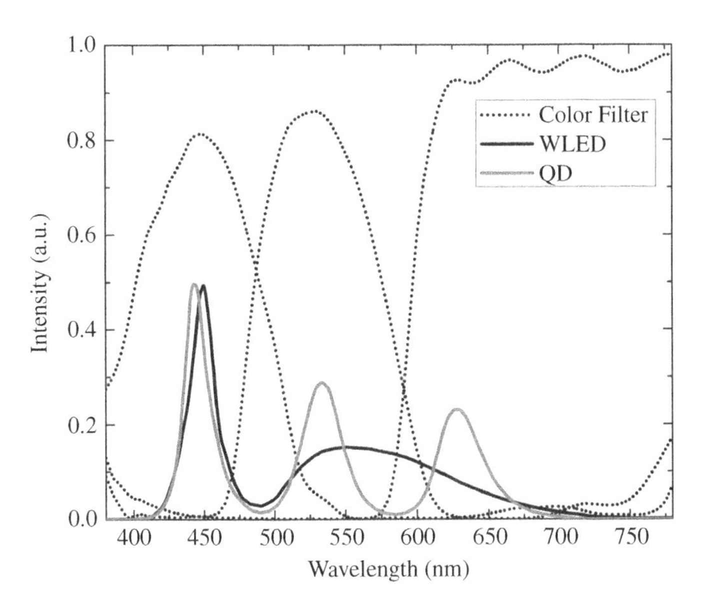

Figure 4.2 Transmission spectra of RGB color filters (dotted lines), and emission spectra of WLED backlight (black line) and blue-LED pumped quantum dots (gray line).

the color purity (or color saturation) of the display [10]. Therefore, the color gamut of a typical transmissive TFT LCD is ~75% of the NTSC (National Television System Committee) standard. Blue-LED pumped quantum dots have narrower emission spectra that are also better matched with the transmission spectra of the color filters, thus, the resulting color gamut reaches ~115% of the NTSC specification [11]. For display devices, a wider color gamut is not the only critical parameter, total light efficiency is equally important, especially for mobile devices (smartphones or tablet computers) powered by battery.

After taking into account the optical losses from polarizers, color filters, and TFT aperture ratio, the overall system optical efficiency is about 7% for a direct-view LCD. If wide view technology is included [12], the total light efficiency is decreased to ~5%. Lower optical efficiency implies a higher power consumption and more heat generation inside the display chassis. For a thin LCD, thermal dissipation is a critical issue. For portable displays, low power consumption is desirable because it lengthens operating battery life. Several approaches have been developed to reduce power consumption, e.g. polarization conversion of the backlight [13] and a two-dimensional local dimming LED backlight [14–16]. Beside HDR, the use of a local dimming mini-LED backlight offers several additional advantages such as 16 bits of grayscale (8 bits from LED backlight and 8 bits from LCD), wider color gamut, ~4× reduction in power consumption, and fast turn-on and off times (~10 ns) for high frame rate and low duty ratio (DR) operation, which help to reduces motion image blurs [17, 18]. Some technological issues which remain to be fully overcome include changes in color and power efficiency as the LED junction temperature changes, and increased cost.

4.3 LIQUID CRYSTAL MATERIALS

A liquid crystal (LC) possesses physical properties that are intermediate between those of a crystalline solid and an isotropic liquid. They are fluid like, yet the arrangement of molecules exhibits structural order. Three types of liquid crystals have been discovered so far: (i) thermotropic, (ii) polymeric, and (iii) lyotropic. Thermotropic liquid crystals have been studied extensively and their applications are widespread. Most TFT-LCDs use thermotropic nematic liquid crystals. Liquid crystal polymers have found interesting applications in optical films, electronic devices, and ultra-high-strength materials. Lyotropic liquid crystals are receiving increasing scientific and technological attention because of the way they reflect the unique properties of their constituent molecules.

Although the LC material only occupies a thin layer in a display device, it makes crucial contributions to the device performance. For instance, the device contrast ratio, response time, viewing angle, and operating voltage are all related to the LC material used, and its alignment. The refractive indices and cell gap determine the phase retardation of the LC device. The dielectric constants and elastic constants jointly determine the threshold voltage. The viscosity, elastic constants, cell gap, driving voltage, and temperature determine the response time.

Within the thermotropic LC family, there are three distinct phase structures: smectic, nematic, and cholesteric. Within the smectic branch, ferroelectric liquid crystals (FLCs) exhibit many fascinating properties, such as bistable switching, layered structure, in-plane molecular reorientation, and microsecond response time [19]. FLC has been used in near-to-eye microdisplays [20] where a typical panel size is about 2 cm by 2 cm. Thus, a uniform thin cell gap is easier to maintain. For large panels, the alignment uniformity and mechanical stability are both challenging problems. Thus, the FLC display industry for large area devices is still waiting to take off.

CLCs exhibit a helical structure which reflects color without using color filters or a polarizer if the Bragg reflection is adjusted to lie in the visible spectral region. The contrast ratio of a polarizer-free CLC is ~30 : 1 which is inadequate for high-end displays, such as computers and TVs. A 30 : 1 contrast ratio implies ~3% unwanted light reflection or scattering at each pixel which will degrade the color purity. However, for electronic paper applications, a 30 : 1 contrast is sufficient. A typical newspaper has ~8 : 1 contrast and ~50–60% reflectivity. High quality white paper has ~80% reflectivity and print then provides ~15 : 1 contrast. Therefore, CLC is better suited for such applications. The encapsulated CLC display [21] is thin and flexible. It is an emerging technology for flexible LCDs and electronic books.

In this section, we will focus on the basic molecular structures and physical properties of nematic liquid crystals as used in mainstream TFT-LCDs.

4.3.1 Phase Transition Temperatures

Only a few LC compounds exhibit mesogenic phase at room temperature (~23 °C). The following compound, 4′-pentyl-4-cyanobiphenyl (best known as 5CB) [22] is such an example:

C_5H_{11}—⬡—⬡—CN (I)

Its nematic range is actually from 24 to 35.3 °C, but due to super-cooling of the crystallization transition, 5CB remains liquid at room temperature. Therefore, many of its properties can be studied conveniently at room temperature without a bulky heating device. However, for display applications a wide nematic range (from −40 to 90 °C) is highly desirable. To widen the nematic range, eutectic mixtures are commonly used. To obtain such a wide nematic range and also optimize the physical properties of the material (viscosity, dielectric constants, and elastic constants), a commercial mixture consisting of 10–15 components is not uncommon.

4.3.2 Eutectic Mixtures

Let us use a binary mixture as an example to illustrate the working principles. Figure 4.3 shows the phase diagram of a binary mixture. The mesogenic ranges of components 1 and 2 are shown on the left and right vertical axes, where $T_{\mathrm{mp1,2}}$ represents the melting temperatures and $T_{\mathrm{c1,2}}$ the clearing temperatures of compounds 1 and 2, respectively. The horizontal axis represents the molar concentration (X_2) of component 2. As the concentration of compound 2 increases, the melting of the mixture gradually decreases. At an appropriate molar concentration, the melting point of the mixture reaches a minimum. The mixture formulated according to this composition is called a eutectic. As the concentration of the second compound is increased above that of the eutectic point, the mixture's melting temperature increases gradually. Meanwhile, the clearing point of the mixture is a linear interpolation between those of the two components. That is to say, at the eutectic point the mixture has the lowest melting point and typically the widest mesogenic range.

The optimal mixing ratio of a eutectic mixture is predicted by the Schröder–Van Laar equation, formulated more than a century ago [23, 24]:

$$\ln(X_i) = \frac{\Delta H_i}{R}\left[\frac{1}{T_i} - \frac{1}{T_{\mathrm{mp}}}\right], \tag{4.1}$$

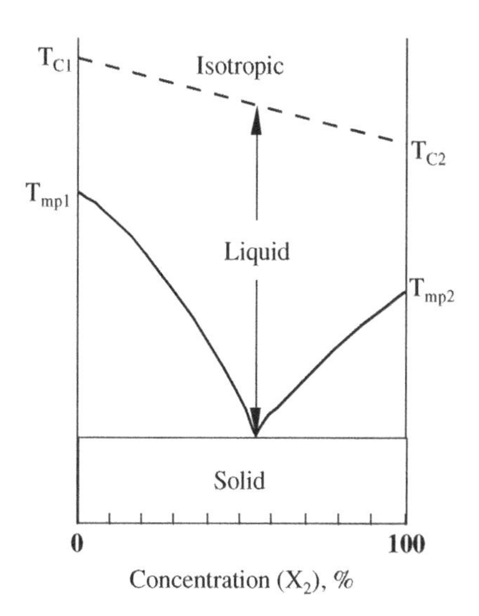

Figure 4.3 Diagram showing eutectic formulation of a binary mixture.

where T_{mp} is the mixture's melting temperature (in unit of K), T_i the melting point, ΔH_i the heat fusion enthalpy (in unit of cal/mol), X_i the mole concentration of the component i, and R is the gas constant (1.98 cal/mol/K). To solve Eq. (4.1), we need an additional boundary condition, which is $\sum X_i = 1$.

In the Schröder–Van Laar equation, the following assumptions are about the compounds which allow them to form an ideal eutectic mixture: (i) The two components crystallize in pure form and do not form mixed crystals, (ii) The liquid phase is a thermodynamically ideal mixture, and (iii) The differences in the heat capacities of the pure components in the molten and crystallized forms, are small. If the compound structures are not too similar to each other, then the calculated melting point from Eq. (4.1) will usually be close to the experimental data.

The clearing temperature (T_c) of the eutectic mixture can be calculated from the clearing point (T_{ci}) of the individual components as:

$$T_c = \sum_i X_i T_{ci}. \tag{4.2}$$

As shown in Figure 4.3, the final clearing temperature of the binary mixture is a weighted linear sum of those of the individual compounds. LC compounds with a high T_{ci} are helpful to enhance the T_c of the mixture. However, their melting points and ΔH values all need to be taken into consideration.

Example 4.1 ***E7 Mixture*** The commercial Merck mixture E7 consists of four components: 5CB, 7CB (4-cyano-4′-*n*-heptyl-biphenyl), 8OCB (4-cyano-4′-*n*-oxyoctyl-biphenyl), and 5CT (4-cyano-4″-*n*-pentyl-*p*-terphenyl) [25]. Their phase transition temperatures and associated heat fusion enthalpies are listed in Table 4.1.

As shown in Table 4.1, compound 8OCB has a smectic-A phase below the nematic phase range. For mixture formulation, this could create some uncertainty about which temperature and heat fusion enthalpy to use in Eq. (4.1). According to experimental observation, a nematic mixture which contains a small percentage of smectic compound will still exhibit only a nematic phase as long as the smectic component is less than ~20%. The calculated melting temperature and clearing point of E7 are −3 and 60 °C respectively. In calculations we use mole concentration, but in real mixture formulations we prefer to use weight percentage (wt%). To convert mole concentration to weight percentage, we use the following formula:

$$(\text{wt }\%)_i = \frac{X_i M_i}{\sum_j X_j M_j} \tag{4.3}$$

In Eq. (4.3), M_i indicates the molecular weight for each individual component. Table 4.1 lists both the molar concentration and weight percentage of the four components used in E7. For comparison, the experimental melting point of E7 is about −10 °C and clearing point is 60.5 °C. Note that the composition in Table 4.1 is the calculated eutectic for E7, rather than the actual composition of the commercial mixture. Here, we just use E7 to elucidate the eutectic mixture design principles.

Table 4.1 Composition of E7.

Compound	Phase transitions	ΔH	mol %	wt%
5CB	K 24 N 35.3 I	4100	49.60	45.53
7CB	K 30 N 42.8 I	6200	28.13	28.74
8OCB	K 54.5 S_A 67.5 N 81 I	5900	14.38	16.28
5CT	K 131 N 240 I	4100	7.89	9.46

K: crystalline phase, N: nematic, S_A: smectic-A, and I: isotropic. The unit of phase transition temperatures is °C and ΔH is cal/mol.

4.3.3 Dielectric Constants

The dielectric constants of a liquid crystal affect the operating voltage and response time of an LCD. For example, in a vertical alignment (VA) cell the threshold voltage (V_{th}) is related to the dielectric anisotropy ($\Delta\varepsilon = \varepsilon_{//} - \varepsilon_{\perp}$) and the bend elastic constant (K_{33}) by the expression $V_{th} = \pi\sqrt{\frac{K_{33}}{\varepsilon_o \Delta\varepsilon}}$ [26]. Thus, a smaller K_{33} and larger $\Delta\varepsilon$ would lower the threshold voltage. However, a large $\Delta\varepsilon$ may also lead to a higher LC viscosity because increasing it generally requires more, or more strongly polar groups in the LC structure.

From the Maier and Meier mean field theory [27], the dielectric anisotropy of a liquid crystal is mainly determined by the molecular dipole moment (μ), its orientation angle (θ) with respect to the principal molecular axis, and the order parameter (S) as follows:

$$\Delta\varepsilon = NhF\{(< \alpha_{//} > - < \alpha_{\perp} >) - (F\mu^2/2kT)(1 - 3\cos^2\theta)S\} \tag{4.4}$$

Here, N is the molecular packing density, $h = 3\varepsilon/(2\varepsilon + 1)$ is the cavity field factor, $\varepsilon = (\varepsilon_{//} + 2\varepsilon_{\perp})/3$ is the averaged dielectric constant, F is the Onsager reaction field [28], $\langle \alpha_{//} \rangle$ and $\langle \alpha_{\perp} \rangle$ are the principal elements of the molecular polarizability tensor.

From Eq. (4.4), for a non-polar compound, $\Delta\varepsilon$ is determined mainly by the differential molecular polarizability, i.e. the first term in Eq. (4.4). In this case, $\mu \sim 0$ and the material dielectric anisotropy is very small ($\Delta\varepsilon < 0.5$). For a polar compound, the dielectric anisotropy depends on the dipole moment, angle θ, temperature (T), and applied frequency. If a LC has more than one dipole, then the resultant dipole moment is their vector sum. From Eq. (4.4), if a polar compound has an effective dipole at $\theta < 55°$, then its $\Delta\varepsilon$ is positive. On the other hand, $\Delta\varepsilon$ becomes negative if $\theta > 55°$.

Fluoro (F) [29], cyano (CN) [22], and isothiocyanato (NCS) [30] are three commonly employed polar groups. Among them, the fluoro group possess a modest dipole moment ($\mu \sim 1.5$ Debye), and contributes to a high resistivity and low viscosity. However, its strong electronegativity reduces the polarizability of electron clouds and, consequently, lowers the compound's birefringence. For direct-view LCDs, the required birefringence is around 0.1, depending on the LC alignment and cell gap (d) employed. On the other hand, cyano and isothiocyanato groups not only exhibit a large dipole moment ($\mu \sim 3.9$ Debye for C≡N and ~3.7 Debye for N=C=S) but also help to extend π-electron conjugation. As a result, their derivatives' birefringence is much higher than for the fluorinated counterpart. High birefringence is favorable for long wavelength applications such as infrared (IR) beam steering, in order to use a thin cell gap to achieve the same phase change while keeping a fast response time [31]. Under strong anchoring conditions, the LC response time is proportional to d^2. However, CN compounds are more viscous than the corresponding NCS and fluoro compounds. Therefore, their major applications are in low-end displays such as wrist watches and calculators where response time and resistivity are not crucial.

Example 4.2 ***Positive*** $\Delta\varepsilon$ ***LCs*** Positive $\Delta\varepsilon$ LCs have been used in twisted nematic (TN) [32] and IPS [33, 34] displays, although IPS can also use negative $\Delta\varepsilon$ LCs. For TFT LCDs, the LC material employed must also possess a high resistivity (>10^{13} Ω cm) in order to reliably hold the charges and avoid image flickering [35]. The resistivity of a LC mixture depends heavily on the concentration of e.g. ionic impurities. The purification process plays an important role in removing ions in order to achieve high resistivity. Fluorinated compounds exhibit a high resistivity and are the natural choice for TFT LCDs [36, 37].

A typical fluorinated LC structure is shown below:

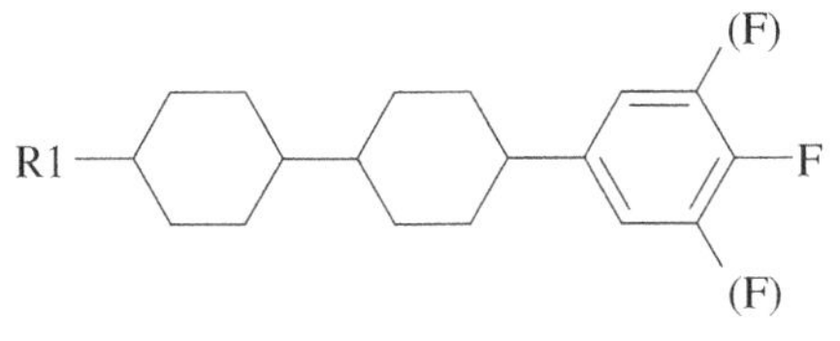

(F)

R1 — F

(F)

(II)

Most liquid crystal compounds discovered so far possess at least two rings, either cyclohexane–cyclohexane, cyclohexane–phenyl, or phenyl–phenyl, and a flexible alkyl or alkoxy chain. The compound shown in structure (II) has two cyclohexane and one phenyl rings. The R_1 group represents a terminal alkyl chain, and a single or multiple fluoro substitutions are located on the phenyl ring. For multiple dipoles, the net resultant dipole moment can be calculated from their vector sum. From Eq. (4.4), to obtain the largest $\Delta\varepsilon$ for a given dipole, the best position for the fluoro substitution is along the principal molecular axis, i.e. in the 4 position. The monofluorinated compound should have $\Delta\varepsilon \sim 5$. To further increase $\Delta\varepsilon$, more fluoro groups can be added. For example, compound (II) with two more fluoro groups in the 3 and 5 positions [38] has a $\Delta\varepsilon$ about 10, but its birefringence will slightly decrease (because of the lower molecular packing density) and its viscosity increases substantially (because of the higher moment of inertia and increased Van der Waals forces between the molecules). The birefringence of compound (II) is around 0.07. If a higher birefringence is needed, the middle cyclohexane ring can be replaced by a phenyl ring. The elongated electron cloud will enhance the birefringence to ~0.12 without increasing the viscosity noticeably.

The phase transition temperatures of a LC compound are difficult to predict before the compound is synthesized. In general, lateral fluoro substitution lowers the melting temperature of the parent compound because increased intermolecular separation leads to weaker molecular association. Thus, a smaller thermal energy is able to separate the molecules which implies to a lower melting point. A drawback of lateral substitution is the increased viscosity which results.

Example 4.3 ***Negative $\Delta\varepsilon$ LCs*** From Eq. (4.4), in order to obtain a negative dielectric anisotropy, the dipoles should be in the lateral (2,3) positions. In the interest of obtaining high resistivity, a lateral difluoro group is a favorable choice. Negative $\Delta\varepsilon$ LCs are useful for vertical alignment device modes [39]. The VA cell exhibits an unrivaled contrast ratio when viewed from the normal direction between two crossed linear polarizers. However, a single domain VA cell has a relatively narrow viewing angle and is only useful for projection displays. For wide-view (WV) LCDs, a multidomain (4 or more domains) vertical alignment (MVA) cell is required.

The following structure is an example of a negative $\Delta\varepsilon$ LC [40]:

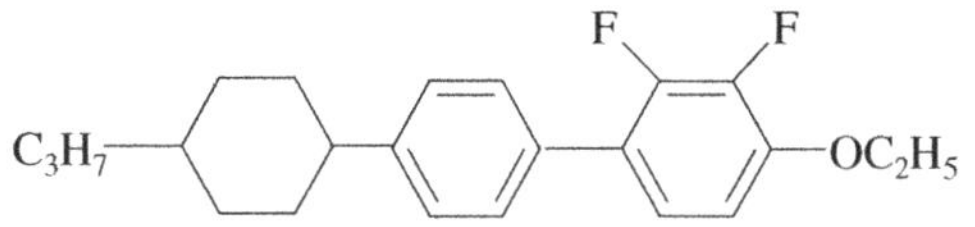

C_3H_7–(cyclohexane)–(phenyl)–(2,3-difluorophenyl)–OC_2H_5

(III)

Compound (III) has two lateral fluoro groups in the (2,3) positions so that the components of their dipoles along the molecular long axis cancel perfectly whereas the perpendicular components add together. Thus, the net $\Delta\varepsilon$ is negative. A typical $\Delta\varepsilon$ of lateral difluoro compounds is −4. The neighboring alkoxy group also has a dipole in the vertical direction. Therefore, it contributes to an increase in the dielectric anisotropy ($\Delta\varepsilon \sim -6$). However, the alkoxy group contributes a higher viscosity than the corresponding alkyl group. To further increase $\Delta\varepsilon$, we could substitute more fluoro groups in the middle phenyl ring or replace one of the fluoro groups with a stronger polar group. Both approaches would cause the rotational viscosity to increase.

4.3.4 Elastic Constants

There are three basic elastic constants (splay K_{11}, twist K_{22}, and bend K_{33}) involved in the electro-optics of a LC cell, their relative importance depending on the molecular alignment [41]. Elastic constants affect a LC device through its threshold voltage and response time. A smaller elastic constant leads to a lower threshold voltage, however, the response time is proportional to a visco-elastic coefficient (γ_1/K_{ii}, where γ_1 is the rotational viscosity) and is therefore increased. Therefore, a proper balance between threshold voltage and response time should be taken into consideration.

Several molecular theories have been developed for correlating the Frank elastic constants with molecular constituents. The commonly employed one is mean-field theory [42, 43]. In the mean-field theory, the three elastic constants are expressed as:

$$K_{ii} = aS^2, \tag{4.5}$$

where a is a proportionality constant and S is the order parameter. For crystalline materials, $S = 1$ and for isotropic materials $S = 0$. A typical nematic LC has $S \sim 0.6$. As the temperature increases, S decreases.

For many LC compounds and mixtures, the magnitude of elastic constants has the following order: $K_{33} \gtrsim K_{11} \approx 2 \times K_{22}$. Therefore, the nature of the LC realignment caused by an applied voltage, also plays an influential role on response time. For example, a VA cell in which the main distortion is director bend, with its threshold determined by K_{33}, should have a faster response time than the IPS or FFS cells whose distortions are dominated by K_{22}, provided that other parameters such as cell gap and viscosity remain the same. Usually, lateral difluoro substitution of the LC increases the rotational viscosity because of the increased molecular width and moment of inertia.

4.3.5 Rotational Viscosity

Viscosity, especially the rotational viscosity (γ_1), plays a crucial role in the LC response time. The response time of a nematic LC device is linearly proportional to γ_1 [44]. The rotational viscosity of an LC mixture depends on the detailed molecular composition, structure, intermolecular association, and temperature. A bulkier and higher molecular weight compound tends to have a higher viscosity. As the temperature increases, viscosity decreases rapidly. Several theories, with a basis in fundamental physics or semi-empirical principles, have been developed in an attempt to account for the origin of the LC viscosity [45, 46]. However, owing to the complicated anisotropic attractive and steric repulsive interactions among LC molecules, these theoretical results are not completely satisfactory [47, 48].

A general temperature dependent rotational viscosity can be expressed as:

$$\gamma_1 = bS\exp(E_a/kT), \tag{4.6}$$

where b is a proportionality constant which takes into account the molecular shape, dimensions, and moment of inertia, S is the order parameter, E_a is the activation energy for molecular rotation, k is the Boltzmann constant, and T is the absolute temperature. When the temperature is not too close to the clearing point (T_c), order parameter can be approximated as follows: [49]

$$S = (1 - T/T_c)^{\alpha}. \tag{4.7}$$

In Eq. (4.7), α is a material parameter. Overall, rotational viscosity is a complicated function of molecular shape, moment of inertia, activation energy, and temperature. Among these factors, the activation energy and temperature are the most crucial ones [50]. The activation energy depends on the detailed inter-molecular interactions. An empirical rule is that for every 10°–15° of temperature rise, the rotational viscosity decreases by ~2×.

From the molecular structural viewpoint, a linear LC molecule is more likely to have a low viscosity [51]. However, all other properties need to be taken into account too. For instance, a linear structure may lack flexibility and lead to a higher melting point. Within a homologous series, a longer alkyl chain will in general (except for the even-odd effect) have a lower melting temperature. However, its moment of inertia is increased. As a result, the homologue with a longer chain length is likely to exhibit a higher viscosity.

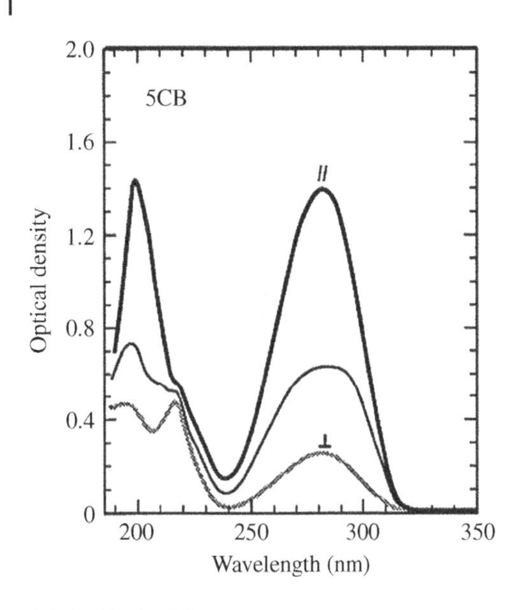

Figure 4.4 The measured polarized absorption spectra of 5CB. The middle trace is for unpolarized light. $\lambda_1 \sim 200$ nm and $\lambda_2 \sim 282$ nm.

4.3.6 Optical Properties

The major optical absorption bands of the LC compounds used in displays, occur in the ultraviolet (UV) and infrared (IR) regions. The $\sigma \rightarrow \sigma^*$ electronic transitions take place in the vacuum UV (100–180 nm) region whereas $\pi \rightarrow \pi^*$ electronic transitions occur in the UV (180–400 nm) region. Figure 4.4 shows the measured polarized UV absorption spectra of 5CB [52]. The λ_1 band which is centered at ~200 nm consists of two closely overlapped bands. The λ_2 band shifts to ~282 nm. The λ_0 band should occur in the vacuum UV region ($\lambda_0 \sim 120$ nm) which is not shown in the figure.

4.3.7 Refractive Indices

Refractive index has a great impact on LC devices. Almost every electro-optic effect in LC modulators, no matter whether amplitude or phase modulation is concerned, involves a refractive index change. An aligned LC exhibits anisotropic properties, including dielectric, elastic, and optical anisotropies. Let us take a homogeneous, planar alignment as an example [53]. Assume that linearly polarized light is incident on the LC cell along the normal direction. If the polarization axis is parallel to the LC alignment axis (i.e. along the LC director which represents an average molecular distribution axis), then the light experiences the extraordinary refractive index, n_e. If the polarization is perpendicular to the LC director, then the light sees the ordinary refractive index n_o. The difference between n_e and n_o is called the birefringence, defined as $\Delta n = n_e - n_o$. Refractive indices are dependent on the wavelength and temperature. For a full-color LCD, RGB color filters are employed. Thus, the refractive indices at these wavelengths need to be known in order to optimize the device performance. Moreover, about 50% of the backlight is absorbed by the polarizer. The absorbed light is converted to heat and causes the LCD panel's temperature to increase. As the temperature increases, the birefringence decreases gradually. The following sections will describe how the wavelength and temperature affect the LC refractive indices.

4.3.7.1 Wavelength Effect

Based on the electronic absorption spectrum, a three-band model which takes one $\sigma \rightarrow \sigma^*$ transition (the λ_0-band) and two $\pi \rightarrow \pi^*$ transitions (the λ_1- and λ_2-bands) into consideration has been developed. In the three band model, the refractive indices (n_e and n_o) are expressed as follows [54, 55]:

$$n_{e,o} \approx 1 + g_{0e,o}\frac{\lambda^2 {\lambda_0}^2}{\lambda^2 - {\lambda_0}^2} + g_{1e,o}\frac{\lambda^2 {\lambda_1}^2}{\lambda^2 - {\lambda_1}^2} + g_{2e,o}\frac{\lambda^2 {\lambda_2}^2}{\lambda^2 - {\lambda_2}^2}, \tag{4.8}$$

where $g_{0e,o}$, $g_{1e,o}$, and $g_{2e,o}$ represent the proportionality constants of the λ_0- λ_1- and λ_2-band, respectively. The three-band model clearly describes the origins of the refractive indices of LC compounds. However, a commercial mixture usually consists of several compounds with different molecular structures in order to obtain a wide nematic range. The individual λ_i's are therefore different. Under such a circumstance, Eq. (4.8) would have too many unknowns to quantitatively describe the refractive indices of a LC mixture.

In the off-resonance region, the three right-hand terms in Eq. (4.8) can be expanded in a power series up to the λ^{-4} term to form the extended Cauchy equations for describing the wavelength-dependent refractive indices of *anisotropic* LCs [56, 57]:

$$n_{e,o} \approx A_{e,o} + \frac{B_{e,o}}{\lambda^2} + \frac{C_{e,o}}{\lambda^4}. \tag{4.9}$$

In Eq. (4.9), $A_{e,o}$, $B_{e,o}$, and $C_{e,o}$ are three Cauchy coefficients. Although Eq. (4.9) is derived based on a LC compound, it can be extended easily to include eutectic mixtures by taking the weighted sum of contributions from each compound. From Eq. (4.9), if we measure the refractive indices at three wavelengths, the three Cauchy coefficients ($A_{e,o}$, $B_{e,o}$, and $C_{e,o}$) can be obtained from a fit to the experimental results. Once these coefficients are determined, the refractive indices at any wavelength can be calculated. From Eq. (4.9), both refractive indices and birefringence decrease as the wavelength increases. In the long wavelength (IR and millimeter wave) region, n_e and n_o are reduced to A_e and A_o, respectively. The coefficients A_e and A_o are constants; they are independent of wavelength, but depend on the temperature. That means, in the IR region the refractive indices are insensitive to wavelength, except for the resonance enhancement effect near the local molecular vibration bands. This prediction is consistent with many experimental observations [58].

Figure 4.5 depicts the wavelength-dependent refractive indices of E7 at $T = 25\,°C$. Open squares and circles represent the n_e and n_o of E7 in the visible region while the downward- and upward-triangles indicate measured data at $\lambda = 1.55$ and $10.6\,\mu m$, respectively. Solid curves are fits to the experimental n_e and n_o data in the visible spectrum using the extended Cauchy equations (Eq. (4.9)). The fitting parameters are as follows: ($A_e = 1.6933$, $B_e = 0.0078\,\mu m^2$, $C_e = 0.0028\,\mu m^4$) and ($A_o = 1.4994$, $B_o = 0.0070\,\mu m^2$, $C_o = 0.004\,\mu m^4$). In Figure 4.5, the extended Cauchy model is extrapolated to the near- and far-infrared regions. The extrapolated

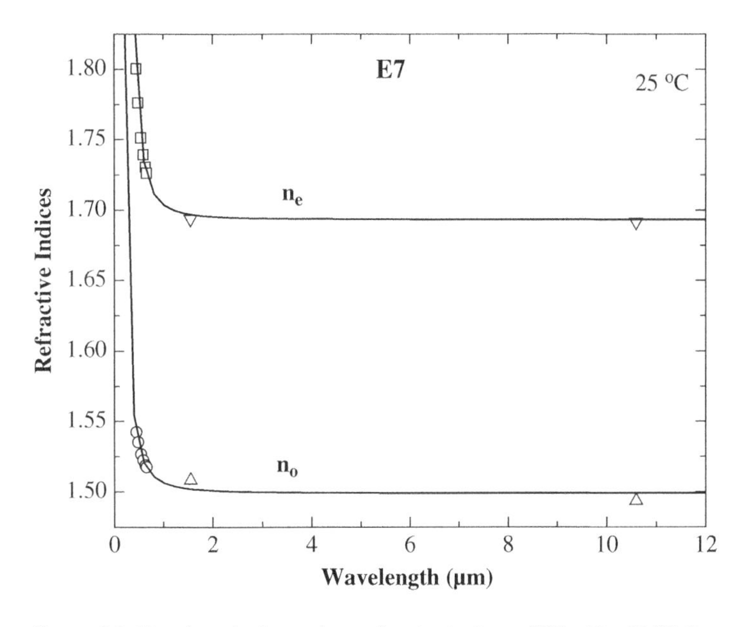

Figure 4.5 Wavelength-dependent refractive indices of E7 at $T = 25\,°C$. Open squares and circles are the n_e and n_o measured in the visible spectrum. Solid lines are fittings to the experimental data measured in the visible spectrum by using the extended Cauchy equation (Eq. (4.9)). The downward- and upward-triangles are the n_e and n_o measured at $T = 25\,°C$ and $\lambda = 1.55$ and $10.6\,\mu m$, respectively.

lines almost exactly pass through the experimental data points measured at $\lambda = 1.55$ and $10.6\,\mu m$. The largest difference between the extrapolated and experimental data is only 0.4%.

Equation (4.9) applies equally well to both high and low birefringence LC materials in the off-resonance region. For low birefringence ($\Delta n < 0.12$) LC mixtures, the λ^{-4} terms are insignificant and can be omitted and the extended Cauchy equations may be simplified to [59]:

$$n_{e,o} \approx A_{e,o} + \frac{B_{e,o}}{\lambda^2}. \tag{4.10}$$

Thus, n_e and n_o each has only two fitting parameters. By measuring the refractive indices at two wavelengths, we can determine $A_{e,o}$ and $B_{e,o}$. Once these two parameters are determined, n_e and n_o can be calculated at any wavelength of interest. Because most TFT LC mixtures have $\Delta n \sim 0.1$, the two-coefficient Cauchy model is adequate to describe the refractive index dispersions. Although the extended Cauchy equation fits experimental data well [60], its physical origin is not clear. A greater physical meaning can be attached to the three-band model which takes the major electronic transition bands into consideration.

4.3.7.2 Temperature Effect

The effect of temperature is particularly important for projection displays [5]. Due to the heating effect of the lamp, the temperature of the display panel could reach 50 °C. It is important to know the LC properties at the anticipated operating temperature beforehand.

Birefringence is defined as the difference between the extraordinary and ordinary refractive indices, $\Delta n = n_e - n_o$, and the average refractive index $\langle n \rangle$ is defined as $\langle n \rangle = (n_e + 2n_o)/3$. Based on these two definitions, n_e and n_o can be rewritten as:

$$n_e \approx < n > + \frac{2}{3}\Delta n, \tag{4.11}$$

$$n_o \approx < n > - \frac{1}{3}\Delta n. \tag{4.12}$$

To describe the temperature dependent birefringence, the Haller approximation (Eq. (4.7)) can be employed when the temperature is not too close to the clearing point:

$$\Delta n(T) = (\Delta n)_o (1 - T/T_c)^{\alpha}, \tag{4.13}$$

In Eq. (4.13), $(\Delta n)_o$ is the LC birefringence in the perfectly ordered crystalline state (or in the LC state at 0 K), the exponent α is a material constant, and T_c is the clearing temperature of the LC material under investigation. On the other hand, the average refractive index decreases linearly with increasing temperature as [61]:

$$< n > = A - BT, \tag{4.14}$$

because the LC density decreases with increasing temperature. By substituting Eqs. (4.14) and (4.13) back to Eqs. (4.11) and (4.12), a four-parameter model for describing the temperature dependence of the LC refractive indices is given as [62]:

$$n_e(T) \approx A - BT + \frac{2(\Delta n)_o}{3}\left(1 - \frac{T}{T_c}\right)^{\alpha}, \tag{4.15}$$

$$n_o(T) \approx A - BT - \frac{(\Delta n)_o}{3}\left(1 - \frac{T}{T_c}\right)^{\alpha}, \tag{4.16}$$

The parameters $[A, B]$ and $[(\Delta n)_o, \alpha]$ can be obtained separately by two-stage data fits. To obtain $[A, B]$, one can fit the average refractive index $\langle n \rangle = (n_e + 2n_o)/3$ as a function of temperature using Eq. (4.14). To find $[(\Delta n)_o, \alpha]$, one can fit the birefringence data as a function of temperature using Eq. (4.13). Therefore, these two sets of parameters can be obtained separately from the same set of refractive indices using different fits.

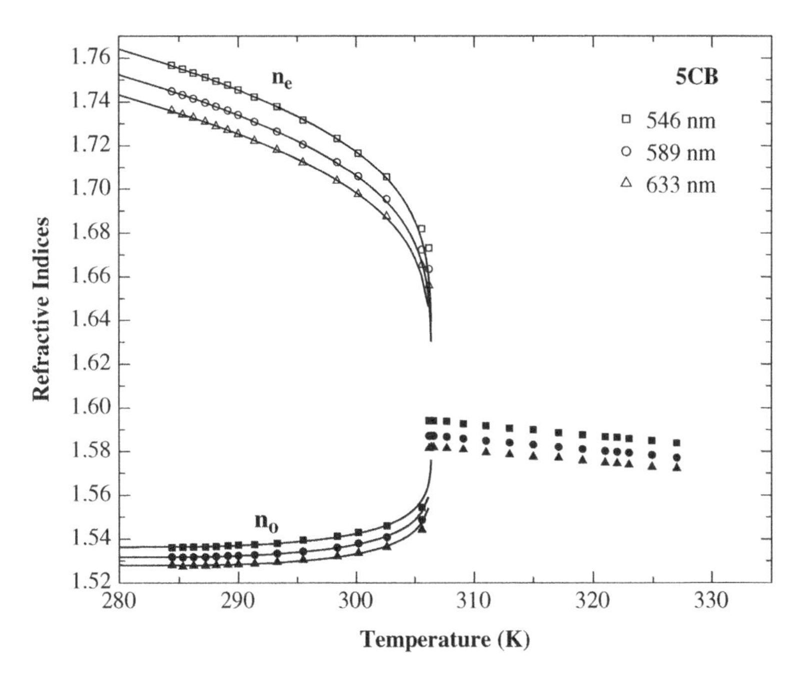

Figure 4.6 Temperature-dependent refractive indices of 5CB at $\lambda = 546$, 589, and 633 nm. Squares, circles, and triangles are experimental data for refractive indices measured at $\lambda = 546$, 589, and 633 nm, respectively.

Figure 4.6 is a plot of the temperature dependent refractive indices of 5CB at $\lambda = 546$, 589, and 633 nm. As the temperature increases, n_e decreases, but n_o gradually increases. In the isotropic state, $n_e = n_o$ and the refractive index decreases linearly as the temperature increases. This correlates with the density effect.

4.4 LIQUID CRYSTAL ALIGNMENT

The liquid crystal mixtures developed for displays are anisotropic liquids at room temperature. In order to obtain useful electro-optic properties, they need to be confined and aligned between two glass or plastic substrates. The substrates are usually coated with a thin (~80 nm) alignment layer in order to align the LC molecules. Four alignment methods which have been commonly employed are mechanically rubbed polyimide (PI), ion beam etched PI, evaporated SiO_x [63], and photoalignment. PI has commonly been used in large screen displays because of its fabrication simplicity. The inorganic SiO_x layer is widely used in projection displays because it is robust and can withstand high intensity illumination from a high power lamp. Photoalignment is becoming mainstream technology for high resolution LCDs. To achieve photo-induced LC alignment, the substrate is first coated with a thin photosensitive polymer. After exposure to polarized UV light, an anisotropic interfacial interaction is generated which provides orientational ordering to the LC molecules. In comparison with rubbing method, photoalignment offers several outstanding advantages, as it is contactless, readily forms multi-domain alignment for wide view devices, and shows excellent alignment uniformity for high resolution devices [64].

Figure 4.7 shows rubbing and ion beam etching induced LC alignment. In mechanical buffing, a nylon cloth is used. The LC directors tilt up along the rubbing direction, as shown in Figure 4.7a [65]. The pretilt angle depends on the PI material and rubbing strength. For a typical rubbing process, the pretilt angle is around 3°–5°. On the contrary, after ion beam etching, the LC directors tilt up in the direction toward the source of the etching beam, as shown in Figure 4.7b. An important function of pretilt angle is to bias the LC director to have a uniform reorientation direction when an external field is applied [66]. Without this pretilt angle,

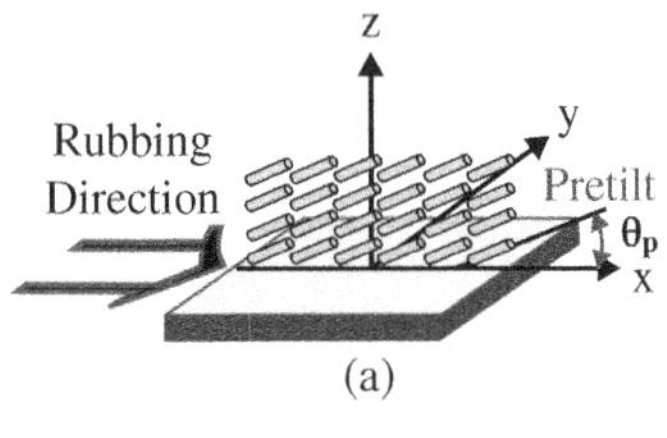

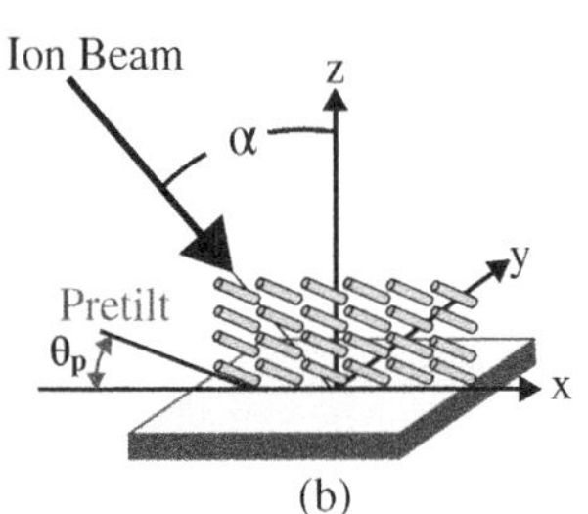

Figure 4.7 (a) Rubbing and (b) ion beam etching induced pretilt angle on a PI coated substrate. Source: Redrawn from Ref. 58.

the LC molecules may not switch in the same direction over the device area, which can cause non-uniform appearance, light scattering and slow response.

Several LC cell geometries have been developed depending on the rubbing direction and pretilt angle of each substrate. When two substrates are rubbed in antiparallel directions so that their pretilt angles (typically $\sim 3°-5°$) are in opposite directions, a homogeneous cell is formed. Similarly to a homogeneous cell, if the PI has a large pretilt angle, say $85°-90°$, a homeotropic (also known as vertical alignment) cell is formed [39]. When two low tilt substrates are rubbed in orthogonal directions, a twisted-nematic (TN) cell is formed [32]. If a chiral dopant is added to the LC material a twist angle larger than 90° can be obtained, and the cell is then called super twisted nematic (STN) [67]. Some polymer-dispersed liquid crystals do not need surface treatment. Due to phase separation, they form liquid crystal droplets [68].

Four types of LC alignment have been widely used in the display industry. They are the 90° TN cell, a homogeneous low tilt cell which is used in IPS and FFS devices, MVA devices which uses a patterned homeotropic (high tilt) alignment, and the bend or π-cell which uses parallel rubbing with pretilt angle intermediate between the homogenous and homeotropic devices. Besides LC alignment, the electrode configuration also determines the performance of a LCD, especially its viewing angle. In the following sections, we will describe these four widely employed LC alignments.

4.5 HOMOGENEOUS CELL

Homogeneous alignment has been used in IPS, FFS, and phase-only modulators, depending on the electric field direction. In IPS and FFS cells, interdigitated electrodes are formed on the bottom substrate such that the electric field is mainly in the transverse direction. However, in phase-only spatial light modulators [69], the electrodes are on both substrates and the electric field is in longitudinal direction. In-plane molecular reorientation leads to a wide viewing angle and is suitable for direct-view displays. On the other hand, the out-of-plane reorientation induced by the longitudinal field has a narrow viewing angle. Such a spatial light modulator has been used for augmented reality and head-up displays [70] and for laser beam steering [71]. To widen the viewing angle for direct-view displays, a special compensation film needs to be used. In this section, we will describe the basic electro-optics of a homogeneous cell using longitudinal electric field. We will defer the description of wide-viewing angle devices (IPS and FFS cells) to later sections.

In a homogeneous cell (also known as electrically controlled birefringence [ECB] cell), the top and bottom substrates are rubbed in antiparallel directions (x and $-x$) to generate opposite pretilt angles θ_p, as depicted

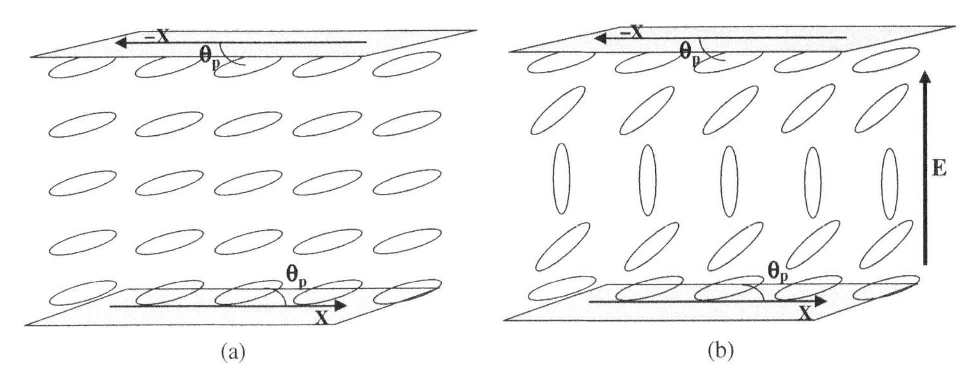

Figure 4.8 LC director profile in a homogeneous cell. (a) V = 0, and (b) V ≫ Vth.

in Figure 4.8. If the LC alignment directions are parallel and the pretilt angles are in the same direction, then a π-cell is formed [72]. We will discuss the electro-optics of the π-cell in Section 4.10. As shown in Figure 4.9b, when the applied voltage is well above the Freederisckz transition threshold [73] ($V_{th} = \pi\sqrt{K_{11}/(\varepsilon_o \Delta\varepsilon)}$), the LC director close to the central plane of the cell will be reoriented almost along the electric field direction, provided that the employed LC has a positive dielectric anisotropy. The existence of this threshold-like voltage (there is no true threshold because this is a pretilted cell so has a torque on the director under arbitrarily low fields) is because the electric field-induced torque has to overcome the restoring elastic torque in order to reorient the LC directors. If the anchoring energy is weak, then the threshold voltage will decrease and the overall response time (rise + decay) becomes slower [74]. As shown in Figure 4.8b, the applied electric field (E) is in the longitudinal direction (defined as the z-axis), therefore, the LC directors tilt out of the plane. The LC layers closest to the substrates are anchored by the alignment layers and do not reorient under normal strengths of applied field. The pretilt angle ($\theta_p \sim 3°$) gives a pre-determined direction for the LC directors to follow. Without this pretilt angle, the LC directors do not know in which initial direction to rotate and this would cause a patchy optical appearance and light scattering. If the LC has a negative $\Delta\varepsilon$, then the electric field cannot reorient the LC molecules. To drive the negative $\Delta\varepsilon$ LC in a homogeneous cell, a transverse electric field (or so-called fringe field) has to be used. This mechanism will be discussed in Sections 4.7 and 4.8.

For a given voltage, as shown in Figure 4.8b, the LC director distribution inside the cell is not uniform, but is symmetric about the mid-layer of the cell, assuming the pretilt angles on both surfaces are equal. Figure 4.9 shows the LC director configuration at different voltages, normalized to V_{th}. Below V_{th}, the LC directors are slightly reoriented due to the 3° pretilt angle. As the voltage increases to 3–4x the threshold, the majority of the LC has been reoriented. The middle layer ($z/d = 0.5$) has already been rotated by more than 80°. Further increase of the applied voltage to 10 V_{th} only reorients the LC layers closer to the surfaces; the bulk layers have already been completely reoriented.

4.5.1 Phase Retardation Effect

When a plane wave is incident normally on a planar aligned uniaxial liquid crystal layer sandwiched between two polarizers, the extraordinary and ordinary components of the outgoing beam will have experienced a relative phase retardation (δ) due to their different propagation speeds inside the LC medium:

$$\delta = \frac{2\pi d}{\lambda}(n_e - n_o) = 2\pi d\Delta n/\lambda, \tag{4.17}$$

where d is the cell gap, Δn is the birefringence, and λ is the wavelength. When a homogeneous cell is sandwiched between two polarizers, the normalized light transmittance is governed by the following equation:

$$T = \cos^2\chi - \sin 2\beta \sin 2(\beta - \chi)\sin^2(\delta/2). \tag{4.18}$$

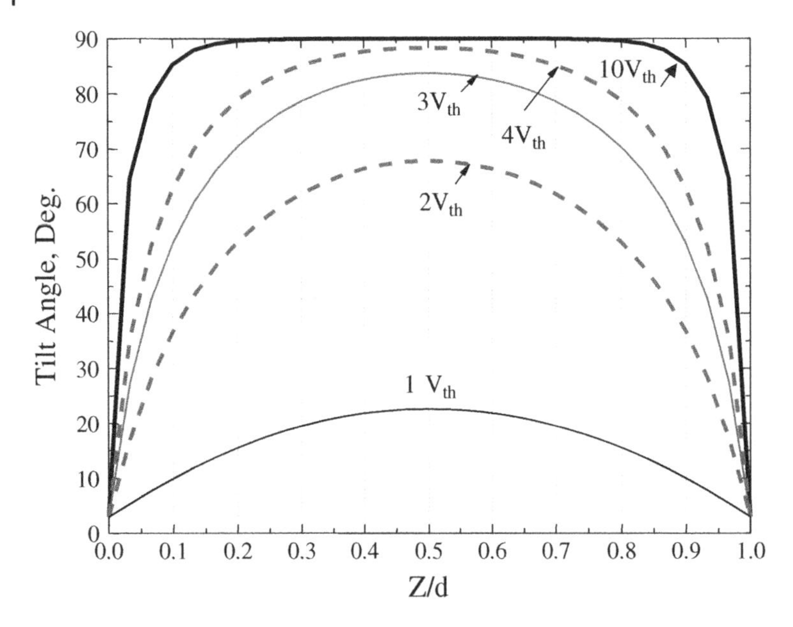

Figure 4.9 LC director distribution profile of a homogeneous cell under different voltages.

Here χ is the angle between polarizer and analyzer, β is the angle between the polarizer and the LC director, and δ is the retardation as above. For the simple case when $\beta = 45°$ and the two polarizers are crossed ($\chi = 90°$), the normalized light transmittance simplifies to:

$$T_{\perp} = \sin^2(\delta/2). \tag{4.19}$$

For a homogeneous cell, the effective phase retardation depends on the wavelength, temperature, and applied voltage. On applying a voltage to a planar LC layer with positive dielectric anisotropy, the effective birefringence and the phase retardation is decreased. In a high voltage regime where $V \gg V_{th}$, LC director is aligned nearly normal to the substrates except in the boundary layers. The effective phase retardation is then small owing to this vanishing effective birefringence. However, the boundary layers are quite resistant to being completely reoriented by the external field. This implies that a good dark state is difficult to achieve from a homogeneous cell without a phase compensation film.

4.5.2 Voltage Dependent Transmittance

Figure 4.11 depicts the voltage-dependent light transmittance of a homogeneous cell between crossed polarizers. Here the optical losses from polarizers, substrate surfaces, and ITO are all neglected. The LC cell parameters are listed as follows: LC mixture MLC-6297-000, elastic constants $K_{11} = 13.4\,\text{pN}$, $K_{22} = 6.0\,\text{pN}$, and $K_{33} = 19.0\,\text{pN}$, dielectric constants $\varepsilon_{//} = 10.5$ and $\Delta\varepsilon = 6.9$, cell gap $d \sim 4.3\,\mu\text{m}$, $\Delta n = 0.125$, 0.127 and 0.129 for $R = 650$ nm, $G = 550$ nm, and $B = 450$ nm, pretilt angle $\theta_p \sim 2°$. From Figure 4.11a, the bright state transmission varies between the RGB colors by around 10%. Due to the small $d\Delta n$ value (275 nm which is $\sim\lambda/2$), the transmittance in the high voltage region decreases monotonically, but it is difficult to get a common dark state for the RGB colors. This is because the residual phase retardation of the boundary layers as shown in Figure 4.9 is additive (parallel) rather than subtractive (orthogonal). One method to obtain a common dark state for a homogeneous cell is to add a uniaxial compensation film (to be discussed in Section 4.7.4) whose optic axis is orthogonal to that of the LC director at each surface. Figure 4.10b shows the results of a uniaxial film-compensated homogeneous cell whose $d\Delta n = 368$ nm and where the compensation film's $d\Delta n = -96$ nm.

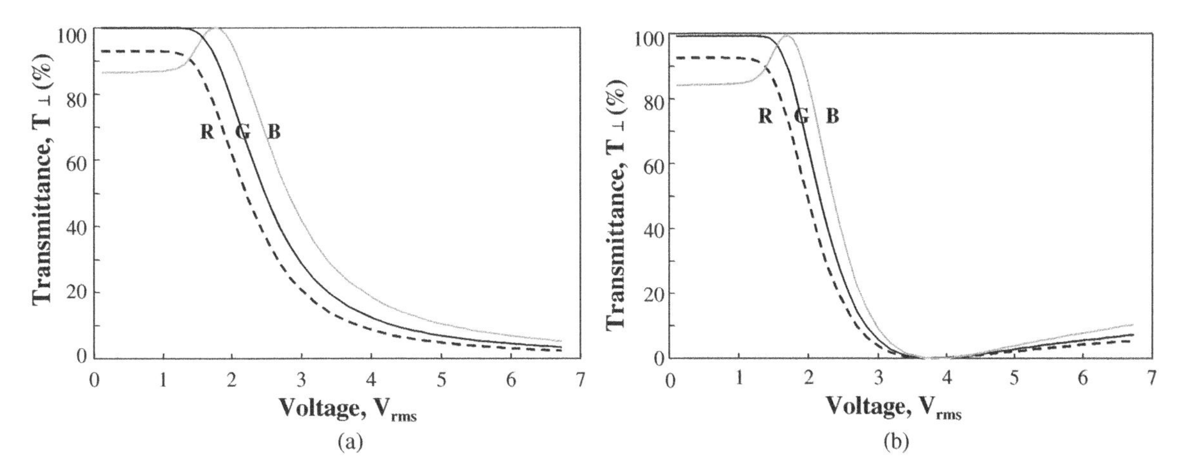

Figure 4.10 (a) VT curve of a homogeneous cell with dΔn = 275 nm, and (b) a uniaxial film-compensated homogeneous cell (cell dΔn = 368 nm and film dΔn = 96 nm but oriented to be orthogonal to the LC directors).

The dark state is optimized at a reduced voltage of ~4 V_{rms}. This dark state voltage is dependent on the $d\Delta n$ values of the LC cell and compensation film employed. For a smaller $d\Delta n$ in the compensation film, the dark state voltage will occur at higher voltage and its width (ΔV while keeping a good dark state) will be broader.

4.6 TWISTED NEMATIC (TN)

The 90° twisted-nematic (TN) cell has been used extensively for small size displays and notebook computers where the viewing angle is not too critical. Figure 4.11 shows the LC director configurations of the normally white (NW) TN cell in the voltage-off (left) and voltage-on (right) states.

The top LC alignment is parallel to the optic axis of the top polarizer while the bottom LC directors are rotated 90° and parallel to the optic axis of the bottom analyzer. This is called e-mode operation. Another configuration is called o-mode, which is to orient the transmissive axis of the polarizer orthogonal to the front LC director [75]. When the $d\Delta n$ of the LC layer satisfies the Gooch-Tarry first minimum condition [76], the incoming linearly polarized light appears to follow the molecular twist and be transmitted through the crossed analyzer. Although this waveguiding effect is slightly dependent on the wavelength, the transmittance variation between the RGB wavelengths is less than 8%. Thus, such a 90° TN cell is basically a broadband half-wave plate. In the voltage-on (~5 V_{rms}) state, the LC director is reoriented perpendicular to the substrates, except in the boundary layers. The incoming light experiences little phase change and is absorbed by the analyzer, resulting in a dark state. The beauty of the TN cell is that the boundary layers are orthogonal so that their residual phase in the high voltage state compensates for each other. But also as important, the boundary layers are effectively optically thin uniaxial slabs aligned with the adjacent polarizer film, so by Eq. (4.18) with $\beta = 0$, do not contribute to any stray transmission. As a result, the dark state occurs at a relatively low voltage. A typical contrast ratio of an active matrix TN LCD is about 1000 : 1. If the twist angle departs from 90°, the dark state will be degraded and operating voltage increased accordingly.

4.6.1 Optical Transmittance

To compare different operating modes, let us focus on the normalized transmittance by ignoring the optical losses from polarizers, ITO ($n \sim 1.8$) layers, and the interfacial reflections from the substrates. The normal-

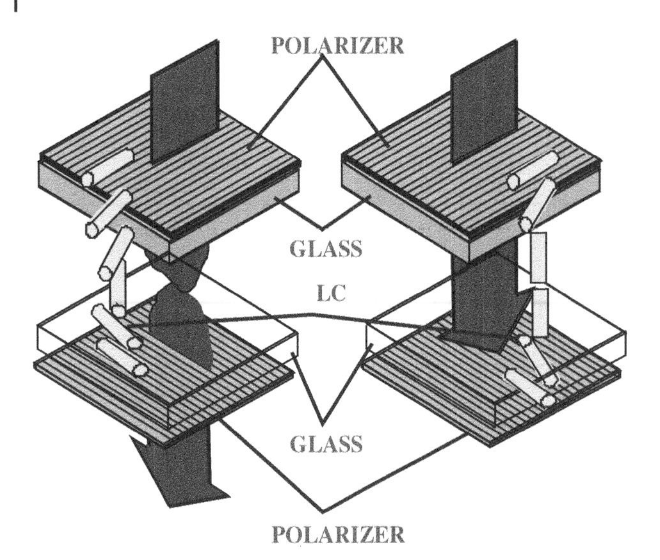

Figure 4.11 LC and polarizer configurations of a 90° TN cell. Left: V = 0, and right $V \gg V_{th}$.

ized transmittance ($T_\perp$ without any voltage) of a TN cell can be described by the following Jones matrices as $T_\perp = |M|^2$ [77]:

$$M = |\cos\beta \quad \sin\beta| \begin{vmatrix} \cos\phi & -\sin\phi \\ \sin\phi & \cos\phi \end{vmatrix} \begin{vmatrix} \cos X - i\dfrac{\Gamma}{2}\dfrac{\sin X}{X} & \phi\dfrac{\sin X}{X} \\ -\phi\dfrac{\sin X}{X} & \cos X + i\dfrac{\Gamma}{2}\dfrac{\sin X}{X} \end{vmatrix} \begin{vmatrix} -\sin\beta \\ \cos\beta \end{vmatrix} \tag{4.20}$$

Here β is the angle between the polarization axis and the front LC director, ϕ is the twist angle, $X = \sqrt{\phi^2 + (\Gamma/2)^2}$, $\Gamma = 2\pi\ d\ \Delta n/\lambda$, and d is the cell gap. By simple algebraic calculations, the following analytical expression for $|M|^2$ is derived:

$$|M|^2 = T_\perp = \left(\frac{\phi}{X}\cos\phi\sin X - \sin\phi\cos X\right)^2 + \left(\frac{\Gamma}{2}\frac{\sin X}{X}\right)^2 \sin^2(\phi - 2\beta) \tag{4.21}$$

Equation (4.21) is a general formula describing the light transmittance of a TN cell (without voltage) as a function of twist angle, beta angle, and $d\Delta n/\lambda$. For a 90° TN cell, $\phi = \pi/2$ and Eq. (4.21) simplifies to:

$$T_\perp = \cos^2 X + \left(\frac{\Gamma}{2X}\cos 2\beta\right)^2 \sin^2 X. \tag{4.22}$$

Equation (4.22) has a special solution when $\cos^2 X = 1$. When $\cos X = \pm 1$ (i.e. $X = m\pi$; $m =$ integer), then $\sin X = 0$ and the second term in Eq. (4.22) vanishes. Therefore, $T_\perp = 1$, independent of β. By setting $X = m\pi$ and knowing that $\Gamma = 2\pi d\Delta n/\lambda$, the Gooch–Tarry's condition is found as follows:

$$\frac{d\Delta n}{\lambda} = \sqrt{m^2 - \frac{1}{4}}. \tag{4.23}$$

For the lowest order $m = 1$, $d\Delta n/\lambda = \sqrt{3}/2$. This is the Gooch–Tarry first minimum condition for the 90° TN cell. Here, the first minimum condition refers to the smallest $d\Delta n$ value for a given wavelength to achieve $T_\perp = 1$. For the second order $m = 2$ and $d\Delta n/\lambda = \sqrt{15}/2$. The second minimum condition is used only for low-end displays such as wrist watches and calculators because a large cell gap is easier to fabricate and the cyano-biphenyl LCs are less expensive. For notebook TFT-LCDs, the first minimum is preferred because it requires a thinner cell gap which leads to a faster response time and wider viewing angle.

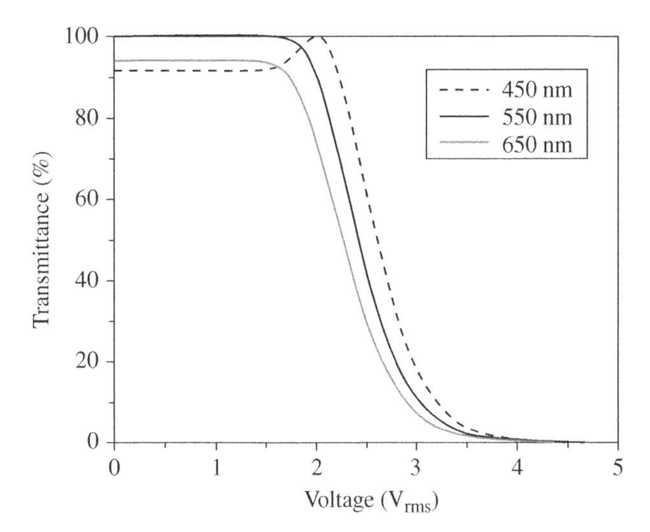

Figure 4.12 Voltage-dependent transmittance of a normally white 90° TN cell. dΔn = 480 nm.

Figure 4.12 depicts the normalized voltage-dependent light transmittance ($T_{\perp}$) of the 90° TN cell at three primary wavelengths: $R = 650$ nm, $G = 550$ nm, and $B = 450$ nm. Since the human eye has its greatest sensitivity at green wavelengths, we normally optimize the cell design at $\lambda \sim 550$ nm. From Eq. (4.22), the first $T_{\perp} = 1$ occurs at $d\Delta n \sim 480$ nm. For a 5 μm cell gap, the required birefringence is $\Delta n \sim 0.096$. From Figure 4.12, the wavelength effect on the transmittance at $V = 0$ is below 8%. Therefore, the TN cell can be treated as an "achromatic" half-wave plate.

The response time of a TN LCD depends on the cell gap and the γ_1/K_{11} of the LC mixture employed. For a 5 μm cell gap, the optical response time is ~20–30 ms. At $V = 5\ V_{rms}$, the contrast ratio (CR) reaches ~1000 : 1. These performance figures, although not perfect, are acceptable for notebook computers. A major drawback of the TN cell is its relatively narrow viewing angle and grayscale inversion which originates from the LC director tilting out of the plane. Because of this molecular tilt, the viewing angle in the vertical direction is narrow and asymmetric, and has grayscale inversion [78].

4.6.2 Viewing Angle

In a TN cell, the applied voltage is in the longitudinal direction. As a result, the bulk LC directors tilt toward the electric field direction, as shown in Figure 4.13a. This out of plane orientation leads to an asymmetric viewing angle in the vertical viewing direction, as illustrated in Figure 4.13b. The cross-section of the refractive index ellipsoid viewed from this up-down direction is quite asymmetric. Therefore, its light transmittance through crossed polarizers would be different from different azimuth directions leading to an asymmetric viewing angle. On the other hand, the viewing angle in the horizontal direction is wider and more symmetric.

Figure 4.14 shows the viewing angle characteristics of a TN LCD. The middle photo represents the normal view, the right and left are horizontal, and top and bottom are vertical views. The horizontal view is indeed fairly symmetric, but the vertical view is much worse. A severe grayscale inversion occurs in the bottom-up direction (bottom photo).

To extend TN LCDs to large screen monitors or TVs, the viewing angle has to be improved. A convenient way to widen the viewing angle of a TN LCD is to use phase compensation films instead of implementing multi-domain structures [79]. The films can be laminated to the inner side of the polarizers. However, due to the asymmetric view of the TN cell, the required compensation film also needs asymmetric phase retardation in the vertical direction.

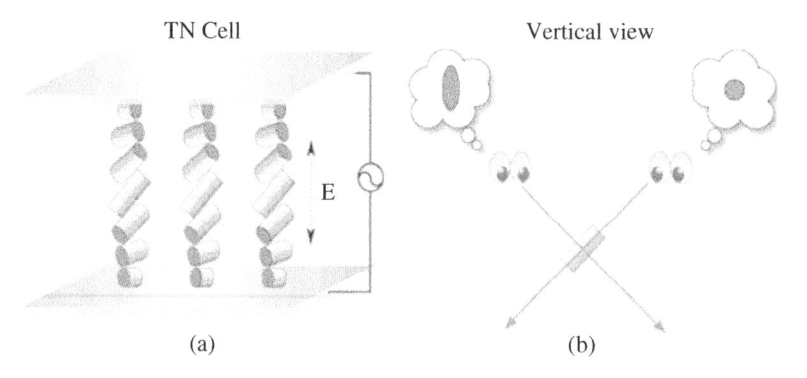

Figure 4.13 (a) LC tilt angles of a TN cell in a voltage-on state, and (b) viewing disparity from vertical direction.

Figure 4.14 Viewing angle of a TN LCD. The middle image is a normal view. Right and left: horizontal view. Top and bottom: vertical view. Grayscale inversion occurs at the bottom-up viewing direction. Source: Courtesy of Dr. Y. Saitoh of Fuji Film.

4.6.3 Film-Compensated TN

In the on-state of a TN cell, the LC directors in the upper half are reoriented along the rubbing direction with almost no twist and the lower half-cell has a similar structure with the director plane orthogonal to that of the upper half. Thus, a uniform phase compensation film, such as a uniaxial A-plate, cannot compensate the upper and lower parts simultaneously. Instead, a pair of wide-view films needs to be used separately on both sides of the TN LC cell in order to compensate each of the half layers. Fuji Photo has developed discotic LC films for widening the viewing angle of TN cells [80]. The molecular structures of the wide-view (WV) discotic materials are shown in Figure 4.15.

A discotic material (triphenylene derivatives) is coated on an alignment layer on a TAC (tri-acetyl cellulose) substrate. The discotic material has a hybrid alignment structure and three important features. (i) It has π

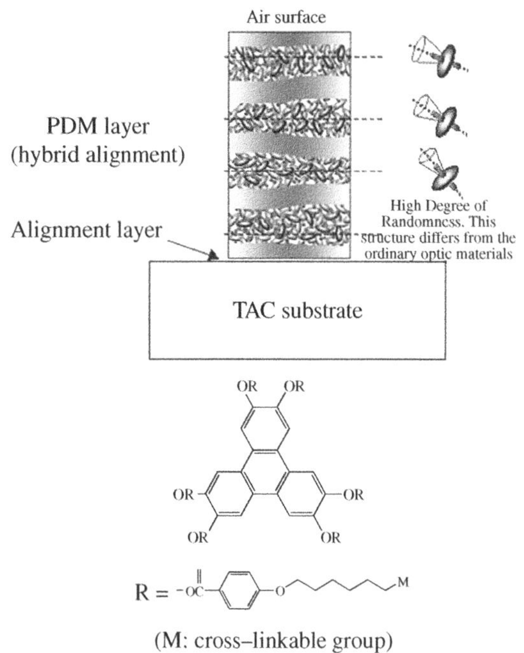

Figure 4.15 Structure of the WV film and the employed discotic compound. PDM represents a polymerized discotic material.

electron clouds distributed over a disc-like shape, which gives it a high negative birefringence. (ii) It takes on discotic nematic (N_D) phase at a lower temperature than the temperature at which the TAC substrate starts to deform. This feature allows a uniform monodomain film to be coated over a large area without defects. (iii) It has cross-linkable groups on all six side chains to make the completed film durable. When heated, the discotic material forms a discotic nematic (N_D) phase. In the vicinity of the alignment layer, the discotic molecules tend to align with their molecular plane almost parallel to the alignment layer surface and have a few degrees of pretilt angle in the plane of the rubbing direction of the alignment layer surface. On the other hand, in the vicinity of the air interface the discotic molecules tend to align with their molecular plane almost perpendicular to the air surface. With this pinned alignment at each interface, the discotic material exhibits a hybrid alignment structure in the N_D phase. When cured by UV light, the discotic material is polymerized and the hybrid alignment structure of the polymerized discotic material (PDM) layer is fixed even after it is cooled down to room temperature. Each film has a hybrid aligned structure in which the director continuously changes through the PDM layer thickness without any twist while the direction of each discotic molecule fluctuates. This hybrid aligned structure includes both splay and bend deformations.

The wide-view Fuji film remarkably improves the viewing angle of TN LCDs without reducing the light transmittance or degrading the image quality. A TN LCD with ~80° viewing cone at $CR > 10:1$ has been demonstrated [81]. A significant advantage of the film-compensated TN LCD is that no change in the panel process is required because the conventional polarizer is simply replaced by a new polarizer laminated with the compensation film. The discotic film is also cost effective. These features have allowed TN LCDs to penetrate into the larger size display market segment, for example up to 20–26 in. diagonal. However, grayscale inversion can still be observed in large panel film-compensated TN LCDs which ultimately limits their competitiveness with IPS and MVA devices.

4.7 IN-PLANE SWITCHING (IPS)

To overcome the narrow viewing angle of a TN LCD, a device mode called IPS which had been proposed in the 1970s [33, 82] was implemented in TFT-LCDs in the 1990s [83, 84]. Interdigitated electrodes are located on

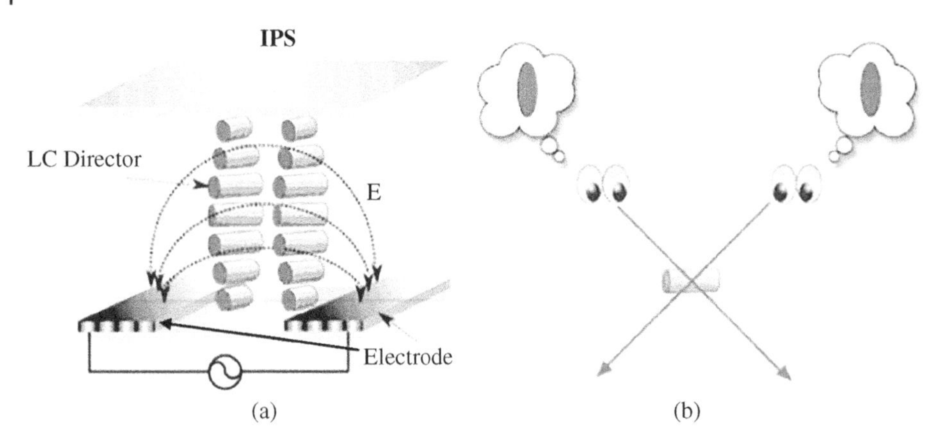

Figure 4.16 (a) IPS electrodes, electric fields, and LC director orientations in a voltage-on state, (b) viewing angle characteristics of in-plane rotation of LC molecules.

the lower substrate and the top substrate has no electrodes. The resulting electric field is curved by the fringing fields at the electrode edges, but its main component is in the plane of the cell. As shown in Figure 4.16, when a field is applied the LC directors are rotated in the device plane which results in a wide viewing angle.

4.7.1 Device Structure

In an IPS mode display, the interdigitated electrodes are fabricated on the same substrate and an LC with positive dielectric anisotropy is aligned homogeneously with a rubbing angle offset by ~10° with respect to the stripe electrodes. The transmission axis of the polarizer can be set to be parallel (e-mode) or perpendicular (o-mode) to the LC director and the analyzer is crossed with respect to the polarizer. The in-plane electric field induced by the electrodes twists the LC directors as shown in Figure 4.17. The incoming linearly polarized light from the polarizer experiences a phase retardation so that its polarization state is changed, resulting in light transmission through the crossed analyzer. However, a strong vertical electric field with only a small horizontal component exists above the electrode surfaces, so the LC director in these regions mainly tilts rather than twisting. As a result, the transmittance above the electrodes is greatly reduced. Overall, the conventional IPS mode has a light efficiency ~75% of that of a TN LCD, when a positive $\Delta\varepsilon$ LC material is used. Using a negative $\Delta\varepsilon$ liquid crystal in the IPS mode can increase the light efficiency to ~85%, but the on-state voltage depends on the electrode gap and the $\Delta\varepsilon$ of the employed LC material [85]. For low power consumption, the preferred operating voltage is lower than $5\,V_{rms}$.

4.7.2 Voltage-Dependent Transmittance

Figure 4.18 depicts the voltage-dependent light transmittance at RGB wavelengths corresponding to the IPS device structure shown in Figure 4.17. The threshold is $\sim 1.5\,V_{rms}$ and maximum transmittance occurs at $\sim 5\,V_{rms}$ for all the wavelengths. Due to absorption, the maximum transmittance of the two polarizers (without the LC cell) is 35.4%, 33.7%, and 31.4% for RGB wavelengths, respectively.

4.7.3 Viewing Angle

Figure 4.19 shows the calculated isocontrast contours of an IPS LCD without any compensation films. The 10 : 1 isocontrast field extends beyond a 60° viewing cone. For mobile displays, this viewing angle is adequate because the display is mainly viewed by a single user. However, for large size displays such as TVs, this viewing angle is still inadequate and compensation films are required. A properly compensated IPS LCD can have a contrast ratio > 100 : 1 beyond 80° viewing cone.

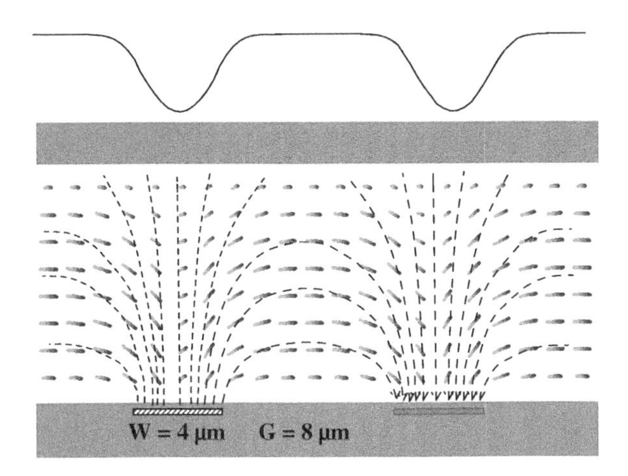

Figure 4.17 The LC director distribution, electric field profile (dashed lines), and corresponding light transmission of an IPS cell at $V = 5\ V_{rms}$. LC: MLC-6686, $\Delta\varepsilon = 10$, electrode width $w = 4\ \mu m$, gap $G = 8\ \mu m$, and cell gap $d = 3.6\ \mu m$.

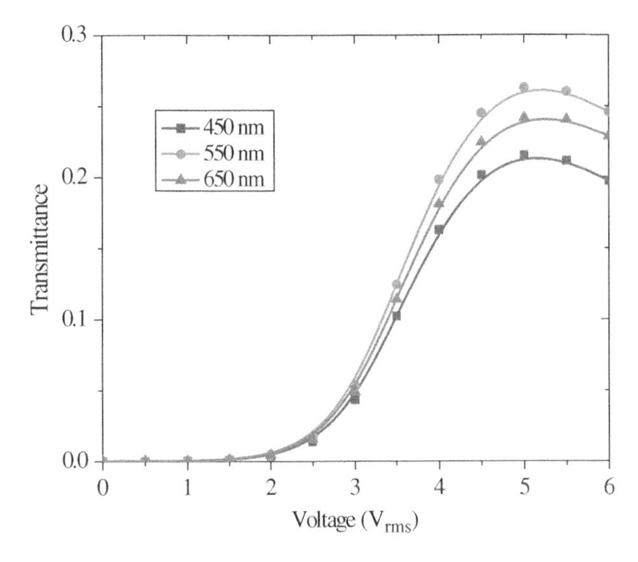

Figure 4.18 Voltage-dependent light transmittance of the IPS LCD. The device structure is shown in Figure 4.17.

4.7.4 Phase Compensation Films

In an IPS LCD, the LC director is aligned parallel (or perpendicular) to the optic axis of the entrance polarizer. At normal incidence, the LC layer in the voltage-off state does not modulate the polarization state of the incident linearly polarized light. As a result, a good dark state is achieved because the outgoing linearly polarized light is efficiently absorbed by the crossed analyzer. However, at oblique angles the incident light leaks through the crossed polarizers, especially at the bisectors. This light leakage arises from two factors: First, the absorption axes of the crossed polarizers are no longer orthogonal to each other under off-axis oblique view, thus, the extinction ratio of the crossed-polarizers decreases and light leakage occurs. Second, obliquely incident linearly polarized light sees the birefringence of the LC layer and becomes elliptically polarized after traversing through the cell. Consequently, the analyzer cannot completely absorb the elliptically polarized light leading to light leakage at off-axis angles of view. This light leakage in the dark state degrades the contrast ratio at oblique viewing angles. To suppress light leakage at oblique angles and further widen the viewing

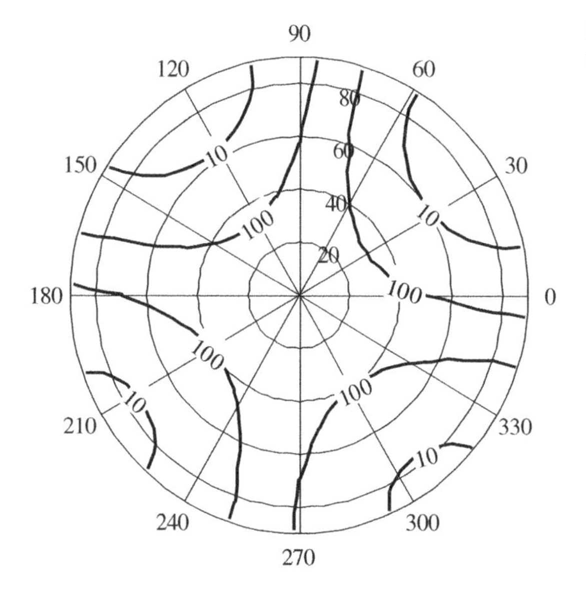

Figure 4.19 Simulated isocontrast contours of the IPS LCD without any compensation films.

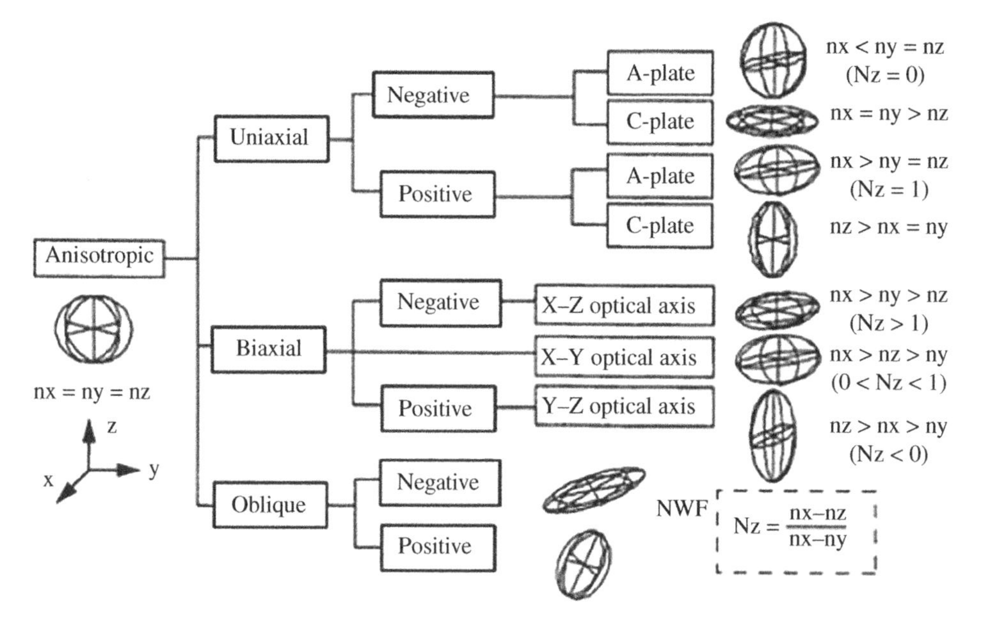

Figure 4.20 Different types of compensation films used for wide view LCDs.

angle, several phase compensation schemes using uniaxial films [86–88] and biaxial films [89–91] have been proposed.

Figure 4.20 shows some commercially available compensation films, classified by their refractive indices. Different LC modes need different types of compensation films in order to obtain a satisfactory compensation effect. For example, the IPS mode needs a biaxial film with $n_x > n_z > n_y$ [92], while VA and OCB modes need a compensation film with $n_x > n_y > n_z$ [93].

A uniaxial film is an anisotropic birefringent film with only one optic axis. On the basis of their optic axis orientation, uniaxial films can be classified into A-films and C-films. The optic axis of an A-film is aligned in the plane of the film, i.e. the *XY* plane, while a C-film's optic axis is perpendicular to the film surface (along the

Z-axis). The commonly used $\lambda/2$ and $\lambda/4$ plates are examples of A-plates. They are usually made of stretched polymers or reactive mesogen films. Their optic axis is defined by mechanical stretching the film along either the X or Y axis. On the other hand, a C-plate is isotropic in the XY plane but its optical anisotropy occurs along the Z axis.

4.8 FRINGE FIELD SWITCHING (FFS)

To overcome the dead zones that occur above the electrodes in the IPS device, a FFS display [94] has been developed. FFS LCDs have a similar structure to IPS mode devices, but have the significant difference that in the FFS cell, the spacing of the in-plane electrodes is less than the thickness of the display cell. This means that the in-plane electric field is concentrated, close to the lower substrate. The bulk of the LC layer experiences a low field which reduces the tendency of dark lines to appear over the electrode areas. FFS LCDs have been widely used in smartphones and tablet devices. A wide viewing angle allowing many simultaneous observers, high resolution to reduce the screen-door effect, high transmittance for low power consumption and long battery life, and pressure-resistance for integration with a touch screen are the key requirements. Both positive and negative dielectric anisotropy ($\Delta\varepsilon$) liquid crystals (LCs) can be used in FFS LCDs. The pros and cons are described as follows. For the FFS with a positive $\Delta\varepsilon$ LC, called p-FFS, it is relatively easy to obtain large $\Delta\varepsilon$ (~10) while keeping a low viscosity. A high $\Delta\varepsilon$ helps to reduce the operating voltage, while a low viscosity helps to shorten the response time. However, p-FFS displays exhibit some problems: (i) the peak transmittance is limited to ~88%; (ii) the voltage-dependent transmittance (VT) curves do not overlap well for RGB colors. Thus, three gamma curves are required, which increases the complexity of the driving electronics; (iii) the electro-optic effect is sensitive to the cell gap, and (iv) a small but noticeable image flicker occurs because of a splay-induced flexoelectric effect. In an FFS cell, the electric field is strong and non-uniform in both lateral and longitudinal directions. As a result, the LC director experiences splay and bend, which in turn causes a non-negligible flexoelectric polarization [95, 96].

4.8.1 Device Configurations

Figure 4.21 depicts the device structure (side-view), LC director distribution, and electric field distribution of a FFS cell in the voltage-on state. The bottom substrate comprises a common electrode (ITO), a thin dielectric passivation layer, pixel electrodes, and then a thin LC alignment layer. The passivation layer lies between the common and the pixel electrodes, providing a built-in storage capacitor (C_{st}) for each pixel. Thus, no external storage capacitor is needed, which increases the aperture ratio and optical efficiency. This feature is particularly important for high pixel density displays for smartphones and virtual reality headsets. On the top substrate, there is a thin LC alignment layer but no electrode. In the voltage-off state, the homogeneously-aligned LC directors are perpendicular to the optical axis of the polarizer. As a result, the incoming linearly polarized light experiences no phase retardation after traversing through the LC layer. The outgoing light remains linearly polarized and is blocked by the crossed analyzer, leading to a good dark state. As the voltage increases, strong electric fields are generated between the pixel electrodes and the common electrode. Because of the narrow electrode width ($W \approx 3\,\mu m$) and short electrode gap ($G \approx 3\,\mu m$), the dead zones as observed in IPS (Figure 4.17) are greatly suppressed. Therefore, the transmittance of FFS LCD is about 10–15% higher than that of IPS LCD with the same electrode width and gap.

The LC rubbing angle (or photoalignment direction) can vary from 0° to about 10° with respect to the stripe pixel electrodes. A larger LC alignment angle leads to a faster rise time (the decay time remains unchanged), but higher operation voltage because the remaining phase retardation is smaller. A special case exists when the rubbing angle is exactly 0°. Because of molecular rotation symmetry, the response time is improved by ~4×, but the major tradeoff is decreased transmittance because of virtual walls which form in the switched structure

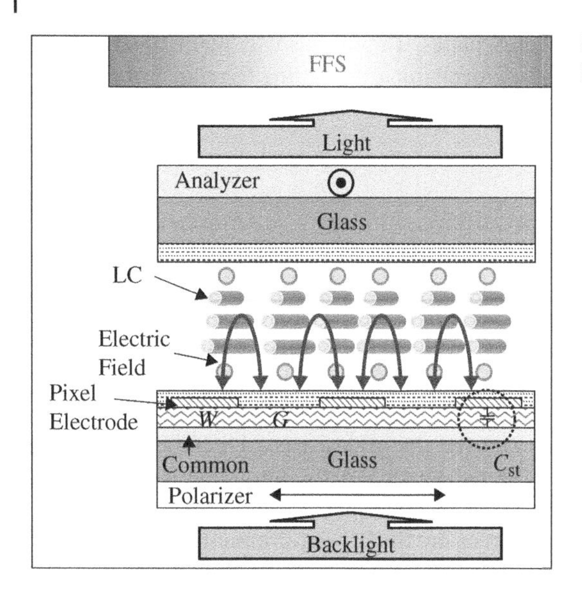

Figure 4.21 Device configuration, electric field distributions, and LC director deformation of a FFS structure in a voltage-on state.

[97–99]. To provide a reasonable compromise between transmittance, operating voltage, and response time, a typical rubbing angle is 7°.

In the IPS mode, the gap between electrodes is larger than the cell gap (d). Thus, the horizontal component of the electric field is dominant between electrodes, through the full thickness of the LC layer. However, in the FFS mode the electric field is concentrated near the surface and decreases rapidly through the LC bulk, with increasing distance from the lower plate. The LC director is strongly anchored on both surfaces. Therefore, in the voltage-on state, the LC director is actually twisted over a short length scale, and then gradually untwists as a function of distance from the bottom to the top surfaces [100]. Due to this double-twist nature, the voltage-dependent transmittance curves are not too sensitive to the wavelength. Moreover, the electric field generated between the pixel electrodes and common electrode is strong enough to reorient the LC directors above the pixel electrodes. Thus, the transmission dips (dead zones) which are observed in an IPS cell are much weaker in the FFS cell.

4.8.2 n-FFS versus p-FFS

Both positive and negative $\Delta\varepsilon$ LC materials can be employed in a FFS cell [101, 102]. To make a fair comparison between n-FFS and p-FFS, we set the electrode width $W = 2.5\,\mu m$, electrode gap $G = 3.5\,\mu m$, and pretilt angle (2°), while the rubbing angle is 7° for p-FFS and 83° for n-FFS with respect to the direction of the strip pixel electrodes. The passivation layer between the pixel and common electrodes is Si_3N_4 whose thickness is $d_p = 150$ nm and the dielectric constant is $\varepsilon_p = 7$. The cell is sandwiched between two crossed linear polarizers, and the transmission axis of the bottom polarizer is parallel to the rubbing direction. The preferred $d\Delta n$ value to achieve peak transmittance at $\lambda = 550$ nm and fast response time is ~320 nm for n-FFS and ~340 nm for p-FFS. This difference results from a more efficient LC director reorientation in the n-FFS cell [103].

Figure 4.22 depicts the simulated voltage-dependent transmittance (VT) curves of a single-domain (1D) n-FFS (dark line) and two p-FFS cells (gray and dashed lines). From Figure 4.22, the transmittance of the n-FFS cell rises more quickly and reaches its peak transmittance (93.5% vs. 87.1%) at a lower voltage ($V_p = 4.8$ versus 5.5 V_{rms}) than the corresponding p-FFS with the same $|\Delta\varepsilon|$ value. To reduce the V_p of p-FFS to 4.8 V, we have to increase the $\Delta\varepsilon$ value to 6.2 (gray line). This is because the LC directors in n-FFS are reoriented more uniformly in terms of twist and tilt angles, resulting in a higher effective birefringence. This more uniform LC reorientation also explains why the required $d\Delta n$ (= 320 nm) for n-FFS is smaller than that of p-FFS (340 nm). The n-FFS has a weaker flexoelectric effect than p-FFS [104]. However, p-FFS with a lower $+\Delta\varepsilon$ LC leads to

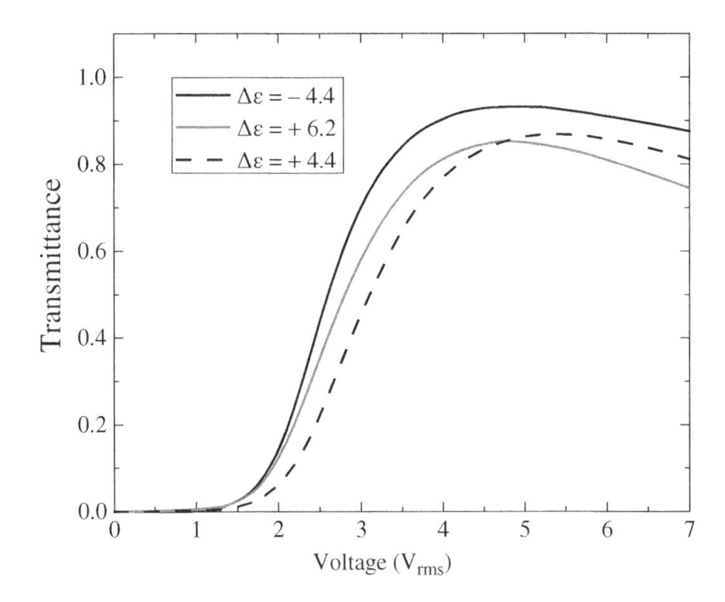

Figure 4.22 Simulated VT curves of n-FFS with $\Delta\varepsilon = -4.4$ and p-FFS with $\Delta\varepsilon = 4.4$ and 6.2.

a higher peak transmittance. As a result, there is crossover point as shown in Figure 4.22 where the gray and the dashed lines have same transmittance (85.3% at $V_p = 4.8V_{rms}$). A lower $\Delta\varepsilon$ LC usually possesses lower γ_1 and smaller activation energy, resulting in faster response time, especially at low temperature [105]. However, grayscale inversion can be seen in 1D n-FFS devices in the high transmittance region, but it is less evident for 1D p-FFS. Therefore, if we want to use n-FFS for practical applications we have to consider a 2D structure. This is indeed used in high-end LCD-based smartphones.

Another major difference between n-FFS and p-FFS lies in their response times. In an IPS or FFS cell, the electric field, which contains both horizontal and vertical components, is not uniform between the electrode gaps. Thus, it is difficult to derive an analytical equation which describes the LC response time. A semi-empirical equation has been found to correlate with the decay time as a function of cell gap, rotational viscosity (γ_1), and twist elastic constant (K_{22}) [106]:

$$\tau_d = A \cdot \frac{\gamma_1 d^2}{K_{22}\pi^2}, \tag{4.24}$$

here A is a proportionality constant. In a range of device structures, materials, and rubbing angles investigated, A is only sensitive to the anchoring energy. Under strong anchoring conditions, $A \approx 1.238$ is basically a constant.

From Eq. (4.24), an LC with a small visco-elastic constant helps to improve the response time. As discussed earlier, both positive and negative $\Delta\varepsilon$ LC materials can be used in FFS LCDs. Moreover, low $|\Delta\varepsilon|$ LCs exhibit three attractive features: (i) low viscosity as Figure 4.23 shows, (ii) small flow activation energy, which leads to a smaller increase in viscosity at low temperature, and (iii) reduced flexoelectric effect because the LC tilt angles are more spatially uniform [104]. However, a major drawback is the increased switching voltage. The optimal $|\Delta\varepsilon|$ value depends on the intended operation voltage. For both p-FFS and n-FFS, the strategy is to use the minimum acceptable $|\Delta\varepsilon|$ in order to obtain the advantages of ultra-low viscosity and small activation energy. Beside a low viscosity, K_{22} also plays an important role because the response time is proportional to γ_1/K_{22}.

Figure 4.23 depicts the relationship between $|\Delta\varepsilon|$ and rotational viscosity for some positive $\Delta\varepsilon$ LC and negative $\Delta\varepsilon$ LC mixtures. Generally speaking, for the same $|\Delta\varepsilon|$ value, positive LCs exhibit a 2× lower viscosity than their negative counterparts. Let us take $\Delta\varepsilon = 5$ as an example. For a positive $\Delta\varepsilon$ LC, we can use one

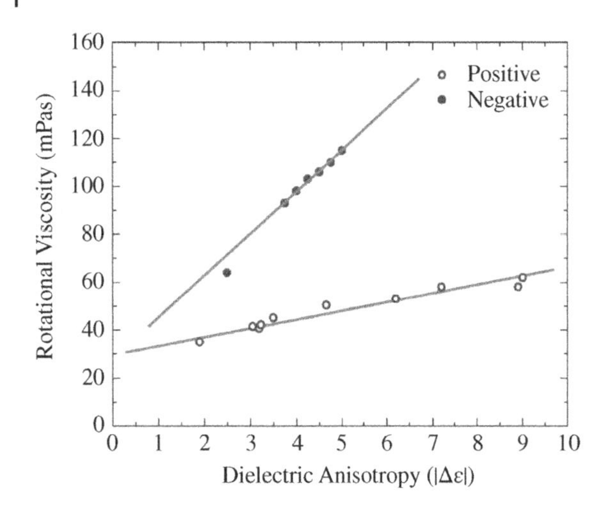

Figure 4.23 Rotational viscosity versus $|\Delta\varepsilon|$ of some LC mixtures with $\Delta n \approx 0.1$ and clearing point around 80 °C.

polar fluoro group along the principal molecular axis, of an LC structure such as alkyl-biphenyl. But to obtain $\Delta\varepsilon = -5$, we have to use two fluoro groups in the (2,3) lateral positions of the right phenyl ring. The lateral fluoro substitutions increase the rotational viscosity significantly.

4.9 VERTICAL ALIGNMENT (VA)

Vertical alignment (VA), also called homeotropic alignment [39, 107], is also commonly used in direct-view transmissive displays and reflective projection displays. VA exhibits the highest contrast ratio among all the LC modes developed, moreover, its contrast ratio is insensitive to the incident light wavelength, LC layer thickness, and operating temperature. Both projection [108, 109] and direct view displays using VA cells [110, 111] have been demonstrated. For projection displays, a single domain VA is employed in reflective LCoS devices. But for direct-view LCDs, multi-domain VA structures have to be used in order to obtain a wide viewing angle.

4.9.1 Voltage-Dependent Transmittance

Figure 4.24 shows the voltage-dependent transmittance of a VA cell with $d\Delta n$ = 350 nm between crossed polarizers. Here, a single domain VA cell employing the Merck high resistivity MLC-6608 LC mixture is considered. Some physical properties of MLC-6608 are listed as follows: $n_e = 1.562$, $n_o = 1.479$ (at $\lambda = 546$ nm and $T = 20$ °C) [112]; clearing temperature $T_c = 90$ °C; dielectric anisotropy $\Delta\varepsilon = -4.2$, and rotational viscosity $\gamma_1 = 186$ mPas at 20 °C. In principle, to obtain 100% transmittance for a transmissive VA cell only requires that $d\Delta n \sim \lambda/2$. Since the human eye is most sensitive to green light ($\lambda \sim 550$ nm), the required $d\Delta n$ is around 275 nm. However, this is the minimum $d\Delta n$ value required because under such condition the 100% transmittance would occur at $V \gg V_{th}$. Due to the finite voltage swing available from an active matrix TFT (preferred to be below 6 V_{rms}), the required $d\Delta n$ should be increased to ~0.6λ, i.e. $d\Delta n \sim 330$ nm.

Figure 4.24 shows that an excellent dark state is obtained at normal incidence. As the applied voltage exceeds the Freedericksz threshold voltage ($V_{th} \sim 2.1\ V_{rms}$), the LC director is reoriented by the longitudinal electric field resulting in light transmission through the crossed analyzer. As shown, the RGB wavelengths reach their peak transmissions at different voltages, blue at ~4 V_{rms}, and green at ~6 V_{rms}. The on-state dispersion is less critical to the contrast ratio than the dark state transmission. A small amount of light leakage in the dark state would degrade the contrast ratio significantly, but minor changes in the luminance of the bright state are less noticeable.

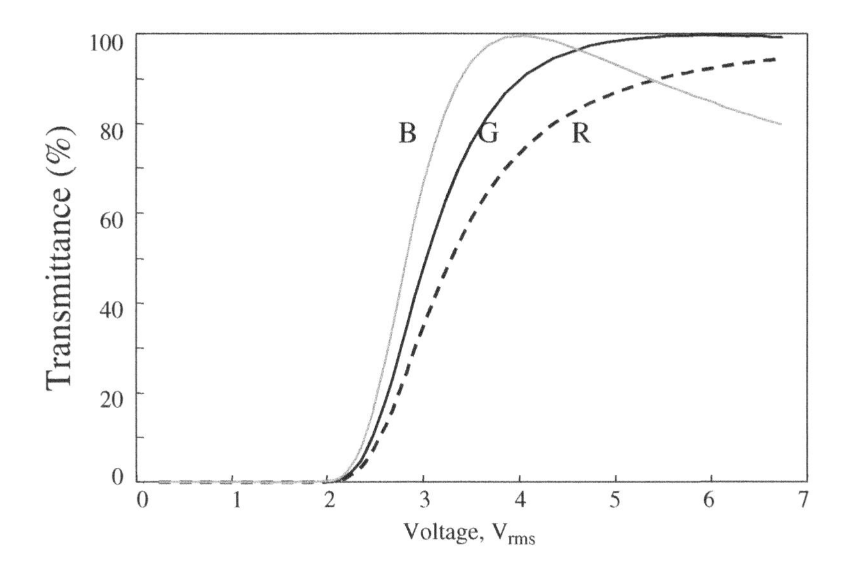

Figure 4.24 Voltage-dependent normalized transmittance of a VA cell. LC: MLC-6608. dΔn = 350 nm. R = 650 nm, G = 550 nm, and B = 450 nm.

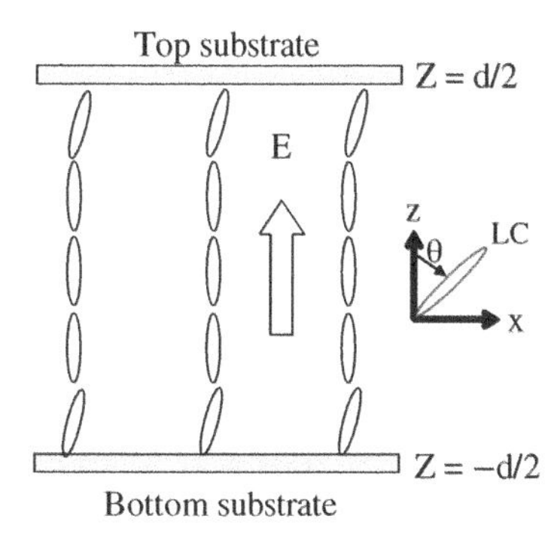

Figure 4.25 Schematic drawing of a VA LC cell with a pretilt angle and boundary conditions.

4.9.2 Response Time

Figure 4.25 shows a single-domain VA LC layer sandwiched between two parallel substrates where the bottom and top substrates are located at $z = -d/2$ and $+d/2$, respectively. The z-axis is normal to the plane of the substrates, and the electric field E is along the z-axis. When backflow and inertial effects are ignored, the Erickson–Leslie equation which describes the dynamics of the LC director has the following form [113, 114]:

$$(K_{11}\sin^2\theta + K_{33}\cos^2\theta)\frac{\partial^2\theta}{\partial Z^2} + (K_{33} - K_{11})\sin\theta\cos\theta\left(\frac{\partial\theta}{\partial Z}\right)^2 + \varepsilon_o\Delta\varepsilon E^2 \sin\ \theta\cos\theta = -\gamma_1\frac{\partial\theta}{\partial t} \tag{4.25}$$

In Eq. (4.25), γ_1 is the LC rotational viscosity, K_{11} and K_{33} represent the splay and bend elastic constants, respectively, $\varepsilon_o\Delta\varepsilon E^2$ is the change in electric field energy density under director reorientation, $\Delta\varepsilon$ is the LC dielectric anisotropy, and θ is the tilt angle defined as the angle between the z-axis and the LC director.

In general, Eq. (4.25) can only be solved numerically. However, under a small angle approximation ($\sin\theta \approx \theta$) [115] and a single elastic constant approximation ($K_{33} \approx K_{11}$), the Erickson–Leslie equation can be simplified to:

$$K_{33}\frac{d^2\theta}{dz^2} + \varepsilon_o\Delta\varepsilon E^2\theta = -\gamma_1\frac{\partial\theta}{\partial t}. \tag{4.26}$$

When both top and bottom substrates have the same alignment treatment, Eq. (4.26) has following simple solution:

$$\theta = \theta_m \cos(\alpha z)\cdot \exp(-t/\tau). \tag{4.27}$$

At a given voltage, θ_m represents the maximum tilt angle in the center of the LC cell ($\theta|_{z=0} = \theta_m$). When the pretilt angle θ_p is zero and the anchoring energy is strong, the following boundary conditions hold:

$$\theta_{Z=-\frac{d}{2},\frac{d}{2}} = \theta_p = 0. \tag{4.28}$$

Equations (4.27) and (4.28) lead to the following analytical solutions for the decay time (τ_d) and rise time (τ_r):

$$\tau_d = \tau_o = \frac{\gamma_1 d^2}{K_{33}\pi^2}, \tag{4.29}$$

$$\tau_r = \frac{\tau_o}{\left|\left(\frac{V}{V_{th}}\right)^2 - 1\right|}. \tag{4.30}$$

In Eq. (4.29), τ_o is called the free relaxation time; i.e. the decay time in the absence of any bias voltage, and in Eq. (4.30) the threshold voltage is defined as $V_{th} = \pi\sqrt{K_{33}/(\varepsilon_o \mid \Delta\varepsilon \mid)}$. It should be mentioned that the τ_d and τ_r described here represent the LC director's response time. To obtain the optical response time, we need to convert the LC director reorientation to transmittance [116]. In general, the optical response time is about two times faster than the LC director response time, shown in Eqs. (4.29) and (4.30).

If the pretilt angle deviates from zero, then we have:

$$\theta_{Z=-\frac{d}{2},\frac{d}{2}} = \theta_p \neq 0. \tag{4.31}$$

Equation (4.27) should satisfy the boundary conditions described by Eq. (4.31) at $Z = -d/2$ and d/2. From Eqs. (4.27) and (4.31), we find that the parameter α has following form:

$$\alpha = \frac{2}{d} cos^{-1}\left(\frac{\theta_p}{\theta_m}\right). \tag{4.32}$$

Based on Eq. (4.26), the modified response times that take the effect of a finite pretilt angle into consideration are derived as follows [117]:

$$\tau_d^* = \tau_o^* = \frac{\gamma_1}{\alpha^2 K_{33}}, \tag{4.33}$$

$$\tau_r^* = \frac{\gamma_1}{|\varepsilon_o|\Delta\varepsilon|E^2 - \alpha^2 K_{33}|}. \tag{4.34}$$

Under normal operating conditions, the maximum tilt angle is much larger than the pretilt angle, i.e. $\theta_m \gg \theta_p$. Under such a condition, the $\cos^{-1}()$ term in Eq. (4.32) can be approximated as:

$$cos^{-1}\left(\frac{\theta_p}{\theta_m}\right) \sim \frac{\pi}{2} - \frac{\theta_p}{\theta_m}, \tag{4.35}$$

and the rise and decay times are as follows:

$$\tau_d^* = \tau_o^* = \frac{\gamma_1}{\alpha^2 K_{33}} = \frac{\gamma_1 d^2}{4K_{33}\left(\frac{\pi}{2} - \frac{\theta_p}{\theta_m}\right)^2}, \tag{4.36}$$

$$\tau_r^* = \frac{\gamma_1}{\left|\varepsilon_o|\Delta\varepsilon|E^2 - \frac{4K_{33}}{d^2}\left(\frac{\pi}{2} - \frac{\theta_p}{\theta_m}\right)^2\right|}. \tag{4.37}$$

Strictly speaking, the LC threshold voltage no longer exists if the pretilt angle is non-zero, although threshold-like behavior in the voltage-dependent transmittance still appears if the pretilt is not too large. For simplicity, let us assume the threshold voltage still exists. Under such a condition, Eq. (4.37) can be simplified as:

$$\tau_r^* = \frac{\tau_o^*}{\left|\left[\frac{V}{\left(1-\frac{2\theta_p}{\pi\theta_m}\right)V_{th}}\right]^2 - 1\right|}. \tag{4.38}$$

As expected, Eqs. (4.36) and (4.38) reduce to Eqs. (4.29) and (4.30) when the pretilt angle is zero. Equations (4.36) and (4.38) suggest that the LC response time is also dependent on the ratio of θ_p/θ_m where θ_m depends on the applied voltage. For the case of small pretilt angles, this term is negligible because $\theta_p/\theta_m \to 0$.

4.9.3 Overdrive and Undershoot Addressing

From Eq. (4.30), the rise time depends on the applied voltage (V), especially near the threshold region. Let us use a normally black VA cell as an example. Typically, the cell is biased at a voltage (V_b) which is slightly below V_{th} in order to reduce the delay time incurred during the rise period and to keep a high contrast ratio. For some intermediate gray levels, the applied voltage is only slightly above V_{th}. Under such circumstances, the rise time would be very slow. To overcome the slow rise time, we could apply a higher voltage for a short period and then hold the transmittance at the desired gray level, as shown in Figure 4.26. This is the so-called overdrive voltage method [118]. Meanwhile, during the decay period, the voltage is turned off for a short period and

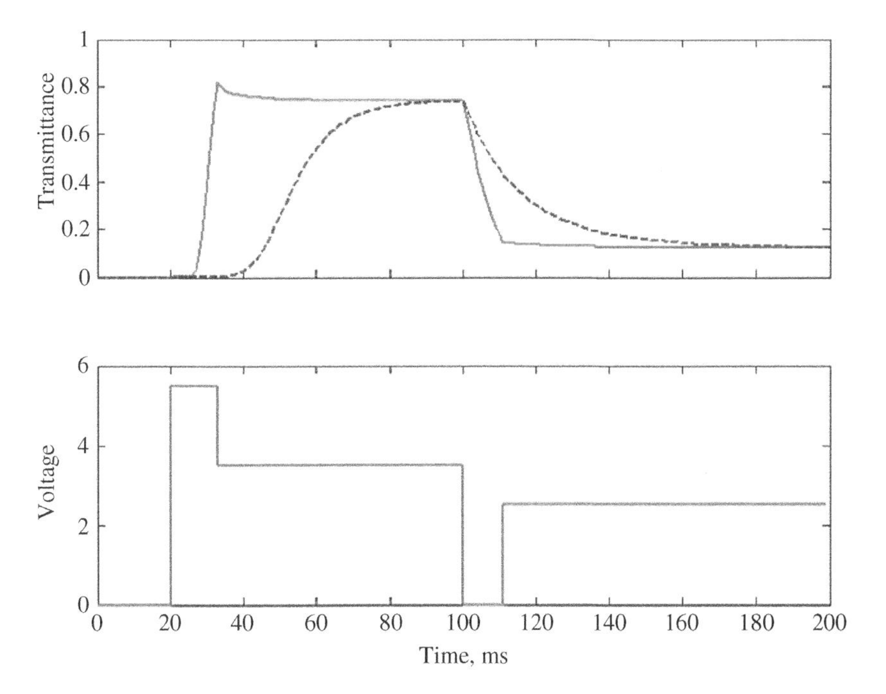

Figure 4.26 The overdrive and undershoot voltage method for speeding up LC rise and decay times. Top traces are the optical response and bottom traces are the corresponding voltage waveforms. The dashed lines represent a normal driving and the solid lines are with overdrive and undershoot voltages.

then a small holding voltage is applied to keep the LC at the desired gray level. This is the undershoot effect [119]. With voltage overdrive and undershoot, the LC response time can be reduced by ~2–3×, depending on the applied voltage.

4.9.4 Multi-domain Vertical Alignment (MVA)

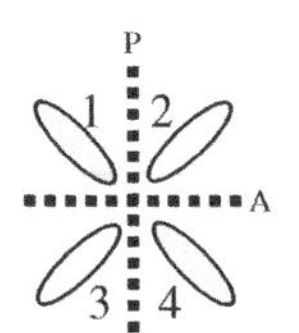

Figure 4.27 LC orientation in a four-domain structure. P: polarizer, and A: analyzer.

Single domain VA devices have a narrow viewing angle and can only be used in projection displays. For direct-view displays, at least four domains are required in order to eliminate grayscale inversion and widen the viewing angle. Assume each domain is located in one quadrant, as shown in Figure 4.27. Here, P and A represent the optic axes of the polarizer and analyzer, respectively. In order to obtain maximum transmittance, the LC directors in each domain should be oriented at 45° with respect to the polarizer's axis.

Fujitsu has developed a protrusion-type MVA [110, 120] and Samsung has developed patterned VA (PVA) [121, 122] using slits to generate fringing fields. The operating mechanisms are alike, but PVA does not require any physical protrusions so its contrast ratio is higher. Figure 4.28 shows the schematic structure of PVA. In the voltage-off state, as shown in Figure 4.28a, the LC directors are aligned perpendicular to the substrates. Because there is no physical protrusion, a very good dark state can be obtained. In a voltage-on state, as depicted in Figure 4.28b, the fringe fields from the top and bottom slits create two opposite tilted domains as highlighted in the dashed circles. By using zigzag electrodes with a 90° tilt angle, a four domain VA can be formed. With a combination of A-plate and C-plate compensators, both MVA and PVA can achieve more than 100 : 1 contrast ratio over an 85° viewing cone.

As shown in Figure 4.28, the PVA cell has no pretilt angle. The four domains are induced by the fringe electric fields. The response time, especially the rise time, is relatively slow. To improve the response time, it is highly desirable to create a pretilt angle in each domain for the MVA to guide the LC reorientation direction. Based on this concept, the surface polymer sustained vertical alignment (PS-VA) technique has been developed [123]. A very small percentage (~0.2 wt%) of reactive mesogen monomer and photoinitiator are mixed in a negative $\Delta\varepsilon$ LC host and injected into a PVA-type LCD panel. While a voltage is applied to generate four tilt domains, UV light is used to cure the monomers. As a result, the monomers polymerize and are adsorbed onto the surfaces. These cured polymers, although present at low density, will provide a pretilt angle within each domain to guide the LC reorientation. Thus, the rise time is reduced by nearly 2× while the decay time remains more or less unchanged [124].

Generally speaking, multi-domain VA shows a higher contrast ratio than IPS or FFS, but only within the central 20° viewing cone, as shown in Figure 4.29. Beyond this region, its contrast ratio decreases more quickly

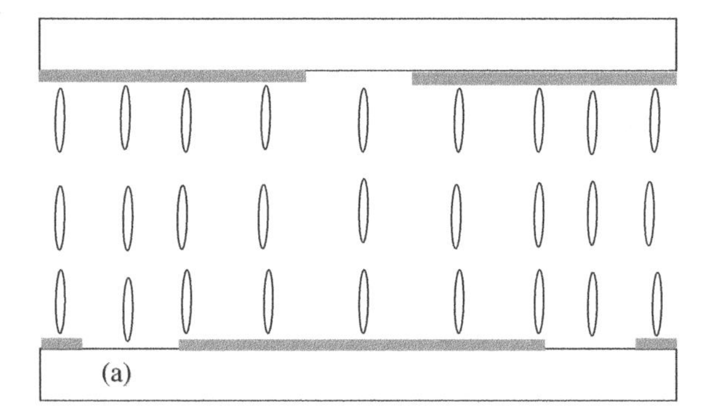

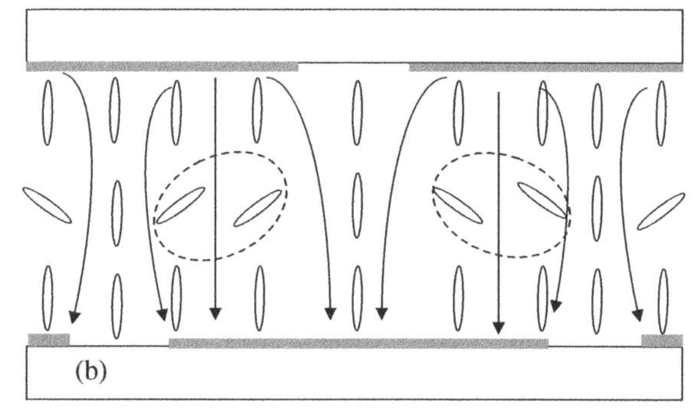

Figure 4.28 (a) LC directors of PVA at V = 0, and (b) LC directors of PVA in a voltage-on state. The fringe fields generated by the top and bottom slits create two domains with opposite tilt directions in this cross-section. When zigzag electrodes are used, four domains are generated.

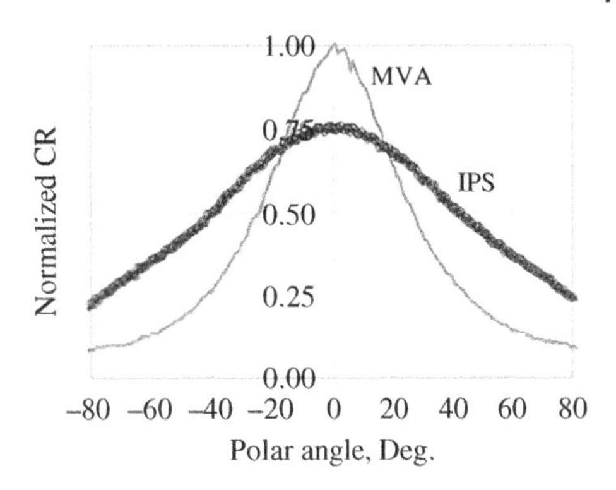

Figure 4.29 Comparison of normalized contrast ratio of MVA and IPS.

than that of IPS. Meanwhile, IPS has a weaker color shift (color change at oblique angle compared to the normal) than multi-domain VA [125]. Therefore, each technology has its own advantages and shortcomings. For touch panels, FFS mode shows no ripples and is a preferred approach. With the advance of local dimming mini-LED backlight, both FFS, IPS and MVA can achieve 1 000 000 : 1 dynamic contrast ratio.

4.10 AMBIENT CONTRAST RATIO

Contrast ratio is a key display metric affecting the image quality. For an emissive display, e.g. organic light-emitting diode (OLED), the CR can reach 1 000 000 : 1, whereas for a non-emissive LCD, its CR is limited to about 5000 : 1 for a commercial MVA LCD TV. As a result, it is generally perceived that OLEDs exhibit a much higher contrast ratio than LCD. Indeed, this is true under dark ambient viewing conditions. However, in most application scenarios, whether in an office or outdoors, ambient light is unavoidable. Thus, the performance of LCDs and OLED displays under different ambient lighting conditions is a practically important issue. In the presence of surrounding light, the ambient contrast ratio (ACR) rather than the dark room contrast ratio is more representative. In fact, ACR has been widely used to evaluate the sunlight readability of LCDs and OLED displays [126, 127].

4.10.1 Modeling of Ambient Contrast Ratio

ACR is an important parameter to quantitatively evaluate the sunlight readability of display devices. It is generally defined as [128]:

$$\mathrm{ACR} = \frac{L_{\mathrm{on}} + L_{\mathrm{ambient}} \cdot R_{\mathrm{L}}}{L_{\mathrm{off}} + L_{\mathrm{ambient}} \cdot R_{\mathrm{L}}}, \tag{4.39}$$

where L_{on} (L_{off}) stands for the on-state (off-state) luminance value of an LCD or OLED, L_{ambient} is ambient luminance (i.e. illuminance divided by π), and R_{L} is the luminous reflectance defined as:

$$R_{\mathrm{L}} = \frac{\int_{\lambda_1}^{\lambda_2} V(\lambda)S(\lambda)R(\lambda)d\lambda}{\int_{\lambda_1}^{\lambda_2} V(\lambda)S(\lambda)d\lambda}, \tag{4.40}$$

where $V(\lambda)$ is the human eye sensitivity function, $R(\lambda)$ is the spectral reflectance of the display device, and $S(\lambda)$ is the spectrum of the ambient light (a CIE standard D65 source is used here).

4.10.2 Ambient Contrast Ratio of LCD

Figure 4.30 shows the schematic diagram for analyzing the ACR of an LCD, where the main reflection occurs at the first interface (R_1). The ambient light entering the LCD panel is mostly absorbed by the crossed polarizers

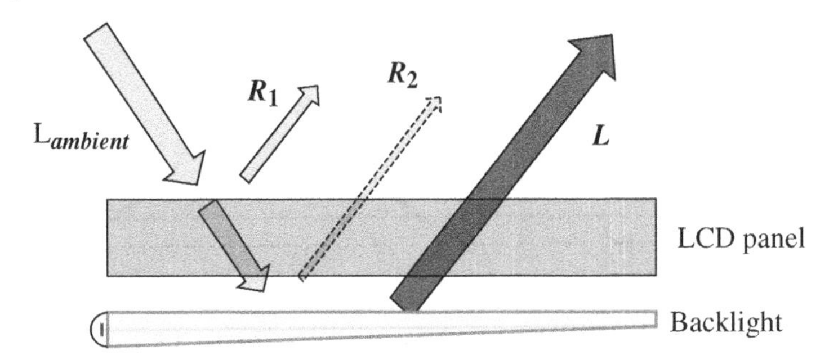

Figure 4.30 Schematic diagram for analyzing the ACR of an LCD.

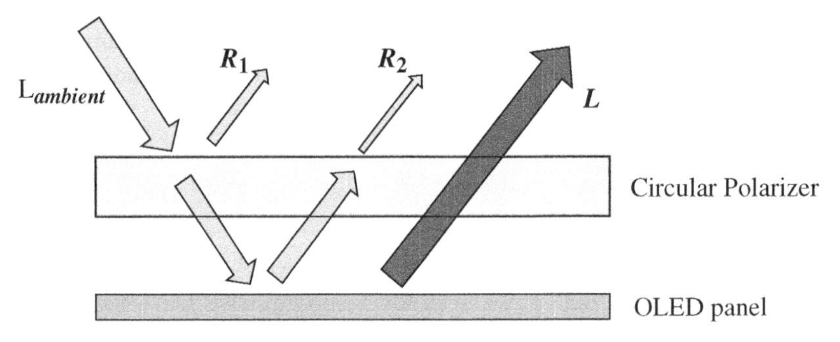

Figure 4.31 Schematic diagram for analyzing the ACR of an OLED display.

and other optical components. Typically, R_2 is smaller than 1% and can be neglected. Then the ACR of the whole device can be simplified as:

$$\mathrm{ACR}_{\mathrm{LCD}}(\theta,\phi) = \frac{L_{\mathrm{on}}(\theta,\phi)+R_1}{L_{\mathrm{off}}(\theta,\phi)+R_1}; \quad R_1 = L_{\mathrm{ambient}} \cdot R_{\mathrm{L\_surface}}(\theta,\phi). \tag{4.41}$$

In Eq. (4.41), θ and ϕ represent the polar angle and azimuthal angle, respectively. For a display device, it is desirable to analyze the ACR for the entire viewing zone.

4.10.3 Ambient Contrast Ratio of OLED

Unlike an LCD, a conventional OLED uses metal (e.g. Ag or Al) as one electrode (i.e. cathode); hence, the OLED itself is a highly reflective device. To suppress ambient light reflection, a broadband circular polarizer (CP) is commonly used, as Figure 4.31 depicts. The conventional circular polarizer comprising a linear polarizer and a quarter-wave ($\lambda/4$) film is a narrowband device. To broaden the spectral bandwidth, we need to laminate a $\lambda/2$ plate between the linear polarizer and the $\lambda/4$ plate. However, such a circular polarizer has a relatively narrow acceptance angle. At large oblique angles, light leakage is quite severe. Thus, in addition to surface reflection, circular polarizer light leakage (R_2 in Figure 4.31) should also be considered for OLED displays:

$$\begin{aligned} ACR_{\mathrm{OLED}}(\theta,\phi) &= \frac{L_{\mathrm{on}}(\theta,\phi)+R_1+R_2}{L_{\mathrm{off}}(\theta,\phi)+R_1+R_2}, \\ R_1 &= L_{\mathrm{ambient}} \cdot R_{L\_\mathrm{surface}}(\theta,\phi), \\ R_2 &= L_{\mathrm{ambient}} \cdot [1 - R_{\mathrm{L\_surface}}(\theta,\phi)] \cdot R_{\mathrm{L\_OLED}}(\theta,\phi). \end{aligned} \tag{4.42}$$

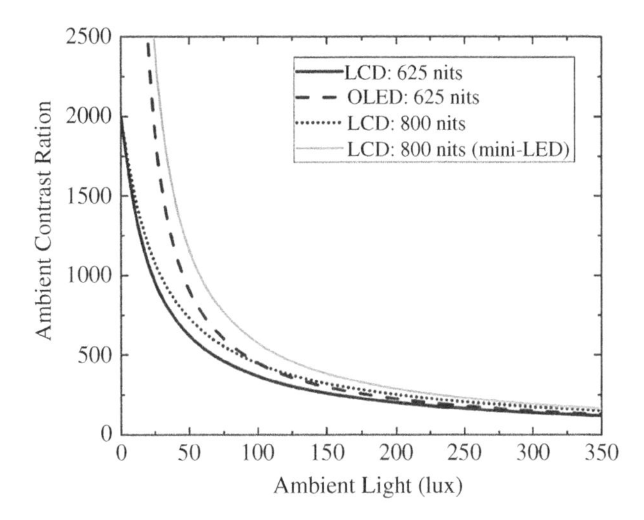

Figure 4.32 Calculated ACR as a function of different ambient lighting conditions for LCD and OLED based smartphones.

4.10.4 Simulated ACR for Mobile Displays

Let us first investigate how the ambient light affects the ACR of LCDs and OLEDs. Next, we will explore ACR at different viewing angles, represented by an *ambient isocontrast contour*.

For mobile displays, such as smartphones, an anti-reflection coating may not be used due to the requirements of touch functionality. Therefore, the outer surface of the display is a cover glass. For simplicity, let us assume 4.4% luminous reflectance for both LCDs and OLEDs. For LCD smartphones, FFS is commonly used. Let us assume its CR = 2000 : 1 and peak brightness = 625 nits. For OLED, we assume its peak brightness is also 625 nits and CR = 10^6 : 1. Then we calculate the ACR under different ambient lighting conditions. Results are plotted in Figure 4.32.

As expected, when the ambient light is dim, OLED shows a much higher ACR than LCD. As the ambient light gets brighter, both ACR curves decline sharply and then converge gradually. If we slightly boost the LCD peak brightness to 800 nits, two ACR curves would crossover at 90 lx. It means the OLED exhibits a higher ACR than the LCD below 90 lx, but the situation is reversed above 90 lx. The mini-LED backlit LCD can provide 10 000 s of local dimming zones to suppress the halo (blooming) effect and can achieve CR ~ 10^6 : 1 while keeping a high brightness (>1000 nits). Under such a condition, the LCD shows superior performance than OLED in all ambient lighting conditions, as Figure 4.32 shows.

4.10.5 Simulated ACR for TVs

For large-sized TVs, we often use remote control to select programs. Because no touch functionality is required, we can apply an anti-reflection coating to the display panel to reduce ambient light reflection. Let us assume a single-layer magnesium fluoride (MgF_2) anti-reflection coating is used and its luminous reflectance is 1.5% at normal angle. Meanwhile, a TV is usually powered by an electrical outlet so that its peak brightness can be boosted by an embedded adaptive brightness control sensor. Nowadays, a state-of-the-art LCD TV can deliver >2000 nits of peak brightness, while OLED has about 1000 nits. In terms of static CR, a MVA LCD is assumed to be 5000 : 1, while OLED is 10^6 : 1. With all these data, we can calculate an ACR for both LCD and OLED TVs. Similarly, as Figure 4.33 depicts, the OLED exhibits a higher ACR in the low illuminance region (dark room), but declines sharply as the ambient light increases. The crossover point is 72 lx, which is a typical family room lighting level. For a mini-LED backlit LCD (assuming the peak luminance = 1500 nits and CR = 10^6 : 1), the performance is improved substantially, enabling a high ACR in all lighting scenarios.

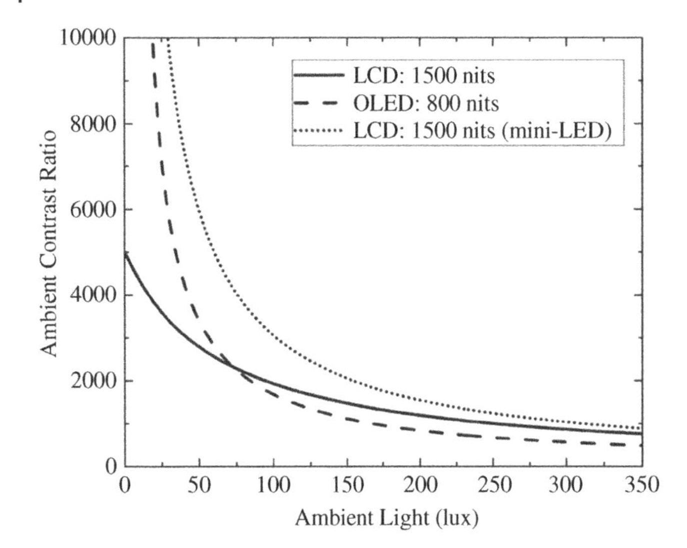

Figure 4.33 Calculated ACR as a function of different ambient lighting conditions for LCD and OLED TVs.

4.10.6 Simulated Ambient Isocontrast Contour

So far, we only concentrated on the ACR at normal angle. Next, we examine the ACR at different viewing angles. Before that, we have to elucidate the device parameters for both LCD and OLED. As discussed above, two LCD modes are used in our simulation: FFS for smartphones and MVA for TVs. For both LCD modes, the parameters are listed as follows: the polarizer and analyzer are 24-μm thick with $n_o = 1.5$, $k_o = 0.000306$, $n_e = 1.5$, and $k_e = 0.019027$, where k_o and k_e are imaginary refractive index values. Compensation films are implemented to suppress the color shift and gamma shift at large oblique angles. Also, the depolarization effects from the TFT array, LC layer, and color filters, have been considered in order to better present realistic cases [129].

For an OLED, a broadband circular polarizer is used. The optical configuration is shown in Figure 4.34a. The parameter of the linear polarizer is the same as that used in LCDs. For the half-wave plate, a 183.33 μm thick positive A-film is used with $n_o = 1.5095$ and $n_e = 1.511$ at $\lambda = 550$ nm. The $\lambda/4$ plate uses the same A-film, but is 2× thinner (91.67 μm). Next, we calculate its light leakage at different wavelengths and different angles. Results are shown in Figure 4.34b,c. At normal angle, the light leakage is less than 1% in the visible region (450–700 nm); the broadband feature is indeed verified. As the viewing angle increases, light leakage gradually increases and rises to almost 40%. This will undoubtedly affect the final perceived ACR at oblique viewing directions.

4.10.6.1 Mobile Displays

For smartphones, we compare an FFS LCD with an OLED display, and assume the cover glass is BK-7. Figure 4.35a shows the calculated luminous reflectance of BK-7 at different viewing directions. When the polar angle is less than 45°, the luminance reflectance stays below 5%; but it increases rapidly as the viewing angle increases further.

To evaluate the isocontrast contour, we need to know the brightness distribution. An OLED is self-emissive and its angular distribution is much broader than an LCD, as Figure 4.35b depicts. For instance, at a 30° viewing angle, the OLED brightness only decreases by ~20%, whereas LCD brightness decreases by nearly 50%.

Using all this information, we can calculate the ACR of an LCD and OLED at different viewing angles. At 500 lx (office lighting condition), LCD and OLED show quite similar *ambient isocontrast contour* pattern, as Figure 4.36a,b depict. In theory, the OLED has a broader angular distribution, and it should perform better

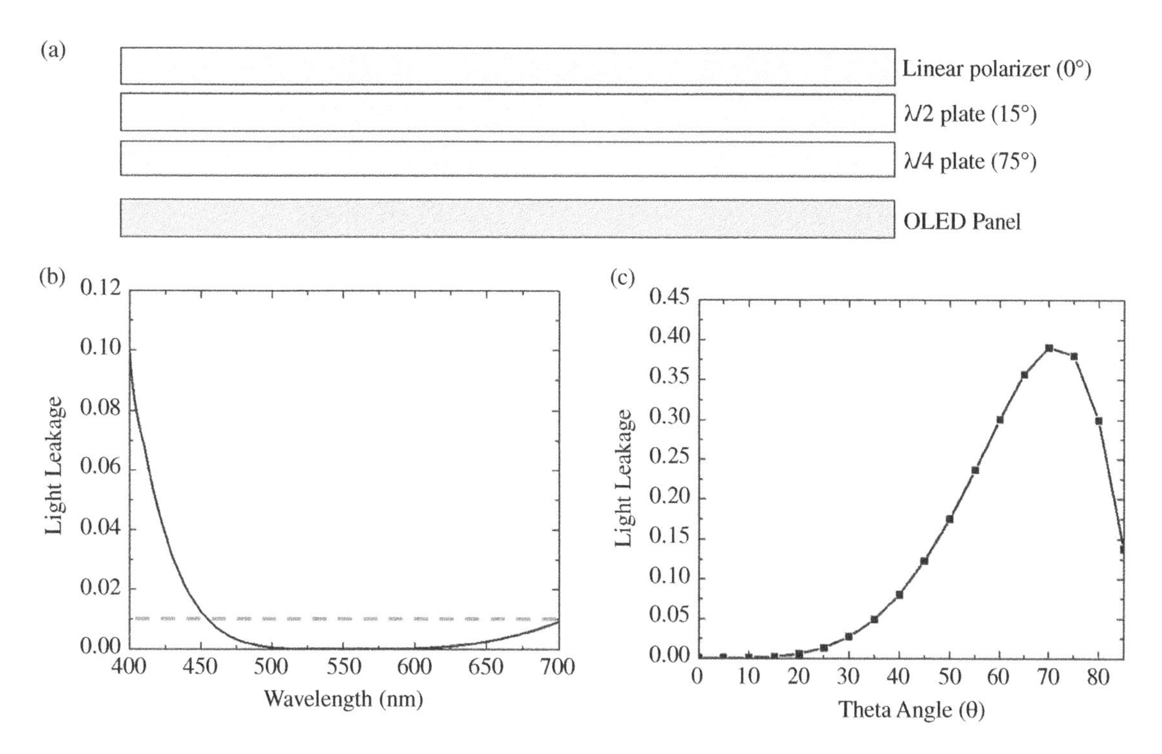

Figure 4.34 (a) Schematic diagram of optical configuration of broadband circular polarizer; (b) Calculated light leakage at different wavelengths at normal angle ($\theta = 0°$, $\phi = 0°$); and (c) Calculated light leakage at different polar angles ($\phi = 0°$).

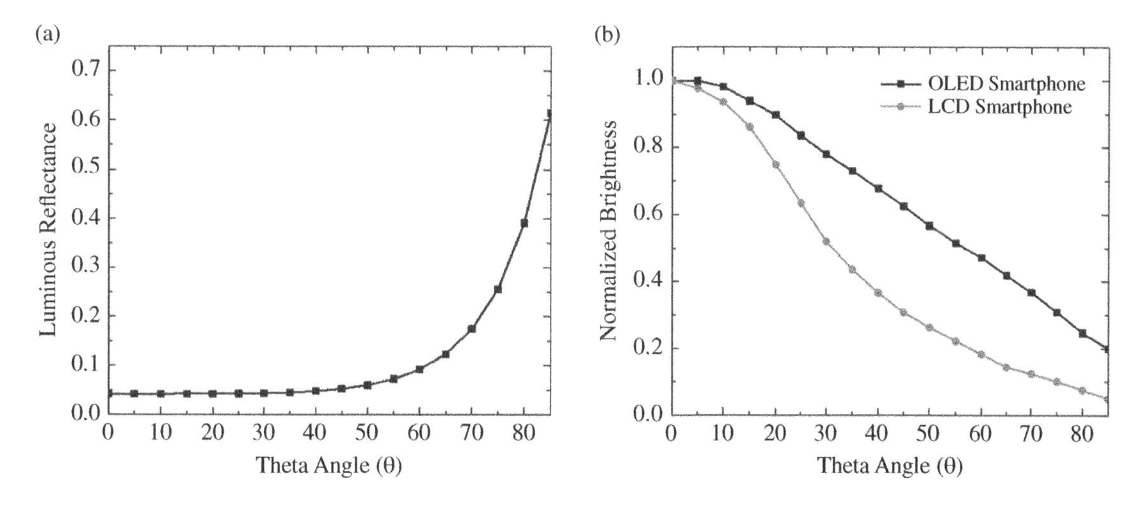

Figure 4.35 (a) Calculated luminous reflectance of BK-7 cover glass at different angles, and (b) normalized brightness distribution of LCD and OLED smartphones.

at large angles. But this advantage is evened out by the light leakage of the circular polarizer. Also, from these two figures, most of the viewing zone shows ACR $\geq 5:1$, which is adequate for normal reading. As the ambient light increases to 5000 lx (outdoor with moderate overcast sky), LCD and OLED show much reduced but still quite similar ACR pattern. When ACR < 2, the display is hardly readable. Therefore, from Figure 4.36c,d, the workable zone for both LCDs and OLEDs is limited to $\pm 50°$ in an overcast day.

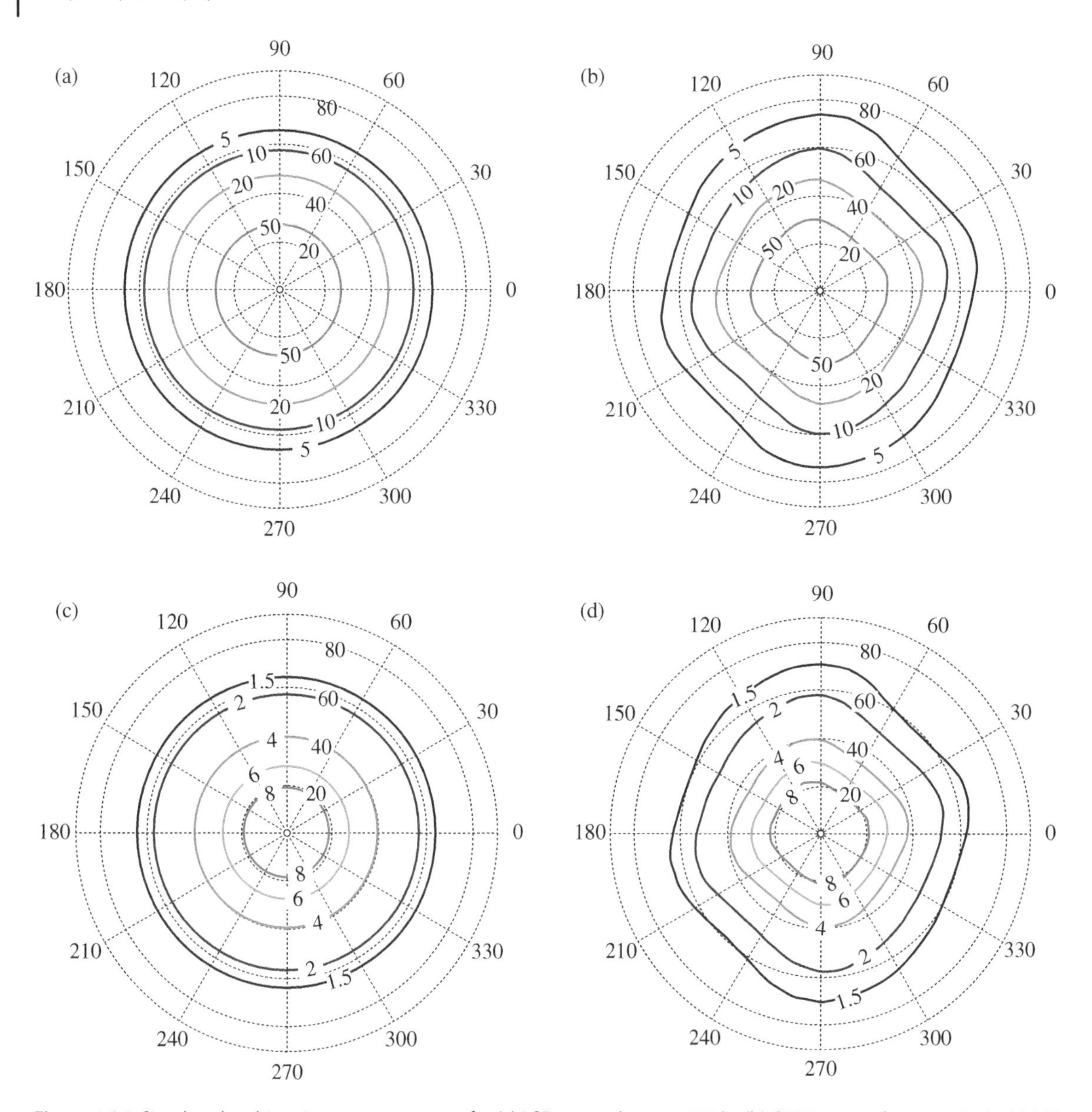

Figure 4.36 Simulated ambient isocontrast contour for (a) LCD smartphone at 500 lx, (b) OLED smartphone at 500 lx, (c) LCD smartphone at 5000 lx, and (d) OLED smartphone at 5000 lx.

4.10.6.2 Large-Sized TVs

To provide an antireflection coating for TVs, several approaches can be employed [130]. Currently, single-layer MgF_2 is a favored choice due to its simple configuration, low cost and fairly good performance. Figure 4.37a shows the calculated luminous reflectance of antireflection-coated BK-7 at different angles. Within 45°, the R_L value remains less than 2%, which is ~2.5× lower than that of a bare BK-7 glass. Also, the brightness decay for LCDs and OLEDs is considered, as shown in Figure 4.37b. Unlike smartphones, a wide angle of view is more critical for TVs. As a result, its luminance distribution is broader; the OLED TV brightness decrease at $\theta = 30°$ is less than 10%, while for LCD it is ~35%.

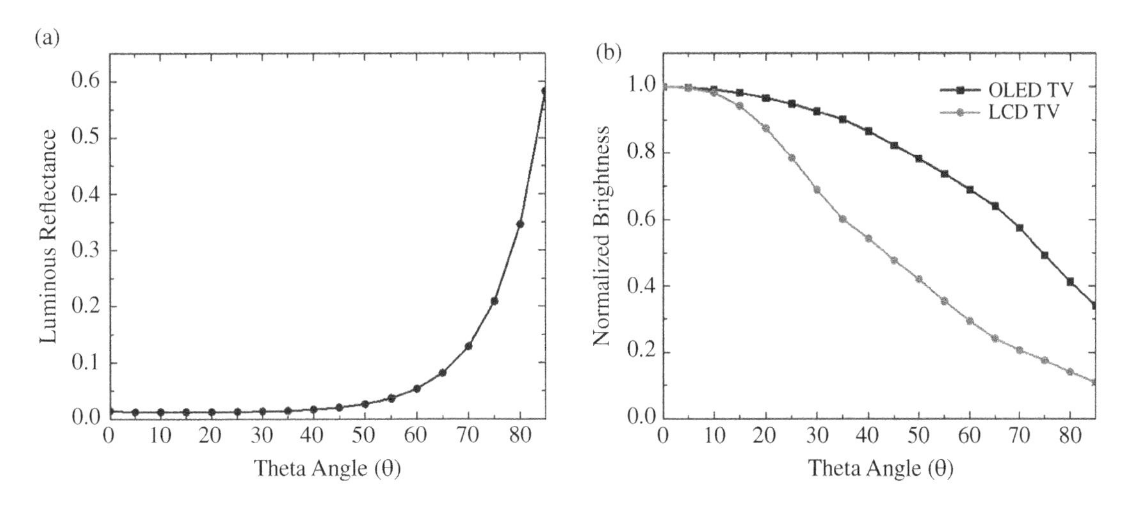

Figure 4.37 (a) Calculated luminous reflectance of AR-coated BK-7 cover glass at different angles, and (b) normalized brightness of LCD and OLED TVs.

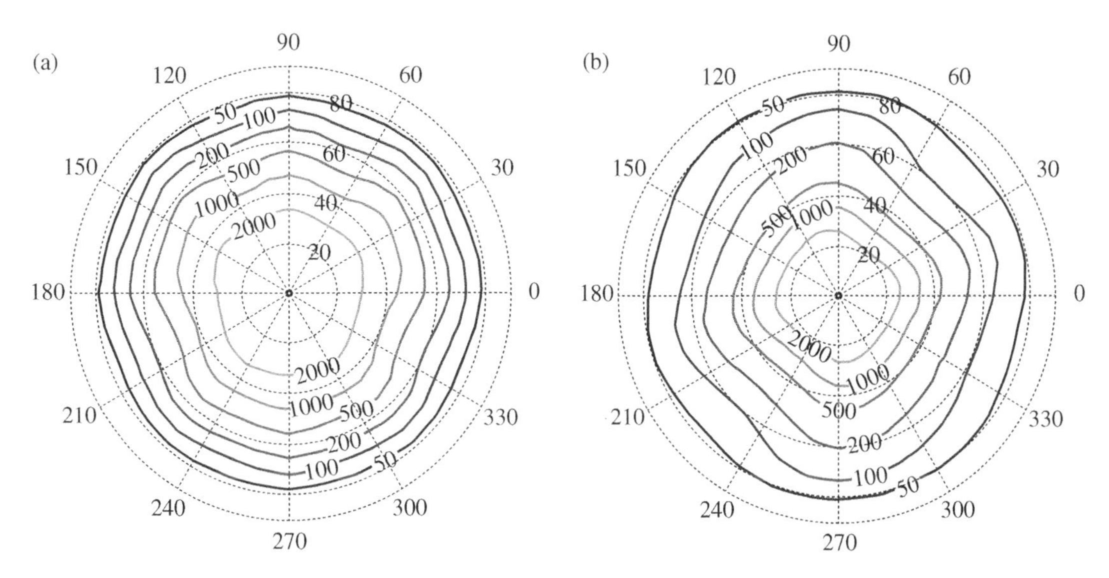

Figure 4.38 Simulated ambient isocontrast contour for (a) LCD TV (peak brightness ~1500 nits) at 50 lx, and (b) OLED TV (peak brightness ~800 nits) at 50 lx.

Figure 4.38 depicts the ambient isocontrast contour under 50 lx of ambient light (a typical lighting condition in living rooms). From Figure 4.38, firstly, both LCD and OLED TVs can provide reasonably good performance (ACR ≥ 50 : 1) over nearly the entire viewing zone (±80°). In the central region, the LCD shows superior ACR to OLED. For example, for the LCD panel the viewing zone with ACR ≥ 1000 : 1 extends to over ±40°; whereas for OLED it is ±30°. This is mainly because LCD exhibits a much higher peak brightness than OLED (1500 nits versus 800 nits).

4.10.7 Improving LCD's ACR

As discussed above, ACR is jointly determined by several factors, such as ambient light illuminance, surface reflection, and display luminance and light leakage. To improve ACR for LCDs and OLEDs we should employ

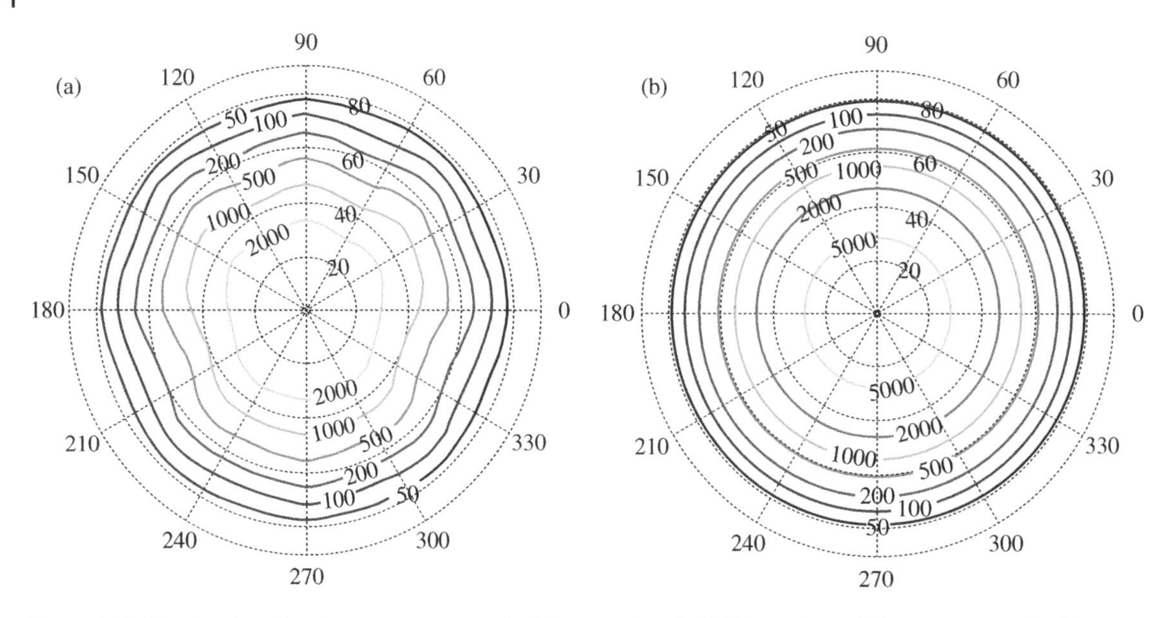

Figure 4.39 Simulated ambient isocontrast contour for (a) conventional LCD TV at 50 lx, and (b) new mini-LED backlit LCD TV at 50 lx.

different strategies. For LCDs, high brightness is their strength, which is helpful for keeping a relatively high ACR at strong ambient lighting conditions. But under low ambient light, the LCD needs to improve its CR by suppressing the light leakage in the dark state. Two approaches can be used to dramatically increase the dynamic range: a mini-LED backlight and a dual-panel structure. The mini-LED backlit can provide over 10 000 local dimming zones to suppress the halo effect and achieve 1 000 000 : 1 contrast ratio. On the other hand, the dual-panel LCD consists of a lower resolution (e.g. 1920 × 1080) black-and-white panel with a high resolution (3840 × 2160) full-color panel. The black-and-white panel provides two-million local dimming zones. As a result, the dynamic CR can reach $10^6:1$. Here, we compare the viewing angle performance between conventional LCD and mini-LED backlit LCD. Their ambient isocontrast contours are plotted in Figure 4.39. Such a mini-LED backlit LCD TV can get more than 2× higher ACR at normal direction (7312.5 versus 2931.3), and the high ACR (e.g. 2000 : 1) region is widened.

4.10.8 Improving OLED's ACR

OLEDs exhibit a true black state, leading to excellent ACR in a dark ambient. But this outstanding feature gradually deteriorates as the ambient light increases, due to its limited luminance. To improve the peak brightness, both materials and device configurations can be considered [131]. Another limiting factor is the circular polarizer. To suppress the light leakage off-axis, here we propose a new configuration by replacing two uniaxial films with two biaxial films [132], as shown in Figure 4.40a. The physical parameters for these two films are: biaxial film #1: $d = 78.57\,\mu\text{m}$, $n_x = 1.5124$, $n_y = 1.5089$, $n_z = 1.50978$ @ 550 nm, and biaxial film #2: $d = 39.29\,\mu\text{m}$, $n_x = 1.5124$, $n_y = 1.5089$, $n_z = 1.51055$ @ 550 nm. Clearly, compared to the conventional circular polarizer (Figure 4.40b), the new circular polarizer shows a much lower light leakage (Figure 4.40c). Within ±40°, it is less than 2%. The highest light leakage is about 10%.

Figure 4.41 shows the simulated ACR of an OLED using the new circular polarizer. The viewing angle is widened noticeably, especially in the central region. The zone with ACR ≥ 500 approaches ±60°. But with a conventional circular polarizer, it is limited to ±40°.

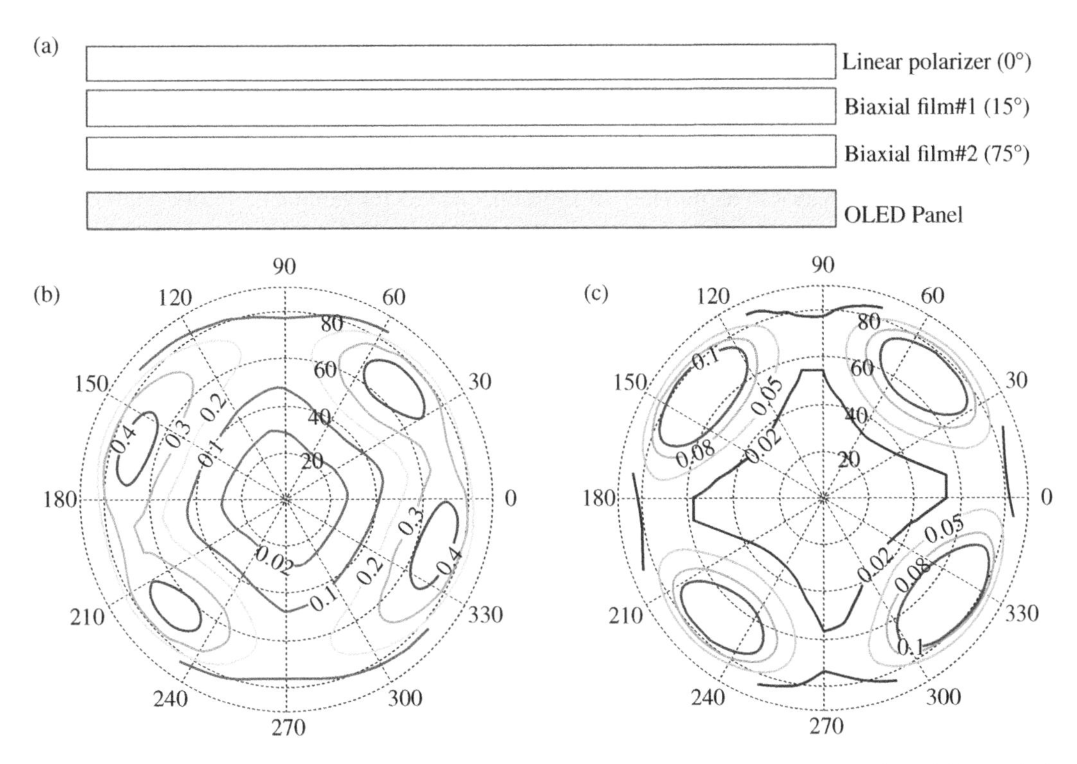

Figure 4.40 (a) Schematic diagram of a broadband and wide-view circular polarizer with two biaxial films; Calculated light leakage for (b) conventional broadband circular polarizer, and (c) new broadband circular polarizer.

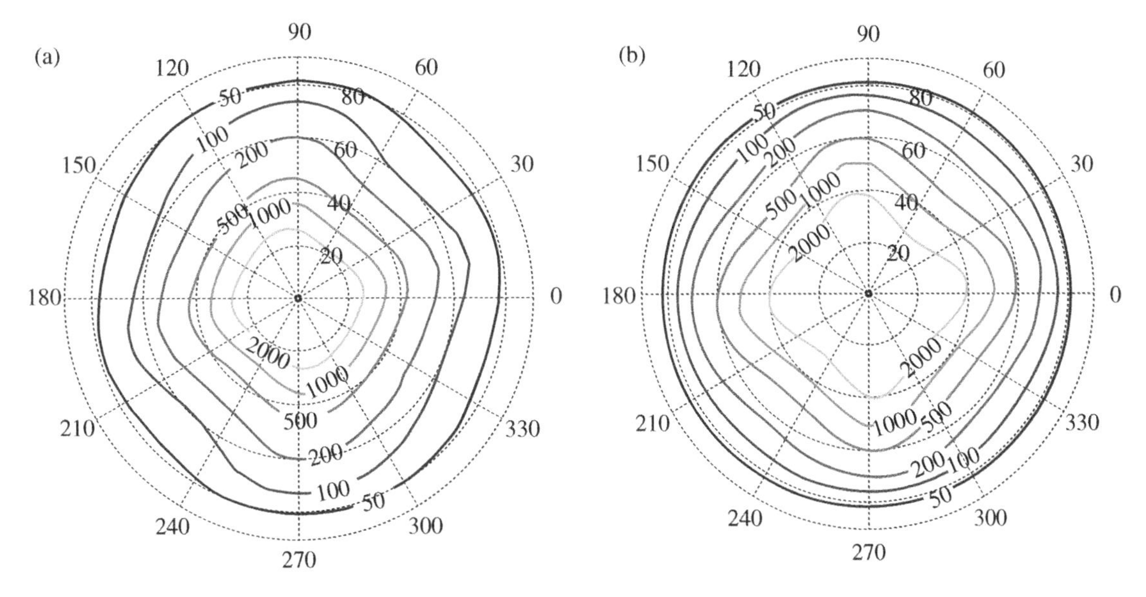

Figure 4.41 Simulated ambient isocontrast contour for (a) an OLED TV at 50 lx with conventional broadband circular polarizer, and (b) an OLED TV at 50 lx with new broadband circular polarizer.

4.11 MOTION PICTURE RESPONSE TIME (MPRT)

The TFT LCD was first introduced in the mid-1980s [133]. After about three decades of extensive material research, device innovation, and heavy investment in advanced manufacturing technologies, TFT LCDs have become ubiquitous in our daily lives [134]. Its widespread applications range from TVs, monitors, personal computers, tablets, to smartphones, just to name a few. In addition, displays for gaming monitors, augmented reality, and virtual reality systems are a rapidly growing category, which demand a higher resolution density, more vivid colors, higher ACR, and no noticeable image blur. Lately, "LCD versus OLED, who wins?" is a heated topic of debate [135, 136]. Each technology has its own pros and cons. Generally speaking, LCD leads in lifetime, peak brightness, and cost; it is comparable to OLED in resolution and pixel density, power consumption, ACR and viewing angle, but inferior to OLED in black state quality, panel flexibility, color gamut, and response time. Therefore, the LCD camp has devoted extensive efforts to narrowing their performance gap relative to OLED displays, by means including quantum-dot backlights [137, 138] to achieve wider color gamut and lower power consumption, and local dimming [139, 140] to enhance the dynamic contrast ratio to 1 000 000 : 1. The remaining grand challenge for LCDs is response time; especially nematic LCDs suffer ~100× slower response time than OLED (~0.1 ms). Thus, it is commonly perceived that LCDs exhibit more severe image blur than OLEDs for fast-moving objects [141]. To improve LC response time, several approaches have been investigated, e.g. polymer-stabilized blue phase LCs [142, 143], low viscosity nematic LCs [144–146], and ferroelectric LCs [147]. Nevertheless, it remains challenging for nematic LCs to achieve ~0.1 ms response time while keeping a low operating voltage (5 V).

Both TFT LCDs and OLEDs are holding-type displays, that means, the displayed image is held on by TFTs in a given frame time. As a result, they both suffer from different degrees of image blur, depending on the frame rate and response time. That is to say, an OLED display could still exhibit motion blur even if its response time were zero [148]. To quantify image blur, the motion picture response time (MPRT) has been proposed [149]. Unlike LCDs and OLEDs, the CRT (cathode ray tube) is an impulse-type display, whose MPRT ≈ 1.5 ms and is nearly free from motion blur [150]. The required MPRT depends on the specific application. For example, if a smartphone or monitor is mainly intended for static images, then a slow MPRT does not affect the display performance. But to clearly display an object moving at a speed $v = 960$ pixels per second, ideally the required MPRT should be less than 1 ms in order to avoid image blur. The minimum acceptable MPRT is 2 ms. At the present time, most LCD and OLED TVs are operated at 120 Hz (i.e. MPRT~6.66 ms), thus image blur remains noticeable for fast-moving objects. It is strongly desirable to reduce the MPRT to ~1.5 ms (or shorter) to eliminate motion blur.

The image blur of a TFT LCD (or OLED) has been analyzed and a simple equation correlating MPRT (in ms) with LC (OLED) response time (τ in ms) and frame time ($T_f = 1000/f$; where f is the frame rate in Hertz) has been derived as follows [18]:

$$\mathrm{MPRT} \approx \sqrt{\tau^2 + (0.8T_\mathrm{f})^2}. \tag{4.43}$$

From Eq. (4.43), MPRT is governed by two important parameters: LC (or OLED) response time and the TFT sample and hold time.

Figure 4.42 depicts the calculated MPRT versus LC (OLED) response time at four frame rates: $f = 60$, 120, 240, and 480 Hz. There are three important trends: (i) If the LC response time is not fast enough, say $\tau = 10$ ms, then increasing the frame rate from 60 to 120 Hz (vertical arrows) makes a big improvement in MPRT, but on further increasing the frame rate to 240 and 480 Hz the improvement is less obvious. This prediction is consistent with experimental observations [151]. (ii) At a given frame rate, say 120 Hz, as the LC response time decreases, MPRT decreases almost linearly and then gradually saturates. Note that the MPRT for $\tau = 2$ ms is only 4% longer than that for $\tau = 0$. Therefore, if an LCD's response time is 2 ms, then its MPRT is comparable to that of an OLED, even if the OLED's response time is assumed to be 0. (iii) As the TFT frame rate increases, the limiting MPRT (assuming $\tau = 0$; open circles) decreases linearly, because the limiting MPRT $= 0.8T_\mathrm{f}$.

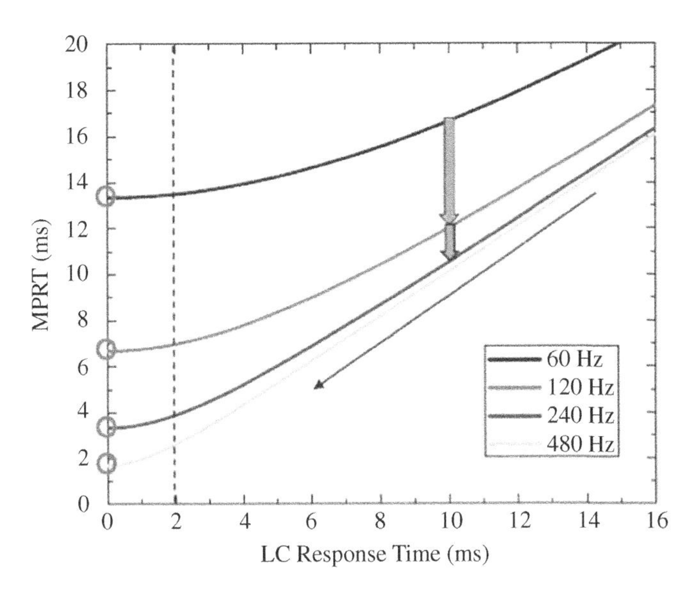

Figure 4.42 (a) LC response time versus MPRT at four specified frame rates. Solid lines represent the calculated results from Eq. (4.41). Open circles represent results for OLED by assuming its $\tau = 0$.

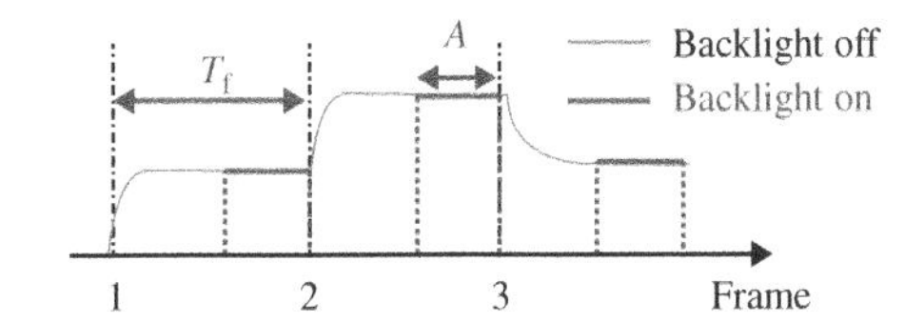

Figure 4.43 Definition of backlight duty ratio: A is the duration of backlight-on in a given frame time (T_f).

Beside the LC response time, another factor affecting motion blur and optical efficiency is backlight modulation, as depicted in Figure 4.43, where A stands for the time that the backlight (e.g. LED) is turned on in one frame time (T_f).

The duty ratio (DR) is defined as:

$$\text{DR} = \frac{A}{T_\text{f}}. \tag{4.44}$$

If the LC response time is fast (e.g. $\tau \leq 2$ ms), then the LC director can achieve its final gray-level configuration before the backlight is turned on. Under such a condition, MPRT can be expressed as:

$$\text{MPRT} \approx 0.8 \times T_\text{f} \times \text{DR} = 800 \times \text{DR}/f. \tag{4.45}$$

From Eq. (4.45), we can achieve a fast MPRT by reducing the duty ratio or by increasing the frame rate.

Figure 4.44 shows simulation results of LC response time dependent MPRT with different duty ratios. Here, we assume the frame rate is $f = 144$ Hz, which is the frame rate employed in some commercial products such as gaming monitors. The limiting MPRT (i.e. for $\tau = 0$) reduces linearly as the backlight duty ratio decreases. The reasons are twofold: (i) The slow transition part of the LC response is obscured by the delayed backlight illumination, and (ii) the sample-and-hold effect is suppressed because such an operation mode is similar to the CRT's impulse driving. As a matter of fact, to suppress image blur Sony's OLED TVs have employed 50% duty ratio [152], and LG's OLED head-mounted displays have used 20% duty ratio [153], because MPRT decreases linearly with the duty ratio. To minimize an LCD's motion blur for high-speed gaming or sports, the targeted MPRT is 1.5 ms, similar to a CRT. As Figure 4.44 shows, if we raise the frame rate to 144 Hz and reduce the duty ratio to 20%, then the MPRT is ~1.1 ms. A low duty ratio helps to shorten MPRT, but the major tradeoff is decreased luminance. To compensate for the luminance loss, we can boost the current in the LED backlight.

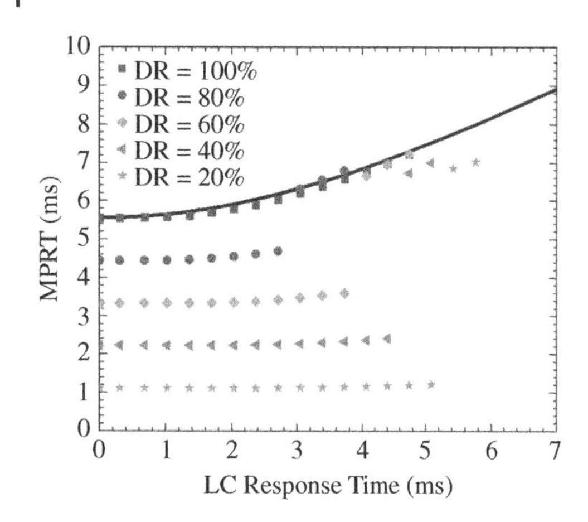

Figure 4.44 LC response time versus MPRT with different duty ratios at $f = 144$ Hz.

For OLED, in principle we can use the same impulse driving. However, high current impulse driving of OLEDs leads to a substantial efficiency roll-off [154] and lifetime degradation [155]. Similarly for LCDs, high current driving of blue LEDs also results in a droop effect [156], i.e. the internal quantum efficiency declines as the current density increases. However, the impact of the droop effect in LEDs is substantially weaker than the lowered efficiency and compromised lifetime in OLEDs. That is to say, the OLED is much more vulnerable than LCD to the impulse driving.

The third approach to achieve much faster MPRT is to combine a high frame rate with backlight modulation. From Eq. (4.45), if we increase the frame rate to 240 Hz while keeping the duty ratio at ~45%, then we can also achieve MPRT ≈ 1.5 ms. However, increasing the frame rate would cause two undesirable effects: reduced TFT charging time and increased electric power consumption. The former is particularly problematic for a high resolution (e.g. 8K) display.

4.12 WIDE COLOR GAMUT

The backlight is a critical component for the non-emissive LCD because it affects the color gamut, optical efficiency, dynamic range, and viewing angle. White light-emitting diodes (called 1pc-WLED) based on a blue LED-pumped yellow phosphor (YAG:Ce^{3+}) have been widely used in the backlight unit because of their high efficiency, long lifetime, low cost, and simple optical configuration [17]. However, the yellow spectrum generated by the phosphor is relatively broad, leading to a rather narrow (75% of NTSC) color gamut. In the past two decades, the color gamut evaluation metric has gradually advanced from sRGB to NTSC, and recently to the Rec.2020 standard (for ultra-high definition TVs), whose coverage of the color space is nearly twice as great in area as that of sRGB [157, 158]. Thus, a new backlight technology offering vivid colors is in great demand.

Using discrete RGB LEDs is a straightforward approach to expand the color gamut, despite the drawback that three sets of LED driving circuits are required [159–161]. Moreover, the choice of high efficiency green LEDs is still limited [162]. This is known as the green gap in LED technology [163]. Another option to widen the color gamut is to use a two-phosphor-converted WLED (2pc-WLED). The key advantages are long term stability, high efficiency, and low cost [164, 165]. However, the narrowest available full width at half maximum (FWHM) of the green phosphor (β-sialon:Eu^{2+}) is still as broad as 55 nm [166, 167]. The red phosphor KSF ($K_2SiF6:Mn^{4+}$) has five emission peaks and the individual FWHM of each band is quite narrow, and the average peak wavelength centers at ~625 nm, which is close to the optimum red wavelength (633 nm) for the Rec.2020 color space [168, 169]. However, the main factor limiting the color gamut of a display device is crosstalk between the color filters.

Lately, the quantum dot (QD)-enhanced backlight has found widespread applications because of the following outstanding features: (i) its central emission wavelength can be tuned by controlling the size of the nanoparticles, (ii) its FWHM is around 20–30 nm, being mainly determined by the particle size uniformity, (iii) its photoluminescence efficiency is high, and (iv) its device configuration is simple [137, 170, 171]. Briefly speaking, the QD backlight uses a blue LED to excite green/red colloidal nanoparticles, generating a white light with three well-separated RGB peaks. Therefore, three highly saturated primary colors can be obtained [138, 172, 173].

4.12.1 Material Synthesis and Characterizations

Since their discovery in the 1980s, colloidal QDs have been explored extensively for their scientific interest and for potential applications [174–176]. In general, they are nanometer sized (e.g. 2 ~ 10 nm) semiconductor particles governed by quantum confinement effects. Unlike a bulk material, these semiconductor nanoparticles exhibit unique optical and electrical properties dictated by their size, shape, and the quantum physics that arises at the nanoscale.

Briefly speaking, their outstanding features can be summarized in three aspects: (i) large freedom in tailoring the emission wavelengths. As the Brus equation suggests [177]:

$$E^* \cong E_g + \frac{\hbar\pi^2}{2R^2}\left[\frac{1}{m_e} + \frac{1}{m_h}\right], \tag{4.46}$$

where E_g is the bandgap of the bulk semiconductor, R is the particle radius, and m_e and m_h are the effective masses of electrons and holes, respectively. The effective bandgap, and hence the fluorescent emission wavelength, of a QD system is dependent on the particle size. For example, a 2 nm diameter CdSe QD would emit blue light, while an 8 nm CdSe QD would emit deep red. In principle, we can obtain any visible color by simply controlling the particle size during the synthesis process. (ii) Vivid colors. Through sophisticated chemical synthesis techniques, QD's particle size can be controlled precisely and uniformly [178]. The corresponding FWHM of Cd-based QDs is ~25–30 nm [171]. With some modifications to the process, 10 nm FWHM colloidal particles have been reported in the form of platelets [179]. Such a narrow emission linewidth would undoubtedly produce an exceedingly wide color gamut. (iii) Excellent quantum yield and stability. This is attributed to the unique core-shell structure of QDs [180]. Shells as well as surrounding organic ligands work as a protection layer and provide necessary processability. Both the efficiency and lifetime are improved compared to core-only systems.

Several classes of QD materials have been synthesized and studied; they can be roughly divided into two groups: cadmium-based QDs and cadmium-free QDs [181–183]. Here, we choose two commonly used CdSe and InP QDs as representatives from each group for describing their characteristics.

A. *Cadmium based QDs*
   The II–VI semiconductor CdSe is the most developed and well characterized QD material system. Its bulk bandgap is 1.73 eV ($\lambda = 716$ nm). According to Eq. (4.46), its emission spectrum can be adjusted to cover the entire visible region by tailoring the particle size, as illustrated in Figure 4.45a. Meanwhile, when synthesized by the well-established hot-injection technique [184], Cd-based QDs exhibit a narrow FWHM (20 ~ 30 nm) and high luminescent quantum efficiency (>95%). Figure 4.45b depicts the typical emission spectra for green and red CdSe QDs as well as a high power InGaN blue LED. With commercially available color filters, a color gamut spanning 90% of Rec. 2020 has been realized [138, 185]. Such a high quality QD material seems to be a perfect choice for display applications. Indeed, Cd-based QDs have already been employed in some commercial products. However, owing to its toxicity, there is increasing demand for manufacturers to eliminate cadmium, along with other heavy metals, from consumer products. For example, in 2003 the European Union issued a directive known as the Restriction of Hazardous Substances (RoHSs), wherein the maximum cadmium content in any consumer electronic product is limited to 100 ppm. Therefore, heavy metal-free or low Cd content QDs have become the new trend for display applications [186].

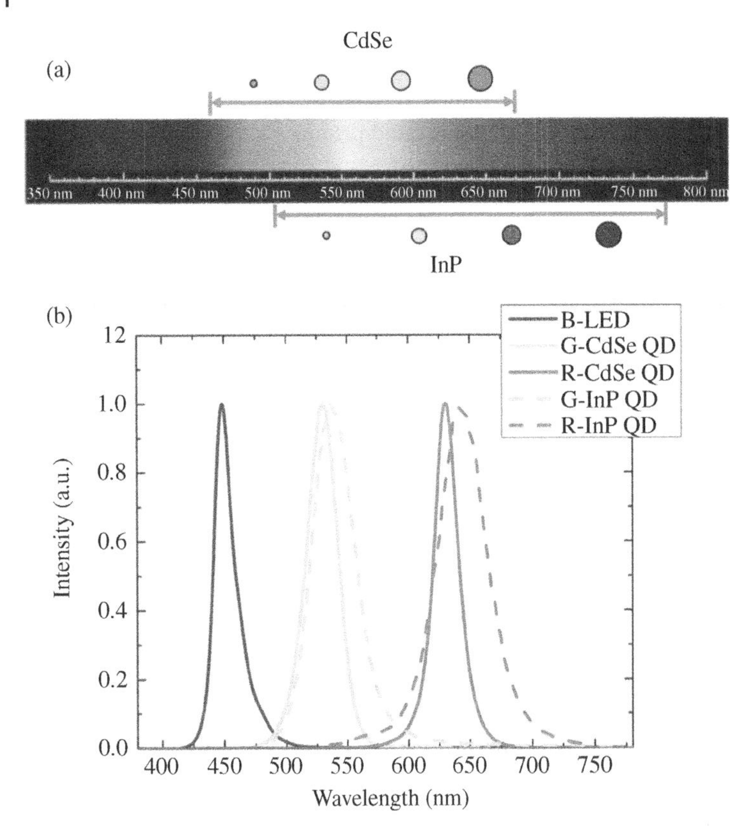

Figure 4.45 (a) Potential emission spectral range of CdSe and InP QDs; (b) Typical emission spectra for green and red quantum dots using CdSe (solid line) and InP (dashed line). Blue LED with emission peak $\lambda = 450$ nm is also shown here.

B. *Cadmium-free QDs*

Among several Cd-free QD candidates, InP is a viable alternative for visible light emission [187–190]. Its bandgap in the bulk material is 1.35 eV, which is smaller than that of CdSe. Thus, to reach the same emission wavelength, the core size of InP QDs has to be smaller than that of CdSe (Figure 4.45a). Smaller bandgap and smaller particle size lead to a much stronger confinement effect. Therefore, the emission spectrum of InP QDs is more sensitive to particle size variation. As a result, its FWHM is somewhat broader (>40 nm) (Figure 4.45b), corresponding to 70–80% Rec. 2020 color gamut, depending on the color filters employed. Moreover, its quantum yield and stability are slightly inferior to those of Cd-based QDs, which has been attributed as mostly due to the immature chemical synthesis method [191, 192]. In 2015, Nanoco claimed to have improved the InP lifetime to over 30 000 hours using their molecular seeding synthesis method [193]. If the FWHM can be further reduced, InP QDs will be more attractive for display applications. Another approach is to combine red InP QD with green Cd-based QD [194]. This hybrid approach not only retains 90% Rec. 2020 color gamut but the reduced cadmium content also complies with the RoHS regulation.

4.12.2 Device Configurations

Three QD backlight geometries have been developed: (i) on-chip, (ii) on-edge, and (iii) on-surface, as Figure 4.46 depicts [195]. Each design has its own pros and cons, and should be chosen carefully depending on the application requirements.

A. *On-chip geometry*
This design (Figure 4.46a) is fully compatible with current backlight systems and it uses the smallest amount of QD materials. We simply replace the YAG:Ce^{3+} yellow phosphor with green and red QD mixtures in the LED housing. However, due to the high light flux and high LED junction temperature (~150 °C), the QD's lifetime and stability could be compromised [196]. Another concern is the packaging issue because QDs are sensitive to oxygen and moisture. Hermetic sealing is necessary, which in turn increases the total cost and design complexity. Nevertheless, encouraging progress has been achieved. For example, some on-chip QDs show no evident degradation in 3000 hours under a flux condition of 52 W/cm^2 at 85 °C and 85% relative humidity [197, 198].

B. *On-edge geometry*
While the problems of the on-chip design remain to be fully solved, the on-edge geometry (also known as the QD rail) offers an alternative approach, especially for large size TVs. Compared to the on-chip design, the QD rail (Figure 4.46b) offers a much longer lifetime because it is located at a greater distance from the blue LED. Meanwhile, the QD material consumption is still acceptable. However, QD rail assembly is challenging. This is because QDs act as an isotropic emitter; light is emitted in all directions. To avoid optical loss, we need to collect the back-scattered light and divert it into the LGP [199]. To avoid degradation, the QD rail should also be sealed in vacuum.

C. *On-surface geometry*
At present time, the on-surface design (also known as a quantum dot enhancement film, QDEF) is the most commonly used geometry [200]. As Figure 4.46c depicts, the QDEF is laminated on top of the LGP surface, which is spatially decoupled from the LED heat source. The operating temperature should be close to room temperature. As a result, both reliability and long-term stability are enhanced significantly. In fact,

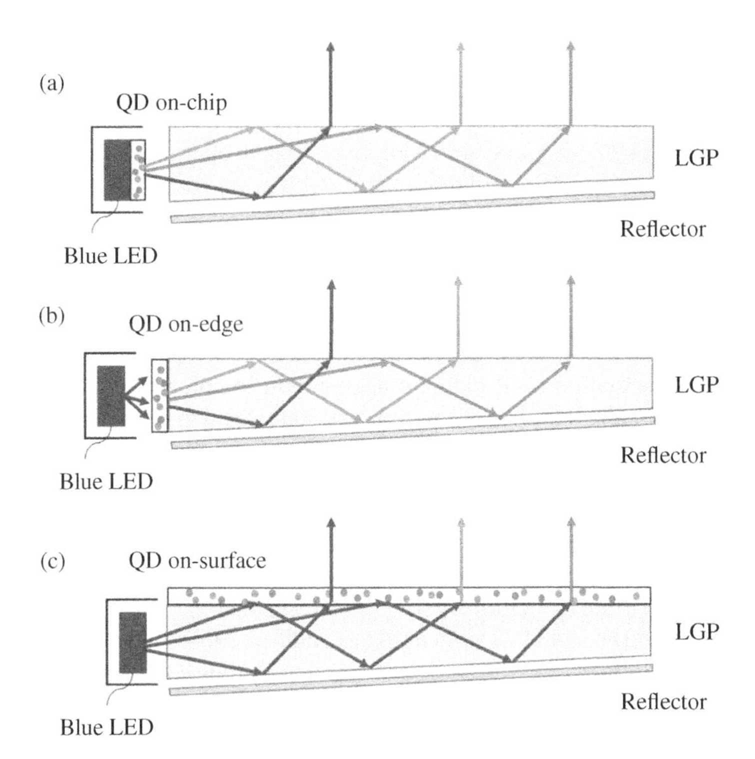

Figure 4.46 Schematic diagram for three device geometries implementing QD materials. (a) QD is placed within an LED package. (b) QD is placed between LED and light guide plate (LGP), or known as quantum rail. (c) QD is placed on the top surface of LGP, or known as quantum dot enhancement film (QDEF).

a lifetime over 30 000 hours has been achieved according to accelerated aging tests [201]. One drawback of the QDEF is that it consumes more QD materials, especially for large screen TVs. As the production capacity of QDEF keeps growing, the cost should decrease accordingly.

4.13 HIGH DYNAMIC RANGE

HDR requires a peak luminance over 1000 nits, black state lower than 0.01 nit, more than 10 bits of gray level, and a wide color gamut [202]. In other words, the contrast ratio is over 100 000 : 1. For an OLED display, it is fairly easy to get a true black state, but to obtain a brightness over 1000 nits would lead to compromised lifetime [203]. On the contrary, it is relatively easy to boost an LCD's peak brightness to 1000 nits, but to lower the dark state to <0.01 nits is challenging. A typical CR of a MVA LCD is ~5000 : 1, which is about 20× lower than what HDR demands. To reduce the dark state light leakage, a local dimming technique has been widely applied [139, 204]. For OLED displays, each pixel is driven individually from its black state to the fully bright state. But for LCD, the commonly employed local dimming method is to divide the LED backlight into hundreds of individually controlled zones, which is still far from pixel-by-pixel dimming. Such an approach improves the contrast ratio, but the backlight illumination can still leak through the LC panel if the displayed scenes have sharp boundaries between bright and dark areas. This phenomenon is known as the halo effect [205–207]. The magnitude of the halo is governed by the LCD's native contrast ratio and the number of local dimming zones.

If a LCD's CR can be improved by 20×, then more gray levels can be displayed. For a standard dynamic range (SDR) display, an 8-bit signal depth is commonly used. However for HDR, at least 10-bit depth is required [208, 209]. The increased bit depth helps to enhance image quality, but may also introduce an additional burden on the driving schemes. Let us assume the minimal voltage interval allowed between two adjacent gray levels is 5 mV, which is limited by the TFT array. Under such conditions, to achieve 10-bit (i.e. 1024 gray levels), 12-bit, and 14-bit image depth, the required on-state voltage would be 5.12, 20.48, and 81.92 V, respectively. Therefore, it is challenging for the present TFT-LCD to achieve 12-bit grayscales. Moreover, higher operating voltage leads to higher electronic power consumption. To achieve HDR and more than 12-bit grayscales, mini-LED backlit LCD and dual-panel LCD have been developed.

4.13.1 Mini-LED Backlit LCDs

Conventional LCDs exhibit a limited contrast ratio because of the depolarization effects from non-uniform LC alignment, light scattering from LC director fluctuations and pigment aggregation in color filters, and diffraction from the pixelated electrodes. To suppress the light leakage in the dark state, local dimming with a spatially segmented mini-LED backlight has been developed. Each segment, called a local dimming zone, is controlled independently. With 10-bit backlight modulation, the CR can increase from 1000 ~ 5000 : 1 to ~1 000 000 : 1. A schematic mini-LED backlit LCD is shown in Figure 4.47. For discussion purposes, let us assume each mini-LED has a square shape. The emitted light propagates some distance (e.g. through the adhesive layer) before reaching the diffuser. This distance and the scattering strength of the diffuser need to be optimized so that the outgoing light is spatially uniform before entering the LCD panel.

To illustrate the light modulation process of mini-LED backlit LCDs, let us use a candle picture as Figure 4.48 shows as an example. Here, the backlight consists of 12 × 24 local dimming zones and each zone contains 6 × 6 mini-LEDs in order to achieve a desired luminance. According to the image content, the mini-LEDs in each dimming zone are pre-determined to show different gray levels, as Figure 4.48a depicts. After passing through the diffuser, the outgoing light spreads out uniformly before reaching the LCD panel (Figure 4.48b). The gray level of each LCD pixel is controlled by a TFT, and each color filter only transmits the designated color. Finally, a full-color image is generated, as Figure 4.48c shows.

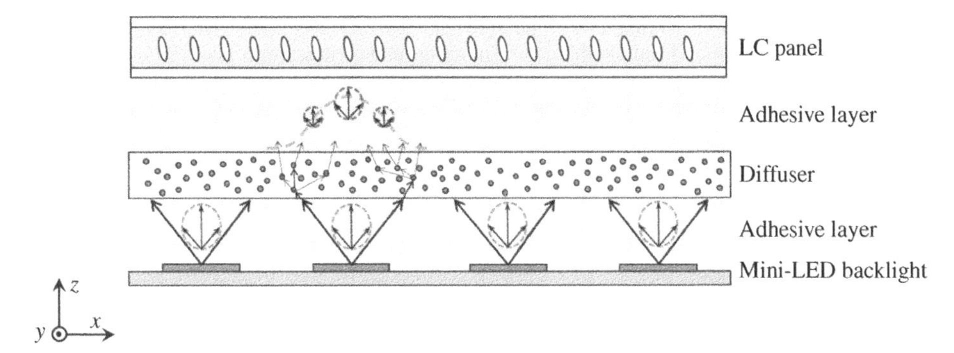

Figure 4.47 Schematic diagram of mini-LED backlit LCD.

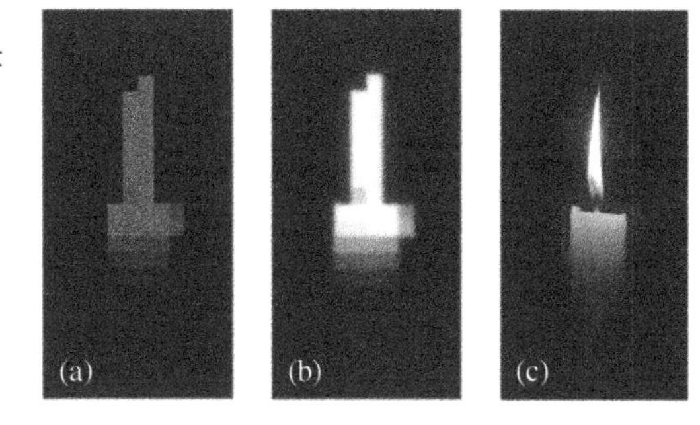

Figure 4.48 Light modulation process of mini-LED backlit LCD: (a) mini-LED backlight modulation; (b) luminance distribution of the light incident on the LC layer, and (c) displayed image after LCD panel.

The mini-LED backlight enables a new LCD with high peak luminance, excellent dark state, and thin form factor, and in the meantime suppresses the undesired halo effect and clipping effect. The halo effect is the light leakage from bright objects to adjacent dark areas, while the clipping effect originates from insufficient luminance in a local dimming zone when the adjacent zones are dimmed [210]. Conventional edge-lit LCDs feature thin profile, but the LGP is relatively thick if a high-luminance large-area LED array is adopted [211, 212]. On the other hand, conventional direct-lit LCDs with fewer LEDs can provide high luminance and HDR, but a relatively long travel distance is needed to ensure good backlight uniformity [14, 139]. In comparison, the small chip size and large number of mini-LEDs allows the light to spread out evenly so that the required optical distance between LED and diffuser is shorter.

The system configuration of a mini-LED backlit LCD determines the severity of the halo effect and clipping effect, and affects the total thickness of the backlight unit. The number of local dimming zones and the LCD's CR, have the dominant impacts on the performance of local dimming. However, between two comparable panels, sometimes the one with fewer local dimming zones can exhibit a better performance, which is contrary to the general trend. This conflict comes from different optical designs, where LED light expansion and local light confinement also jointly contribute to the final local dimming performance. In the following, we will discuss the influence of each factor and then suggest the corresponding optimization strategies. The following discussions are based on a 6.4 in. smartphone placed at a 25 cm viewing distance, but these results can be scaled up and applied to large-size panels as well.

A mini-LED BLU can effectively suppress the halo effect if the LCD contrast ratio and the density of local dimming zones are properly chosen [213]. By simulating the displayed images of a mini-LED backlit LCD with different system configurations and conducting subjective experiments, it is found that the peak signal-to-noise ratio in the CIE 1976 ($L^*a^*b^*$) color space (LabPSNR) can be used as a metric to evaluate the halo effect. When LabPSNR >47.4 dB, less than 5% of people could differentiate the displayed image on a mini-LED backlit LCD from the original picture. Figure 4.49 shows the correlation between the LCD CR and the local dimming zone number. The black dashed line indicates LabPSNR = 47.4 dB; above this level the halo

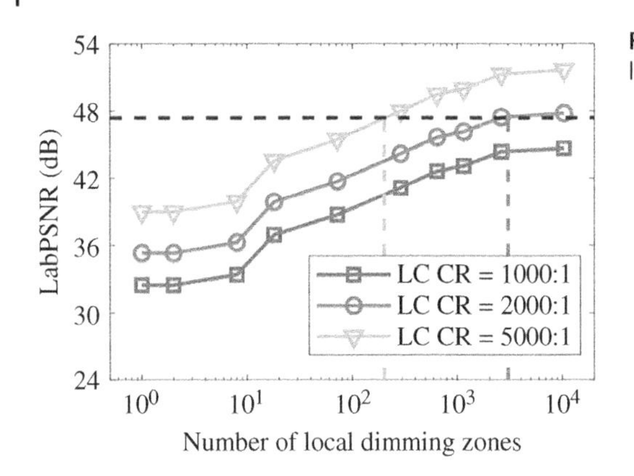

Figure 4.49 Simulated LabPSNR for HDR display systems with various local dimming zone numbers and contrast ratio.

effect is unnoticeable. From Figure 4.49, we find that ~3000 local dimming zones is required for a FFS LCD with CR = 2000 : 1, and ~200 zones is required for a MVA LCD with CR = 5000 : 1. However, if an LCD's CR is lower than 1000 : 1, then even 10 000 zones is still inadequate. For practical applications, we have to consider the decreased contrast ratio as the viewing angle increases, as Figure 4.49 depicts. These guidelines are consistent with several prototypes which have been demonstrated [214–216].

4.13.2 Dual-Panel LCDs

The dual-panel approach combines a high-resolution full-color display panel (say, 8K LCD #2) with a low-resolution black-and-white local dimming panel (say 2K LCD #1), as Figure 4.50 depicts [140]. Let us assume the contrast ratio of the two LCD panels is CR1 and CR2, respectively, then the effective contrast ratio of cascaded display system should be CR1*CR2. A typical CR for a FFS LCD is ~2000 : 1 and for TN is ~800 : 1, thus ideally the combined CR of a dual panel should be 1 600 000 : 1. In experiments, a contrast ratio over 1 000 000 : 1 and 16-bit signal depth (8 bits from LCD #1 and 8 bits from LCD #2) at only 5 V

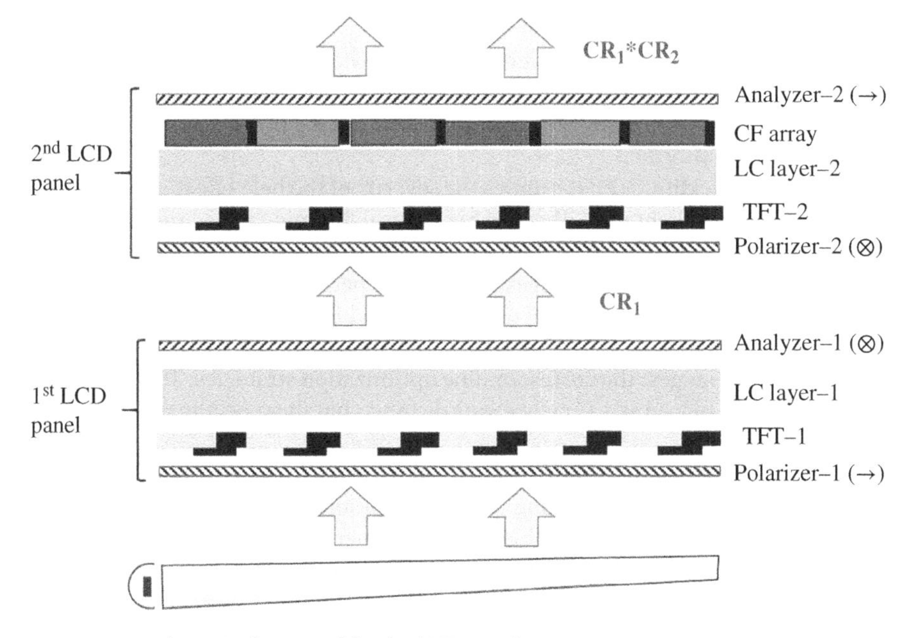

Figure 4.50 Schematic diagram of the dual LCD panels.

on-state voltage has been demonstrated. The major advantage of this dual-panel design is that LCD#1 can easily provide ~2 million (1920 × 1080) local dimming zones for achieving a comparable black state with the pixel-level dimming OLED panel. In practice, the resolution of the LCD #1 can be further reduced, as long as the halo effect is not noticeable, which depends on the eye's acuity. Of course, there are some drawbacks, such as a ~30% reduced optical efficiency (assuming the aperture ratio of LCD #1 is 70%), increased panel weight, Moiré pattern, parallax error effect at off-axis viewing, and increased cost.

4.14 FUTURE DIRECTIONS

Recently, LCDs have faced a strong challenge from OLEDs, especially in small and medium panels. Figure 4.51 shows a spider chart comparison of nine performance metrics: cost, lifetime, peak brightness, color, resolution density, power consumption, MPRT, contrast ratio (CR), and panel flexibility. From Figure 4.51, we can find that LCD is leading in cost, lifetime, and peak brightness. LCD is comparable to OLED in color (with quantum dots), resolution density, power consumption, and MPRT. However, LCD is trailing behind OLED in CR, and flexibility.

1. Cost: As the manufacturing technology advances, the production cost of both LCD and OLED will continue to decrease. However, LCD still has a competitive edge in material cost, equipment cost, and manufacturing yield.
2. Lifetime: LCD is voltage driven, while OLED is current driven. Most LCDs only need one TFT per pixel as a simple voltage switch, but OLED needs multiple (usually more than five) TFTs (including switching, driving, and compensation circuits) in order to provide a stable current. Overall, OLED is more vulnerable to moisture, oxygen, electric current, and temperature than LCD [217]. Especially, the lifetime of blue OLEDs is the shortest among the three primary colors, so differential aging is a big concern. However, the life cycle time of a smartphone is typically aimed for two to three years, in which case OLED lifetime is still acceptable. But for TVs, the expected operational lifetime is 50 000 hours. In this case, LCD is a more favorable choice.

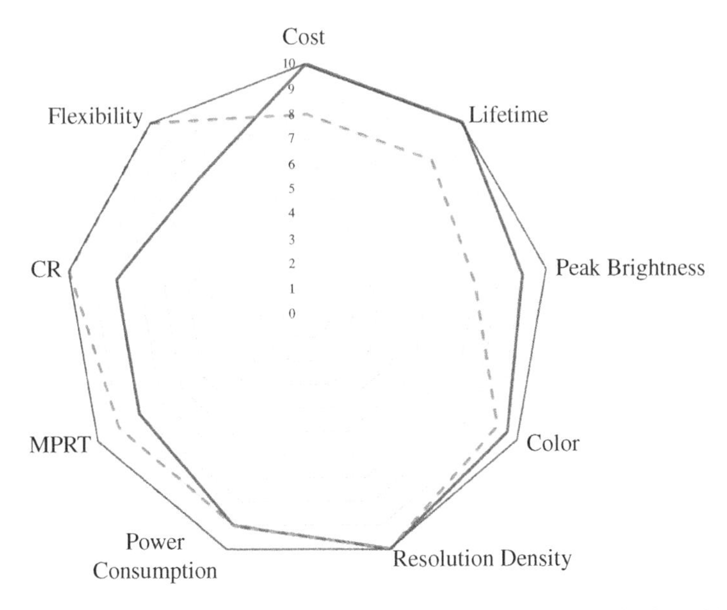

Figure 4.51 Performance comparison of LCD (solid line) versus RGB OLED (dashed lines) in nine categories.

3. Peak brightness: The brightness of an LCD is jointly determined by the LED backlight and the LCD panel's optical throughput. For a given LCD panel efficiency, say 5%, we can still boost the LED backlight to obtain high peak brightness, say >2000 nits. In principle, we can do the same for OLED, except that high current impulse driving of OLED would lead to substantial efficiency roll-off and compromised lifetime.
4. Color: The LCD with 1pc WLED (blue LED + yellow phosphor) backlight shows ~75% NTSC (or ~50% Rec.2020) color gamut, which is inferior to OLEDs [218]. By adopting quantum dots or an ultra-narrow (FWHM ≈ 10 nm) LED backlight, the color gamut greatly improves to 90% Rec.2020. Similar progress has occurred in the OLED camp as well. Some RGB OLED displays have employed a longer wavelength red OLED ($\lambda \approx 625$ nm), and as a result, its color gamut increases to 84% Rec.2020. However, for TVs using white OLED, the crosstalk of color filters is the limiting factor. The ultimate goal for both LCD and OLED displays is to be comparable to laser displays.
5. Resolution density: The resolution density of a display panel is determined by its pixel dimensions. Both active matrix LCDs and OLEDs are based on TFTs, thus the resolution density is determined by how small a pixel can be fabricated. A high mobility silicon, such as low temperature poly-silicon or crystalline silicon, helps reduce the TFT area for a higher light throughput. Both LCD and OLED can then achieve high resolution density. If an LCD has submillisecond response time, then it would enable field sequential color (FSC) operation with suppressed color breakup. In an FSC display using RGB LEDs, the color filters can be eliminated, resulting in tripled optical efficiency and resolution density.
6. Power consumption: The power consumption of an OLED display depends heavily on the image content, according to the average pixel luminance level (APL). APL = 0 denotes a completely black screen, i.e. every pixel is off, while APL = 1 indicates a completely white screen, i.e. every pixel is on. For a conventional LCD without local dimming, its power consumption is independent of the displayed image. Studies show that APL ~30% is the crossover point for white OLED TVs with color filters, i.e. OLED wins if APL < 30%, but LCD wins if APL > 30%. For TVs, the average APL is about 50%. On the contrary, for RGB-based OLEDs in which no color filter is needed, the crossover point increases to about 60%. This explains why the RGB-OLED is gaining momentum for smartphone applications. Similarly to OLED, the mini-LED backlit LCD's power consumption is also dependent on the displayed image content. A smaller APL helps to reduce the power consumption because fewer mini-LED zones will be turned on.
7. MPRT: A detailed discussion of MPRT has been given in Section 4.11. Both TFT-LCDs and OLEDs are holding type devices. To achieve a CRT-like display with MPRT ≈ 1.5 ms, impulse driving for OLEDs or backlight modulation for LCDs is needed. With a low viscosity liquid crystal and low duty ratio, it is possible for an LCD to achieve a comparable MPRT with OLED. However, the OLED still has a major advantage over LCDs at low temperature, where the LC's viscosity increases exponentially.
8. Contrast ratio: It is fairly easy for OLEDs to achieve CR > 1 000 000 : 1 in a dark room, but it is difficult for LCDs. The mini-LED backlight and double-panel approaches open a new gateway for LCDs to close this gap. Sunlight readability is a general concern for both LCDs and OLEDs, especially for automotive displays. With ambient lighting conditions, high luminance is more important than high contrast ratio.
9. Flexible and foldable displays: These are OLED's major advantages [219]. However, OLEDs still need a circular polarizer (CP) in order to obtain a high ACR. Therefore, an ultra-thin CP is critically needed for flexible OLEDs [220]. Flexible LCDs, although not addressed here in detail, are also emerging. Some plastic LCD prototypes [221] and LCDs with organic TFTs (called OLCD) [222] and polymer-stabilized spacers for controlling the cell gap uniformity during bending have been developed. Potential applications are foreseeable in smartphones and foldable display devices.
10. Transparent LCDs: In 2017, Japan Display Inc. demonstrated a 4 in. transparent LCD prototype with 80% transparency [223]. The LCD mode employed is a polymer network liquid crystal (PNLC). As a result, the response time is fast enough (<2 ms) to enable FSCs (180 Hz) with edge-lighting provided by RGB LEDs. In the voltage-off state, the PNLC layer is highly transparent, while in a voltage-on state it scatters light strongly. The device contrast ratio is around 16 : 1, and the color gamut is about 112% NTSC. Such a transparent LCD opens up new applications for augmented reality and automotive displays, etc.

Homework Problems

4.1 An eutectic mixture is comprised of following two compounds:

(1) C_3H_7 NCS

K 39 N 41.3 I; $\Delta H = 4300$ cal/mol

(2) C_3H_7 NCS

K 66 N 190 I; $\Delta H = 3000$ cal/mol

(a) Calculate its nematic range (in °C)

(b) Calculate the molecular weight of compounds (1) and (2).

(c) If we want to prepare 10 g of this mixture, how many grams of each compound should be used?

4.2 In a fast response LC phase modulator, we have to consider birefringence and visco-elastic coefficient together. A figure-of-merit (FoM) is defined as:

$$\mathrm{FoM} = \frac{K_{11}(\Delta n)^2}{\gamma_1}$$

where K_{11} is the splay elastic constant, Δn is the birefringence, and γ_1 is the rotational viscosity. Assume $K_{11} \sim S^2$; $\Delta n \sim S$; $\gamma_1 \sim S \cdot \exp(E/kT)$ and $S = (1 - T/T_c)^{\alpha}$

(a) Prove that FoM exhibits a maximum at the optimal operating temperature (T_{op})

(b) Derive the analytical expression for T_{op}. Estimate T_{op} if activation energy $E = 0.35$ eV, $\alpha = 0.18$ and $T_c = 100$ °C, and Boltzmann constant $k = 0.0861$ meV/K.

(c) Explain why FoM has a maximum at T_{op}.

4.3 At $T = 20$ °C, an LC mixture has refractive indices as follows: $(n_e, n_o) = (1.5733, 1.4859)$ at $\lambda = 450$ nm and $(1.5565, 1.4751)$ at $\lambda = 633$ nm. What are the extrapolated (n_e, n_o) at $\lambda = 1550$ nm?

4.4 A homogeneous cell is useful as a tunable phase retardation plate. The following chart plots the voltage-dependent transmittance of a homogeneous LC cell at $\lambda = 633$ nm. The polarizers are crossed and the angle between the front polarizer and the LC rubbing direction is $\beta = 45°$.

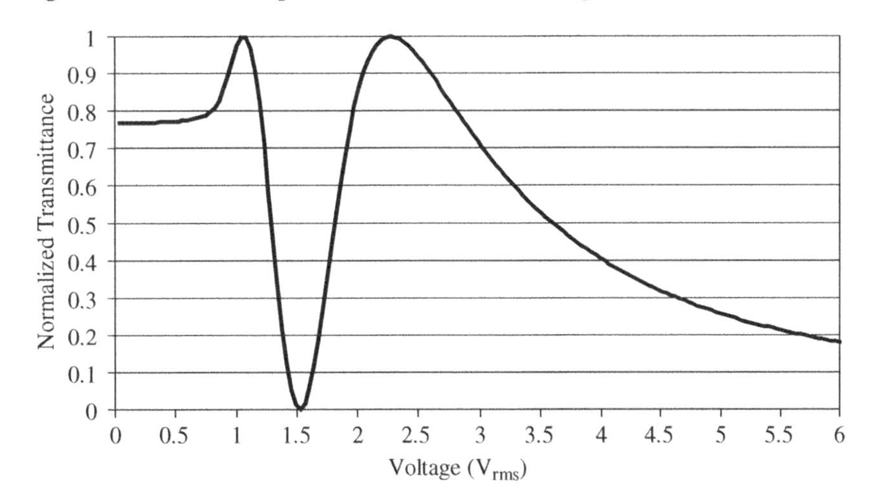

(a) If the cell gap is $d = 5\,\mu m$, what is the birefringence of the LC?
(b) At what voltages is the output beam (before the analyzer) circularly polarized?
(c) If we want to switch the outgoing beam from circular to linear polarization, which voltages do we use in order to obtain the fastest response time?
(d) Plot the VT curve using the same cell between crossed polarizers if $\beta = 0$.

4.5 A student prepared three cells: VA, 90° TN, and homogeneous cell and filled them with liquid crystal, but forgot to put on labels. Can you help the student to identify which cell is which by using a white light table and two linear polarizers?

4.6 In-Plane-Switching (IPS) and Multi-domain Vertical Alignment (MVA) are the two major approaches for wide-view LCD TVs. Compare their pros and cons.

4.7 Similar to a homogeneous cell, a VA cell can be used for phase-only modulation. From the viewpoint of LC molecular structures, explain why the homogeneous cell is a favored choice.

4.8 The extinction ratio of two crossed polarizers can exceed $10^5:1$ easily. However, the contrast ratio of a VA cell is about 5000 : 1 and for IPS is about 2000 : 1. What causes the degradation of the LCD's contrast ratio?

4.9 Let us assume OLED has 1 μs response time and LCD has 2 ms response time. Calculate the MPRT of OLED and LCD at 120 Hz frame rate. How to obtain CRT-like MPRT, which is 1.5 ms?

4.10 From Figure 4.49, the required local dimming zone number for the fringing-field switching LCD with CR = 2000 : 1 is ~3000. But in reality, the actual zone number is often doubled. Why? (*Hint: we need to consider the viewing angle effect of contrast ratio, as* Figure 4.29 *shows*).

4.11 Discuss the pros and cons of mini-LED backlit LCDs and double-panel LCDs.

4.12 Quantum dots have been integrated into the backlight unit to increase the color gamut of an LCD. Can quantum dots replace the pigment color filters?

References

1 Armitage, D., Underwood, I., and Wu, S.T. (2006). *Introduction to Microdisplays.* Wiley.
2 Khan, A., Schneider, T., Montbach, E. et al. (2007). Recent progress in color flexible reflective cholesteric displays. *SID Symp. Dig.* 38: 54.
3 Bae, J.H., Jang, S.J., Choi, Y.S. et al. (2007). The stabilized bistable LC mode for flexible display. *SID Symp. Dig.* 38: 649.
4 Kress, B.C. and Cummings, W.J. (2017). Towards the ultimate mixed reality experience: HoloLens display architecture choices. *SID Symp. Dig.* 48: 127.
5 Stupp, E.H. and Brennesholtz, M. (1998). *Projection Displays.* Wiley.
6 Wu, S.T. and Yang, D.K. (2001). *Reflective Liquid Crystal Displays.* Wiley.
7 Ge, Z.B. and Wu, S.T. (2010). *Transflective Liquid Crystal Displays.* Wiley.
8 Crawford, G.P. (2005). *Flexible Flat Panel Displays.* Wiley.
9 Huang, Y., Tan, G., Gou, F. et al. (2019). Prospects and challenges of mini-LED and micro-LED displays. *J. Soc. Inf. Disp.* 27: 387.
10 Chen, H., Zhu, R., He, J. et al. (2017). Going beyond the limit of an LCD's color gamut. *Light Sci. Appl.* 6: e17043.
11 Luo, Z., Chen, Y., and Wu, S.T. (2013). Wide color gamut LCD with a quantum dot backlight. *Opt. Express* 21: 26269.

12 Yang, D.K. and Wu, S.T. (2014). *Fundamentals of Liquid Crystal Devices*, 2e. Wiley.
13 J. M. Jonza, M. F. Weber, A. J. Ouderkirk, and C. A. Stover, "Polarizing beam-splitting optical component", US Patent 5,962,114 (1999).
14 de Greef, P. and Hulze, H.G. (2007). Adaptive dimming and boosting backlight for LCD-TV systems. *SID Symp. Dig.* 38: 1332.
15 Chen, H., Sung, J., Ha, T., and Park, Y. (2007). Locally pixel-compensated backlight dimming for improving static contrast on LED backlit LCDs. *SID Symp. Dig.* 38: 1339.
16 Lin, F.C., Liao, C.Y., Liao, L.Y. et al. (2007). Inverse of mapping function method for image quality enhancement of high dynamic range LCD TVs. *SID Symp. Dig.* 38: 1343.
17 Anandan, M. (2008). Progress of LED backlights for LCDs. *J. Soc. Inf. Disp.* 16: 287.
18 Peng, F., Chen, H., Gou, F. et al. (2017). Analytical equation for the motion picture response time of display devices. *J. Appl. Phys.* 121, 023108.
19 Goodby, J.W. (1991). *Ferroelectricity Liquid Crystals: Principles, Properties and Applications*. Routledge.
20 Wand, M., Thurmes, W.N., Vohra, R.T., and More, K.M. (1997). Advances in ferroelectric liquid crystals for microdisplay applications. *SID Symp. Dig.* 27: 157.
21 Yang, D.K., Lu, Z.J., Chien, L.C., and Doane, J.W. (2003). Bistable polymer dispersed cholesteric reflective display. *SID Symp. Dig.* 34: 959.
22 Gray, G., Harrison, K.J., and Nash, J.A. (1973). New family of nematic liquid crystals for displays. *Electron. Lett* 9: 130.
23 Schroder, L. (1893). *Z. Phys. Chem.* 11: 449.
24 Van Laar, J.J. (1908). *Z. Phys. Chem.* 63: 216.
25 Bedjaoui, L., Gogibus, N., Ewen, B. et al. (2004). Preferential solvation of the eutectic mixture of liquid crystals E7 in a polysiloxane. *Polymer* 45: 6555.
26 Deuling, H.J. (1978). Solid State Physics. Suppl. 14. In: *Liquid Crystals* (ed. L. Liebert). New York: Academic.
27 Maier, W. and Meier, G. (1961). A simple theory of the dielectric characteristics of homogeneous oriented crystalline-liquid phases of the nematic type. *Z. Naturforsch. Teil A* 16: 262.
28 Tironi, I.G., Sperb, R., Smith, P.E., and van Gunsteren, W.F. (1995). A generalized reaction field method for molecular dynamics simulations. *J. Chem. Phys.* 102: 5451.
29 Schadt, M. (1992). Field-effect liquid-crystal displays and liquid-crystal materials – key technologies of the 1990s. *Displays* 13: 11.
30 Dabrowski, R. (1990). Isothiocyanates and their mixtures with a broad range of nematic phase. *Mol. Cryst. Liq. Cryst.* 191: 17.
31 He, Z., Gou, F., Chen, R. et al. (2019). Liquid-crystal beam steering devices: principles, recent advances and future developments. *Crystals* 9: 292.
32 Schadt, M. and Helfrich, W. (1971). Voltage-dependent optical activity of a twisted nematic liquid crystal. *Appl. Phys. Lett.* 18: 127.
33 Soref, R.A. (1973). Transverse field effect in nematic liquid crystals. *Appl. Phys. Lett.* 22: 165.
34 Oh-e, M. and Kondo, K. (1995). Electro-optical characteristics and switching behavior of the in-plane switching mode. *Appl. Phys. Lett.* 67: 3895.
35 Xu, D., Peng, F., Chen, H. et al. (2014). Image sticking of liquid crystal displays with lateral electric fields. *J. Appl. Phys.* 116, 193102.
36 Tarao, R., Saito, H., Sawada, S., and Goto, Y. (1994). Advances in liquid crystals for TFT displays. *SID Tech. Dig.* 25: 233.
37 Geelhaar, T., Tarumi, K., and Hirschmann, H. (1996). Trends in LC materials. *SID Tech. Dig.* 27: 167.
38 Goto, Y., Ogawa, T., Sawada, S., and Sugimori, S. (1991). Fluorinated liquid crystals for active matrix displays. *Mol. Cryst. Liq. Cryst.* 209: 1.
39 Schiekel, M.F. and Fahrenschon, K. (1971). Deformation of nematic liquid crystals with vertical orientation in electric fields. *Appl. Phys. Lett.* 19: 391.

40 R. Eidenschink and L. Pohl, US patent 4,415,470 (1983).
41 de Gennes, P.G. and Prost, J. (1993). *The Physics of Liquid Crystals*, 2e. Oxford.
42 Maier, W. and Saupe, A. (1960). A simple molecular statistical theory for nematic liquid crystal phase, part II. *Z. Naturforsch. Teil A* 15: 287.
43 Gruler, H. (1975). The elastic constants of a nematic liquid crystal. *Z. Naturforsch. Teil A* 30: 230.
44 Jakeman, E. and Raynes, E.P. (1972). Electro-optic response times of liquid crystals. *Phys. Lett. A* 39: 69.
45 Imura, H. and Okano, K. (1972). Temperature dependence of the viscosity coefficients of liquid crystals. *Jpn. J. Appl. Phys.* 11: 1440.
46 Diogo, A.C. and Martins, A.F. (1981). Thermal behavior of the twist viscosity in a series of homologous nematic liquid crystals. *Mol. Cryst. Liq. Cryst.* 66: 133.
47 Belyaev, V.V., Ivanov, S., and Grebenkin, M.F. (1985). Temperature dependence of rotational viscosity of nematic liquid crystals. *Sov. Phys. Crystallogr.* 30: 674.
48 Wu, S.T. and Wu, C.S. (1990). Rotational viscosity of nematic liquid crystals. *Liq. Cryst.* 8: 171.
49 Haller, I. (1975). Thermodynamic and static properties of liquid crystals. *Prog. Solid State Chem.* 10: 103.
50 Osipov, M.A. and Terentjev, E.M. (1989). Rotational diffusion and rheological properties of liquid crystals. *Z. Naturforsch. Teil A* 44: 785.
51 Wu, S.T. and Wu, C.S. (1990). Experimental confirmation of Osipov-Terentjev theory on the viscosity of liquid crystals. *Phys. Rev. A* 42: 2219.
52 Wu, S.T., Ramos, E., and Finkenzeller, U. (1990). Polarized UV spectroscopy of conjugated liquid crystals. *J. Appl. Phys.* 68: 78–85.
53 Wu, S.T., Efron, U., and Hess, L.D. (1984). Birefringence measurement of liquid crystals. *Appl. Opt.* 23: 3911.
54 Wu, S.T. and Wu, C.S. (1989). A three-band model for liquid crystal birefringence dispersion. *J. Appl. Phys.* 66: 5297.
55 Wu, S.T. (1991). A semi-empirical model for liquid-crystal refractive index dispersions. *J. Appl. Phys.* 69: 2080.
56 Wu, S.T., Wu, C.S., Warenghem, M., and Ismaili, M. (1993). Refractive index dispersions of liquid crystals. *Opt. Eng.* 32: 1775.
57 Li, J. and Wu, S.T. (2004). Extended Cauchy equations for the refractive indices of liquid crystals. *J. Appl. Phys.* 95: 896.
58 Wu, S.T., Efron, U., and Hess, L.D. (1984). Infrared birefringence of liquid crystals. *Appl. Phys. Lett.* 44: 1033.
59 Li, J. and Wu, S.T. (2004). Two-coefficient Cauchy model for low birefringence liquid crystals. *J. Appl. Phys.* 96: 170.
60 Mada, H. and Kobayashi, S. (1976). Wavelength and voltage dependences of refractive indices of nematic liquid crystals. *Mol. Cryst. Liq. Cryst.* 33: 47.
61 Li, J., Gauza, S., and Wu, S.T. (2004). High temperature-gradient refractive index liquid crystals. *Opt. Express* 12: 2002.
62 Li, J. and Wu, S.T. (2004). Temperature effect on liquid crystal refractive indices. *J. Appl. Phys.* 96: 19.
63 Cognard, J. (1982). Alignment of nematic liquid crystals and their mixtures. *Mol. Cryst. Liq. Cryst.* Suppl. 1: 1.
64 Chigrinov, V.G., Kozenkov, V.M., and Kwok, H.S. (2008). *Photoalignment of Liquid Crystalline Materials: Physics and Applications*. Wiley.
65 Stöhr, J., Samant, M.G., Lüning, J. et al. (2001). Liquid crystal alignment on carbonaceous surfaces with orientational order. *Science* 292: 2299.
66 Scheffer, T.J. and Nehring, J. (1977). Accurate determination of liquid-crystal tilt bias angles. *J. Appl. Phys.* 48: 1783.
67 Scheffer, T. and Nehring, J. (1990). Twisted nematic and super-twisted nematic mode LCDs. In: *Liquid Crystals Applications and Uses*, vol. 1, Ch. 10 (ed. B. Bahadur), 231–274. Singapore: World Scientific.

68 Drzaic, P.S. (1995). *Liquid Crystal Dispersions*. World Scientific.
69 Chen, H.M., Yang, J.P., Yen, H.T. et al. (2018). Pursuing high quality phase-only liquid crystal on silicon (LCoS) devices. *Appl. Sci.* 8: 2323.
70 Huang, Y., Liao, E., Chen, R., and Wu, S.T. (2018). Liquid-crystal-on-silicon for augmented reality displays. *Appl. Sci.* 8: 2366.
71 McManamon, P.F., Dorschner, T.A., Corkum, D.L. et al. (1996). *Proc. IEEE* 84: 268.
72 Bos, P.J. and Koechler, K.R. (1984). The pi-cell: a fast liquid crystal optical switching device. *Mol. Cryst. Liq. Cryst.* 113: 329.Beran
73 Freedericksz, V. and Zolina, V. (1933). Forces causing the orientation of an anisotropic liquid. *Trans. Faraday Soc.* 29: 919.
74 Jiao, M., Ge, Z., Song, Q., and Wu, S.T. (2008). Alignment layer effects on thin liquid crystal cells. *Appl. Phys. Lett.* 92, 061102.
75 Lien, A., Takano, H., Suzuki, S., and Uchida, H. (1991). The symmetry property of a 90° twisted nematic liquid crystal cell. *Mol. Cryst. Liq. Cryst.* 198: 37.
76 Gooch, C.H. and Tarry, H.A. (1975). The optical properties of twisted nematic liquid crystal structures with twisted angles $\leq 90°$. *J. Phys. D* 8: 1575.
77 Wu, S.T. and Wu, C.S. (1999). Mixed-mode twisted-nematic cell for transmissive liquid crystal display. *Displays* 20: 231.
78 Mori, H., Itoh, Y., Nishiura, Y. et al. (1997). *Jpn. J. Appl. Phys.* 36: 143.
79 K. H Yang. Int'l Display Research Conf. p. 68 (1991).
80 Mori, H., Nagai, M., Nakayama, H. et al. (2003). Novel optical compensation method based upon a discotic optical compensation film for wide-viewing-angle LCDs. *SID* 34: 1058.
81 Mori, H. (2005). The wide view film for enhancing the field of view of LCDs. *J. Disp. Technol.* 1: 179.
82 Soref, R.A. (1974). Field effects in nematic liquid crystals obtained with interdigital electrodes. *J. Appl. Phys.* 45: 5466.
83 R. Kiefer, B. Weber, F. Windscheid, and G. Baur, "In-plane switching of nematic liquid crystals", Japan Displays'92, p.547 (1992).
84 M. Oh-e, M. Ohta, S. Arantani, and K. Kondo, "Principles and characteristics of electro-optical behavior with in-plane switching mode", Asia Display'95, p.577 (1995).
85 Ge, Z., Zhu, X., Wu, T.X., and Wu, S.T. (2006). High-transmittance in-plane-switching liquid-crystal displays using a positive-dielectric-anisotropy liquid crystal. *J. SID* 14: 1031.
86 Chen, J., Kim, K.H., Jyu, J.J. et al. (1998). Optimum film compensation modes for TN and VA LCDs. *SID Tech. Dig.* 29: 315.
87 Anderson, J.E. and Bos, P.J. (2000). Methods and concerns of compensating in-plane switching liquid crystal displays. *Jpn. J. Appl. Phys., Part 1* 39: 6388.
88 Hong, Q., Wu, T.X., Zhu, X. et al. (2005). Extraordinarily high-contrast and wide-view liquid-crystal displays. *Appl. Phys. Lett.* 86, 121107.
89 Saitoh, Y., Kimura, S., Kusafuka, K., and Shimizu, H. (1998). Optimum film compensation of viewing angle of contrast in in-plane-switching-mode liquid crystal display. *Jpn. J. Appl. Phys., Part 1* 37: 4822.
90 T. Ishinabe, T. Miyashita, T. Uchida, and Y. Fujimura, "A wide viewing angle polarizer and a quarter-wave plate with a wide wavelength range for extremely high quality LCDs," *Proc. 21st Int'l Display Research Conference (Asia Display/IDW'01)*, 485 (2001).
91 Ishinabe, T., Miyashita, T., and Uchida, T. (2002). Wide-viewing-angle polarizer with a large wavelength range. *Jpn. J. Appl. Phys., Part 1* 41: 4553.
92 Pasqual, F.D., Deng, H., Fernandez, F.A. et al. (1999). Theoretical and experimental study of nematic liquid crystal display cells using the in-plane-switching mode. *IEEE Trans. Electron Devices* 46: 661.
93 Ohmuro, K., Kataoka, S., Sasaki, T., and Koite, Y. (1997). *SID Tech. Dig.* 26: 845.
94 Lee, S.H., Lee, S.L., and Kim, H.Y. (1998). Electro-optic characteristics and switching principle of a nematic liquid crystal cell controlled by fringe-field switching. *Appl. Phys. Lett.* 73: 2881.

95 Meyer, R.B. (1969). Piezoelectric effects in liquid crystals. *Phys. Rev. Lett.* 22: 918.
96 Tan, G., Lee, Y.H., Gou, F. et al. (2017). Review on polymer-stabilized short-pitch cholesteric liquid crystal displays. *J. Phys. D: Appl. Phys.* 50, 493001.
97 Choi, T.H., Oh, S.W., Park, Y.J. et al. (2016). Fast fringe-field switching of a liquid crystal cell by two-dimensional confinement with virtual walls. *Sci. Rep.* 6, 27936.
98 Chen, H., Tan, G., Huang, Y. et al. (2017). A low voltage liquid crystal phase grating with switchable diffraction angles. *Sci. Rep.* 7, 39923.
99 Matsushima, T., Seki, K., Kimura, S. et al. (2018). New fast response in-plane switching liquid crystal mode. *J. Soc. Inf. Disp.* 26: 602.
100 Ge, Z., Wu, S.T., Kim, S.S. et al. (2008). Thin cell fringe-field-switching liquid crystal display with a chiral dopant. *Appl. Phys. Lett.* 92, 181109.
101 Ge, Z., Zhu, X., Wu, T.X., and Wu, S.T. (2006). High transmittance in-plane-switching liquid crystal displays. *J. Disp. Technol.* 2: 114.
102 Yun, H.J., Jo, M.H., Jang, I.W. et al. (2012). Achieving high light efficiency and fast response time in fringe field switching mode using a liquid crystal with negative dielectric anisotropy. *Liq. Cryst.* 39: 1141.
103 Chen, Y., Luo, Z., Peng, F., and Wu, S.T. (2013). Fringing-field switching mode with a negative dielectric anisotropy liquid crystal. *J. Disp. Technol.* 9: 74.
104 Chen, H., Peng, F., Hu, M., and Wu, S.T. (2015). Flexoelectric effect on image flickering of a liquid crystal display. *Liq. Cryst.* 42: 1730.
105 Chen, H., Peng, F., Luo, Z. et al. (2014). High performance liquid crystal displays with a low dielectric constant material. *Opt. Mater. Express* 4: 2262.
106 Xu, D., Peng, F., Tan, G. et al. (2015). A semi-empirical equation for the response time of in-plane switching liquid crystal and measurement of twist elastic constant. *J. Appl. Phys.* 117, 203103.
107 Kahn, F.J. (1972). Electric-field-induced orientational deformation of nematic liquid crystals. *Appl. Phys. Lett.* 20: 199.
108 Grinberg, J., Bleha, W.P., Jacobson, A.D. et al. (1975). Photoactivated birefringence liquid crystal light valve for color symbology display. *IEEE Trans. Electron Devices* ED-22: 775.
109 Sterling, R.D. and Bleha, W.P. (2000). D-ILA technology for electronic cinema. *SID Tech. Dig.* 31: 310.
110 Takeda, A., Kataoka, S., Sasaki, T. et al. (1998). A super-high-image-quality multi-domain vertical alignment LCD by new rubbing-less technology. *SID Tech. Dig.* 29: 1077.
111 Oh-e, M., Yoneya, M., and Kondo, K. (1997). Switching of negative and positive dielectric anisotropic liquid crystals by in-plane electric fields. *J. Appl. Phys.* 82: 528.
112 Li, J., Wen, C.h., Gauza, S. et al. (2005). Refractive indices of liquid crystals for display applications. *J. Disp. Technol.* 1: 51.
113 Erickson, J.L. (1961). Conservation laws for liquid crystals. *Trans. Soc. Rheol.* 5: 23.
114 Leslie, F.M. (1968). Some constitutive equations for liquid crystals. *Arch. Ration. Mech. Anal.* 28: 265.
115 Jakeman, E. and Raynes, E.P. (1972). Electro-optic response times in liquid crystals. *Phys. Lett. A* 39: 69.
116 Wang, H., Wu, T.X., Zhu, X., and Wu, S.T. (2004). Correlations between liquid crystal director reorientation and optical response time of a homeotropic cell. *J. Appl. Phys.* 95: 5502.
117 Nie, X., Xianyu, H., Lu, R. et al. (2007). Pretilt angle effects on liquid crystal response time. *J. Disp. Technol.* 3: 280.
118 Wu, S.T. and Wu, C.S. (1988). Small angle relaxation of highly deformed nematic liquid crystals. *Appl. Phys. Lett.* 53: 1794.
119 Wu, S.T. (1990). A nematic liquid crystal modulator with response time less than 100 μs at room temperature. *Appl. Phys. Lett.* 57: 986.
120 Ohmuro, K., Kataoka, S., Sasaki, T., and Koike, Y. (1997). Development of super-high-image-quality vertical alignment-mode LCD. *SID Tech. Dig.* 28: 845.
121 Kwag, J.O., Shin, K.C., Kim, J.S. et al. (2000). Implementation of new wide viewing angle mode for TFT-LCDs. *SID Tech. Dig.* 31: 256.

122 Kim, S.S. (2005). The world's largest (82-in) TFT LCD. *SID Tech. Dig.* 36: 1842.

123 Hanaoka, K., Nakanishi, Y., Inoue, Y. et al. (2004). A new MVA-LCD by polymer sustained alignment technology. *SID Tech. Dig.* 35: 1200.

124 Kim, S.G., Kim, S.M., Kim, Y.S. et al. (2007). Stabilization of the liquid crystal director in the patterned vertical alignment mode through formation of pretilt angle by reactive mesogen. *Appl. Phys. Lett.* 90, 261910.

125 Hong, H.K., Shin, H.H., and Chung, I.J. (2007). In-plane switching technology for liquid crystal display television. *J. Disp. Technol.* 3: 361.

126 Singh, R., Narayanan Unni, K.N., and Solanki, A. (2012). Improving the contrast ratio of OLED displays: an analysis of various techniques. *Opt. Mater.* 34: 716.

127 Tan, G., Zhu, R., Tsai, Y.S. et al. (2016). High ambient contrast ratio OLED and QLED without a circular polarizer. *J. Phys. D: Appl. Phys.* 49, 315101.

128 Chen, H., Tan, G., and Wu, S.T. (2017). Ambient contrast ratio of LCDs and OLED displays. *Opt. Express* 25: 33643.

129 Chen, H., Tan, G., Li, M.C. et al. (2017). Depolarization effect in liquid crystal displays. *Opt. Express* 25: 11315.

130 Raut, H.K., Ganesh, V.A., Nair, A.S., and Ramakrishna, S. (2011). Anti-reflective coatings: a critical, in-depth review. *Energy Environ. Sci.* 4: 3779.

131 Müllen, K. and Scherf, U. (2006). *Organic Light Emitting Devices: Synthesis, Properties and Applications*. Wiley.

132 Hong, Q., Wu, T.X., Lu, R., and Wu, S.T. (2005). A wide-view circular polarizer consisting of a linear polarizer and two biaxial films. *Opt. Express* 13: 10777.

133 Ishii, Y. (2007). The world of the TFT-LCD technology. *J. Disp. Technol.* 3: 351.

134 Schadt, M. (2009). Milestone in the history of field-effect liquid crystal displays and materials. *Jpn. J. Appl. Phys.* 48, 03B001.

135 Ukai, Y. (2013). TFT-LCDs as the future leading role in FPD. *SID Symp. Dig. Tech. Pap.* 44: 28.

136 Yoon, J.K., Park, E.M., Son, J.S. et al. (2013). The study of picture quality of OLED TV with WRGB OLEDs structure. *SID Symp. Dig. Tech. Pap.* 44: 326.

137 Shirasaki, Y., Supran, G.J., Bawendi, M.G., and Bulović, V. (2013). Emergence of colloidal quantum-dot light-emitting technologies. *Nat. Photonics* 7: 13.

138 Luo, Z., Xu, D., and Wu, S.-T. (2014). Emerging quantum-dots-enhanced LCDs. *J. Disp. Technol.* 10: 526.

139 Chen, H., Ha, T.H., Sung, J.H. et al. (2010). Evaluation of LCD local-dimming-backlight system. *J. Soc. Inf. Disp.* 18: 57.

140 Chen, H., Zhu, R., Li, M.C. et al. (2017). Pixel-by-pixel local dimming for high-dynamic-range liquid crystal displays. *Opt. Express* 25: 1973.

141 Chen, H., Peng, F., Gou, F. et al. (2016). Nematic LCD with motion picture response time comparable to organic LEDs. *Optica* 3: 1033.

142 Kikuchi, H., Yokota, M., Hisakado, Y. et al. (2002). Polymer-stabilized liquid crystal blue phases. *Nat. Mater.* 1: 64.

143 Huang, Y., Chen, H., Tan, G. et al. (2017). Optimized blue-phase liquid crystal for field-sequential-color displays. *Opt. Mater. Express* 7: 641.

144 Chen, H., Hu, M., Peng, F. et al. (2015). Ultra-low viscosity liquid crystal materials. *Opt. Mater. Express* 5: 655.

145 Peng, F., Huang, Y., Gou, F. et al. (2016). High performance liquid crystals for vehicle displays. *Opt. Mater. Express* 6: 717.

146 Peng, F., Gou, F., Chen, H. et al. (2016). A submillisecond-response liquid crystal for color sequential projection displays. *J. Soc. Inf. Disp.* 24: 241.

147 Srivastava, A.K., Chigrinov, V.G., and Kwok, H.S. (2015). Ferroelectric liquid crystals: excellent tool for modern displays and photonics. *J. Soc. Inf. Disp.* 23: 253.

148 Ito, H., Ogawa, M., and Sunaga, S. (2013). Evaluation of an organic light-emitting diode display for precise visual stimulation. *J. Vis.* 13: 1.

149 Kurita, T. (2001). *SID Symp. Dig. Tech. Pap.* 32: 986.

150 Sluyterman, A. (2006). What is needed in LCD panels to achieve CRT-like motion portrayal? *J. Soc. Inf. Disp.* 14: 681.

151 Emoto, M., Kusakabe, Y., and Sugawara, M. (2014). High-frame-rate motion picture quality and its independence of viewing distance. *J. Disp. Technol.* 10: 635.

152 Igarashi, Y., Yamamoto, T., Tanaka, Y. et al. (2003). Proposal of the perceptive parameter motion picture response time (MPRT). *SID Symp. Dig. Tech. Pap.* 34: 1039.

153 Vieri, C., Lee, G., Balram, N. et al. (2018). An 18 megapixel 4.3-in. 1,443-ppi 120-Hz OLED display for wide field-of-view high-acuity head-mounted displays. *J. SID* 26: 314.

154 Murawski, C., Leo, K., and Gather, M.C. (2013). Efficiency roll-off in organic light-emitting diodes. *Adv. Mater.* 25: 6801.

155 Féry, C., Racine, B., Vaufrey, D. et al. (2005). Physical mechanism responsible for the stretched exponential decay behavior of aging organic light-emitting diodes. *Appl. Phys. Lett.* 87, 213502.

156 Verzellesi, G., Saguatti, D., Meneghini, M. et al. (2013). Efficiency droop in InGaN/GaN blue light-emitting diodes: physical mechanisms and remedies. *J. Appl. Phys.* 114, 071101.

157 ITU-R Recommendation BT.709–5, "Parameter values for the HDTV standards for production and international programme exchange," (2002).

158 Masaoka, K., Nishida, Y., Sugawara, M., and Nakasu, E. (2010). Design of primaries for a wide-gamut television colorimetry. *IEEE Trans. Broadcast.* 56: 452.

159 Harbers, G. and Hoelen, C. (2001). High performance LCD backlighting using high intensity red, green and blue light emitting diodes. *SID Symp. Dig. Tech. Pap.* 32: 702.

160 Chiu, H.J. and Cheng, S.J. (2007). LED backlight driving system for large-scale LCD panels. *IEEE Trans. Ind. Electron.* 54: 2751.

161 Wu, C.Y., Wu, T.F., Tsai, J.R. et al. (2008). Multistring LED backlight driving system for LCD panels with color sequential display and area control. *IEEE Trans. Ind. Electron.* 55: 3791.

162 Schubert, E.F., Gessmann, T., and Kim, J.K. (2005). *Light Emitting Diodes*. Wiley.

163 Bulashevich, K.A., Kulik, A.V., and Karpov, S.Y. (2015). Optimal ways of colour mixing for high-quality white-light LED sources. *Phys. Status Solidi A* 212: 914.

164 Xie, R.J., Hirosaki, N., and Takeda, T. (2009). Wide color gamut backlight for liquid crystal displays using three-band phosphor-converted white light-emitting diodes. *Appl. Phys. Express* 2, 022401.

165 Oh, J.H., Kang, H., Ko, M., and Do, Y.R. (2015). Analysis of wide color gamut of green/red bilayered freestanding phosphor film-capped white LEDs for LCD backlight. *Opt. Express* 23: A791.

166 Wang, L., Wang, X., Kohsei, T. et al. (2015). Highly efficient narrow-band green and red phosphors enabling wider color-gamut LED backlight for more brilliant displays. *Opt. Express* 23: 28707.

167 Hirosaki, N., Xie, R.J., Kimoto, K. et al. (2009). Characterization and properties of green-emitting β-SiAlON:Eu^{2+} powder phosphors for white light-emitting diodes. *Appl. Phys. Lett.* 86, 211905.

168 Adachi, S. and Takahashi, T. (2008). Direct synthesis and properties of K_2SiF_6:Mn^{4+} phosphor by wet chemical etching of Si wafer. *J. Appl. Phys.* 104, 023512.

169 Murphy, J.E., Garcia-Santamaria, F., Setlur, A.A., and Sista, S. (2015). PFS, K_2SiF_6:Mn^{4+}: the red-line emitting LED phosphor behind GE's TriGain Technology™ platform. *SID Symp. Dig. Tech. Pap.* 46: 927.

170 Jang, E., Jun, S., Jang, H. et al. (2010). White-light-emitting diodes with quantum dot color converters for display backlights. *Adv. Mater.* 22: 3076.

171 Coe-Sullivan, S., Liu, W., Allen, P., and Steckel, J.S. (2013). Quantum dots for LED downconversion in display applications. *ECS J. Solid State Sci. Technol.* 2: R3026.

172 Bourzac, K. (2013). Quantum dots go on display. *Nature* 493: 283.

173 Steckel, J.S., Ho, J., Hamilton, C. et al. (2015). Quantum dots: the ultimate down-conversion material for LCD displays. *J. Soc. Inf. Disp.* 23: 294.

174 Goldstein, L., Glas, F., Marzin, J.Y. et al. (1985). Growth by molecular beam epitaxy and characterization of InAs/GaAs strained-layer superlattices. *Appl. Phys. Lett.* 47: 1099.

175 Murray, C., Norris, D.J., and Bawendi, M.G. (1993). Synthesis and characterization of nearly monodisperse CdE (E = sulfur, selenium, tellurium) semiconductor nanocrystallites. *J. Am. Chem. Soc.* 115: 8706.

176 Bruchez, M., Moronne, M., Gin, P. et al. (1998). Semiconductor nanocrystals as fluorescent biological labels. *Science* 281: 2013.

177 Brus, L. (1986). Electronic wave-functions in semiconductor clusters – experiment and theory. *J. Phys. Chem.* 90: 2555.

178 Steckel, J.S., Colby, R., Liu, W. et al. (2013). Quantum dot manufacturing requirements for the high volume LCD market. *SID Symp. Dig. Tech. Pap.* 44: 943.

179 Ithurria, S., Bousquet, G., and Dubertret, B. (2011). Continuous transition from 3D to 1D confinement observed during the formation of CdSe nanoplatelets. *J. Am. Chem. Soc.* 133: 3070.

180 Dabbousi, B.O., Rodriguez-Viejo, J., Mikulec, F.V. et al. (1997). (CdSe) ZnS core-shell quantum dots: synthesis and characterization of a size series of highly luminescent nanocrystallites. *J. Phys. Chem. B* 101: 9463.

181 Anc, M.J., Pickett, N.L., Gresty, N.C. et al. (2013). Progress in non-Cd quantum dot development for lighting applications. *J. Solid State Sci. Technol.* 2: R3071.

182 Yang, X.Y., Zhao, D.W., Leck, K.S. et al. (2012). Full visible range covering InP/ZnS nanocrystals with high photometric performance and their application to white quantum dot light-emitting diodes. *Adv. Mater.* 24: 4180.

183 Kim, H., Han, J.Y., Kang, D.S. et al. (2011). Characteristics of CuInS2/ZnS quantum dots and its application on LED. *J. Cryst. Growth* 326: 90.

184 Hines, M.A. and Guyot-Sionnest, P. (1996). Synthesis and characterization of strongly luminescing ZnS-capped CdSe nanocrystals. *J. Phys. Chem.* 100: 468.

185 Zhu, R., Luo, Z., Chen, H. et al. (2015). Realizing Rec. 2020 color gamut with quantum dot displays. *Opt. Express* 23: 23680.

186 Pickett, N.L., Gresty, N.C., and Hines, M.A. (2016). Heavy metal-free quantum dots making inroads for consumer applications. *SID Symp. Dig. Tech. Pap.* 47: 425.

187 Yang, X., Divayana, Y., Zhao, D. et al. (2012). A bright cadmium-free, hybrid organic/quantum dot white light-emitting diode. *Appl. Phys. Lett.* 101, 233110.

188 Lim, J., Park, M., Bae, W.K. et al. (2013). Highly efficient cadmium-free quantum dot light-emitting diodes enabled by the direct formation of excitons within InP@ZnSeS quantum dots. *ACS Nano* 7: 9019.

189 Kim, Y., Greco, T., Ippen, C. et al. (2013). Indium phosphide-based colloidal quantum dot light emitting diodes on flexible substrate. *Nanosci. Nanotechnol. Lett.* 5: 1065.

190 Lee, S.H., Lee, K.H., Jo, J.H. et al. (2014). Remote-type, high-color gamut white light-emitting diode based on InP quantum dot color converters. *Opt. Mater. Express* 4: 1297.

191 Yang, S.J., Oh, J.H., Kim, S. et al. (2015). Realization of InP/ZnS quantum dots for green, amber and red down-converted LEDs and their color-tunable, four-package white LEDs. *J. Mater. Chem. C* 3: 3582.

192 Ippen, C., Greco, T., Kim, Y. et al. (2015). Color tuning of indium phosphide quantum dots for cadmium-free quantum dot light-emitting devices with high efficiency and color saturation. *J. Soc. Inf. Disp.* 23: 285.

193 Pickett, N.L., Harris, J.A., and Gresty, N.C. (2015). Heavy metal-free quantum dots for display applications. *SID Symp. Dig. Tech. Pap.* 46: 168.

194 Lee, E., Wang, C.K., Hotz, C. et al. (2016). Greener quantum-dot enabled LCDs with BT. 2020 color gamut. *SID Symp. Dig. Tech. Pap.* 47: 549.

195 Chen, H., He, J., and Wu, S.T. (2017). Recent advances in quantum-dot-enhanced liquid crystal displays. *IEEE J. Sel. Top. Quantum Electron.* 23, 1900611.

196 Zhao, Y., Riemersma, C., Pietra, F. et al. (2015). High-temperature luminescence quenching of colloidal quantum dots. *ACS Nano* 6: 9058.

197 Kurtin, J., Puetz, N., Theobald, B. et al. (2014). Quantum dots for high color gamut LCD displays using an on-chip LED solution. *SID Symp. Dig. Tech. Pap.* 45: 146.

198 Chen, W., Hao, J., Qin, J. et al. (2016). Luminescent nanocrystals and composites for high quality displays and lighting. *SID Symp. Dig. Tech. Pap.* 47: 556.

199 Twietmeyer, K. and Sadasivan, S. (2016). Design considerations for highly efficient edge-lit quantum dot displays. *J. Soc. Inf. Disp.* 24: 312.

200 Chen, J., Hardev, V., Hartlove, J. et al. (2012). A high-efficiency wide-color-gamut solid-state backlight system for LCDs using quantum dot enhancement film. *SID Symp. Dig. Tech. Pap.* 43: 895.

201 Thielen, J., Lamb, D., Lemon, A. et al. (2016). Correlation of accelerated aging to in-device lifetime of quantum dot enhancement film. *SID Symp. Dig. Tech. Pap.* 47: 336.

202 Seetzen, H., Heidrich, W., Stuerzlinger, W. et al. (2004). High dynamic range display systems. *ACM Trans. Graphics* 23: 760.

203 Oh, C.H., Shin, H.J., Nam, W.J. et al. (2013). Technological progress and commercialization of OLED TV. *SID Symp. Dig. Tech. Pap.* 44: 239.

204 Lin, F.C., Huang, Y.P., Liao, L.Y. et al. (2008). Dynamic backlight gamma on high dynamic range LCD TVs. *J. Disp. Technol.* 4: 139.

205 Seetzen, H., Whitehead, L.A., and Ward, G. (2003). A high dynamic range display using low and high resolution modulators. *SID Symp. Dig. Tech. Pap.* 34: 1450.

206 Shu, X., Wu, W., and Forchhammer, S. (2013). Optimal local dimming for LC image formation with controllable backlighting. *IEEE Trans. Image Process.* 22: 166.

207 Hoffman, D.M., Stepien, N.N., and Xiong, W. (2016). The importance of native panel contrast and local dimming density on perceived image quality of high dynamic range displays. *J. Soc. Inf. Disp.* 24: 216.

208 Kwon, J.U., Bang, S., Kang, D., and Yoo, J.J. (2016). The required attribute of displays for high dynamic range. *SID Symp. Dig. Tech. Pap.* 47: 884.

209 Zhu, R., Chen, H., and Wu, S.T. (2017). Achieving 12-bit perceptual quantizer curve with liquid crystal display. *Opt. Express* 25: 10939.

210 Kim, S., An, J.-Y., Hong, J.-J. et al. (2009). How to reduce light leakage and clipping in local-dimming liquid-crystal displays. *J. Opt. Soc. Am. A* 17: 1051.

211 Shirai, T., Shimizukawa, S., Shiga, T. et al. (2006). RGB-LED backlights for LCD-TVs with 0D, 1D, and 2D adaptive dimming. *SID Intl. Symp. Dig. Tech. Pap.* 37: 1520.

212 Hulze, H.G. and de Greef, P. (2009). Power savings by local dimming on a LCD panel with side lit backlight. *SID Intl. Symp. Dig. Tech. Pap.* 40: 749.

213 Tan, G., Huang, Y., Chen, M.C. et al. (2018). High dynamic range liquid crystal displays with a mini-LED backlight. *Opt. Express* 26: 16572.

214 Zheng, B., Deng, Z., Zheng, J. et al. (2019). An advanced high-dynamic-range LCD for smartphones. *SID Intl. Symp. Dig. Tech. Pap.* 50: 562.

215 Wu, Y.-E., Lee, M.H., Lin, Y.-C. et al. (2019). Active matrix mini-LED backlights for 1000PPI VR LCD. *SID Intl. Symp. Dig. Tech. Pap.* 50: 566.

216 Masuda, T., Watanabe, H., Kyoukane, Y. et al. (2019). Mini-LED backlight for HDR compatible mobile displays. *SID Intl. Symp. Dig. Tech. Pap.* 50: 390.

217 Pang, H., Michalski, L., Weaver, M.S. et al. (2014). Thermal behavior and indirect life test of large-area OLED lighting panels. *J. Solid State Light.* 1: 7.

218 Chen, H., Lee, J.H., Lin, B.Y. et al. (2018). Liquid crystal display and organic light-emitting diode display: present status and future perspectives. *Light Sci. Appl.* 7: 17168.

219 Watanabe, K., Iwaki, Y., Uchida, Y. et al. (2016). A foldable OLED display with an in-cell touch sensor having embedded metal-mesh electrodes. *J. Soc. Inf. Disp.* 24: 12.

220 Goto, S., Miyatake, M., and Saiki, Y. (2016). A novel ultra-thin polarizer to achieve thinner and more-flexible displays. *SID Symp. Dig. Tech. Pap.* 47: 510.

221 Chiu, P.-H., Li, W.-Y., Chen, Z.-H. et al. (2016). Roll TFT-LCD with 20R curvature using optically compensated colorless-polyimide substrate. *SID Symp. Dig. Tech. Pap.* 47: 15.

222 Harding, M.J., Horne, I.P., and Yaglioglu, B. (2017). Flexible LCDs enabled by OTFT. *SID Symp. Dig. Tech. Pap.* 48: 793.

223 Okuyama, K., Nakahara, T., Numata, Y. et al. (2017). Highly transparent LCD using new scattering-type liquid crystal with field sequential color edge light. *SID Symp. Dig. Tech. Pap.* 48: 1166.

5

Light-Emitting Diodes

5.1 INTRODUCTION

The operating principle of light-emitting diodes (LEDs) is that electrons and holes are injected into a single-crystal semiconductor, where these charge carriers recombine and generate photons. The emission wavelength of an LED is determined by the material used, the design of the active-layer, and the device structure, and can cover the whole visible range. Although electroluminescence (EL) in a semiconductor (SiC) was first observed in the early 1900s [1], the commercial realization of visible LEDs did not begin until around 60 years later. In the 1960s, visible-LEDs which emitted red light were manufactured using a GaAsP active layer on a GaAs substrate [2]. One of the major problems of this system comes from a lattice mismatch between the GaAs (lattice constant = 5.65 Å) and GaAsP (lattice constant = 5.45 Å) which results in misfit dislocations and in turn increases the non-radiative recombination rate. The emission wavelength of GaAsP LEDs can be blue-shifted by increasing the percentage of phosphorus. However, a transition from a direct to indirect bandgap (which will be discussed in Section 5.2) occurs when the P mole fraction reaches 50%, which decreases the efficiency. By incorporating selected isoelectronic impurities (N and Zn: O) into indirect bandgap materials (such as GaP or GaAsP), trap levels are formed inside the forbidden bandgap, and the resulting LEDs exhibit shorter emission wavelengths (orange, yellow, or even green). LEDs using these principles became available around the 1970s [3]. Another important material system is AlGaAs on a GaAs substrate (1980), which exhibits the advantage of a low lattice mismatch since the lattice constants are nearly the same for AlAs (5.66 Å) and GaAs (5.65 Å) [4]. Hence, a high quality crystalline epitaxial film can be obtained, thereby lowering the non-radiative recombination rate. In addition, this near-perfect lattice-matching means that the layer structure can easily be engineered into structures such as heterojunctions and quantum wells to improve carrier confinement and fine-tune the emission wavelength. With an increasing concentration of aluminum, the emission wavelength blueshifts. However, a direct–indirect transition occurs at 621 nm when the aluminum content is 45%, which limits the operation of AlGaAs/GaAs LEDs to the red spectral region. Then, in 1990, the quaternary alloy, AlGaInP, was introduced, bringing the advantages of: (i) a lattice which can be matched to GaAs by tuning the Al/Ga ratio, and (ii) a shorter direct–indirect transition wavelength (555 nm; yellowish green) than GaAsP and AlGaAs systems. The AlGaInP system is currently the standard material used in "long wavelength" (red, orange, and yellow) visible LEDs [5].

GaN, InGaN, and AlGaN (Group III-nitrides) are material systems, which cover the short wavelength range of the visible spectrum from the UV and blue, to green. Three major problems of these material systems come from the lack of any lattice-matched substrate, the difficulty in achieving p-type doping, and the difficulty in incorporating indium. By pre-growth of a buffer layer (e.g. AlN or GaN) at low temperature, it is possible to form an epitaxial nitride film on the lattice-mismatched substrate, sapphire (Al_2O_3) [6]. Acceptors (which should provide p-type doping) in nitride materials are easily passivated by hydrogen atoms introduced during epitaxial film growth. Hence, a high temperature annealing step under nitrogen ambient is needed to activate

Introduction to Flat Panel Display, Second Edition. Jiun-Haw Lee, I-Chun Cheng, Hong Hua, and Shin-Tson Wu.

the acceptors and achieve an "effective" p-type nitride material [7]. Because GaN emits in the UV, high quality InGaN is needed to give blue emission. However, due to the high vapor pressure of indium at the growth temperature of GaN (~1000 °C), incorporation of In at the required concentration is difficult, a difficulty which was solved by re-design of the equipment used for epitaxial growth [8]. In recognition of their realization of blue GaN-LEDs, the 2014 Nobel Prize in Physics was awarded to the three major inventors, Professor Isamu Akasaki, Professor Hiroshi Amano, and Professor Shuji Nakamura. In order to generate white light, two- or three-color mixing is needed. Typical approaches include: (i) the all semiconductor method using different color LEDs, and (ii) down-conversion materials (such as phosphors) optically pumped by a short wavelength (UV or blue) LED. Figure 5.1 shows the time evolution of LED performance. We can see that the power efficiencies of LEDs have made huge progress for the last several decades and are already higher than those of incandescent and fluorescent lamps [9].

To achieve a high performance LED [10], it is necessary to: (i) inject and transport the carriers into the active layer for recombination without injection barriers and with little ohmic loss, (ii) have a much higher radiative than non-radiative recombination rate, and (iii) couple the light out of the device efficiently. High resistance of the semiconductor results in higher operating voltage and lower efficiency. An ohmic contact, rather than a Schottky barrier, is needed for efficient carrier injection. Hence, a p–n diode structure with low resistance is preferred for LED application. Since the radiative recombination process in LEDs is a Langevin-type process, the recombination rate distribution is proportional to the product of the electron and hole volume concentrations. Hence, it is advantageous to confine the carriers in the active region by using heterojunction, electron blocking layer (EBL), quantum well, wire, and dot structures.

Once electron–hole pairs recombine, they may relax their energy radiatively by generation of photons or non-radiatively by heat generation. Single crystalline semiconductor materials are needed for LED applications since defects in a semiconductor typically act as non-radiative centers and cause a decrease

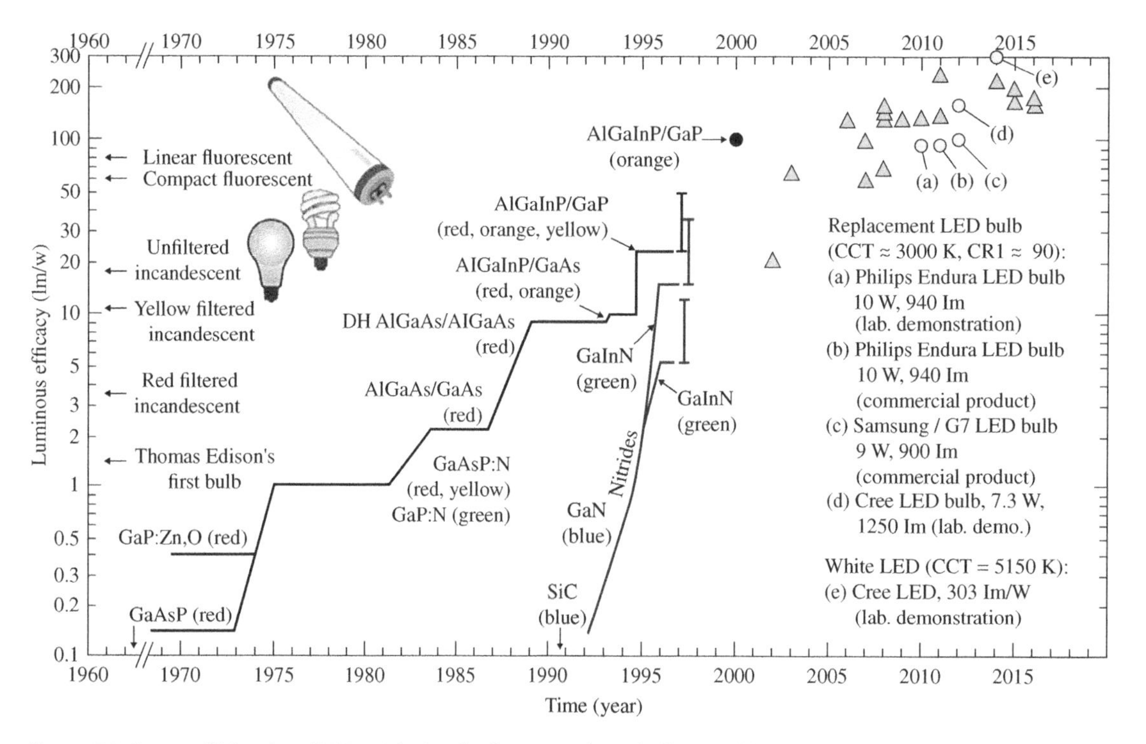

Figure 5.1 Power efficiencies of LEDs with time for "long wavelength (GaAsP, AlGaAs, and AlInGaP)," and "short wavelength (nitride)" materials, and for white lighting systems.

in light emission. The bandgap of the semiconductor material within the active layer determines the emission wavelength from LEDs. The active layer structure also plays an important role in determining emission spectra. Quantum-well, wire, and dot structures shift the emission wavelength from the bandgap to the quantized states which allows some flexibility in tuning the emission wavelength. Also, the density of states (DOSs) of low-dimensional structures (quantum-well, wire, and dot structures) are different from that of the bulk material, which means that the full-width at half-maximum (FWHM) of the emission peak and the temperature dependence of the emission wavelength are altered.

Once photons are generated in the active layer of an LED, the next objective is to radiate the photons out of the semiconductor material efficiently. According to the usual properties of spontaneous emission, the photon emission is isotropic. Photons may then experience absorption (by defect states), reabsorption (band-to-band or free carrier) and scattering (by defects or interfaces) when propagating in the semiconductor materials. If the substrate of the LED has a smaller bandgap (e.g. GaAs with $E_g = 1.424$ eV, corresponding to 870 nm in the near infrared region) than the emitted light, absorption by the substrate (theoretically 50%) needs to be taken into consideration. When the photons propagate out of the semiconductor into the epoxy layer which is used for packaging the LEDs, total internal reflection occurs due to the refraction index difference between the semiconductor ($n = 3.5$) and the epoxy ($n = 1.5$). From a simple calculation using Snell's law, the critical angle is about 25° and light with an angle of incidence from the semiconductor greater than this cannot pass into the epoxy layer. Those photons will be reflected back into the semiconductor and may be absorbed by defects or the substrate, or propagate inside the semiconductor as the waveguided mode. To improve the emission efficiency, many LED structures have been proposed: (i) surface texturing and a reflector have been used to disrupt the waveguide effect and reflect the light out; (ii) a transparent substrate (e.g. GaP) can be employed to decrease the substrate absorption, and (iii) the reflective metal electrode can be replaced with a transparent electrode such as ITO.

Crystalline semiconductor layers with p–n structures and an active region, several μm in thickness are grown on single-crystal substrates, using a so-called epitaxial process [11]. Liquid phase epitaxy (LPE) and vapor phase epitaxy (VPE) are common techniques. A VPE system with organometallic material sources, called metallorganic chemical vapor deposition (MOCVD), is one of the most popular methods for epitaxial growth. Many important material characteristics are determined during the epitaxial growth process, such as doping concentrations, doping profile, defect density, and alloy compositions. Other requirements for LEDs with high power efficiency rely on suitable device design and fabrication techniques. Contact resistance between the electrode and the semiconductor should be as low as possible, and can be minimized by thermal annealing. In addition, the layout of the electrical contacts should be optimized to achieve a uniform current density combined with a high outcoupling efficiency. An etch process is needed to confine the current flow and increase the extraction efficiency. However, the periodic lattice is damaged at the etched face which increases the non-radiative recombination rate. Suitable thermal annealing can effectively passivate the dangling bonds and restore device efficiency. After the LED device is fabricated, suitable packaging is essential in real applications to: (i) protect the LED chips mechanically and environmentally, (ii) improve the extraction efficiency, and (iii) provide a thermal dissipation path for high power applications.

LEDs have many applications in the display industry. For applications in traffic signals, LEDs have higher power efficiency and hence lower power consumption than the conventional filtered incandescent lamps. Longer lifetime is another important advantage which reduces the maintenance cost. Electronic signage and very large area displays (over several hundred inches diagonal) provide another specialist market for LED displays. The conventional backlight used for LCDs was previously a cold cathode fluorescent lamp (CCFL), which contains Hg and can therefore be harmful to the environment. The position of the peaks in the CCFL spectrum, displaced from the ideal wavelengths of the primaries, limits the color gamut. Otherwise, if the color filter were re-engineered to have narrow transmission peaks at the best primary wavelengths, the continuum spectrum of the lamp would lower the power efficiency. LEDs replace the CCFL as the backlight sources for LCDs, and bring environmentally friendly characteristics, wide color gamut, long lifetime, and fast response. As well as backlight units for LCDs, LEDs can be also used for general lighting due to their high power efficiency and long lifetime. LEDs are also rugged due to their all solid-state nature. The operating voltage is quite low (less than 5 V) compared to other lighting technologies, which means it is easy to operate and safer. By reducing the size of LED chips to the subpixel size, it is possible to construct complete displays from millions of miniaturized LEDs, a technology referred to as "micro-LED." Due to their high efficiency, high brightness and fast response time, it is possible for these devices to satisfy the requirements for applications from small panels (such as projectors and watches) to larger displays (such as TVs). By transferring micro-LEDs from the semiconductor to another substrate, flexible, wearable, and even stretchable displays can be achieved using micro-LED technology.

In this chapter, we will first describe the material systems used for visible LEDs. Then the electrical (diode) and optical (light-emitting) characteristics are introduced, followed by an account of device fabrication. Finally, LED applications in displays are discussed.

5.2 MATERIAL SYSTEMS

The materials systems used in LEDs determine their emission wavelengths and radiative recombination efficiencies. To emit photons with the desired wavelength, an epitaxial active layer (which may be a binary, ternary, or quaternary compound-semiconductor) with a certain bandgap should be selected (a direct bandgap is much preferred). Then, a suitable substrate is needed, which has to be lattice-matched to the epitaxial layer, to reduce the non-radiative recombination rate.

Elemental semiconductors of group IV such as Ge (germanium), Si (silicon), and C (diamond) are indirect bandgap materials which are not suitable for LED applications. Figure 5.2 shows the relations of the bandgap and emission wavelength versus the lattice constant for the III–V compound semiconductors. The points are binary materials and the lines connecting the points are ternary materials. Alloying two ternaries which differ in one atom type (e.g. GaInP and AlInP) creates a quaternary material (e.g. AlGaInP). We can see a general trend in Figure 5.2 that as the lattice constant increases, the bandgap decreases and the emission wavelength increases. Since the binding energy (which determines the bandgap) is higher for smaller atomic numbers, compound semiconductors composed of lighter atoms exhibit a larger bandgap. So, for example, the bandgap of GaAs < GaP < GaN since the order in atomic size is As > P > N. The same principle is generally true when replacing the group III material. For example, the bandgap sequence InN < GaN < AlN follows a similar trend since the atom size is Al < Ga < In. The As- and P-based compound semiconductors (e.g. GaP, GaAsP, AlGaAs, GaAs, InGaAsP) can be operated in the red, orange, yellow, and even green region. To push the emission toward shorter wavelengths, nitrides which contain the smallest group V atom are used.

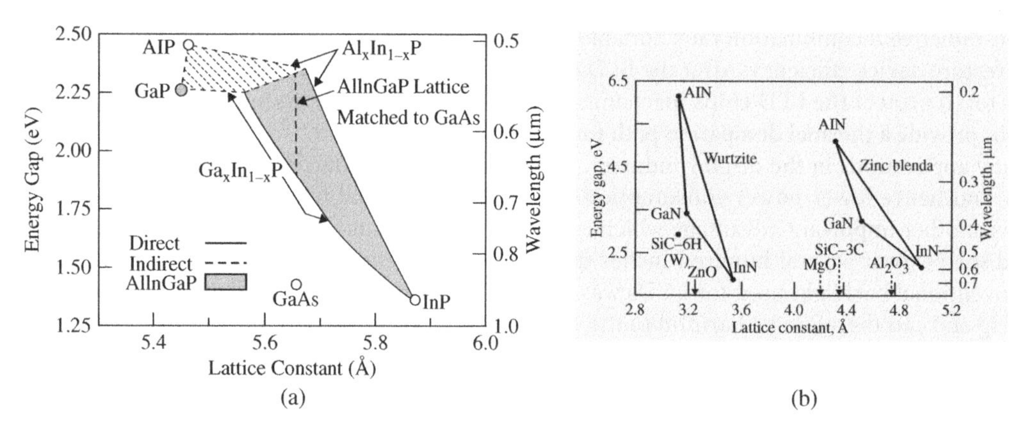

Figure 5.2 Bandgap and emission wavelength versus lattice constant for the III–V semiconductors of (a) AlInGa-P and (b) AlInGa-N material systems [12, 13].

The lattice constants of the substrate and epitaxial layers should be kept as equal as possible. "Lattice mismatch," defined as the difference in these lattice constants normalized relative to that of the substrate, $\Delta a/a_0$, should typically be as low as $\leq 0.1\%$ to allow high-quality epitaxial layer growth. When the lattice constants of the epitaxial layer and the substrate are not matched, residual strain results in deformation of the epitaxial layer, since it is much thinner than the substrate. If the accumulated strain is too high, dislocations are formed which increase the non-radiative recombination rate. Hence, the lattice-matching condition should be addressed carefully when selecting the material system. Common substrate materials are GaAs, InP, and GaP. From Figure 5.2, one can see that GaP and InP exhibit extremely high and low bandgap values, respectively. For these two compounds, no suitable ternary and quaternary P-based materials can be found which are lattice-matched to the substrate. GaAs is a suitable substrate choice since it is lattice matched to AlGaAs and $(AlGa)_{0.5}In_{0.5}P$. However, one has to note that the bandgap of GaAs is 1.424 eV, corresponding to 870 nm in the infrared region, which means it absorbs visible light and hence reduces the device efficiency. On the other hand, for the nitride-based materials, no suitable III–V semiconductor materials can be found as the substrate, a fact which delayed the development of short wavelength LEDs until the 1990s. However, by pre-growth of low-temperature GaN with a wurtzite structure on the lattice mismatched substrate Al_2O_3, high efficiency green and blue LEDs can be fabricated, as will be discussed in the following sections.

As discussed in Chapter 3, semiconductors can be divided into direct and indirect bandgap materials. In Figure 5.2 compositions providing direct and indirect bandgaps are designated by solid and dashed lines, respectively. Figure 5.3 shows schematic band structures of direct and indirect bandgap semiconductor materials. In a direct bandgap material the energy minimum of the conduction band coincides with the maximum of the valence band, in momentum space. Since the momentum of a photon is so small compared to those of electrons and holes, the optical transition is typically vertical in the band diagram as shown in Figure 5.3. In the case of an indirect bandgap condition, as shown in Figure 5.3b, electrons accumulate at the lowest energy valley of the conduction band, which does not coincide with the maximum of the valence band. Hence, momentum cannot be conserved during direct emission of a photon. "Phonons" – lattice vibrations with low energy but large momentum–must then participate in the light emission process. Typically, direct bandgap materials exhibit much higher radiative recombination rates than indirect ones. As the bandgap of the As- and P-materials is increased, such as by increasing the Al ratio of AlGaAs and the $(InGa)_{0.5}Al_{0.5}P$, a direct–indirect bandgap transition occurs, which results in the efficiency decreasing and limits EL emission at longer wavelengths. Note that the transition between direct and indirect bandgap does not bring about an abrupt change in efficiency. In general, electrons tend to reside near the lowest energy point in momentum space, as determined by the Boltzmann distribution. In semiconductor compositions which correspond to a direct bandgap, but close to the direct–indirect transition, some electrons can occupy states corresponding to an indirect minimum (see Figure 5.3c) when the energy difference between the direct and indirect minima is small enough.

Hence, as shown in Figure 5.2a, $(AlGa)_{0.5}In_{0.5}P$ systems are latticed matched to GaAs and are direct bandgap semiconductors at compositions emitting from the red to amber spectral regions, so are suitable for long wavelength visible LEDs. The absorbing GaAs substrate can be removed and replaced by a transparent GaP substrate after the epitaxial growth to further improve the external quantum efficiency (EQE). On the other hand, moving from short to longer wavelengths, InGaN with increasing In concentrations shows a high efficiency emission from the blue toward the green regions of the spectrum. Hence, the blue-green III–N and amber–red III–P LEDs can cover the whole visible spectrum.

Since the III–N and III–P materials are typically grown on different substrates (i.e. sapphire and GaAs, respectively), it is difficult to emit broadband white light from a single LED chip. There are several ways to obtain white light emission, such as (i) by assembling multiple LEDs in a module, or (ii) using phosphors which absorb LED blue light and emit yellow light [14, 15]. By using multiple LEDs, the electrical and optical characteristics (such as efficiency, driving voltage, and emission wavelength) of each device can be optimized individually. On the other hand, the cost is high and the module is bulky since more than one LED is used. Besides, after long term operation, there may be color shift problems due to the different lifetime performance of different LEDs. For the phosphor-coated white LED, the power efficiency is theoretically lower than for the

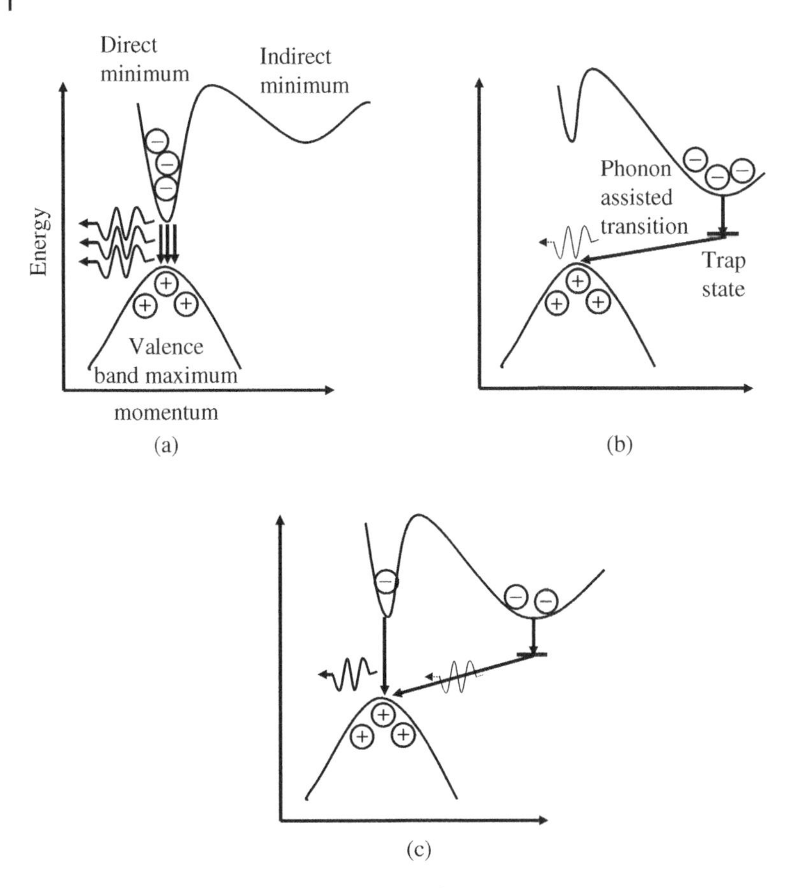

Figure 5.3 Energy versus momentum curves for semiconductors with (a) direct, (b) indirect, and (c) near direct-indirect bandgaps.

multi-LED technique since the emission wavelength of the phosphor is longer than the absorption wavelength. Advantages of this technique include low cost, compact form, and stable CIE coordinates. As discussed in Chapter 2, a "white" source color requires specific CIE coordinates located near the center of the chromaticity diagram. However, for the light source of a reflective display, the color of the LEDs should not only be white, but a high color rendering index (CRI) value is also important to reproduce colors as they would appear under the illumination of a blackbody radiator.

5.2.1 AlGaAs and AlGaInP Material Systems for Red and Yellow LEDs

As shown in Figure 5.2, GaAs is almost lattice matched to AlAs, which means that the ternary system AlGaAs can be epitaxially grown on a GaAs substrate with low defect density. It also provides great flexibility in designing epi-layer structures, such as heterojunction and quantum-well structures (see Sections 5.3.4 and 5.3.5), in order to obtain higher efficiency and control the emission wavelength [16]. The energy gap of $Al_xGa_{1-x}As$ can be approximated by the equations [11]:

$$E_g(\text{eV}) = 1.424 + 1.247x;\ x < 0.45\ \text{(direct bandgap)} \tag{5.1}$$

$$E_g\ (\text{eV}) = 1.9 + 0.125x + 0.143x^2; 0.45 < x < 1\ \text{(indirect bandgap)} \tag{5.2}$$

From the above equations, one can deduce that the cross-over point of the direct–indirect bandgap occurs at 1.985 eV, which corresponds to red emission at 624 nm. It limits AlGaAs devices to operating in the red to infrared emission region although the internal quantum efficiency (IQE) of the AlGaAs/GaAs system can be as high as 99%. To obtain shorter emission wavelengths, III–P material systems are used to replace the III–As ones. To fit the lattice constant of the conventional GaAs substrate, a quaternary ($(Al_xGa_{1-x})_{0.5}In_{0.5}P$) rather than ternary material is needed. The energy bandgap can be tuned from red to amber efficiently by varying the x value, which is indicated by the vertical line labeled "AlGaInP/GaAs" in Figure 5.2. A more precise expression for the lattice match condition for $(Al_xGa_{1-x})_yIn_{1-y}P$ on GaAs can be expressed as:

$$y = 0.616/(1 - 0.027x) \tag{5.3}$$

The y-value ranges from 0.516 to 0.525 and the approximate value of 0.5 is typically used. The bandgap of the AlGaInP can then be represented by [17]:

$$E_g\,(\text{eV}) = 1.900 + 0.61x;\; x < 0.58\,(\text{direct bandgap}) \tag{5.4}$$

$$E_g\,(\text{eV}) = 2.204 + 0.085x;\; 0.58 < x < 1\,(\text{indirect bandgap}) \tag{5.5}$$

As we can see from Figure 5.2, the wide bandgap materials GaP, AlP and their ternaries are indirect semiconductors. Hence, there is a crossover point from direct to indirect when x increases to 0.65, which corresponds to a bandgap of 2.3 eV or 540 nm (green emission). However, an efficiency drop is observed when the wavelength is shorter than 590 nm due to accumulation of electrons in an indirect minimum. So, AlGaInP material systems can provide efficient emission in a range from red to amber. The refractive indexes of AlGaAs and AlGaInP systems are around 3.2–3.6, depending on the composition and wavelength. Typically, a smaller bandgap corresponds to a larger refractive index, as can be derived from the Kramers–Kronig relation [18].

5.2.2 GaN-Based Systems for Green, Blue, UV and UV LEDs

Typically, single crystal semiconductor substrates are sliced from ingots which are obtained by a melting and re-crystallization process. For III–V compounds, group V materials exhibit much higher vapor pressure than those of group III at the same temperature, which increases the difficulty of fabricating substrates of III–V compound semiconductors compared with group IV Si-ingot fabrication. Besides, unlike As and P which exist as solids at room temperature and 1 atm, N_2 is a gas which means that the vapor pressure of nitrides is much higher than compounds of As and P and makes it very difficult to obtain a nitride-based substrate (e.g. ~16 000 atm and ~1500 °C to form single-crystal GaN bulk) [19].

Due to the lack of any conventional semiconductor substrate which is lattice matched to the III-nitrides, insulating sapphire (Al_2O_3) is commonly used as a substrate even though the lattice mismatch is as high as 16% to GaN. Direct epitaxy of nitrides on the sapphire substrate results in island formation rather than planar growth, combined with a high dislocation density, as shown in Figure 5.4. By introducing an AlN or GaN buffer layer grown at low temperature (LT), a usable nitride epi-layer can be obtained since it: (i) provides nucleation sites for epitaxial growth at the same crystal orientation which reduces the dislocation defect densities, and (ii) decreases the interfacial free energy between the substrate and the epitaxial layer which makes 2D growth preferable [6]. However, the dislocation density is still much higher than that in the III–P system.

The bandgap of a $In_xGa_{(1-x)}N$ ternary alloy is roughly given by

$$E_g\,(\text{eV}) = x\,E_g, \text{InN} + (1 - x)\,E_g, \text{GaN} - (1 - x)xb \tag{5.6}$$

where E_g,InN and E_g,GaN are the bandgaps of InN and GaN, respectively. Reference values of these bandgaps are 1.95 and 3.40 eV for InN and GaN, depending on the strain. The bowing factor, b, has a typical value of 1.00 eV [20]. The reader should note that III-nitride semiconductors exhibit direct bandgaps for the whole ranges of composition. By tuning the x value from 0.15 to 0.45, LEDs emitting from the UV to the green can be fabricated. In nitride LEDs, indium atoms tend to form clusters in the epitaxial layer which results in some inevitable composition fluctuation [21]. Its consequence can be observed in spectral broadening of the photo- or electro-luminescence which is called "alloy broadening." However, these In-clusters also effectively confine

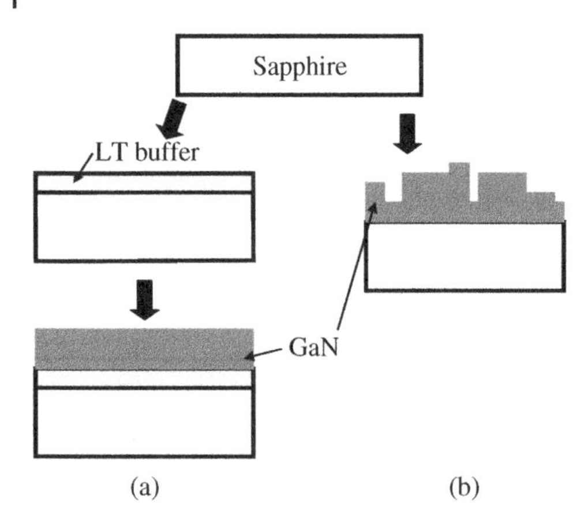

Figure 5.4 Growth of III-nitrides (a) with and (b) without LT buffer layers.

the carriers which then cannot diffuse easily, to be trapped by non-radiative recombination centers. Hence, although the dislocation density of InGaN is as high as $10^9\,cm^{-2}$, compared to $<10^3\,cm^{-2}$ in the III–P system, acceptable efficiency can be achieved [22]. At high indium concentrations, the metallic In-clusters act as quenchers which results in low internal quantum efficiencies of nitride-LEDs for long wavelengths (green, yellow, and red). Electric polarization effects occur in GaN-based LEDs, which can be categorized as spontaneous polarization and strain-induced piezoelectric polarization, and affect their electrical and optical characteristics. Although the crystal structure of GaN can be wurtzite or zinc blende, as shown in Figure 5.2b, high efficiency LEDs usually use the wurtzite structure, as shown in Figure 5.5a [23]. In this structure, due to the polarity of the bond between Ga and N, a horizontal electric polarization is induced, which is called a spontaneous polarization. Besides, a lattice mismatch occurs not only between the substrate and the epitaxial layer, but also between different epitaxial layers. This results in strain between different layers (e.g. quantum well and barrier, see Section 5.3.5) and hence a built-in potential inside the GaN LED along the crystal growth direction,

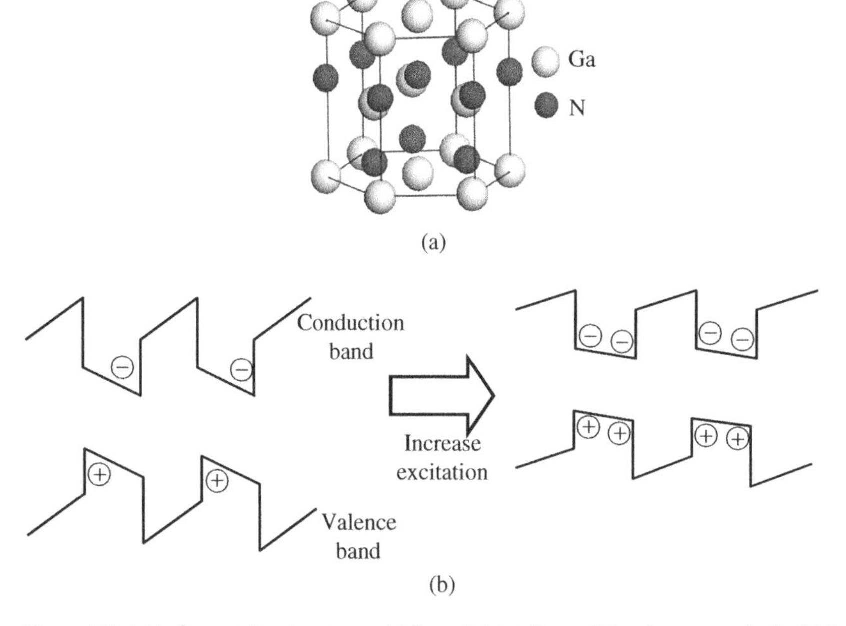

Figure 5.5 (a) Left: wurtzite structure, middle and right: Ga- and N-polar, respectively. (b) Band structure shift of III-nitride under low and high excitation.

which is called a piezoelectric polarization. The combination of these two polarization effects (spontaneous and piezoelectric) in nitride LEDs results in a spatial misalignment of the electron- and hole-wavefunctions, which reduces the efficiency, as shown in the left figure of Figure 5.5b [24]. Besides, because electrons and holes accumulate at the lowest and highest energy level of the conduction and valence band, respectively, the emission wavelength is longer than the bandgap. As the pumping intensity (optical or electrical) is increased, as shown in the right figure of Figure 5.5b, the emission spectrum of the LED blue-shifts due to screening of the built-in electric field by excess carriers. As in the case of III–P materials, it is also possible to use quaternary AlInGaN to control the strain and fine tune the emission characteristics [25].

Mg is typically used for p-type doping of nitride materials. Since NH_3 is used as the nitrogen source, Mg–H is formed during epitaxial growth. This impurity has a high electronic conductivity and results in low hole concentrations. By using high temperature annealing ($> 700\,°C$) for de-hydrogenation, it is possible to achieve a reasonable hole-concentration and resistivity for LED (or even laser diode) applications [26].

Example 5.1 Determine the composition of $In_xGa_{1-x}N$ in the active layer of LEDs with emission wavelengths at 550 and 450 nm, used for green and blue LEDs, respectively.

Answer

For green and blue with emissions at 550 and 450 nm:

$$E_g(\text{eV}) = \frac{1240}{550\text{ nm}} = 2.2545\ E_g(\text{eV}) = \frac{1240}{550\text{ nm}} = 2.2545;\ E_g(\text{eV}) = \frac{1240}{650\text{ nm}} = 2.7556$$

$$E_g(\text{eV}) = \frac{1240}{650\text{ nm}} = 2.7556$$

x-values of ternary alloy $In_xGa_{1-x}N$ can be calculated from Eq. (5.6)

$$In_xGa_{1-x}N\ (550\text{ nm})\ E_g(\text{eV}) = \frac{1240}{550\text{ nm}} = 1.95x + 3.4(1-x) - x(1-x)E_g(\text{eV}) = \frac{1240}{550\text{ nm}}$$
$$= 1.95x + 3.4(1-x) - x(1-x);\ x = 0.6291$$

$$In_xGa_{1-x}N\ (450\text{ nm})\ E_g(\text{eV}) = \frac{1240}{450\text{ nm}} = 1.95x + 3.4(1-x) - x(1-x)E_g(\text{eV}) = \frac{1240}{450\text{ nm}}$$
$$= 1.95x + 3.4(1-x) - x(1-x);\ x = 0.2997$$

5.2.3 White LEDs

To generate white light from LEDs, there are at least two different techniques, as shown in Table 5.1. The first is using multiple LED chips. As mentioned before, since the integration of different substrates is difficult, the physical size of this white-light module cannot be very small. The second method is to use the (blue or UV) LED to excite phosphors. These techniques provide different spectral compositions of the white light, also shown in Table 5.1. Due to the lack of broadband emission from a single LED, the white light from either technique consists of two, three, or even more colors. From colorimetry, white light from mixing two colors exhibits the highest maximum attainable luminous efficiency (in terms of lm/W). However, to achieve a high CRI value in light source applications, three or more colors which provide a pseudo-broadband emission are needed.

Table 5.1 White light generation techniques.

	Multiple LEDs	LED pumps phosphor
Two-color mixing	Y+B LEDs	B-LED pumps Y-phosphors
Three-color mixing	R+G+B LEDs	UV-LED pumps white phosphors (RGB mixture)

It is straightforward to mix two or more LEDs with different wavelengths to obtain white light, following the colorimetric principles described in Chapter 2. The efficiency of this technique is typically higher than the LED/phosphor method since wavelength conversion which wastes energy is not needed. However, use of more LED chips means higher cost and bulk. Due to the differing non-linearity of the intensity–current relation in different LEDs, the color of this white light is also not stable to changes in drive current. With an increasing device temperature, LED intensity decreases and the spectrum redshifts which also results in a color shift of the white light. After long term operation, the different operating lifetimes of each LED type also results in a color change. For the two-color mixing technique, two complementary colored LEDs (typically blue and yellow) are used, as shown in Figure 5.7. Since the sensitivity of the human eye is highest at 555 nm and decreases at both sides in the photopic region, it is reasonable that there exists an optimized value for the maximum attainable luminous efficiency of two complementary colors with different wavelengths, which can be derived from color science and engineering, as described in Chapter 2. With longer or shorter wavelength combinations, a decrease of the luminous efficiency must result, which implies that two-color mixing with narrow emission spectra is the most efficient way to generate white light. Under the assumptions of: (i) a combination of two monochromatic complementary colors, (ii) 100% electrical to optical energy conversion, the upper theoretical limit of the white light generation efficiency is over 400 lm/W. Hence, such a white light source is very suitable for "white" display applications, such as pedestrian traffic signals, due to its low power consumption. However, it is not suitable for illumination applications since the CRI value is typically low, due to its narrowly peaked spectrum and two-color combination. To obtain a high CRI value, a broadband emission is usually needed, which also implies a trade-off between the CRI value and the efficiency. So, it can be expected that the maximum attainable luminous efficiency for three-color mixing is somewhat lower but with a higher CRI value, which is more suitable for lighting applications. It is also possible to increase the number of LED colors to 4 or 5 to obtain a larger color gamut, but the trade-off is lower efficiency and higher cost. Another important application for three-color LEDs sources, is in LCD backlights. For this application, LEDs with narrow emission spectra are preferred as they can provide a larger color gamut.

Another common way to generate white light is by using a blue or a UV LED to pump phosphors. Such a method is relatively simple and low cost since only a single LED is needed. The phosphors can be coated inside the LED package during the encapsulation process. Hence, small physical dimensions can be also achieved. Since the lifetime of conventional phosphors is long, the color stability of this white LED is also very good. However, compared with the multi-LED scheme, this technique has an inevitable disadvantage of theoretically lower efficiency. As shown in Figure 5.6, a phosphor absorbs photons with high energy in the violet/UV region

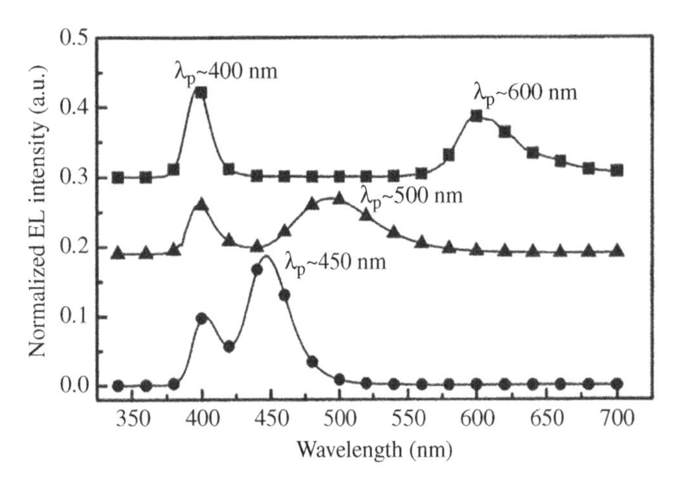

Figure 5.6 Emission spectra of UV (λp~400 nm) pumping (a) red (λp~600 nm), (b) green (λp~500 nm), and (c) blue (λp~450 nm) phosphors, respectively.

with a peak wavelength of 400 nm and emits lower energy photons at 600, 500, and 450 nm, for red, green, and blue emissions, respectively [27]. Typically, the IQE of the phosphor can be as high as 90%, which means 90 photons are emitted per 100 photons absorbed. However, there is an energy loss between the photons absorbed and those emitted, which is called the Stokes shift. In this case, the energy loss originating from the Stokes shift results in an efficiency decrease of 33.3%, 20%, and 11.1%, for red, green, and blue emissions, respectively. Another possible energy loss comes from light scattered by the phosphor which is reabsorbed by the LED chip, package, and reflectors.

There are many "down-conversion" materials which can absorb blue and UV light and generate photons with longer wavelength, such as organic dyes [28], semiconductors [29], and nanocrystals [30]. However, there are some disadvantages to these materials, such as limited lifetime and/or high cost. Typically, inorganic phosphors are commercially used as the wavelength converters due to their advantages of high stability, defect-free structure, easy fabrication and high-efficiency [15]. Figure 5.7 shows the device structure of a blue LED pumping a yellow phosphor and the resulting emission spectrum. The yellow phosphor was coated onto the LED chip during the packaging process. Part of the blue emission is absorbed by the phosphor which re-emits yellow light. The remaining blue light is emitted from the device and white light can be obtained. The blue emission from the LED is relatively narrow, with a FWHM of 50 nm. On the other hand, the phosphor emission is quite broad, and can be fine-tuned by selection of the phosphor material.

One of the most successful yellow phosphor materials is yttrium aluminum garnet (YAG) doped with Ce^{3+} ions ($Y_3Al_5O_{12}:Ce^{3+}$) [15]. As shown in Figure 5.8, by incorporation of Gd or Ga, the emission peak of the phosphor $(Y_{1-x}Gd_x)_3(Al_{1-y}Ga_y)_5O_{12}:Ce^{3+}$ can be red- or blue-shifted, respectively. The absorption peak of

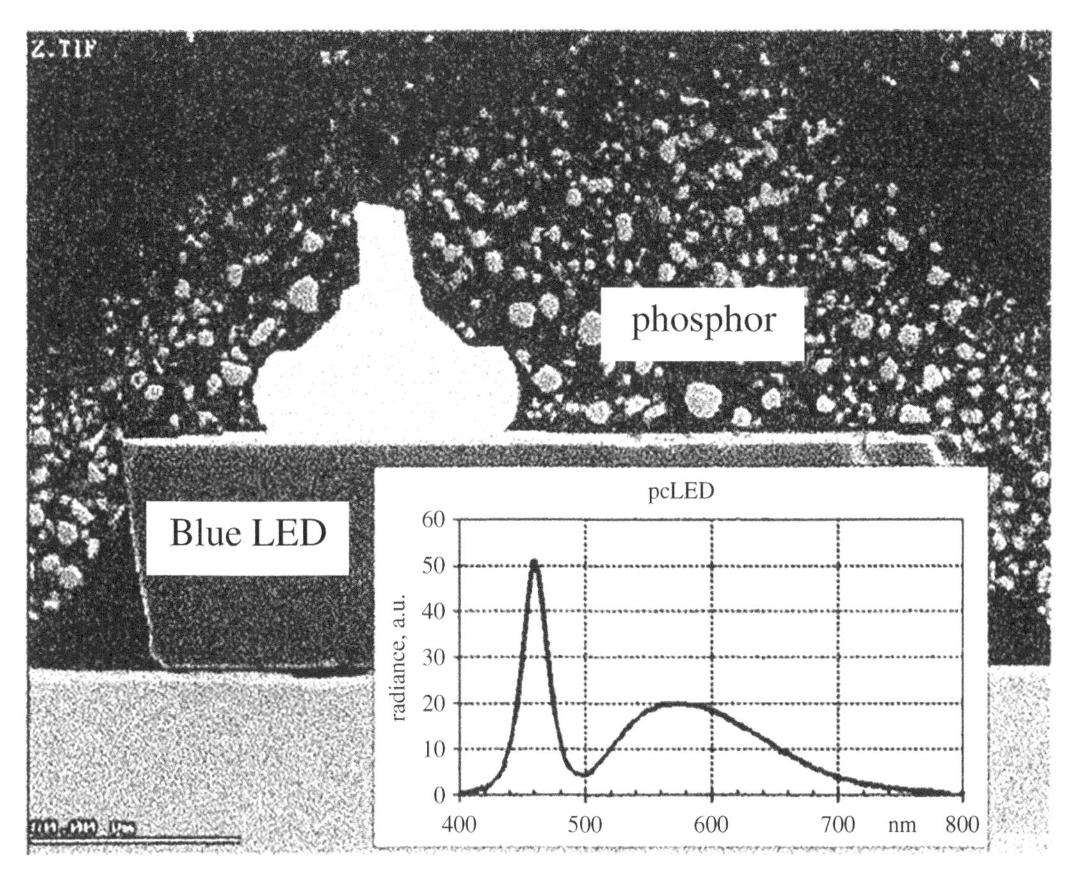

Figure 5.7 White LED structure consisting yellow phosphor pumped by a blue LED and its emission spectra [15].

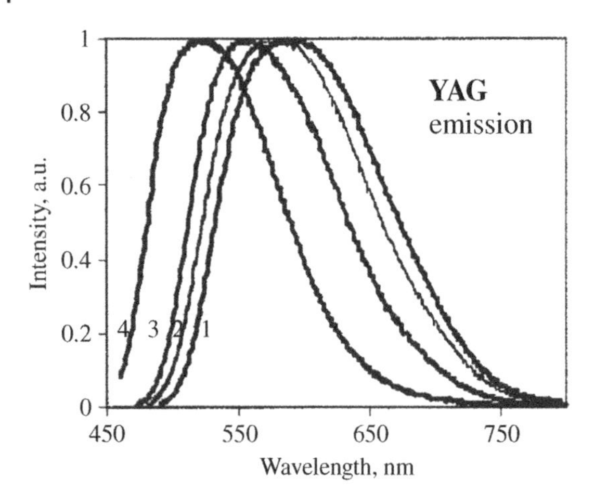

Figure 5.8 Emission spectra of the YAG phosphors with different compositions. "1 → 3": with decreasing Gd concentrations, and "4": with high Ga concentrations[15].

the phosphor is around 460 nm, which is suitable for pumping by a blue LED. The thickness of the phosphor layer is varied to adjust the relative intensity between blue and yellow emission. The peak wavelengths of these two complementary colors are also close to those giving the upper theoretical limiting efficiency of white light generation. In this system, the energy loss from the Stokes shift is also reduced since the blue portion of the light is directly emitted from the LED, rather than from a phosphor. However, the CRI value is typically less than 80, using this two-color mixing. UV-LED pumping of RGB-phosphors can effectively increase the CRI value since the emission spectrum is determined by the broadband emission from the phosphors. If a deep-UV LED is used (200–320 nm), the mature phosphor technology which is typically used in the fluorescent lamp, can be employed. Unfortunately the efficiency of such deep-UV LEDs is still a problem. Note that the peak of the circadian sensitivity function is located at 460 nm, which means that blue light viewed during the night may reduce melatonin levels and cause sleep disturbance, as discussed in Section 2.2. Shifting the blue emission to ~ < 400 nm (violet LED) can avoid the stimulation of the circadian rhythms [31]. However, the large Stokes shift required, is a problem for this approach.

5.3 DIODE CHARACTERISTICS

When a p- and n-layer are placed in contact, a p–n junction forms and shows diode characteristics. Ideally this means that the device behaves as a short circuit and an open circuit under forward and reverse bias conditions, respectively. There are more holes than electrons in the p-type material and correspondingly more electrons in the n-type layer. Hence, holes diffuse from the p- toward the n- side near the p–n junction. Similarly, electrons diffuse from the n-type toward the p-type material. Carriers of opposite charge are concentrated at the junction and recombine as shown in Figure 5.9 [32]. After the carriers' annihilation, they leave behind negative and positive space charge which creates an internal electric field from the n- to the p-side of the junction.

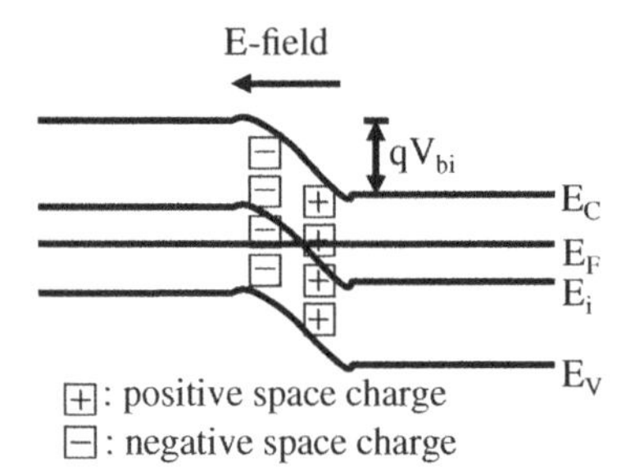

Figure 5.9 Band diagram of a p–n junction under thermal equilibrium.

This internal field results in band bending, as shown in Figure 5.9. By using the Poisson equation, one can obtain the relationship between the donor and acceptor concentrations and the built-in voltage. The region of space charge is nearly free of carriers under thermal equilibrium and is therefore called the "depletion region." Holes and electrons experience the built-in voltage and drift toward the p- and n-sides, respectively. At equilibrium, the drift current balances the diffusion current and no net current flows, as shown in Figure 5.9. Under this condition, one can also note that the Fermi levels at the p- and n-sides are aligned due to the band bending. When a forward bias is applied (a voltage from the p–n side), it opposes and reduces the built-in voltage. The point at which the forward bias exceeds the built-in voltage is called the threshold voltage, and above this value the current density increases exponentially with the applied voltage. The value of threshold voltage of a LED is approximately equal to E_g Volt (which is the bandgap energy in terms of eV) at high doping levels. As the charge carriers pass through the p- and n-type layers into the depletion region for recombination, the resistance of these layers should be as small as possible to reduce ohmic loss in these regions. Besides, the doping levels determine the width of the depletion zone, which in turn affects the recombination zone and recombination efficiency. For this kind of homojunction structure, however, there is still a carrier leakage problem, i.e. some holes and electrons propagate to the cathode and anode, respectively, without recombination due to inefficient carrier confinement. Heterojunction structures and EBLs are used to prevent carrier leakage to the opposite electrodes. Furthermore, quantum structures (well, wire, and dot) are used to confine carriers in a specified recombination region. Energy levels in the quantum structures are discrete, which allows fine-tuning of the emission wavelength.

5.3.1 p- and n-Layer

LEDs are basically formed by a p–n junction. Holes and electrons pass through the p- and n-layers, respectively, and then recombine in the active region. By incorporating group IV (such as: Ge, and Si) and VI (such as: S, Se, and Te) dopants into III–V compounds to replace group III and V atoms respectively, it is possible to achieve n-type doping. Figure 5.10a shows a schematic diagram of Se doped into GaAs and occupying As lattice sites. Since the Se carries one more valence electron than As, it "donates" one more electron to the crystal. This extra electron can be viewed as nearly free, since it cannot form a bond within the crystal. Hence, a donor state is formed slightly below the conduction band in the forbidden gap, for which one can define a donor binding energy (E_D) and treat it as a trap state. Since it is a shallow trap, trapped electrons can be thermally excited into the conduction band as free electrons. Similarly, p-type doping can be achieved by incorporating Cd, Zn, Mg, and Be (group II) and C and Si (group IV) atoms to replace group III and V atoms

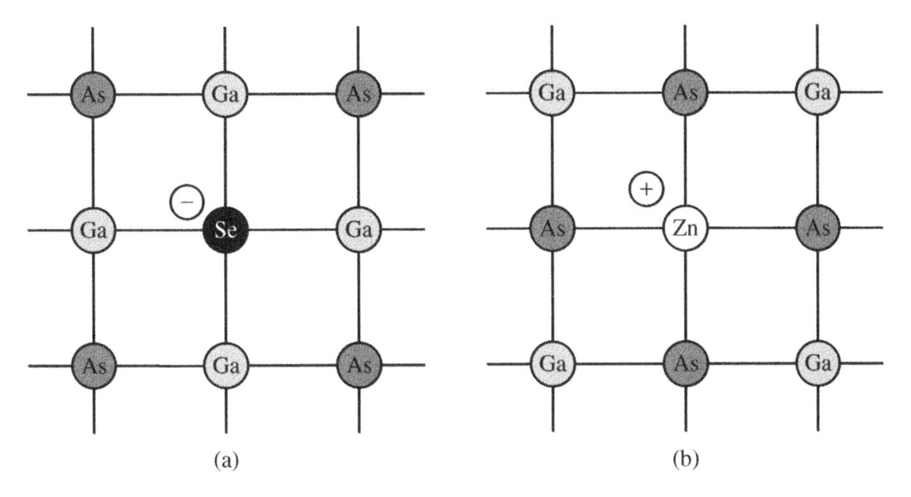

Figure 5.10 Schematic diagram of bonding of GaAs with (a) Se and (b) Zn impurities.

respectively. Figure 5.10b shows an example of Zn doped into GaAs and replacing Ga atoms. Such acceptor doping creates acceptor levels at an energy offset (E_A) above the valence band within the forbidden gap. Electrons in the valence band can be thermally excited to the acceptor state, which also creates a free hole in the valence band.

Typically, the thermal energy required for free-hole creation is much larger than that for free electrons, which means the "ionization" of holes is sometimes not complete and limits the p-doping concentration, especially for the wide-gap nitride semiconductors. For example, the E_A of Mg-doped GaN is 0.2 eV and much higher than the thermal energy (~25 meV at room temperature) which means that only a small proportion of the acceptors are ionized and contribute to the hole carrier concentration. Sometimes, the dopant does not exactly replace the original atoms, which also results in inefficient ionization and hence a low carrier concentration. For example, Mg doped into GaN does not occupy the Ga site as-grown and generates a low hole concentration. After low energy electron beam irradiation (LEEBI), the Mg atoms move to the exact Ga sites and contribute efficiently to raising the hole concentration. Unintended impurities introduced during epitaxy and processing result in defect trap levels which compensate the doping levels. Examples include oxygen contamination of Al-containing compounds and hydrogen passivation of nitride compounds. A suitable post annealing process is required to remove those impurities. Another way to reduce the resistance of the p- and n-layers is to increase the mobility value. Typically, hole mobilities in III–V compounds are much lower than those of electrons due to their larger effective mass. For example, electron and hole mobilities (μ_n and μ_p) are 5000 and 300 $cm^2/V\,s$ in n- and p-type GaAs each with doping concentrations of $1 \times 10^{17}\ cm^{-3}$ at room temperature. There exists a trade-off between the carrier concentration and carrier mobility. With increasing dopant concentration comes increasing scattering of the carriers by the impurities, and the carrier mobility is reduced. Despite this, it is generally practical to reduce the resistance by increasing the carrier concentration.

5.3.2 Depletion Region

In this section, we will quantitatively describe two important parameters of the depletion region, the built-in voltage and the depletion width. As shown in Figure 5.9, the built-in voltage originates from the band bending in the depletion region. On the other hand, in the p- and n-layers, the band is still flat. As shown in Chapter 3, assuming that the acceptors and donors are fully ionized, the hole and electron concentrations can be represented as [32]:

$$p = n_i \exp\left[\frac{E_i - E_{Fp}}{(kT/q)}\right] = N_A p = n_i \exp\left[\frac{E_i - E_{Fp}}{(kT/q)}\right] = N_A \tag{5.7}$$

$$n = n_i \exp\left[\frac{E_{Fn} - E_i}{(kT/q)}\right] = N_D n = n_i \exp\left[\frac{E_{Fn} - E_i}{(kT/q)}\right] = N_D \tag{5.8}$$

where p and n are hole and electron concentrations far away from the depletion region, respectively: n_i is the intrinsic concentration, E_i, E_{Fp}, and E_{Fn} are the Fermi energies of intrinsic, p-doped, and n-doped layers respectively, in units of eV. Note that although the Fermi level appears flat throughout the whole p–n junction as shown in Figure 5.9, the Fermi energy is different with respect to the conduction band and valence bands in p- and n-layers, far away from the depletion region. In other words, the valence and conduction bands bend, so their separation from the Fermi level changes. k is the Boltzmann constant, T is the temperature, N_A and N_D are the acceptor and donor concentrations, respectively. Hence, one can obtain ψ_p, ψ_n, and the built-in voltage, $V_{bi} = (\psi_n - \psi_p)$ as shown in Figure 5.9.

$$\psi_p = -(E_i - E_{Fp}) = -\frac{kT}{q} \ln\left(\frac{N_A}{n_i}\right) \psi_p = -(E_i - E_{Fp}) = -\frac{kT}{q} \ln\left(\frac{N_A}{n_i}\right) \tag{5.9}$$

$$\psi_n = -(E_i - E_{Fn}) = \frac{kT}{q} \ln\left(\frac{N_D}{n_i}\right) \psi_n = -(E_i - E_{Fn}) = \frac{kT}{q} \ln\left(\frac{N_D}{n_i}\right) \tag{5.10}$$

$$V_{bi} = \psi_n - \psi_p = \frac{kT}{q} \ln\left(\frac{N_A N_D}{n_i^2}\right) V_{bi} = \psi_n - \psi_p = \frac{kT}{q} \ln\left(\frac{N_A N_D}{n_i^2}\right) \tag{5.11}$$

One can note that as the dopant concentrations are increased, the V_{bi} increases. In LED applications, heavily pn-doped layers are needed to reduce the device resistance and hence the power consumption, and the Fermi level of the p- and n-layers are close to the valence and conduction bands, respectively. Hence, it follows that the V_{bi} is close to the bandgap of the semiconductor. When a voltage higher than V_{bi} is applied, the direction of band bending is reversed, compared to the zero bias equilibrium case. Under this bias condition, holes and electrons travel easily from the p- and n-type layers to the opposite side of the junction, which results in an exponential increase in current density as described in the next section.

From electrostatic theory, Gauss's law can be represented as:

$$\nabla \cdot (\varepsilon E) = \rho \tag{5.12}$$

or it can be written as Poisson's equation:

$$\nabla^2 V = -\rho/\varepsilon \tag{5.13}$$

where E is the electric field, ε is the electric permittivity, ρ is the charge density, and V is the voltage. Assuming the free carriers are fully ionized in the depletion region, then the distributions of space charge, electric field, and voltage are as shown in Figure 5.11. The electric field distribution can be obtained from:

$$E(x) = \int \rho(x) \mathrm{d}x$$

$$\rho(x) = -qN_A; -x_p < x < 0 + qN_D;\ 0 < x < x_n \tag{5.14}$$

where x_p and x_n are the boundary positions of the depletion region at the p and n sides, respectively. Hence, the maximum electric field (E_{max}), which occurs at the p–n boundary, is:

$$E_{max} = \frac{q}{\varepsilon} N_D x_n = \frac{q}{\varepsilon} N_A x_p E_{max} = \frac{q}{\varepsilon} N_D x_n = \frac{q}{\varepsilon} N_A x_p \tag{5.15}$$

Also note that as

$$x_n + x_p = W \tag{5.16}$$

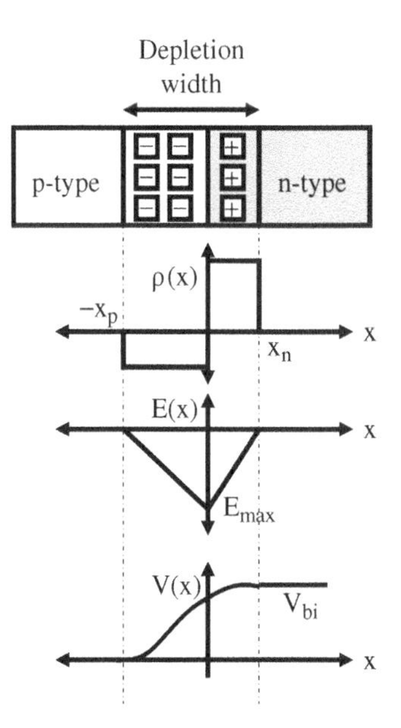

Figure 5.11 Charge density, electric field and voltage distribution of an LED under thermal equilibrium.

where W is the width of the depletion region, under thermal equilibrium the built-in voltage can be given by:

$$V_{\text{bi}} = (1/2)E_{\text{max}}W \tag{5.17}$$

or

$$W = \sqrt{\frac{2\varepsilon}{q}\left(\frac{N_{\text{A}}+N_{\text{D}}}{N_{\text{A}}N_D}\right)V_{\text{bi}}} \tag{5.18}$$

When a forward bias V is applied, the V_{bi} in Eq. (5.18) must be replaced by $(V_{\text{bi}} - V)$, which results in a reduction in the depletion zone width. On the other hand, the depletion region width increases under reverse bias. When the depletion region narrows, it is easier for carriers to cross it, and hence the current density is high. On the other hand, the wider depletion region under reverse bias impedes carrier transport, and these effects underlie the rectification effect of the p–n diode.

Example 5.2 Consider a GaAs p–n diode at room temperature. The intrinsic carrier concentration n_{i} is $1.79 \times 10^6\ \text{cm}^{-3}$ and the doping levels are $N_{\text{A}} = 5 \times 10^{19}\ \text{cm}^{-3}$ in the p-side and $N_{\text{D}} = 1 \times 10^{16}\ \text{cm}^{-3}$ in the n-side. Find the built-in voltage, depletion zone width and maximum electric field. ($\varepsilon_r = 13.18$ for GaAs)

Answer

From (5.9) and (5.10):

$$\psi_{\text{p}} = -\frac{kT}{q}\ln\left(\frac{N_{\text{A}}}{n_{\text{i}}}\right) = -\frac{300 \cdot 1.381 \times 10^{-23}}{1.6 \times 10^{-19}} \cdot \ln\left(\frac{5 \times 10^{19}}{1.79 \times 10^{6}}\right) = -0.8013\ \text{V}$$

$$\psi_{\text{n}} = \frac{kT}{q}\ln\left(\frac{N_{\text{D}}}{n_{\text{i}}}\right) = \frac{300 \cdot 1.381 \times 10^{-23}}{1.6 \times 10^{-19}} \cdot \ln\left(\frac{1 \times 10^{16}}{1.79 \times 10^{6}}\right) = 0.5808\ \text{V}$$

The built-in voltage V_{bi} from (5.11) is given by $V_{\text{bi}} = \psi_{\text{n}} - \psi_{\text{p}} = 1.3821 V_{\text{bi}} = \psi_{\text{n}} - \psi_p = 1.3821$ V
From (5.17) and (5.18)

$$W = \sqrt{\frac{2\varepsilon}{q}\left(\frac{N_{\text{A}}+N_{\text{D}}}{N_{\text{A}}N_{\text{D}}}\right)V_{\text{bi}}} = \sqrt{\frac{2 \cdot 13.18 \cdot 8.85 \times 10^{-14}}{1.6 \times 10^{-19}}\left(\frac{5 \times 10^{19} + 1 \times 10^{16}}{5 \times 10^{19} \cdot 1 \times 10^{16}}\right) \cdot 1.3821}$$

$$\cong 4.48 \times 10^{-5}\text{cm}$$

$$E_{\text{max}} = \frac{2V_{\text{bi}}}{W} \cong 61700 E_{\text{max}} = \frac{2V_{\text{bi}}}{W} \cong 61700\ \text{V/cm}$$

5.3.3 J–V Characteristics

Carrier transport in a semiconductor can be divided into carrier diffusion and drift. Diffusion means that carriers move from high to low concentration regions. Drift current means carrier transport driven by an electric field. The total current density is the sum of the electron and hole contributions. Take the hole current density, Jp, as an example. This can be described by the expressions:

$$J_{\text{p}} = J_{\text{p,diff}} + J_{\text{p,drift}} \tag{5.19}$$

$$J_{\text{p,diff}} = q\text{D}_{\text{p}}(\text{dp}/dx) \tag{5.20}$$

$$J_{\text{p,drift}} = \text{p}q\mu_{\text{p}}\text{E} \tag{5.21}$$

where $J_{\text{p,diff}}$ and $J_{\text{p,drift}}$ are the hole diffusion and drift current densities, respectively. D_{p} is called the diffusivity. μ_{p} is the hole mobility, defined as the velocity (in terms of cm/s) over the electric field (V/cm). Under thermal

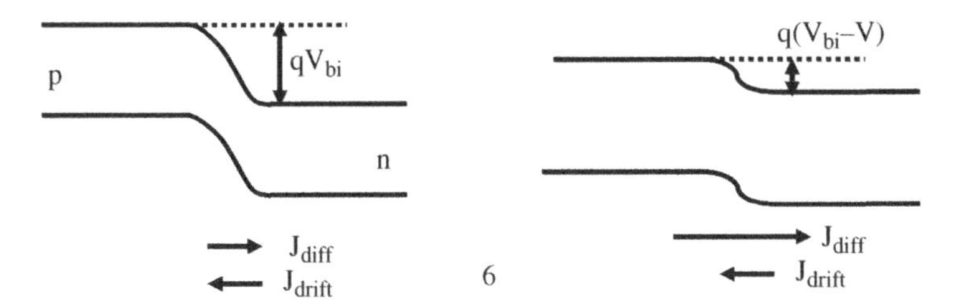

Figure 5.12 Band diagram of a LED under (a) thermal equilibrium and (b) forward bias.

equilibrium, holes diffuse from the p-layer to the n-layer, but this is balanced by the hole drift current from n- to p-side. Since the drift current depends on the electric field, there is no drift current outside the depletion region, where the band is flat. When a forward bias is applied, the energy band offset decreases from the built-in voltage and the current density increases. Obviously, this increase cannot come from any contribution of the drift current since the drift current flows from the n- to the p-side, as shown in Figure 5.12. Hence, the current density increase under forward bias comes from the diffusion current. With an increasing forward bias, the barrier for carrier diffusion decreases which results in a net current from the p- to the n-side under forward bias.

Knowledge of the carrier distribution in a LED, is essential to derive the J–V characteristics. Although there is no explicit formula to describe this distribution, we can still qualitatively describe it as shown in Figure 5.13. p_{p0}, n_{n0}, p_{n0}, and n_{p0} are the hole concentration in p-layer, electron concentration in n-layer, hole concentration in n-layer, and electron concentration in p-layer, respectively, before p–n junction formation. One can note that near the depletion boundary, the minority carrier concentrations p_n, and n_p are higher than p_{n0}, and n_{p0}, respectively, which results in a diffusion-current for holes and electrons in the n- and p-layers. Typically, the J–V characteristics can be approximately represented as:

$$J = q\left[\sqrt{\frac{D_p}{\tau_p}}N_A + \sqrt{\frac{D_n}{\tau_n}}N_D\right]\exp[q(V - V_{bi})/kT] \tag{5.22}$$

where τ_n and τ_p are the respective minority carrier lifetimes. One can note that once the voltage increases above the V_{bi} (approximately the value of E_g in terms of electron volt) an exponential increase in current density is observed. This "threshold" voltage is reasonable. Since the LED emits a photon at a wavelength $\lambda = hc/E_g$, one has to provide a driving voltage of at least $V = E_g/q$.

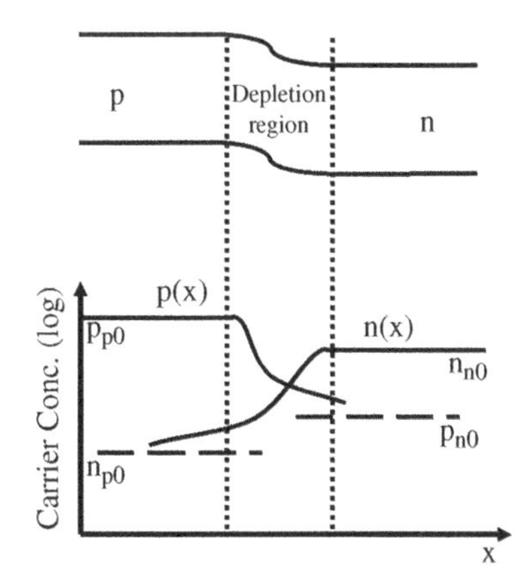

Figure 5.13 Carrier distribution under forward bias.

5.3.4 Heterojunction Structures

Since the bandgap of the III–V semiconductors can be adjusted by varying their composition, materials with different bandgaps can be deposited to provide different layers of a single device during epitaxial growth, creating what is called a heterojunction structure [33]. An LED with a single composition (and therefore bandgap) as described in the previous section is referred to as a "homojunction" device. Heterojuction structures can be used to confine the carriers in the active region to increase their recombination rate. Figure 5.14a shows a single-heterojuction (SH) structure. We can note that, in this LED, the p- and n-type materials are different; the p-layer exhibits a smaller bandgap than n-layer. ΔE_c and ΔE_v denote the difference in energy of the conduction and valence bands respectively, between the two materials. The sum of ΔE_c and ΔE_v is equal to the bandgap difference between the p- and n-type materials. A hole passing from p- to n-type material, encounters an extra barrier height ΔE_v which decreases the hole current into the n-region. On the other hand, for the electron it is energetically favorable to pass from the n- to the p-side, and it may be trapped in the "energy well" formed by ΔE_c. Hence, a high concentration electron hole pairs accumulate at the p-side near the SH interface, increasing the recombination rate and the IQE. Typically, in a heterostructure, the small bandgap material, on the p-side in this case, is designed to act as the emitting layer. In contrast to homojunction devices, the recombination zone of SH-LEDs is mainly determined by the energy difference between the p- and n-layers, and the layer structure. Hence, the p- and n-doping concentrations can be optimized without considering the depletion region width which – as described in previous sections – determines the recombination zone in homojunction LEDs.

Furthermore, one can form heterojunctions in both the p- and n-type materials at the same time, forming a "double heterojunction" (DH) structure, as shown in Figure 5.14b. An active layer with a smaller bandgap is sandwiched between two wide bandgap materials (typically called cladding or confinement layers), which ensures that carriers injected into the active layer are confined in this region to raise the recombination rate. Due to the different carrier mobilities ($\mu_n \gg \mu_p$), electrons can rather easily penetrate the barrier from the active layer to the p-side. Therefore, an EBL with a large bandgap and ΔE_c, is inserted between the active layer and the p-type region as a barrier to block electrons as shown in Figure 5.15 [34].

5.3.5 Quantum-Well, -Wire, and -Dot Structures

By inserting a thin layer with a lower bandgap in the active region of a heterostructure, the emission zone can be spatially defined and hence further increase the recombination rate, which is proportional to the product of the electron and hole concentrations. The reabsorption probability also decreases since the emission zone

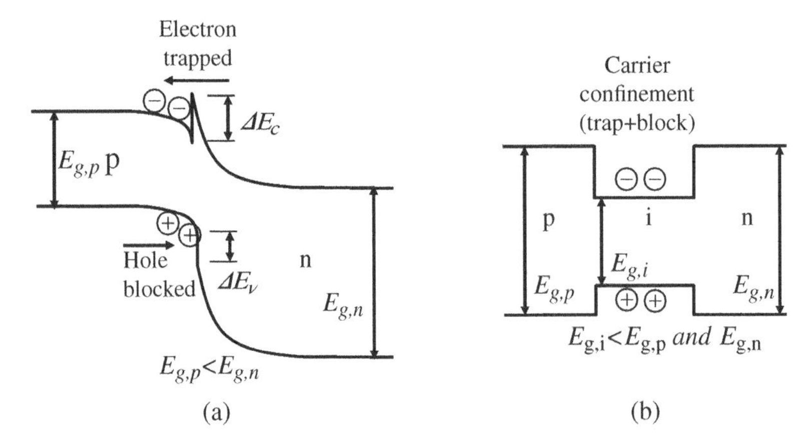

Figure 5.14 Band structures of (a) single-, and (b) double-heterojunction LEDs.

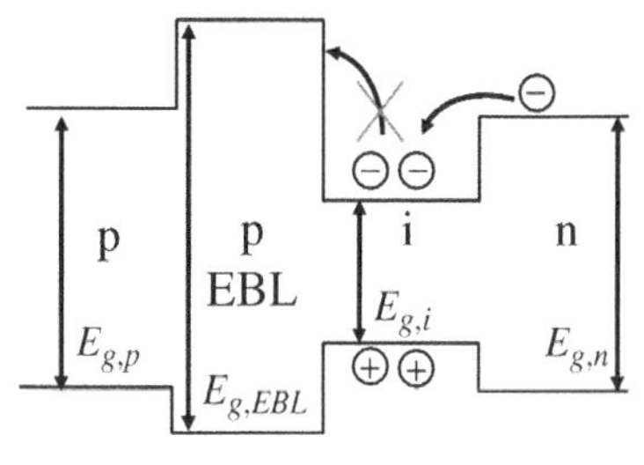

Figure 5.15 Band structures of LED with EBL.

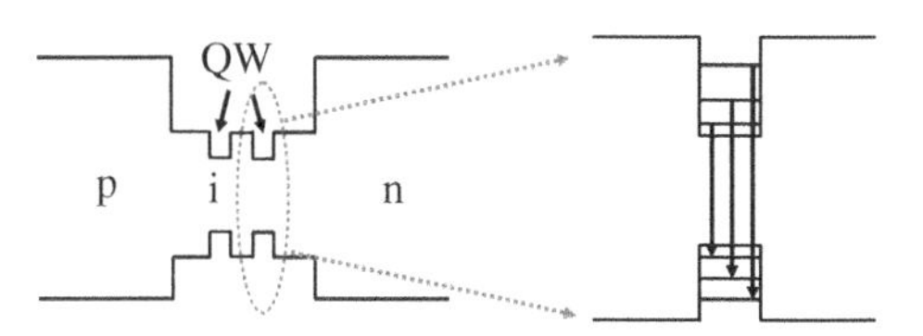

Figure 5.16 Band diagram of a QW structure.

exhibits the smallest bandgap and is itself very thin. Such a layer is typically less than 20 nm thick, which is smaller than the de Broglie wavelength of the electrons and holes. Hence, the energy level becomes quantized, forming what is called a quantum well (QW) structure, as shown in Figure 5.16. The quantized energy levels can be adjusted by changing the composition and width of the QW, which provides a further mechanism for fine-tuning the emission wavelength. Another advantage of the QW structure comes from its ultra-thin nature (compared with the 1 μm thickness of the active layer of the heterostructure) which means that strained layers can be included by epitaxial growth with a lattice mismatch; strain relaxation does not occur in the ultra-thin layer.

The QW structure is an ultra-thin layer, with quantized levels along the epitaxy direction. The composition in the 2D plane perpendicular to the epitaxial growth is homogeneous. As is shown in Figure 5.17, if the low bandgap material can be fabricated in the form of wires or dots, then carriers can be confined in one- and zero-dimensions, respectively, and the resulting structures are called quantum wires and quantum dots (QD) [35]. In contrast to the precise control achieved during epitaxy, fabrication of quantum wires and dots is more difficult since high resolution pattern definition and epitaxial regrowth are needed. One special case occurs in nitride semiconductors. During epitaxy, In clusters form which generate a spontaneous composition fluctuation. Quantum dot structures are therefore formed unintentionally, but also result in a higher recombination efficiency.

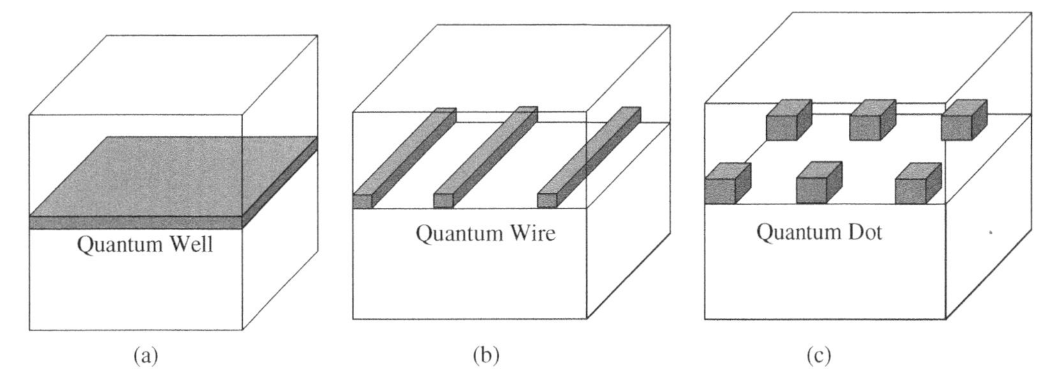

Figure 5.17 Structures of (a) QW, (b) quantum wire, and (c) QD.

5.4 LIGHT-EMITTING CHARACTERISTICS

Once the electron–hole pair recombine, the excited state may relax radiatively or non-radiatively. To minimize the non-radiative relaxation, one of the most important factors is to obtain a high quality epitaxial layer. Since the bandgap originates from the crystalline structure, defects act as quenchers which result in an efficiency decrease. The periodic lattice structure may be damaged during the fabrication process, e.g. at a cleaved facet or etched surface. Annealing processes are then used to reduce the non-radiative recombination rate. Radiative recombination in LEDs is a Langevin process, the rate of which is proportional to the product of the electron and hole concentrations. Under some approximations, a linear relationship between the injection current and light output can be obtained. The emission spectrum from a LED depends on the DOS and the carrier distributions. The DOS functions of the bulk, quantum-well, wire, and dot structures are different causing shifts in the peak wavelength and a change in the FWHM. A spectral shift can be observed at different temperatures and current densities. Once the photons are generated, the next step is for them to radiate out of the semiconductor. Due to the large refractive index (typically over 3) of the semiconductors, total internal reflection limits the extraction efficiency. For III-P LEDs, the generated visible photons may also be absorbed by the GaAs substrate.

5.4.1 Recombination Model

Radiative recombination in LEDs involves holes and electrons. Once these two opposite charged carriers meet in the spatial domain, the recombination process occurs which is a two-particle process and can be described by a Langevin formula:

$$R_{rad} = r_{rad}np \tag{5.23}$$

where R_{rad} is the radiative recombination rate, n and p are electron and hole concentrations, respectively and r_{rad} is called the bimolecular recombination coefficient. When the semiconductor is excited, electrons are promoted from the valence band to the conduction band, which creates the same number of holes in the valence band. That means the numbers of the "excess" holes and electrons are the same, so:

$$\Delta \mathrm{n} = \Delta \mathrm{p} \tag{5.24}$$

where Δn and Δp are excess electron and hole concentrations, respectively. For a p-type doped semiconductor, the hole concentration is much higher than the electron concentration under thermal equilibrium, hence:

$$\mathrm{p}_0 >> \mathrm{n}_0 \tag{5.25}$$

where p_0 and n_0 are the hole and electron concentrations under thermal equilibrium, respectively. Under low-level excitation (when the excess carrier concentration is much less than that of the majority carrier under thermal equilibrium; $\Delta \mathrm{p} \ll \mathrm{p}_0$), we can write:

$$R_{rad} = r_{rad}\mathrm{np} = r_{rad}\mathrm{n}(\mathrm{p}_0 + \Delta \mathrm{p}) \sim r_{rad}\mathrm{np}_0 = \mathrm{n}/\tau_{rad} \tag{5.26}$$

and

$$\tau_{rad} = 1/r_{rad}\mathrm{p}_0 \tag{5.27}$$

where τ_{rad} is called the radiative lifetime. We can see that the radiative lifetime in a doped semiconductor under low-level injection is inversely proportional to the doping concentration. That means recombination is "faster" with higher doping concentrations since there are more majority carriers which can participate in radiative recombination, according to the recombination process.

Energy relaxation by phonon (lattice vibration) generation without light emission, is called non-radiative recombination. In a crystalline semiconductor, the main source of non-radiative recombination comes from defects in the periodic structures. Those defect structures can be quantitatively described as "trap states" inside

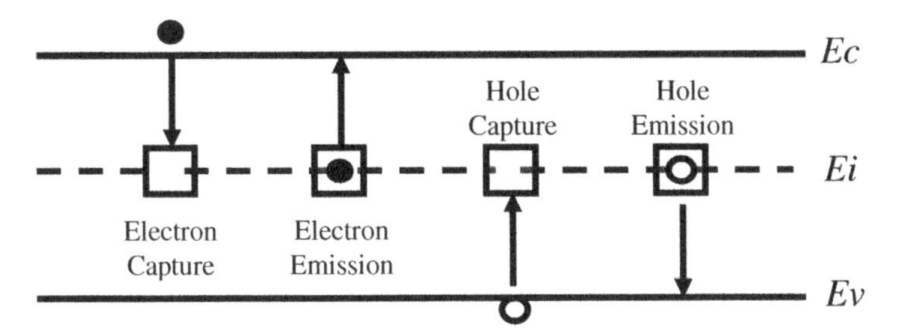

Figure 5.18 Recombination (typically non-radiative) via deep-level traps.

the forbidden bandgap. Electrons (holes) jump between the conduction (valence) band and the trap states via transitions mediated by either phonons or photons. Although the gap states can improve photon emission in some indirect bandgap semiconductors (such as nitrogen doping in GaP), typically they act as luminescence quenchers for most direct bandgap materials.

Quantitative analysis of recombination via traps was accomplished by Schockley, Read, and Hall, and the process they described is called SRH recombination [36, 37]. A schematic illustration is shown in Figure 5.18. Typically, a trap can capture and emit one electron (hole) at a time from or to the conduction (valence) band. Once an electron and a hole are simultaneously captured by a trap before they emit, recombination occurs. If the trap is near the conduction band (i.e. a shallow-trap), electrons are easily captured and released by the trap, but it is difficult for the holes because this is then a multi-phonon process. Only when the trap is located near the mid-gap (i.e. deep-trap condition), is the recombination rate high. For a doped semiconductor (e.g. a p-type material) containing deep-traps, although the derivation of the equation is relatively complex, we can still understand that: (i) since there are many more holes than electrons, recombination takes place rapidly once an electron is captured which means the lifetime due to non-radiative recombination is determined by the minority carrier concentration, and (ii) the traps are typically filled with holes because the traps are distributed near the mid-gap, and therefore the lifetime is independent of hole (majority carrier) concentration. So, we can obtain the relation that:

$$R_{\text{non-rad}} = n/\tau_{\text{non-rad}} \tag{5.28}$$

where R_{rad} and τ_{rad} are the radiative recombination rate and the radiative lifetime, respectively. So, the total recombination rate can be described as:

$$R = R_{\text{rad}} + R_{\text{non-rad}} = (1/\tau_{\text{rad}} + 1/\tau_{\text{non-rad}})\, n = n/\tau \tag{5.29}$$

and τ is called the minority carrier lifetime.

5.4.2 L-J Characteristics

A continuity equation can be used to describe the change in carrier concentration under electrical and optical excitation. For the electron concentration in a p-type semiconductor with a low rate of carrier injection, one can write:

$$\frac{\partial n}{\partial t} = \frac{1}{q}\frac{\partial J_e}{\partial x} + G_0 + G_{\text{ext}} - \frac{n}{\tau} \tag{5.30}$$

where q is the electron charge, J_e is the electron current density and G_0 is the generation rate under thermal equilibrium which is equal to the recombination rate under the same conditions ($R_0 = n_0/\tau$). G_{ext} is the excess generation rate by optical excitation. Eq. (5.30) can also be re-written as:

$$\frac{\partial \Delta n}{\partial t} = \frac{1}{q}\frac{\partial J_e}{\partial x} + G_{\text{ext}} - \frac{\Delta n}{\tau} \tag{5.31}$$

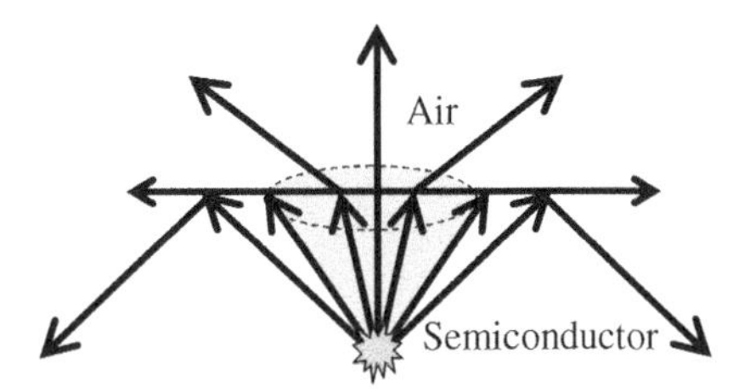

Figure 5.19 Light escape cone of LED due to the total internal reflection.

For a LED under steady state electrical pumping, one can write:

$$0 = \frac{1}{q}\frac{\partial J_e}{\partial x} - \frac{\Delta n}{\tau} \tag{5.32}$$

The optical power (P_{out}) emitted from an LED can be represented as:

$$P_{out} = \eta_{ext} P_{generated} = \eta_{ext} h\upsilon A \int \frac{\Delta n}{\tau_{rad}} dx = \eta_{ext} \frac{h\upsilon A}{q} \int \frac{\tau}{\tau_{rad}} \frac{\partial J_e}{\partial x} dx = \eta_{ext} \frac{h\upsilon A}{q} \frac{\tau}{\tau_{rad}} [J_e(\text{in}) - J_e(\text{out})] \tag{5.33}$$

where η_{ext} is the photon extraction efficiency, $P_{generated}$ is the power generated inside the LED, $h\nu$ is the photon energy, A is device cross-section area, and J_e(in) and J_e(out) are the electron current injected into and flowing out of the device, respectively. By using a thick active layer to provide a wide recombination region or by designing suitable layer structures (e.g. heterojunction or low-dimensional structures) to confine the carriers, carrier leakage can be effectively reduced. If the electrons supplied into the device completely recombine with the same number of holes, then Eq. (5.33) becomes:

$$P_{out} = \eta_{ext} \frac{h\upsilon}{q} \frac{\tau}{\tau_{rad}} I \tag{5.34}$$

where I ($=J^*A$) is electrical current. One can see that the output optical power is proportional to the supply current. To achieve a high optical power output, one has to increase the extraction efficiency (higher η_{ext}), use a wide bandgap material (higher $h\nu$), and reduce the radiative recombination lifetime (smaller τ_{rad}). For a planar LED structure, the extraction efficiency is limited by total internal reflection between the interface of the semiconductor and air governed by Snell's law, as shown in Figure 5.19. To improve light extraction, various corrugated structures have been designed which destroy the planarity of the structure and assist in coupling waveguided light out of the device. Die shaping can also be helpful to redirect the light out of the LED. Packaging with an epoxy dome and reflector not only protects the device but also increases the extraction efficiency. For III–P material systems, since GaAs substrates are optically absorbing, their replacement with a transparent substrate also increases the light output. A detailed discussion of ways to enhance the extraction efficiency will be presented in Sections 5.5.3 and 5.5.4.

5.4.3 Spectral Characteristics

The emission wavelength of an LED is determined by the product of the DOSs and the number of carriers in each energy level. The DOS defines how many states are available for recombination at any given energy. As shown in Figure 5.3, the band diagram has a parabolic shape for both conduction and valence bands. So, the DOS increases from zero at the band edge as the energy is increased. On the other hand, the carrier distribution decreases exponentially as the energy increases. A typical emission spectrum from a LED is shown in Figure 5.20, the emission peak wavelength being a little shorter than the bandgap wavelength. Also, the emission spectrum is asymmetric. There are no allowed states within the bandgap. Hence, the edge on the long wavelength side of the peak is sharp. On the other hand, one can see a smooth curve on the short wavelength side which comes from the exponential tail of the carrier distribution.

For QW and QD structures, the energy is quantized and hence the DOS is different from that in a bulk semiconductor. Figure 5.21 shows the DOS as a function of energy for bulk and quantum structures. For QW

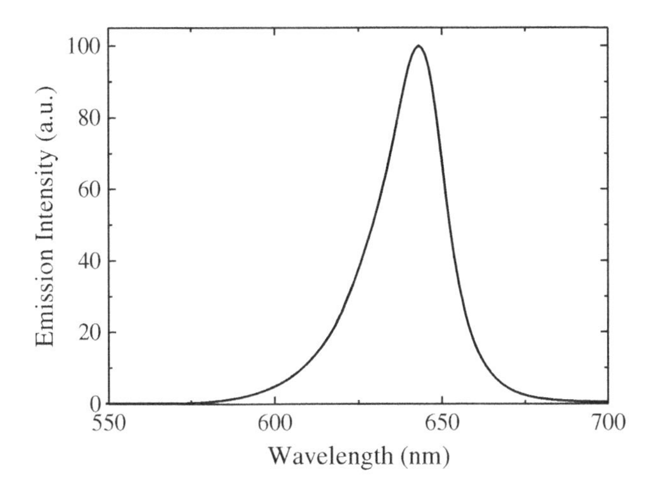

Figure 5.20 Emission spectrum from a LED.

Figure 5.21 DOS of bulk, QW and QD structures.

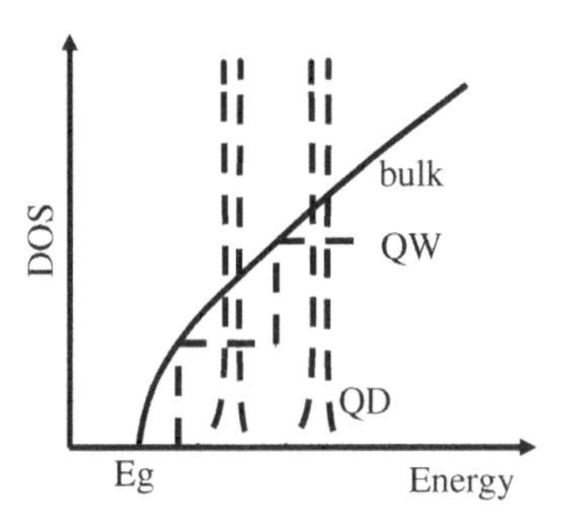

and QD structures, the DOS is a step function and a delta function, respectively, due to energy confinement by the quantized level, which means the emission spectrum of a quantum structure has a narrower FWHM and can provide a larger color gamut in display applications.

When the temperature increases from the normal ambient, the emission intensity of a LED decreases and the spectrum redshifts. Figure 5.22 shows experimental data on the intensity decrease and spectral shift for

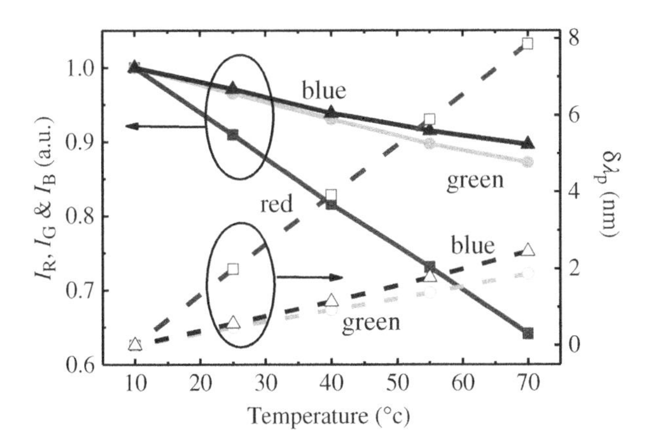

Figure 5.22 Temperature dependence of RGB LEDs.

R, G, and B LEDs at different temperatures. The mechanism of the intensity decrease mainly comes from an increase in the non-radiative recombination rate, which in turn is due to the increase in phonon intensity at higher temperatures. A temperature increase also shifts the bandgap toward lower energy, which causes the spectral redshifts. These two phenomena are most pronounced in the long wavelength LED, as shown in Figure 5.22. Such a temperature dependence of LED performance poses serious problems for display applications. For example, an LED traffic signal might have different colors and a lower intensity at noon than at night, which can cause safety issues. For LEDs used as LCD backlights, a great deal of heat is generated, causing a color shift and color gamut change. Hence, suitable temperature control and compensation circuits are needed to optimize the display performance over a wide temperature range.

Example 5.3 White light is generated by red, green, and blue LEDs with their primary wavelengths at 650 nm, 550 nm and 450 nm at room temperature (25 °C), respectively. However, due to the dissipated heat after 30 minutes operation, the module temperature increases and stabilizes at 85 °C. Assume that these three LEDs emit monochromatic light. For the red LED, the peak wavelength red-shifts 5 nm and the emission intensity decreases 20% under a 30 °C increase in temperature. For green and blue LEDs, the wavelength shift and intensity decay are 2.5 nm and 5% per 30 °C. Find the initial RGB luminance ratio (at 25 °C) of a white light source which emits white light at CIE coordinates of (0.33, 0.33) when operating at 85 °C.

	Red		Green		Blue	
λ (nm)	650	660	550	555	450	455
X	0.725	0.730	0.302	0.337	0.157	0.151
Y	0.275	0.270	0.692	0.659	0.018	0.023
Z	0	0	0.006	0.004	0.825	0.826
$V(\lambda)$	0.107	0.032	0.99 495	1	0.038	0.048

Answer

After a long operating time, the emission wavelength of the red LED shifts and stabilizes at 660 nm.

$$x = \frac{X_r}{X_r + Y_r + Z_r} = 0.73;\quad y = \frac{Y_r}{X_r + Y_r + Z_r} = 0.27;\quad z = \frac{Z_r}{X_r + Y_r + Z_r} = 0 \rightarrow X_r = 2.7Y_r,\quad Z_r = 0$$

Similarly, we can find the relation between X, Y and Z for green (555 nm) and blue (455 nm) LEDs. $\rightarrow X_g = 0.5114\ Y_g$, $Z_g = 0.006\ Y_g \rightarrow X_b = 6.5652\ Y_b$, $Z_b = 35.913\ Y_b$ When white light is obtained from these RGB-LEDs, its X, Y and Z values are:

$$X_w = X_r + X_g + X_b = 2.7Y_r + 0.5114Y_g + 6.5652Y_b$$
$$Y_w = Y_r + Y_g + Y_b$$
$$Z_w = Z_r + Z_g + Z_b = 0.006Y_g + 35.913Y_b$$

For a white light source at (0.33, 0.33)

$$\frac{X_w}{X_w + Y_w + Z_w} = 0.33,\ \frac{Y_w}{X_w + Y_w + Z_w} = 0.33 \text{ and } \frac{Z_w}{X_w + Y_w + Z_w} = 1 - 0.33 - 0.33$$

Hence, one can find that $Y_r{:}Y_g{:}Y_b = 5.02{:}28.857{:}1$ at 85 °C.

In order to compensate for the thermal effect on emission intensity, the initial luminance ratio of the RGB-LEDs at 25C would be $\frac{Y_r}{(1-0.4)} : \frac{Y_g}{(1-0.1)} : \frac{Y_b}{(1-0.1)} = 7.53 : 28.857 : 1$

5.4.4 Efficiency Droop

As shown in Eq. (5.34), as the current density rises, the light output also increases linearly. This suggests that it is possible to use a very small LED to achieve super high light output, just by increasing the current without limit. Of course, this is not possible. For example, a higher current density results in a temperature increase (as shown in Figure 5.22) which decreases the current efficiency of the LED. Even if the temperature effect is excluded, an efficiency decrease is still observed, especially for applications requiring high optical power such as general lighting, using a III–N (blue or UV) LED to pump a phosphor [38]. It is found that the EQE increases then decreases as the current rises, which is called the "droop effect" and is shown in Figure 5.23a. The different curves show the EQE versus current of LEDs at different temperatures; the decrease in EQE with increasing temperature is called "temperature droop." When the temperature is fixed, the EQE increases sharply and decreases gradually. Such a reduction at high current density is called "current droop" [39].

When the current density is high, the excess carrier density is higher than the majority carrier density and hence Eq. (6.26) is modified to:

$$R_{\text{rad}} = r_{\text{rad}}np = r_{\text{rad}}n(p_0 + \Delta p) \sim r_{\text{rad}}n(\Delta p) \tag{5.35}$$

Because the number of the excess electrons and holes is the same, it can also be written as:

$$R_{\text{rad}} \sim r_{\text{rad}}n^2 \tag{5.36}$$

Auger recombination is a multi-electron process, which is frequently observed in semiconductors. Figure 5.23b shows the E–k relation of the semiconductor. When an electron relaxes its energy from the conduction to the valence band, it is possible for its energy to be transferred to a neighboring electron in the conduction band. This promotes the electron to a higher energy in the conduction band as a "hot" electron. In this process, not only energy but also momentum must be conserved. Hence, in Figure 5.23b, one electron goes up-right, and the other goes down-left. Note that there are three carriers involved in this process (an electron recombines with a hole which promotes a second electron). Under high current conditions, the total recombination rate in Eq. (5.29) is usually rewritten as:

$$R_{\text{rad}} = An + Bn^2 + Cn^3 \tag{5.37}$$

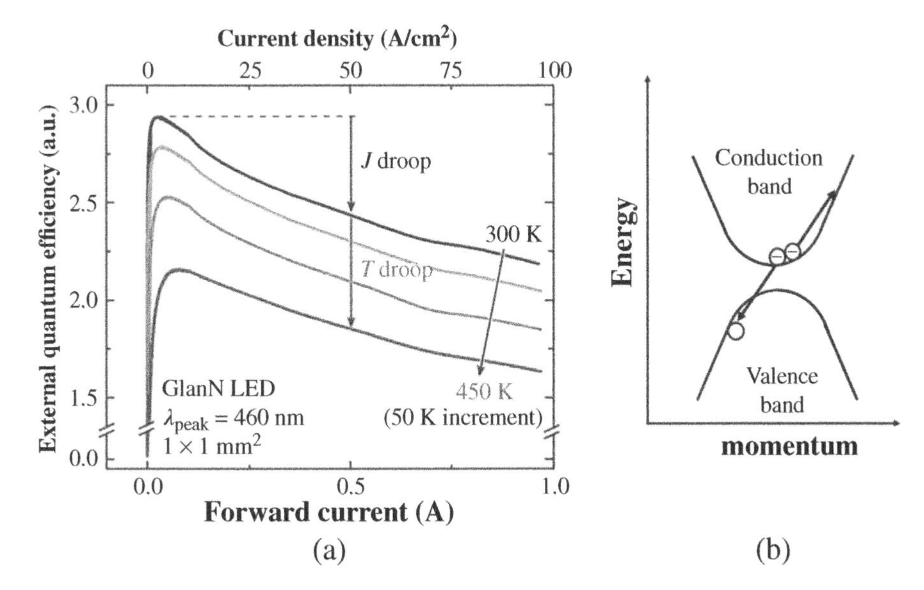

Figure 5.23 (a) EQE versus electrical current for the LED at different temperatures. (b) Schematic diagram of Auger recombination.

where A, Bn, and Cn^2 represent the rates of SRH recombination, radiative recombination, and Auger recombination. The internal quantum efficiency (IQE) is then:

$$\text{IQE} = Bn^2/(An + Bn^2 + Cn^3) = B/(A/n + B + Cn) \tag{5.38}$$

As shown in Eqs. (5.37) and (5.38), when the current density is low (small n), SRH recombination dominates and IQE increases with an increase in the current density. Then, when the current density reaches a certain level (high n), Auger recombination becomes significant and reduces the efficiency again.

In semiconductor materials, the electron mobility is typically higher than that of holes. Hence, under high current conditions, electrons may overflow from the QW region and leak to the p-side which can cause an efficiency droop. Careful design of the LED device is needed to achieve a uniform current density distribution in the "lateral" direction (perpendicular to the p–n junction), and this will be discussed in Sections 6.5.2 and 6.5.3. Otherwise, the region with the highest current density suffers the most serious droop effect, which degrades the efficiency of the LED under high current drive. The III–P material systems are typically not used for general lighting and hence LEDs with super-high optical power are not generally required. Also, due to the long diffusion length, SRH recombination is a more serious problem than efficiency droop. Electron leakage is more serious than Auger recombination for efficiency droop, because the QW structure in III–P LEDs is quite shallow, and the electrons overflow easily at high current density.

On the other hand, for III–N based LEDs, remember that the dislocation density of the (In)GaN material is high (10^8 cm^{-3}). Indium concentration fluctuations play a role in confining carriers for recombination before they are trapped by defect sites. This is an important mechanism for increasing the efficiency of InGaN-LEDs with a high dislocation density. However, at high current densities, indium fluctuation results in localization of the carriers, which results in an increase of local current density and efficiency droop. Carriers may overflow from the In-rich region and be captured by defect states, which also reduces the efficiency. Polarization plays some role in the droop effect in nitride-based OLEDs. The small overlap of the electron–hole wavefunctions in nitride-based QWs, reduces the SRH and recombination efficiency, and so effectively enhances the effects of Auger recombination. The polarization effect at the QW/EBL interface may help electron leakage and retard hole injection, which results in a more serious efficiency droop.

To reduce the efficiency droop (mainly in nitride-based LED), the possible strategies include: (i) to reduce the current density inside the active region, and (ii) to prevent electron leakage outside the active region. A straightforward method to reduce the current density is to increase the QW thickness and number. Reducing polarization in the QW broadens the "effective" active region, and hence decreases the current density. Due to the low conductivity of the p-type GaN layer, insufficient current spreading results in carrier crowding in the lateral direction, which also results in droop effect. To prevent the electron leakage, suitable design of the EBL together with improved hole-injection are necessary. Control of the polarization in the epi-layers can also help for carrier confinement.

5.5 DEVICE FABRICATION

Fabrication of LEDs begins with epitaxial growth on a crystalline substrate. There are several epitaxy techniques, including LPE, VPE, and MOCVD. LPE is the simplest and most economic technique for obtaining a thick epi-layer under thermodynamic equilibrium conditions, but is limited to semiconductor compositions near the lattice-match condition. The fast growth rate (2 μm/min) results in difficulty in obtaining thin layers, such as quantum well structures. On the other hand, VPE provides more flexibility to tune the alloy composition by controlling the gas flow of group III and V materials independently in the vapor phase. According to the chemistry of the vapor sources, there are chloride, hydride, and organometallic sub-types of VPE. Typically, VPE with organometallic sources is referred to as MOCVD and is one of the most common ways for obtaining complex layer thicknesses, composition, and doping profiles. Once a high-performance epi-layer is obtained, the next step is to fabricate a p–n junction. LED fabrication is relatively simple, compared to that of other III–V devices such as laser diodes and transistors. Suitable electrode materials and annealing processes must be used to obtain ohmic contacts and low serial resistance. The electrode layout must be well-designed since the reflective metal blocks light emission. To improve the extraction efficiency, the epitaxial layer should be well designed to allow light emission from an "escape cone" formed by the bottom, top, and side walls. Chip shaping and surface engineering are helpful to redirect the light radiating out of the semiconductor into the epoxy encapsulation layer and prevent light trapping by waveguiding effects. The refractive index of the epoxy should be as high as possible to increase the critical angle at the semiconductor/epoxy interfaces, and the shape of the epoxy is curved to improve light extraction from the epoxy to the air. The package process is important since it not only protects the LED chips and improves the extraction efficiency, but also improves the thermal dissipation which is an important issue for high power applications.

5.5.1 Epitaxy

LPE was the first epitaxial technique for LED mass production, due to its simple process, low cost, and high growth rate. A schematic illustration of the LPE process is shown in Figure 5.24 [40]. Different saturated solutions of the epitaxy materials in a high temperature solvent system, are placed in different source holders (solution bins). The solvent is commonly a molten metal or salt. A GaAs substrate is mounted on a holder which allows the substrate to be brought into contact with different melts by moving the push rod. Each melt allows growth of a layer with a different composition or doping profile. Once the source and the GaAs substrate are brought into contact, the temperature is decreased, and the solute atoms in the liquid phase grow under thermal equilibrium as a monocrystalline layer aligned with the lattice of the crystalline substrate. The thickness of each epi-layer is controlled by the temperature difference maintained between the source solution and the substrate, and the contact time. Since the growth temperature is high (about 700–900 °C), the transfer time between the different melts should be as short as possible to avoid any oxidation occurring, which would be a cause of defect formation at the layer interfaces. LPE has a very high growth rate (up to 2 μm/min) which is suitable for mass production. However, it also raises difficulties for epitaxial growth of thin layers, such quantum well structures.

VPE uses gas sources, rather than the liquid sources of LPE. Hence, it is easier to control the epitaxial thickness and doping profile, and to grow multilayer structures by adjusting the source gas flow. There are different kinds of VPE systems which can be distinguished by the nature of the sources used, such as chloride and hydride VPE, and MOCVD, as shown in Table 5.2. Chloride VPE uses solid and gaseous sources for the group III and V elements, respectively. The group V chloride first reacts with H_2 and forms group V molecules and HCl in gas phase:

$$4VCl_3 + 6H_2 \rightarrow V_4 + 12HCl \tag{5.39}$$

where V denotes one or more group-V elements (e.g. As and P). The resulting HCl then reacts with a solid element or binary compound which forms the source of the group III element and forms a chloride group III

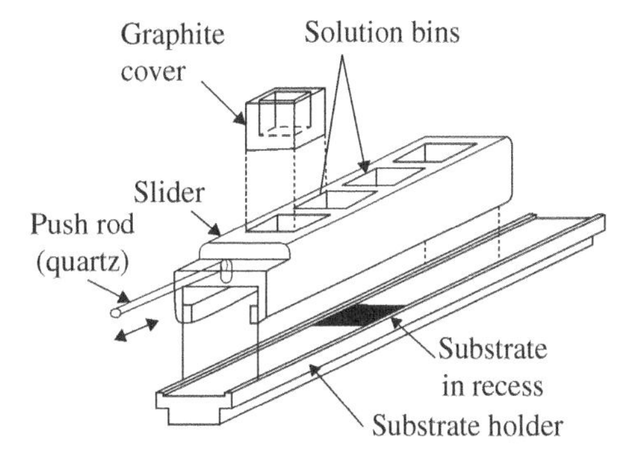

Figure 5.24 Schematic diagram of LPE system.

Table 5.2 Sources used for different VPE systems.

	Group III	Group V	Gas flow
Chloride VPE	Elements (In, Ga...) or binary sources (GaAs, InP, InAs)	Chloride ($AsCl_3$, PCl_3...)	H_2
Hydride VPE	Atoms (Ga, In...)	Hydride (AsH_3, PH_3, NH_3...)	HCl and H_2
MOCVD	Organometallic materials (TMGa, TMAl, TMIn...)	Hydride (AsH_3, PH_3, NH_3...)	H_2, N_2

source. In the case of a binary compound precursor, one can obtain:

$$4\text{IIIV} + 12\,\text{HCl} \rightarrow 4\text{IIICl}_3 + \text{V}_4 + 6\text{H}_2 \tag{5.40}$$

where III means a group-III element (e.g. In or Ga). Note that this reaction is reversible. When the temperature is high (which is the case in the region of the solid source material), the equilibrium tends to shift toward the right-hand side, so the solid group III sources of are etched and form group III gas-phase intermediates. Then, at a lower temperature (where the substrate is placed), the equilibrium shifts to the left which drives formation of an epitaxial layer on the crystalline substrate. Figure 5.25a shows the schematic diagram of a chloride VPE system [41].

One can note that there are successive reactions in chloride VPE, which mean that the amount of group-III intermediate formed is determined by the quantity of group-V source. In hydride VPE, the group V sources are replaced by AsH_3 and PH_3, which are then cracked at high temperature to form As_4 and P_4 [42]. In parallel, HCl gas is injected to react with a group III metal and generate the gaseous source of the group III element:

$$2\text{III} + 6\text{HCl} \rightarrow 2\text{IIICl}_3 + 3\text{H}_2 \tag{5.41}$$

Hence, in hydride VPE (shown in Figure 5.25b) it is possible to control the amounts of the group III and V sources independently, which makes it more flexible for epitaxial growth compared with chloride VPE. The deposition reaction for hydride VPE is the same as in chloride epitaxy. One of the disadvantages of hydride VPE is the use of the toxic source gases arsine (AsH_3) and phosphine (PH_3).

By using metal alkyls, which are liquids or waxy solids at room temperature, the melting points of the group-III sources can be greatly reduced. Epitaxial growth using metal alkyls and hydrides as sources for groups III and V respectively, is called MOCVD, or organometallic VPE (OMVPE), and the reactor configuration is shown in Figure 5.26. A carrier gas, such as a mixture of H_2 and N_2, is passed through the group III alkyls and transports vapor into the growth chamber, so by controlling the gas flow, the delivery rate of the group-III sources can be controlled precisely. Typical group-III sources are trimethylaluminum $Al(CH_3)_3$ (TMAl), trimethylgallium $Ga(CH_3)_3$ (TMGa), and trimethylindium $In(CH_3)_3$. The reaction which occurs in MOCVD can be represented as:

$$x\text{Al(CH}_3)_3 + \text{yGa(CH}_3)_3 + \text{zIn(CH}_3)_3 + \text{VH}_3 \rightarrow \text{Al}_x\text{Ga}_y\text{In}_z\text{V} + 4\text{CH}_4 \tag{5.42}$$

where V denotes a group-V element (e.g. As, P and N). Note that the reaction is irreversible which means MOCVD is an example of non-equilibrium growth. Non-equilibrium growth brings several advantages which include: (i) The possibility to grow abrupt junctions and thin epi-layers, and (ii) greater flexibility in varying the layer composition. These features are very important in LED applications when quantum structures and different emission wavelengths are required [43]. The capability for precise rate control, implies a low growth rate (0.1–10 μm/h) and this can be viewed as a disadvantage of MOCVD.

As shown in Figure 5.4, the growth mode of nitride-based materials onto a sapphire substrate is not planar. Hence, a low temperature buffer layer (of GaN or AlN) is needed, but the resulting dislocation density is still high ($\sim 10^8$ cm^{-2}). To further improve the film quality, the MOCVD equipment must be modified. For example,

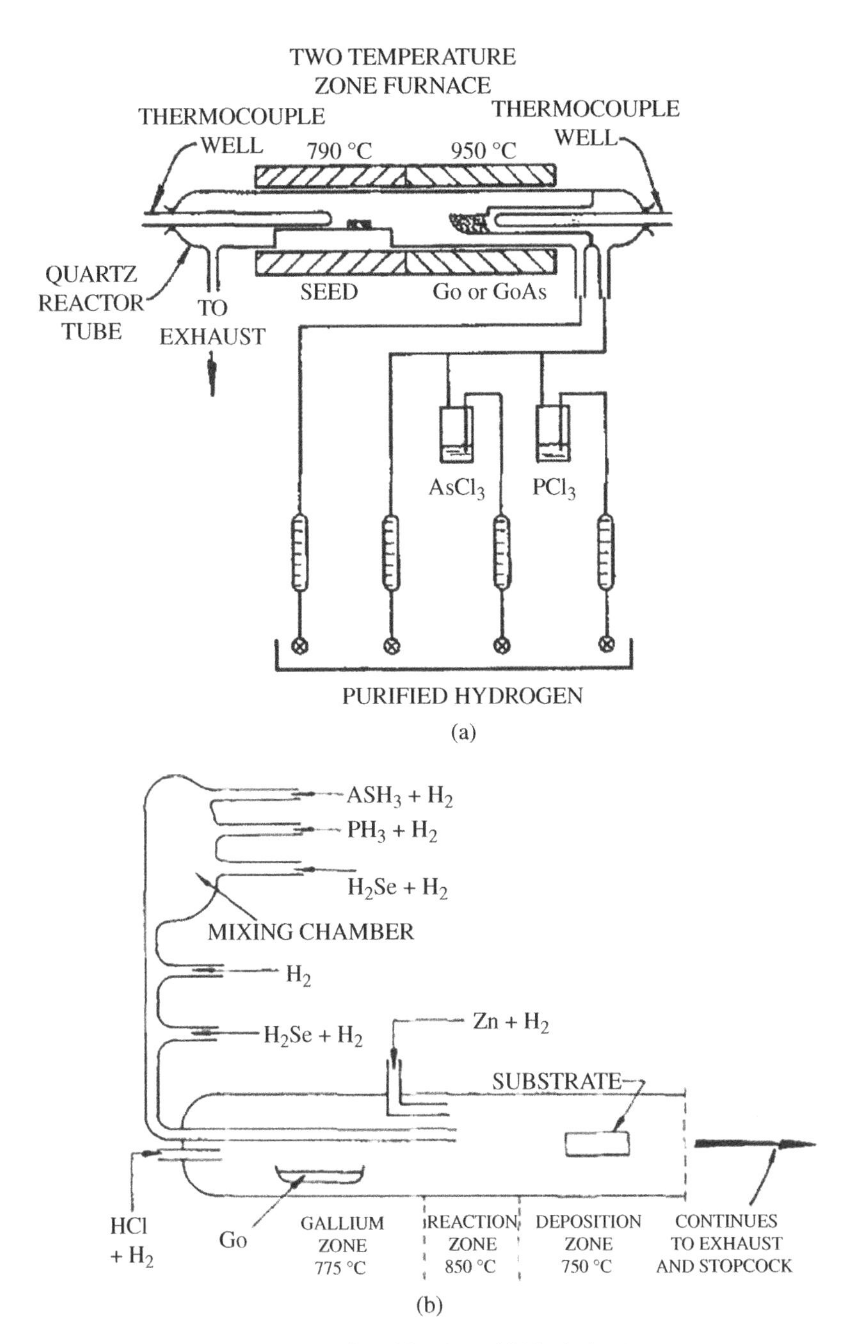

Figure 5.25 Schematic diagram of (a) chloride and (b) hydride VPE systems.

a two-flow system as shown in Figure 5.27a [44] has been introduced. The horizontal main flow carries the reaction gas (H_2, NH_3, and TMGa) onto the surface of the substrate, and an additional vertical sub-flow of N_2 and H_2 is introduced which enhances the reaction at the substrate surface. The outcome is a reduction in 3D growth and dislocation density. Due to the 2D growth characteristics of the two-flow system, it is possible to incorporate indium into GaN. That is an important breakthrough for fabricating blue OLEDs. With careful

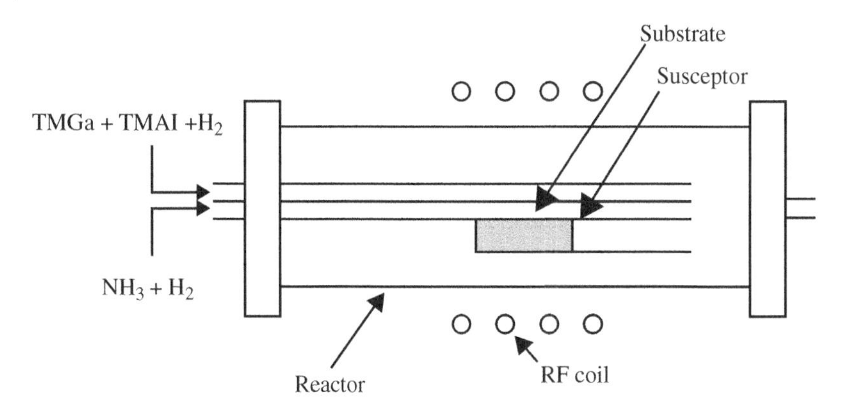

Figure 5.26 Schematic diagram of MOCVD systems.

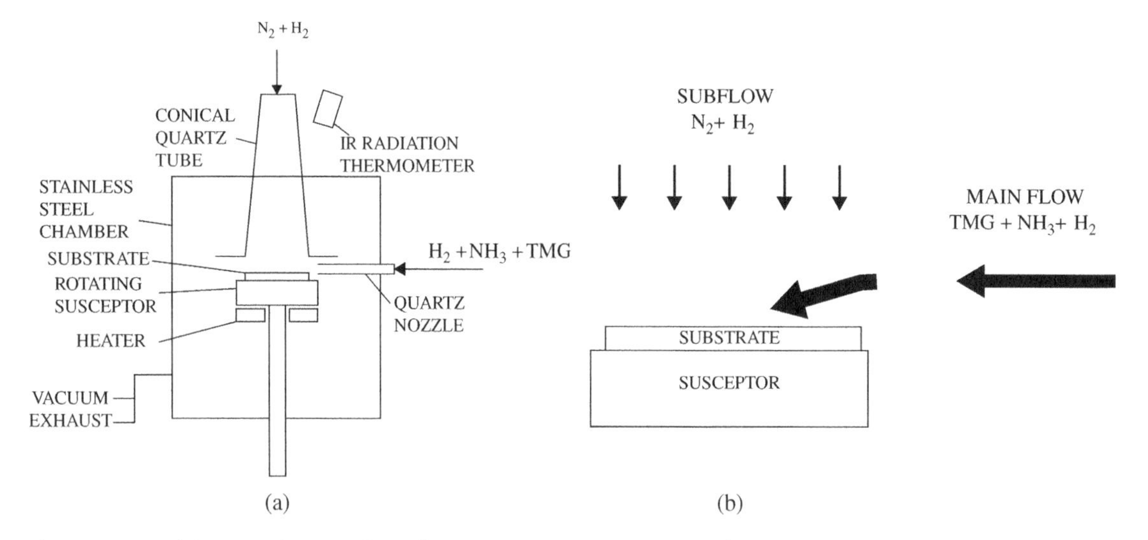

Figure 5.27 (a) Schematic diagram of two-flow MOCVD systems, and (b) gas flow in this system.

design of the gas flow in the chamber, MOCVD reactors with different geometries (such as horizontal and vertical gas flow) have been introduced for mass production.

We can also use the 3D nature of the growth of III-N to obtain a high quality film. A low temperature III-nitride film may be grown and lithographically patterned into a micro- or nano-rod morphology. Then, a second round of epitaxial growth is carried out by high temperature MOCVD. This second III-nitride film begins to grow on the top of the rods. Due to the 3D nature of the growth, it tends to grow laterally and the spacing between the rods allows development of a higher quality film. The high temperature during the second growth also anneals the rods and improves the film quality. This process is called epitaxial lateral overgrowth (ELOG) [45]. Using this technique, the dislocation density can be reduced to lower than 10^7 cm^{-2}. However, additional processing steps are needed: patterning and etching the first epitaxial layer and a second MOCVD growth. An alternative is to use a patterned sapphire substrate (PSS) [46]. Before epitaxial growth by MOCVD, the sapphire substrate is patterned by lithography and an etching process. Different shapes (such as pyramid, cone, and hemisphere) and different dimensions (hundreds nm to ~μm range) have been used. At the beginning of the epitaxial layer growth, deposition is faster in the lateral direction, as is the case in ELOG. Then, the dislocations curve, and the film quality improves in the top region of the epitaxial layer. A useful tool to determine the film quality is to measure the etch pit density (EPD). If the epitaxial film is solution etched, the dissolution rate is faster in the region of a defect, which results in etch pits which are visible under an optical microscope [47]. Due to the non-planar structure of the PSS, light extraction is enhanced by reducing total internal reflection in the active layer, which will be discussed further in Section 5.5.3.

5.5.2 Process Flow and Device Structure Design

Figure 5.28a,b shows typical device structures of III–P and III–N LEDs, respectively. Carriers are injected from the electrodes into the p- and n-layers of the LED, and then recombine in the active region. Photons are generated and are emitted from the LED structure through the top and both sides for the III–P and III–N LEDs, respectively. There are several requirements on the electrode design. First of all, good ohmic contacts are needed between the electrodes and the p- and n-layers for efficient carrier injection. Optimized metal materials and annealing processes can increase the conductivity and reduce the power consumption. Then, the current density should be uniform across the active layer, which means the resistance of the p- and n-layers should be as small as possible. However, it is easy to achieve low resistivity for the p-layer of nitride-based materials. By suitable design of the electrode layouts, carriers can be distributed uniformly across the junction area which results in a dramatic increase in luminous efficiency. A concern for the electrode design is that the area of the metallization should be kept as small as possible since it impedes light extraction. Due to the use of an insulating sapphire substrate, electrodes of nitride LEDs have to be fabricated on the one side of the substrate, which means that mesa etching is required. Then, the LED substrate is diced into individual chips for the packaging process, which will be discussed in Section 5.5.4. Prior to the dicing process, sometimes, a wafer thinning process is used to improve the production yield.

In principle, direct contact of the metal and semiconductor results in a Schottky diode, rather than an ohmic contact. The barrier between the metal work function and the carrier transport band (conduction and valence bands for electrons and holes, respectively) limits the current injected into the LED. This problem can be solved by mixing a dopant metal into the electrode deposition sources and annealing the sample. The dopant metal diffuses into the semiconductor and interacts with the cladding layer to form a thin heavily doped layer. Hence, an ohmic contact between electrode and semiconductor can be achieved. For AlInGaP and AlGaAs systems, Au–Ge and Au–Zn (or Au–Be) alloys are typically used for the n- and p-type electrodes, respectively. For p-type contacts, heavy doping can be included during epitaxial growth (e.g. p + -GaAs, p + -GaP and

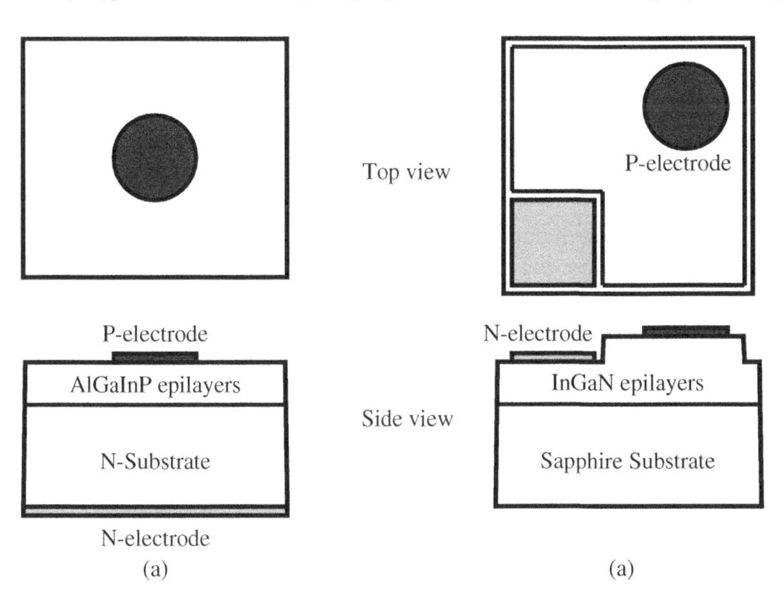

Figure 5.28 Device structures of (a) III-P and (b) III-N LEDs (www.epistar.com.tw).

p + -AlGaAs) to improve the conductivity. However, for nitride LEDs, heavily doped layers are difficult to achieve. Hence, a different approach is used to reduce the barrier height of the Schottky junction by using a metal with a suitable work function. Ti/Al and Ni/Au bilayers are typical metals used for n- and p-type metallization, respectively. Ti reacts with nitrides to form TiN which is conductive. Ni and Au exhibit high work functions thereby reducing the barrier between the electrode and the p-type semiconductor.

Typically, the lateral dimension of LED chips is in the region of hundreds of μm, which is much larger than the thickness of the p-cladding layer. Hence, holes injected from the anode recombine with electrons mostly under the opaque metal region, which results in a low efficiency. Also, since the recombination rate is proportional to the product of electron and hole concentrations, the non-uniformity of the hole distribution results in lower recombination efficiency. One simple remedy is to increase the thickness of the p-cladding layer, to improve the lateral current flow, but this results in a higher serial resistance. Another possible solution is to pattern the electrode. By using "finger" structures to assist the current spreading, the carrier distribution uniformity can be much improved. The area covered by the opaque electrode should be as small as possible since the metal impedes light extraction. For nitride LEDs, since an insulating substrate is used, the anode and the cathode are on the same side of the substrate, which results in even worse situation of current non-uniformity. Hence, both the p- and n-electrode layout must be taken into consideration. Transparent or semi-transparent electrodes such as ITO and thin metals (e.g. Ni/Au or Pt) can be used over the whole area of the device to improve current spreading in the p-type region. On the other hand, for III–P LEDs, ITO can form an ohmic contact only with a GaAs cladding layer. Since GaAs absorbs light in the visible region, the cladding layer should be as thin as possible to reduce the absorption loss.

5.5.3 Extraction Efficiency Improvement

As mentioned in Section 5.4.2, the light extraction efficiency from a LED is limited by internal reflection at the boundaries of the semiconductor planar structure. To improve it, two common methods are used: (i) redirecting the light by optimizing the shape of the entire device, and (ii) using a non-planar structure to couple out waveguided modes. By combining these methods with a properly designed package process (which will be discussed in Section 5.5.4) to increase the critical angle and redirect the light, the extraction efficiency can be effectively increased. For III–P devices, photons will also be absorbed by the GaAs substrate. It is possible to use a distributed-Bragg-reflector (DBR) structure between the substrate and the epi-layer to prevent this absorption and the DBR can act as a wavelength selector at the same time. However, the multi-layer structure increases the complexity and cost of the epitaxial growth. Methods are also available to fabricate transparent substrates, rather than using an absorbing one. Another issue arises from the metal electrodes which may impede the light output. A suitable layout design and device structure are needed to obtain a high extraction efficiency.

Applying Snell's law to the refractive index of the semiconductor (typically n_{semi}~3.5) and the outside medium (typically an epoxy resin used in the package process with, n_{epo}~1.5) provides the critical angle at the interface and hence define the "escape cone" shown in Figure 5.19. For the RI values above, θ~25.38°. Assuming the LED to be a perfect smooth-faced rectangular cuboid and that the light emission is isotropic, the six full escape cones represent the maximum attainable light extraction. In the case that the substrate is optically absorbing, obviously the downward escape cone does not exist. The physical dimensions of the LED are around several hundreds of micrometers in width and length. The thickness of the active layer is around 1 μm. When the thicknesses of the layers on each side of the active layer (typically called cladding layers) is small, photons generated at the active layer experience multiple reflections before escaping out of the LED, and the photons may be absorbed by the metal electrode, active layer, and the absorbing substrate. Hence, the cladding layer should be thick enough (>30 μm) and obviously it must be transparent to open up the four side cones. However, growth of such thick epilayers needs an impractically long time when using the MOCVD technique which is limited to a growth rate of ~1 μm/h.

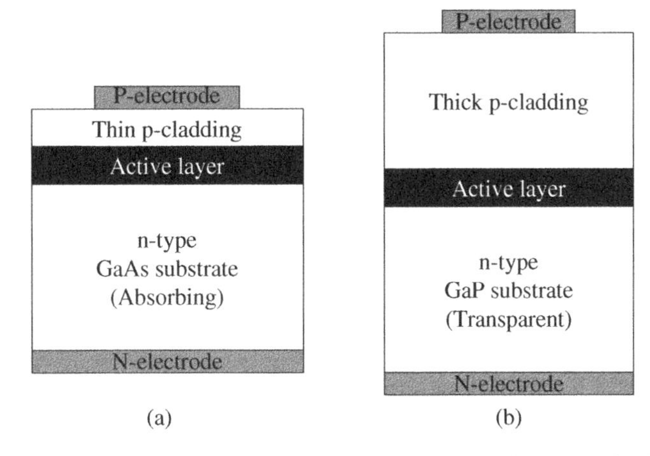

Figure 5.29 (a) Absorbing substrate with a thin cladding layer which has one full (upward) plus four partial (sidewall) escape-cones, and (b) transparent substrate with thick cladding layers which has six large escape-cones.

For the III–P/GaAs system, an upper cladding layer and substrate of transparent GaP is typically used, although there is still a 3.6% lattice mismatch to the GaAs. Using MOCVD, the GaP cladding thickness cannot be more than 15 μm which is not sufficient to open up the complete side cones. However, following MOCVD growth of the active layer, VPE growth can be used to deposit an upper cladding layer with a thickness up to 50 μm. The GaAs substrate can also be removed by an etching process, then the device can be bonded to a GaP substrate by high temperature annealing [48]. The resulting device structure is shown in Figure 5.29. Transparent substrates are intrinsically present in III-nitride devices. However, due to the difficulties in doping technology, it is not easy to provide a thick cladding layer. Also, there are high internal absorption losses in the active layer which attenuate light before it can propagate to an edge and escape from the device. A thick cladding may, therefore, not be very helpful in raising the extraction efficiency, and is normally not used. As has been stated, due to the poor conductivity of the semiconductor (especially when p-doped), a metal "mesh" electrode is typically used to improve the luminance uniformity and this in turn impedes the light emission, as shown in Figure 5.30a. To solve these issues a flip-chip structure has been proposed, as shown in Figure 5.30b [49]. Using an inverted structure which emits light through the transparent sapphire substrate, both good conductivity and high extraction efficiency can be obtained simultaneously. The refractive indexes of GaN and sapphire are 2.45 and 1.78, respectively, which means that light generated within the InGaN active region can

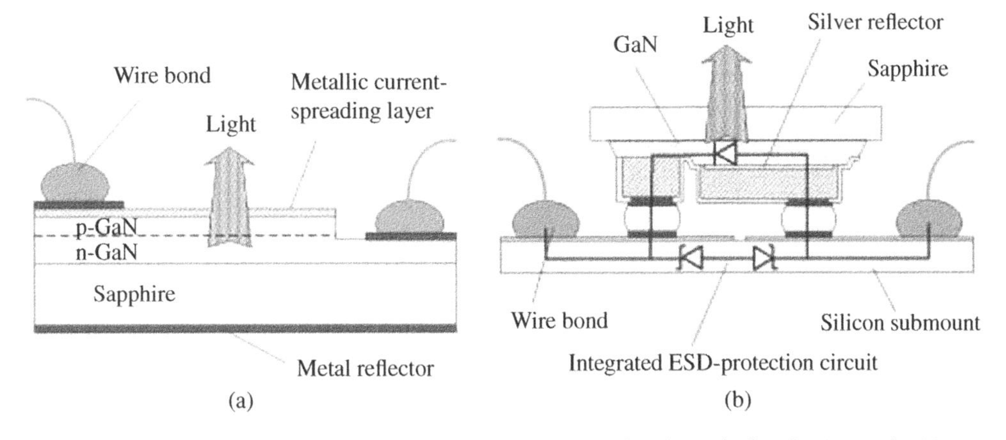

Figure 5.30 (a) Mesh electrode providing a metallic current spreading layer, (b) flip-chip III-nitride LEDs.

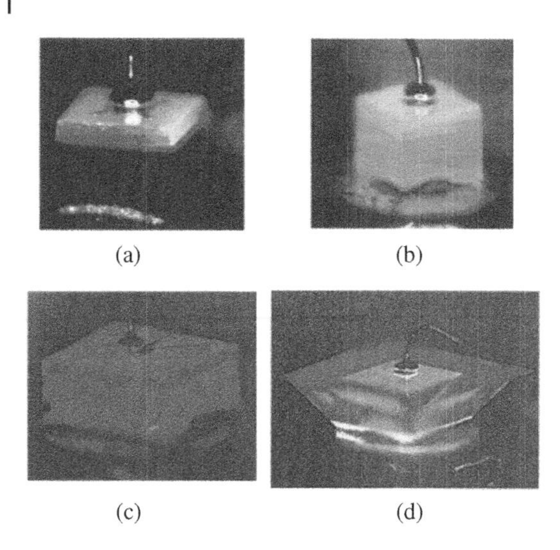

Figure 5.31 Pictures of LED under operation with different geometries: (a) absorbing substrate, (b) transparent substrate, (c) finger electrode arrangement, and (d) shaped die structure.

experience total internal reflection at the GaN/sapphire interface. With a carefully designed PSS (as discussed in Section 5.5.1), the light rays can be redirected to improve the light extraction efficiency, too [50].

The discussion above relates to maximizing the extraction of light from a rectangular cuboid. In practice, it is found that the side cones can extract more light than expected because the edge facets of the structure are not perfectly planar. In addition, changing the shape of the die can further improve the extraction efficiency by redirecting light to favor top-side emission. For example, a trapezoidal structure can reflect the sidewall emission upwards which effectively increases the extraction efficiency. Figure 5.31 shows the evolution of structures giving improved extraction efficiency from (a) a conventional absorbing substrate, to (b) a transparent substrate, (c) a "finger" electrode, and (d) a trapezoidal die structure.

Many efforts have been made to reduce the amount of total internal reflection at the top side of the LED. A straightforward method is to roughen the top surface which results in light scattering, rather than reflection. Periodic photonic micro- and nanostructures have been demonstrated which can couple light out of a waveguided mode by diffraction or plasmonic coupling. Fabrication of suitable structures still raises issues regarding processing, cost and device layout.

5.5.4 Packaging

The packaging process employed during typical semiconductor device fabrication, has the functions of: (i) protecting the LED and its electrodes from the mechanical and environmental damage, (ii) improving the optical extraction efficiency, and (iii) helping to dissipate heat which is especially important for high power applications. Figure 5.32 shows a packaged LED, in which a semiconductor chip with dimensions of several hundreds of μm has been mounted on a reflector cup, which redirects the light upwards. Bonding wire is used to connect the semiconductor to external electrodes which provide the electrical current. Here, a LED with a semiconducting substrate (e.g., AlGaInP on a GaAs substrate) is used as an example. Epoxy resin encapsulates the semiconductor, reflector and bonding wire for protection. The shape of the epoxy is typically designed to be a hemisphere which improves the extraction efficiency at the epoxy-air interface. At the same time, to improve light extraction at the semiconductor-epoxy interface, the refractive index of the epoxy is as high as possible consistent with low absorption in the visible range, thereby maximizing the escape cone. Typical values of the refractive index the epoxy resin range from around 1.5 to 1.8.

To increase the LED optical power, more current must be injected into the device which increases the temperature, and in turn tends to decrease the efficiency and red-shift the emission wavelength. Hence, for high-power LED applications, the thermal resistance of the package has to be taken into consideration in its

Figure 5.32 Illustration of a packaged LED.

design. Typically, a power LED chip is directly mounted onto a heatsink which dissipates heat from its bottom side. Silicone rather than epoxy is also typically used as an encapsulant due to its better thermal stability.

5.6 APPLICATIONS

In the early phase of their development (1960–1980), the applications of LEDs were limited to indicators, alphanumeric displays, and simple dot matrix displays for consumer products or industrial use. The colors of these LEDs were limited to the long wavelength part of the visible range (such as red) due to the lack of suitable semiconductor materials at shorter wavelengths. Recently, because of rapid developments in epitaxial growth of semiconductor materials and device efficiency improvements, high performance LEDs are available which can be used in many applications. In direct-view applications, LEDs are used in traffic signals, electronic signage and ultra-large area (e.g. stadium) displays. The main requirements for direct view displays are high efficiency (for low power consumption) and correct CIE coordinates. LEDs can also be used as backlights in LCD displays, where they bring the advantages of being Hg-free, and providing a high color gamut, long lifetime, and fast response, compared to conventional CCFL backlights. For small LED backlit displays (such as in mobile phones), color gamut is not a major issue. Hence, a white LED (blue + phosphors) can serve as the light source to reduce the cost and minimize the module size. For medium- or large-size LCDs, multi-color LEDs are used to increase the power efficiency and color gamut. For this application, the FWHM of the LEDs should be as small as possible to increase the color gamut value. Changes in the LED output spectrum and intensity due to the ambient temperature and operating time are also important for the display quality. The third category of LED application is in general lighting, which can also be employed as the light source for a reflective display. For this application, not only the CIE coordinates but also a high CRI value of the white LEDs is important to ensure that objects look similar under the LED and under natural light. However, as shown in Section 5.2.3, there is a trade-off between the power efficiency and CRI value.

5.6.1 Traffic Signals, Electronic Signage and Huge Displays

Specifications of the colors, luminous intensities and luminance used for traffic signals are important for safety reasons. Different countries (e.g. United States, Europe, and Japan) have standards which are generally similar but differ in detail. For example, in the United States, the specifications of traffic signals are regulated by the Institute of Transportation Engineers (ITE), which defines acceptable color ranges of five distinct colors for traffic signal lights – red, yellow, blue–green, Portland orange and lunar white. The first three colors are used for the standard red, yellow, and green lights, which indicate "stop," "caution," and "go." Portland orange and lunar white are used for pedestrian control signals, indicating "don't walk" and "walk," respectively.

Using LEDs in traffic signals to replace the conventional filtered incandescent lamp has two obvious advantages, which are lower power consumption and longer lifetime. The power efficiency of a typical incandescent lamp is about 14 lm/W. To obtain a red traffic light, the green and blue portions are filtered out which results in a power efficiency of 3–5 lm/W. When using AlInGaP red LEDs, the power efficiency can be as high as 30 lm/W, which provides a 10-fold reduction in power consumption. In contrast to the 2–3 years lifetime of an incandescent lamp, LEDs offer a 10 year operating lifetime with a consequent saving in maintenance costs. A filtered incandescent traffic signal uses a single bulb with a color filter. On the other hand, LED traffic signals consist of hundreds of LED pixels (~200–700 pixels). Failure of one or several LEDs will not significantly affect the function of the traffic signal. If the bulb of an incandescent traffic signal fails, immediate replacement is needed. Although the unit cost of the LED per lumen of output power is higher than that of the incandescent lamp, the energy and maintenance savings result in lower total cost when using LED technology in traffic signals. LEDs based on AlGaInP are typically used for red, orange, and yellow traffic lights. InGaN LEDs are used for green and white (blue or UV + phosphor) traffic lights.

Besides the price, two disadvantages of LED traffic signals are: (i) light output variation at different ambient temperatures, and (ii) driving circuit complexity. As mentioned in Section 5.4.4, the output intensity decreases and the spectrum redshifts as the temperature is increased. Within a typical range of ambient temperatures (−40 to 55 °C), the CIE shift is not obvious, and will not exceed the acceptable range of tolerance with proper device design. On the other hand, the output intensity variation is obvious, but can be compensated by suitable circuit design. Typically, incandescent lamps are driven by 120 V ac power provided by the traffic signal controller. To use the same controller, an ac–dc converter is needed and a string of LEDs are series connected. A voltage regulator, limiting resistor, or constant current source is needed to provide a uniform light output. To provide some redundancy, which ensures that the random failure of one LED pixel will not affect the functionality of the traffic signal, some parallel connections between different series strings are also needed.

Electronic signage and ultra-large displays are two further important applications for LEDs. Thanks to the rapid improvement of LED technology, the colors available for electronic signage have improved from monochrome (typically red) to multi-color (red, yellow, and green) and finally to full color (red, green, and blue). Compared with conventional electronic signal technologies using small filtered incandescent bulbs, LEDs offer the same advantages noted above, of low power consumption and long lifetime. Using LEDs a higher resolution can be obtained due to their small module size. Since the LED is passivated by the package process, and lacks the fragile glass envelopes and filaments used in bulbs, the LED module is also more rugged. In ultra-large display applications, there will be millions of pixels with three or four LEDs (red, green, and blue) in each. To achieve maximum white light efficiency, sometimes two red LEDs are used in a single pixel. Due to the high brightness of the LEDs, these huge displays can be viewed even under sunlight.

5.6.2 LCD Backlight

Using LEDs to replace the conventional CCFL as the backlight of an LCD has several obvious advantages, which include low power consumption, long lifetime, small size, and ruggedness, as described in the last section. The LED backlight has a much lower driving voltage (<5 V) than the CCFL which is an advantage both for driving-circuit design and safety. Due to the fast response of the LED, it is very easy to switch the LED's backlights on and off to insert "black frames" to eliminate motion blur which arises in hold-type CCFL-LCDs due to their impulse-type characteristics, as discussed in Section 4.11. Another important advantage of LEDs compared to CCFL is that they are Hg-free and environmentally friendly. However, the optical requirements on LEDs used in LCD backlights include: (i) a narrow emission spectrum matched to the required CIE coordinates of each primary color and the transmission curve of the color filter, and (ii) luminance uniformity over the whole panel. Figure 5.33 shows the emission spectra of CCFL and RGB-LEDs, and the transmittance spectra of RGB color filters. The emission spectrum of the CCFL has multiple peaks, of which the sharp emissions at 434, 542, and 610 nm correspond to red, green, and blue colors, respectively. The peaks at 486 and 585 nm lie within the overlapping transmission ranges of the blue/green and green/red color filters respectively and result in a decrease in color saturation. Typically, a CCFL-LCD has a color gamut

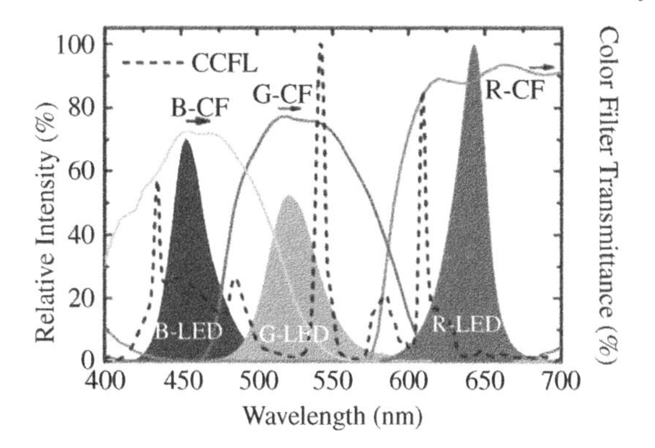

Figure 5.33 Emission spectra of CCFL and RGB-LEDs, and transmission spectra of RGB-color filters.

of 72% NTSC. To enlarge the color gamut, one has to decrease the FWHM of the transmission peaks of the color filter. This requires thicker color filter films which have lower optical transmission and hence give lower overall power efficiency. In contrast to this, RGB-LEDs provide a three-color light source with their FWHMs (typically <50 nm) narrower than those of the color filters which results in a high color saturation and gamut (105% NTSC) [51]. Phosphors and quantum dots pumped by blue LEDs offer another approach to backlight design, as shown in Section 4.12.

When using red, green, and blue LEDs in a backlight, they can be driven individually which makes it possible to emit different colors in series to achieve a filterless LCD, which is called the color sequential or field sequential technique, as shown in Figure 5.34 [52]. In this example, the pixel is intended to display a yellow color, which is a mixture of red and green primary colors. In a conventional LCD, the liquid crystal is switched to transmit light through the red and green subpixels and block light from passing through the blue one. A yellow color is perceived due to the mixing of light from the red and green subpixels. On the other hand, in color sequential driving, blue, green, and red LEDs are switched on sequentially in different time slots. No subpixel or color filter is needed. In this case, the LC pixel is switched to a non-transmissive state when the blue light is turned on, but to a transparent state during the other time slots, allowing red and green light to pass though. The displayed color gamut is defined by the color coordinates of the LEDs, rather than by color filters. Without the color filter, the fabrication process is simpler and panel cost is reduced. Also, since the color filter absorbs 2/3 and transmits only 1/3 of the incident light, the color sequential technique can significantly reduce the power consumption. However, the different colors are lit in series, at a rate which must be fast enough for the eye to fuse them into a uniform perceived resultant. This in turn means that the response of the LC panel must be fast enough. The three-color LED methods described above are suitable for medium- and large-size LCDs, such as in monitor and TV applications. For small-size LCDs (especially for mobile use),

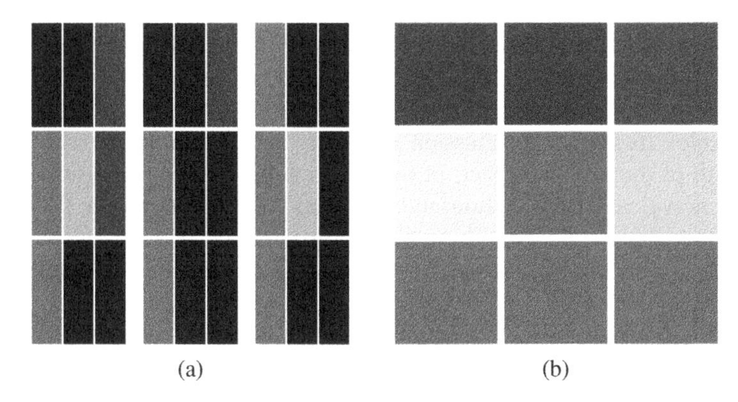

Figure 5.34 Illustrations of (a) conventional driving, and (b) color sequential technique [53].

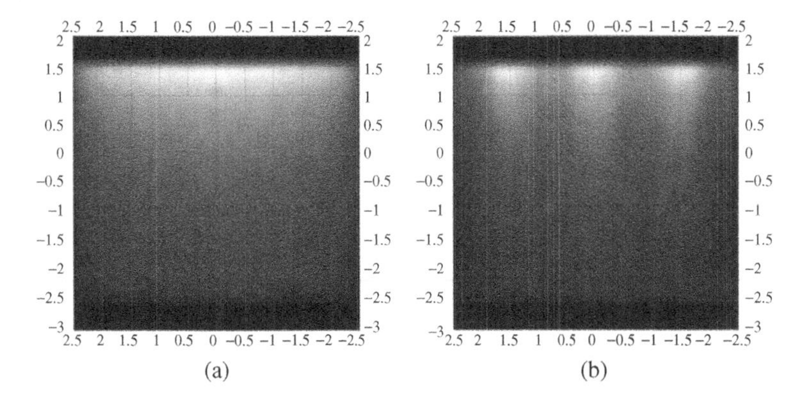

Figure 5.35 Simulation results of luminance distributions of (a) CCFL, and (b) LED backlights.

module size is more critical than color performances and hence white backlight illumination based on blue LEDs plus phosphor are usually used.

As LEDs are point-like light sources, the optical design required to obtain uniform luminance over a panel is more complex than when a CCFL is used. Figure 5.35 shows simulated luminance distributions from backlights using a conventional linear-type CCFL and point-type LEDs. The light source is hidden at the top edge. For the CCFL, the luminance decreases gradually from the top to the bottom. On the other hand, obvious high luminance spots can be observed from the LED chips, which results in a serious non-uniformity. The effect is even more serious for "direct" backlight modules, which are essential components in large LCDs. Careful optical design is needed to achieve a uniform luminance distribution in a LED-backlight system.

The thermal stability and operating lifetime should also be considered. The LCD temperature increases during operation which results in reduced luminance and wavelength redshifts of the LEDs. Since the operational lifetimes of red, green, and blue LEDs are different, differential aging also results in a color shift of the white point after long term operation. Photodetectors can be implemented on the panel to compensate the brightness loss and correct the color performance.

5.6.3 General Lighting

Requirements for general lighting include high power efficiency and high CRI. Hence, a light source with a broadband spectrum is needed. Manmade light sources based on flames, including torches, candles, and gas lighting, were universally used from prehistoric times to well into the nineteenth century and still find some use today. However, the efficiency is low (<1 lm/W), as can be understood from the heat dissipated alongside the light emission. This association explains why people feel "warm" under this kind of yellowish illumination. Typically, this lighting technology involves conversion from chemical to optical energy. Incandescent bulbs and fluorescent lamps are two common lighting sources relying on electrical power. Electric power heats the tungsten filament of an incandescent lamp without causing any chemical reaction. The temperature of the tungsten wire (about 2856 K) determines its color. Its blackbody emission provides a broadband source with very good CRI. The peak emission wavelength of the bulb is in fact, in the infrared rather than the visible region. The lifetime of an incandescent bulb is typically several thousands of hours. During operation, the tungsten will gradually evaporate onto the inner side of the glass thinning the tungsten wire until it burns out. A lower filament temperature results in longer life but shifts the emission peak to longer wavelength and results in lower power efficiency, typically less than 15 lm/W.

The operating principle of a fluorescent lamp is based on the relaxation of excited atoms created in a gas discharge under low pressure. UV emission from a transition from the excited state to the ground state of the gas (e.g. Hg vapor) excites phosphors which then generate visible light. The efficiency of a fluorescent lamp can be several times higher than that of incandescent lamps, which is reflected in the operating temperature of the fluorescent tube being much lower than that of a tungsten bulb. The lifetime of a fluorescent lamp can be as long as several tens of thousands of hours. However, due to the sharp emission peaks in the spectrum of fluorescent lamps, their CRI values typically range from 50 to 80. Due to the rapid improvement in efficiency of LEDs, they can provide an attractive technology for general lighting applications. In order to obtain a high CRI, blue and UV nitride LEDs with phosphors are typically used to broaden the spectrum. Multi-LED modules can be used for higher efficiency. However, since the FWHM of any single LED is about 50 nm, more than three LEDs or a specially engineered LED layer structure may be needed to broaden the spectrum. A long lifetime (predicted >100 000 hours) and good resistance to the environment are two obvious advantages of LED lighting, compared with bulbs and fluorescent tubes, as discussed in Section 5.6.1. LED lighting is also environmentally friendly. Compared with an incandescent bulb, the LED has a much higher efficiency; hence, less electric power is consumed and carbon dioxide emissions are reduced. Although the efficiency of fluorescent lamps is comparable to (or slightly lower than) that of LEDs, the Hg vapor used in fluorescent lamps causes serious environmental concerns.

5.6.4 Micro-LEDs

In a LCD display, LEDs can be used in the backlight system and the LCD provides the optical switch. The transmittance of the LC cell is limited – mainly by the polarizers – to less than 40%, which reduces the power efficiency of the display. As shown in Section 5.6.1, LEDs can also be used for ultra-large displays (over 100 in.) directly without a LC panel. In this application, millions of LEDs are used. Due to the package size of each LED, the pixel size (including red, green, and blue subpixels) is in the range of millimeters and hence the viewing distance of such a display is typically ~10 m or more. In other words, if the pixel size of LED panels can be shrunk, it may be possible to use them directly in mass market displays without an LC switch, an approach which is called "micro-LED."

For ultra-large displays, LED modules are assembled after packaging. However, for micro-LED arrays, to maintain high resolution, pixel integration has to be done before the packaging process. The size of the display is also limited by that of the substrate. Hence, micro-LEDs are most appropriate for small displays with high resolution and high brightness, such as those used in watches or projectors. Integration of drive TFTs is needed to address the LEDs for high-resolution displays. As shown in the bottom figure of Figure 5.36a, the circuits are fabricated on a Si-substrate [54]. The III-nitride LEDs are fabricated on a sapphire substrate and are connected to the circuit substrate in a flip-chip configuration similar to the structure in Figure 5.30b. The top figure of Figure 5.36a shows an SEM image of the "microtube" anisotropic conductor connecting the top and bottom substrate. It is difficult to have a full color panel fabricated on a single substrate. Down-conversion by quantum dots (QD) is a possible solution for achieving full color from a monolithic InGaN LED array on a sapphire wafer. Another possibility to achieve full color is by using a tandem structure. As shown in Figure 5.36b, by stacking the red, green, and blue panels from the bottom to the top, the resolution of the display can be greatly improved [55]. The sidewall reflects the light and confines it to the pixel region which has a dimension of ~10 μm. Light propagating inside the planar structure is coupled out by the vertical waveguide structure with a diameter of ~1 μm.

LEDs must be grown on a crystalline substrate (i.e. GaAs and sapphire for III–P and III–N, respectively), which hinders the application of LEDs to displays larger than the substrate size. Otherwise, a "discrete" method is needed which reduces the resolution (similar to the huge displays described in Section 5.6.1). To surpass the size limitations of the substrate while providing a color display, it is possible to transfer the LEDs to a soft mold, and then to an arbitrary substrate [56]. First, the LEDs were grown on a "native" substrate (GaAs or sapphire). Then, LED mesas were defined by an etching process with high resolution. During epitaxial growth a sacrificial layer which can be selectively etched, was deposited between the substrate and LED. An anchor was formed at the corner of each LED using photoresist to fix the location of the LED. A schematic drawing is shown at the left of Figure 5.37. Here, blue LED arrays were fabricated on the native substrate. Then, a patterned polydimethylsiloxane (PDMS) mold was stamped onto the LED substrate and lifted away selected

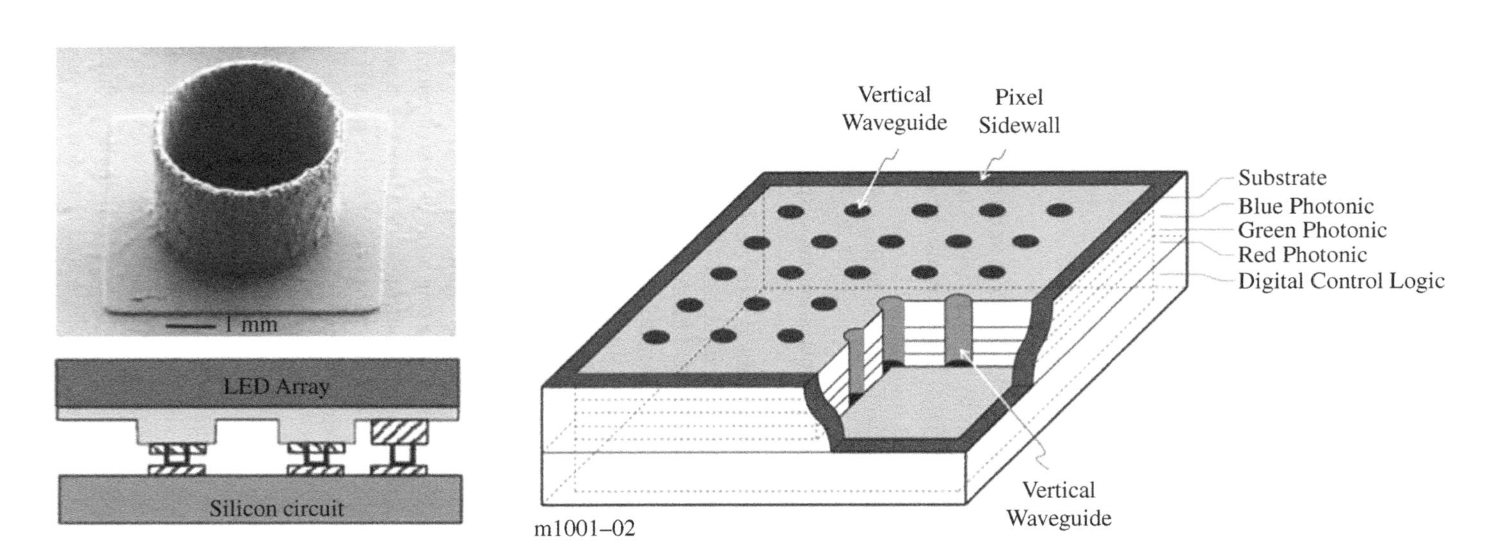

Figure 5.36 (a) Integration of micro-LEDs with a driving circuit on a Si wafer. (b) Tandem micro-LED structure.

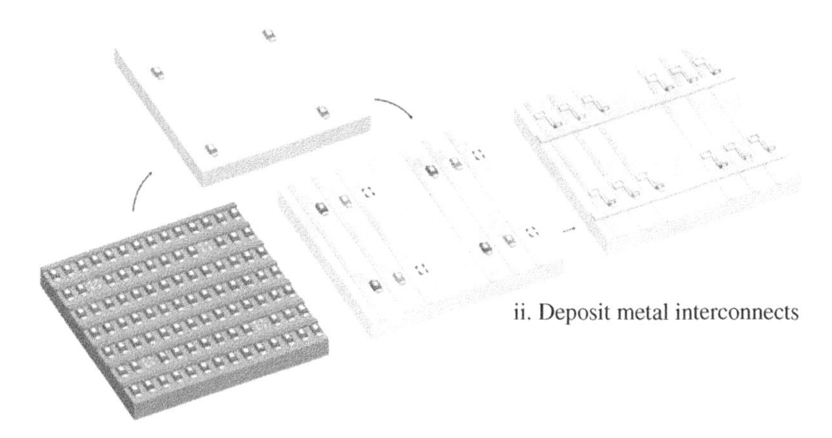

Figure 5.37 Stamping process to selectively transfer micro-LEDs from the native to the destination substrate.

LEDs. So, as shown in the left figure in Figure 5.37, some of the blue LEDs are missing on the native substrate having transferred to the PDMS mold. They can then be transferred to any substrate. For example, the figure shows a destination substrate with drive circuits, and with red and green LEDs already in place. Finally, after all the LEDs were transferred, metal interconnects were deposited to complete the fabrication. Using multiple stamping operations, red, green, and blue LEDs can be placed in suitable positions. There is no limitation on the display size. By fabricating TFTs on the final substrate, AM-LED panels can be achieved.

Homework Problems

5.1 Can one measure the built-in voltage of an LED by using a voltmeter? Why?

5.2 Draw the energy band diagram and current flows (including electron, hole, drift, and diffusion currents) under equilibrium, forward, and reverse bias.

5.3 What is the maximum attainable luminous efficiency of a D65 ($x = 0.313$, $y = 0.329$) light source which operates by mixing light from multiple LEDs (assume each LED emits at a single wavelength)? Is it a di- or tri-chromatic white light? What are the emission wavelengths of the LEDs?

5.4 A p-type semiconductor is uniformly photo-excited at $t < 0$. Then, the excitation is turned off at $t = 0^+$. Find the time-dependent minority carrier concentration $n_p(t)$.

Hint: use the continuity equation: $\frac{\partial n'_{\mathrm{p}}(x,t)}{\partial t} - \frac{1}{q}\frac{\partial J_{\mathrm{e}}(x,t)}{\partial x} = g_{\mathrm{ext}} - \frac{n'_{\mathrm{p}}(x,t)}{\tau_{\mathrm{min}}}$

where $n'_{\mathrm{p}}(x,t)n'_{\mathrm{p}}(x,t)$ is the excess concentration of minority carrier, $J_{\mathrm{e}}(x,t)$ is the electron current density, τ_{min} is the electron carrier lifetime, g_{ext} is the next generation rate due to the excitation source, and n_{p0} is the electron concentration under thermal equilibrium.

References

1 H. J. Round, "A note on carborundum," Electrical World. 19, 309 (1907).

2 Holonyak, N. Jr. and Bevacqua, S.F. (1962). Coherent (visible) light emission from $Ga(As_{1-x}P_x)$ junctions. *Appl. Phys. Lett.* 1: 82.

3 Groves, W.O., Herzog, A.H., and Craford, M.G. (1971). The effect of nitrogen doping on $GaAs_{1-x}P_x$ electroluminescent diodes. *Appl. Phys. Lett.* 19: 184.

4 Rupprecht, H., Woodall, J.M., and Petit, G.D. (1967). Efficient visible electroluminescence at 300 K from $Ga_{1-x}Al_xAs$ p–n junctions grown by liquid-phase epitaxy. *Appl. Phys. Lett.* 11: 81.

5 Kuo, C., Fletcher, R., Osentowski, T. et al. (1990). High performance AlGaInP visible light-emitting diodes. *Appl. Phys. Lett.* 57: 2937.

6 Amano, H., Sawaki, N., Akasaki, I., and Toyoda, Y. (1986). Metalorganic vapor phase epitaxial growth of a high quality GaN film using an AlN buffer layer. *Appl. Phys. Lett.* 48: 353.

7 Nakamura, S., Senoh, M., and Mukai, T. (1991). Highly P-typed Mg-doped GaN films grown with GaN buffer layers. *Jpn. J. Appl. Phys.* 30: L1708.

8 Nakamura, S., Harada, Y., and Seno, M. (1991). Novel metalorganic chemical vapor deposition system for GaN growth. *Appl. Phys. Lett* 58: 2021.

9 Cho, J., Park, J.H., Kim, J.K., and Schubert, E.F. (2017). White light-emitting diodes: history, progress, and future. *Laser Photonics Rev.* 11: 1600147.

10 Schubert, E.F. (2006). *Light-Emitting Diodes*, 2e. Cambridge Univ. Press.

11 Zukauskas, A., Shur, M.S., and Gaska, R. (2002). *Introduction to Solid-State Lighting*. Wiley.

12 Stringfellow, G.B. and Craford, M.G. (eds.) (2000). *High Brightness Light Emitting Diode*, vol. 48. Academic Press.

13 Mueller, G. (ed.) (Volume ed.) (1997). *Electroluminescence I, Semiconductor and Semimetals*, vol. 64. Academic Press.

14 Muthu, S., Schuurmans, F.J.P., and Pashley, M.D. (2002). Red, green, and blue LEDs for white light illumination. *IEEE J. Sel. Top. Quantum Electron.* 8: 333.

15 Mueller-Mach, R., Mueller, G.O., Krames, M.R., and Trottier, T. (2002). High-power phosphor-converted light-emitting diodes based on III-nitrides. *IEEE J. Sel. Top. Quantum Electron.* 8: 339.

16 Stringfellow, G.B. and Craford, M.G. (eds.) (Volume eds.) (1997). *High Brightness Light Emitting Diodes, Semiconductor and Semimetals*, vol. 48. Academic Press.

17 Streubel, K., Linder, N., Wirth, R., and Jaeger, A. (2002). High brightness AlGaInP light-emitting diodes. *IEEE J. Sel. Top. Quantum Electron.* 8: 321.

18 Schubert, M., Woollam, J.A., Leibiger, G. et al. (1999). Isotropic dielectric functions of highly disordered $Al_xGa_{1-x}InP$ ($0 \leqslant x \leqslant 1$) lattice matched to GaAs. *J. Appl. Phys.* 86: 2025.

19 Karpiński, J., Jun, J., and Porowski (1984). Equilibrium pressure of N2 over GaN and high pressure solution growth of GaN. *J. Cryst. Growth* 66: 1.

20 Nakamura, S., Mukai, T., Senoh, M. et al. (1993). $In_xGa_{(1-x)}N/In_yGa_{(1-y)}N$ superlattices grown on GaN films. *J. Appl. Phys.* 74: 3911.

21 Lin, Y.S., Ma, K.J., Hsu, C. et al. (2000). Dependence of composition fluctuation on indium content in InGaN/GaN multiple quantum wells. *Appl. Phys. Lett.* 77: 2988.

22 Lester, S.D., Ponce, F.A., Craford, M.G., and Steigerwald, D.A. (1995). High dislocation densities in high efficiency GaN-based light-emitting diodes. *Appl. Phys. Lett.* 66: 1249.

23 Qin, H., Luan, X., Feng, C. et al. (2017). Mechanical, thermodynamic and electronic properties of wurtzite and zinc-blende GaN crystals. *Materials* 10: 1419.

24 Feng, S.W., Cheng, Y.C., Chung, Y.Y. et al. (2002). Impact of localized states on the recombination dynamics in InGaN/GaN quantum well structures. *J. Appl. Phys.* 92: 4441.

25 Khan, M.A., Yang, J.W., Simin, G. et al. (2000). Lattice and energy band engineering in AlInGaN/GaN heterostructures. *Appl. Phys. Lett.* 76: 1161.

26 Nakamura, S., Fasol, G., and Pearton, S.J. (2000). *The Blue Laser Diode: The Complete Story*. Springer.

27 Sheu, J.K., Chang, S.J., Kuo, C.H. et al. (2003). White-light emission from near UV InGaN–GaN LED chip precoated with blue/green/red phosphors. *IEEE Photonics Technol. Lett.* 15: 18.

28 Xiang, H.F., Yu, S.C., Che, C.M., and Lai, P.T. (2003). Efficient white and red light emission from GaN/tris-(8-hydroxyquinolato) aluminum/platinum(II) meso-tetrakis(pentafluorophenyl) porphyrin hybrid light-emitting diodes. *Appl. Phys. Lett.* 83: 1518.
29 X. Guo, J. Graff, and E. F. Schubert, "Photon recycling semiconductor light emitting diode," IEDM Technical Digest, 600 (1999).
30 Chen, H.S., Yeh, D.M., Lu, C.F. et al. (2006). White light generation with CdSe-ZnS nanocrystals coated on an InGaN-GaN quantum-well blue/green two-wavelength light-emitting diode. *IEEE Photonics Technol. Lett.* 18: 1430.
31 Hurni, C.A., David, A., Cich, M.J. et al. (2015). Bluk GaN flip-chip violet light-emitting diodes with optimized efficiency for high-power operation. *Appl. Phys. Lett.* 106: 031101.
32 Sze, S.M. (2001). *Semiconductor Devices—Physics and Technology*, 2e. Wiley.
33 Chuang, S.L. (1995). *Physics of Optoelectronic Devices*. Wiley.
34 Kim, K.C., Choi, Y.C., Kim, D.H. et al. (2004). Influence of electron tunneling barriers on the performance of InGaN–GaN ultraviolet light-emitting diodes. *Phys. Stat. Sol. (a)* 201: 2663.
35 Peyghambarian, N., Koch, S.W., and Mysyrowicz, A. (1993). *Introduction to Semiconductor Optics*. Prentice-Hall.
36 Shockley, W. and Read, W.T. Jr. (1952). Statistics of the recombinations of holes and electrons. *Phys. Rev.* 87: 835.
37 Hall, R.N. (1952). Electron-hole recombination in germanium. *Phys. Rev.* 87: 387.
38 Verzellesi, G., Saguatti, D., Meneghini, M. et al. (2013). Efficiency droop in InGaN/GaN blue light-emitting diodes: physical mechanisms and remedies. *J. Appl. Phys.* 114: 071101.
39 Meyaard, D.S., Shan, Q., Cho, J. et al. (2012). Temperature dependent efficiency droop in GaInN light-emitting diodes with different current densities. *Appl. Phys. Lett.* 100: 081106.
40 Kupha, E. (1991). Liquid phase epitaxy. *Appl. Phys. A* 52: 380.
41 Finch, W.F. and Mehal, E.W. (1964). Preparation of $GaAs_xP_{1-x}$ by vapor phase reaction. *J. Electrochem. Soc.* 111: 814.
42 Tietjen, J.J. and Amick, J.A. (1966). The preparation and properties of vapor-deposited epitaxial $GaAs_{1-x}P_x$ using arsine and phosphine. *J. Electrochem. Soc.* 113: 724.
43 Hirosawa, K., Hiramatsu, K., Sawaki, N., and Akasaki, I. (1993). Growth of single crystal $Al_xGa_{1-x}N$ films on Si substrates by organometallic vapor phase epitaxy. *Jpn. J. Appl. Phys.* 32: L1039.
44 Nakamura, S., Harada, Y., and Seno, M. (1991). Novel organometallic chemical vapor deposition system for GaN growth. *Appl. Phys. Lett.* 58: 2021.
45 Conroy, M., Zubialevich, V.Z., Li, H. et al. (2015). Epitaxial lateral overgrowth of AlN on self- assembled patterned nanorods. *J. Mater. Chem. C* 3: 431.
46 Wang, M.T., Liao, K.Y., and Li, Y.L. (2011). Growth mechanism and strain variation of GaN material grown on patterned sapphire substrates with various pattern designs. *IEEE Photonics Technol. Lett.* 23: 962.
47 Wuu, D.S., Wang, W.K., Wen, K.S. et al. (2006). Fabrication of pyramidal patterned sapphire substrates for high-efficiency InGaN-based light emitting diodes. *J. Electrochem. Soc.* 153: G765.
48 Kish, F.A., Steranka, F.M., DeFevere, D.C. et al. (1994). Very high-efficiency semiconductor wafer-bonded transparent-substrate $(Al_xGa_{1-x})_{0.5}In_{0.5}P$/GaP light-emitting diodes. *Appl. Phys. Lett.* 64: 2839.
49 Steigerwald, D.A., Bhat, J.C., Collins, D. et al. (2002). Illumination with solid state lighting technology. *IEEE J. Sel. Top. Quantum Electron.* 8: 310.
50 Li, G., Wang, W., Yang, W. et al. (2016). GaN-based light-emitting diodes on various substrates: a critical review. *Rep. Prog. Phys.* 79: 056501.
51 Kakinuma, K. (2006). Technology of Wide Color Gamut Backlight with light-emitting diode for liquid crystal display television. *Jpn. J. Appl. Phys.* 45: 4330.
52 Takahashi, T., Furue, H., Shikada, M. et al. (1999). Preliminary study of field sequential full color liquid crystal display using polymer stabilized ferroelectric liquid crystal display. *Jpn. J. Appl. Phys.* 38: L534.

53 Siemianowski, S., Bremer, M., Plummer, E. et al. (2016). Liquid crystal technologies towards realising a field sequential colour (FSC) display. *SID 16 Digest* 47 (1): 175–178.
54 Templier, F. (2016). GaN-based emissive microdisplays: a very promising technology for compact, ultra-high brightness display systems. *SID 16 Digest* 24 (11): 669–675.
55 El-Ghoroury, H.S., Chuang, C.L., and Alpaslan, Z.Y. (2015). Quantum photonic imager (QPI): a novel display technology that enables more than 3D applications. *SID 15 Digest* 46 (1): 371–374.
56 Meitl, M., Radauscher, E., Bonafede, S. et al. (2016). Passive matrix displays with transfer-printed microscale inorganic LEDs. *SID 16 Digest* 47 (1): 743–746.

6

Organic Light-Emitting Devices

6.1 INTRODUCTION

The operational principles of organic light-emitting devices (OLEDs) are similar to those of the semiconductor light-emitting devices (LEDs) described in Chapter 5, except that the principal materials are organic rather than inorganic semiconductors [1]. When a voltage is applied to such a device, holes and electrons are injected from the anode and cathode, respectively, and then recombine to emit light.

The energy levels in organic materials which can be occupied by charge carriers, correspond to molecular orbitals (MOs) which can each be approximated as a combination of many atomic orbitals (AOs). Electrons fill up the MOs from the lowest energy levels so that a highest occupied molecular orbital (HOMO) can be defined. The MOs having the next higher energy level is called the lowest unoccupied molecular orbital (LUMO). The HOMO and LUMO levels in organic materials correspond in some respects, to the valence and conduction bands in ordinary semiconductor materials. After an electron is excited, optically or electrically, to a higher energy level, it relaxes its energy radiatively or non-radiatively. Due to the difference in spin momentum, there are two kinds of energy levels, singlet and triplet, which have non-degenerate energies. Radiative relaxation of singlet and triplet excited states is called fluorescence and phosphorescence, respectively. Typically, the time constant of radiative recombination from the triplet state is much longer than for the singlet one which results in a low phosphorescence quantum efficiency at room temperature.

Charge carrier injection in organic devices is limited by energy barriers at the metal/organic and organic/organic interfaces which are typically modeled by the Richardson–Schottky (RS) equations for thermionic emission [2]. Organic molecules in OLEDs are generally aggregated in an amorphous form without a well-defined band structure and this results in a quite low charge mobility value, typically less than 10^{-3} cm^2/Vs. Carriers are transported by "hopping" among organic molecules in such a disordered material, which means the mobility value increases with the applied electric field. In organic materials, since the free carrier density under thermal equilibrium is quite low, carrier transport in an organic thin film is usually described by trap-charge limited current (TCLC), and trap-free space-charge limited conduction (SCLC), which is the highest sustainable current in a perfect insulator.

By applying electric field to an anthracene single crystal, Pope et al. observed blue electroluminescence (EL) from organic material in 1965 [3]. However, hundreds of volts of driving potential were required due to the thickness of the organic layer. The quantum efficiency (in terms of emitted photon number per injected carrier) was also very low due to imbalanced charge injection and transport. In 1987, a two-layer device structure was first introduced by Tang et al. who utilized thermal evaporation in a vacuum chamber for deposition of amorphous organic thin films [4]. Figure 6.1 shows the device structure. Since the organic film was as thin as several tens of nanometers, the driving voltage can be less than 10 V. The two organic layers were: (i) aromatic diamine, which was used as the hole-transporting layer (HTL) material, and (ii) aluminum chelate, which was used as the emitting layer (EML) and electron-transporting layer (ETL) material. Electrons and holes were transported through the ETL and HTL respectively, with relatively high electron and hole mobilities

Introduction to Flat Panel Display, Second Edition. Jiun-Haw Lee, I-Chun Cheng, Hong Hua, and Shin-Tson Wu.

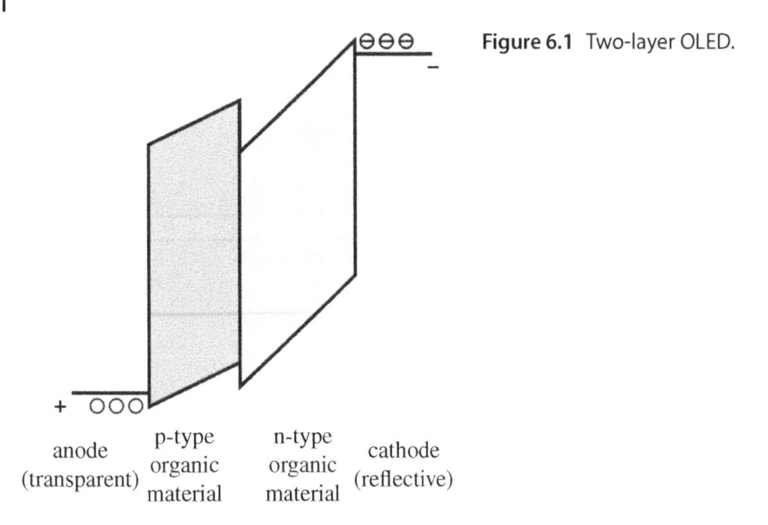

Figure 6.1 Two-layer OLED.

and recombined near the organic/organic interface. This mechanism corresponded in gross detail, with the operation of an ordinary semiconductor LED. Carrier injection into the organic layers was improved by adjusting the workfunctions of the metal electrodes. A high workfunction anode, indium tin oxide (ITO), and low workfunction cathode, Mg:Ag alloy, were suitable choices for effective carrier injection. Since ITO was transparent to visible light, the emitted photons can radiate through the anode and glass substrate, to escape the device structure. To further reduce the driving voltage, improve the quantum efficiency and extend the operational lifetime, multi-layer structures have been proposed, incorporating materials such as a hole-injection layer (HIL), hole-blocking layer (HBL), and electron-injection layer (EIL). The emission wavelength from the OLED can be adjusted by the choice of EML material and device structures. In 1990, Burroughes et al. in Cambridge demonstrated EL from a conjugated polymer, called a polymeric light-emitting device (PLED) [5]. The operating principles of PLED and OLED are basically identical. One major difference comes from the molecular weight and fabrication technology. Conjugated polymers cannot be sublimed under vacuum due to their large molecular weight.

In this chapter, we will first describe the energy diagrams, the photophysical processes, and the electrical properties of organic materials. Then, device structures will be introduced which are exploited to obtain better electrical and optical characteristics of modern OLEDs.

6.2 ENERGY STATES IN ORGANIC MATERIALS

Electrons in isolated atoms occupy AOs, characterized by a set of four quantum numbers which determine the size, shape, relative orientation and energy of each orbital. When two atoms are sufficiently close to one another for chemical bonding to occur, their electrons experience the combined electrostatic attraction of both nuclei. The resulting MOs can in a first approximation, be represented as a sum of the corresponding AOs. The number and nature of the resulting MOs, can be found from an orbital correlation diagram (Figure 6.2a). In general, interaction of two AOs will produce exactly two MOs, one of which has lower energy than the AOs (a bonding MO) and the other with higher energy (antibonding). The bonding MO results from in-phase addition of the AOs. The electron density in this orbital is concentrated between the nuclei, providing an electrostatic attraction which can overcome the internuclear repulsion to form a bond. On the contrary, the antibonding orbital which results from an out of phase combination of AOs has a plane of zero electron density between the nuclei and the resulting electrostatic potential destabilizes the molecule. In the ground state of a simple molecule as shown in Figure 6.2a, the low-lying bonding MO is doubly occupied, while the antibonding MO is vacant.

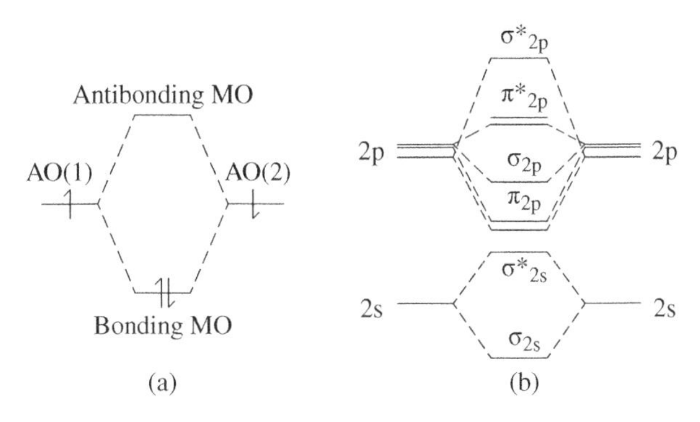

Figure 6.2 Interaction of AOs forming MOs [6].

In bonding between atoms of the second row of the periodic table, which are typical of those making up the skeletons of organic molecules, more AOs are involved, and the correlation diagram is more complex (Figure 6.2b). In particular, the p-orbitals have equal energy in the isolated atoms, but one of these three orbitals can be oriented along the bond axis and then forms bonding and antibonding orbitals with higher energy than the other two. The MO energies indicated by correlation diagrams like Figure 6.2a and b must be recognized as only qualitative guides; the energy of an electron in any particular MO depends on the total occupancy of all the MOs in the molecule.

In practically all important organic molecules, further changes to the geometry and energies of the MOs are caused by the mutual steric repulsion of other atoms, as well as effects such as ring strain and orbital hybridization is usually invoked to describe these effects. Moreover, typical organic semiconductor molecules have many MOs which are distributed over several atoms, e.g. in an aromatic ring structure. Organic semiconductors therefore have a rather large number of MOs, the shapes and energies of which can be influenced by structural modification. These MOs are nevertheless much less numerous than the overlapping levels which form the band structure of a classical semiconductor, and are localized on a single molecule.

In a molecule, the electrons occupy MOs in order from the lowest to higher energy. According to the Pauli exclusion principle, each MO can hold a maximum of two electrons (which must have opposing spins). Typically, in a neutral molecule, bonding MOs are fully or highly occupied, while antibonding MOs are vacant. Non-bonding orbitals (such as "lone pairs") are also common. In any molecule we can identify among the occupied MOs, one which has the highest energy. Similarly, there is an unoccupied MO which has the lowest energy. These are referred to as the HOMO and LUMO (or LVMO; lowest vacant molecular orbital) respectively, as shown in Figure 6.3. The energy associated with transferring an electron from the HOMO level to the LUMO, represents the smallest energy which can form an electronic excited state of the molecule.

Figure 6.3 HOMO and LUMO levels.

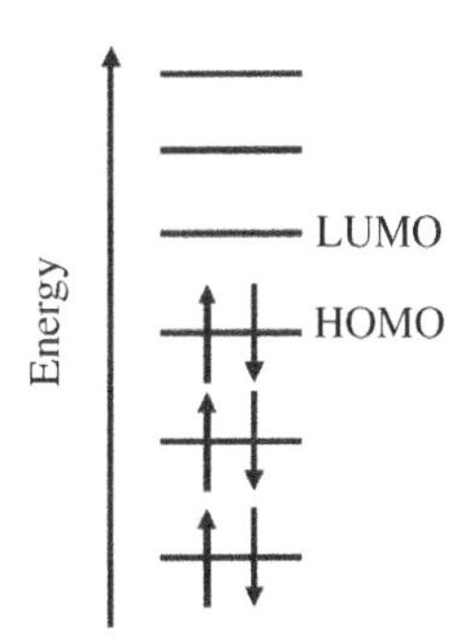

6.3 PHOTOPHYSICAL PROCESSES

Photophysical processes in organic materials include optical absorption and energy relaxation, which follow the Franck–Condon principle. When a molecule absorbs light, it accepts energy from the electromagnetic field. For conjugated molecules and visible/UV light, the absorbed energy results in promotion of an electron from the HOMO to the LUMO level, or possibly (for a photon of sufficiently high energy) to a still higher energy vacant level. The energy may then be relaxed radiatively through light emission (an electronic state transition), or non-radiatively through heat dissipation (vibrational and rotational state transitions). The radiative process is central to the operation of OLEDs and can be subdivided into fluorescence and phosphorescence, which originate from the relaxation of singlet and triplet excited states, respectively. These effects are explored in Section 6.3.2, below. Phosphorescent emission requires a (formally forbidden) spin inversion of an electron which gives the excited state a long lifetime, and hence typically at room temperature phosphorescence is only observable, from special types of molecules. Unfortunately, upon electrical excitation, quantum statistics dictate that the number of triplet excitons generated, is three times that of singlet ones. That means 75% of the excitons do not contribute to light emission in a conventional OLED. A Jablonski diagram is typically used to illustrate the whole range of photophysical processes, such as absorption, electronic state transitions, and energy relaxations in a single molecule. Considering a system with one excited-state molecule and one ground-state molecule, energy transfer from the excited-state molecule to the ground-state molecule may occur. Besides, the two-molecule system may form a complex which exists only in the excited state, which is called excimer (excited dimer) and exciplex (excited complex) when two molecules are identical and different, respectively.

6.3.1 Franck–Condon Principle

Photon energy absorbed by a molecule may excite electronic, vibrational and rotational transitions. Typically, the light absorption and emission in organic materials between LUMO and HOMO levels in UV and visible range, i.e. several electron volts, is most important. Vibrational states, which have energy separations of several tenths of an electron volt, are superimposed on each electronic state and there are selection rules for allowed energy transitions which can be deduced from the absorption and emission spectra. Typically, rotational states are too close in energy, (several hundredths of an electron volt), to be resolved at room temperature. Vibrational states are also not usually well resolved at room temperature in electronic spectra of condensed phases but often partly resolved and are mainly important in determining the shapes of bands, Stokes shift, as discussed later. Figure 6.4 shows potential energy curves of some molecular electronic states with superimposed

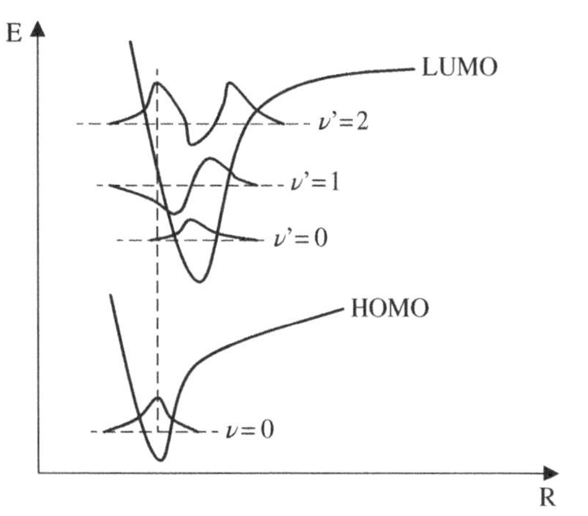

Figure 6.4 Potential curves of electronic states with vibrational energy levels and wavefunctions.

vibrational energy levels and wavefunctions. Rotational modes are usually strongly suppressed in solids. Vibrational modes are excited to some degree, but at room temperature, molecules are predominantly in the ground vibrational state.

Since the light frequency and electronic reconfiguration rate (~10^{15} s^{-1}) for electronic excitations are much faster than the vibrational motion of atomic nuclei (~10^{13} s^{-1}), the electronic transition occurs along vertical line between potential curves, without any configuration change of the molecules. The configuration and wavefunction for the nuclei should be the same before and immediately after an electronic transition, and the transition probability depends on the overlap integral of the respective wavefunctions. As shown in Figure 6.4, when a molecule is in its ground electronic state and ground vibrational state ($v = 0$), it is difficult to excite it to the $v' = 0$ state, since the wavefunction overlap between these two vibrational modes is small. $v = 0\ v' \rightarrow v' = 1$ is possible due to the partial overlap of the wavefunctions. $v = 0v' \rightarrow v' = 2$ is preferred since the wavefunction overlap reaches a maximum in this case. It is called Franck–Condon principle to explain the intensity of vibronic transition between two electronic states.

When the molecule absorbs light, the electrons are promoted to higher electronic and vibrational states, and transition to the ground vibrational state of the excited electronic state ($v' = 0$). This vibrational ground state will be the initial state in the transition leading to light emission, which must also follow the Franck–Condon principle. Hence, the absorption has a higher energy and a shorter wavelength than the emission, as shown in Figure 6.5. The difference in wavelength (or energy) between the absorption and emission peaks is called the "Stokes shift." Also, the shape of the absorption spectrum is often approximately a "mirror image" of the emission peak, since the vibrational modes in the two electronic states are typically similar. Figure 6.5c shows the fluorescence (left) and absorption (right) spectra of oligophenylenevinylenes (nPVs) with different chain lengths in solution in dioxane at room temperature. The structures of nPV are also shown.

6.3.2 Fluorescence and Phosphorescence

According to Pauli's exclusion principle, two electrons in the same orbital must have opposite spins. Those two electrons are called "paired electrons" and the state is called the "singlet ground state," denoted by S_0 in Figure 6.6. Once the molecule absorbs the light, the two electrons retain their different spins in different energy levels and therefore form a "singlet excited state" (S_1). Since two electrons in the excited states are in different orbitals, Pauli's exclusion principle no longer applies, which means these two electrons can also have the same spin, forming a "triplet state" (T_1). The energy of a singlet state can be calculated from quantum mechanics and is higher than that of the corresponding triplet state. It follows that it is energetically preferable for the excited molecule to transform from the S_1 state to the T_1 state, together with a spin inversion and loss of thermal energy. This process is called "intersystem crossing." Note that the triplet state must be an excited state to conform with Pauli's exclusion principle. So, T_1 cannot relax back to S_0 without another spin inversion. In other words, $T_1 \rightarrow S_0$ is "forbidden" since the wavefunction overlap is zero if considering the spin term and only singlet excitons can contribute to light emission.

Formally, electrons with spin-up and -down have a spin angular momentum with either a positive or negative component in the z-direction, respectively. The spin angular momentum is non-zero in x–y plane, as shown in Figure 6.7a. When there are two electrons in separate electronic states, their spins can either add together or cancel, corresponding to singlet or triplet states, respectively. As shown in Figure 6.7b, there are three possible arrangements for the triplet and only one for a singlet state, according to the principles of quantum statistics. In an electrically driven OLED structure, carrier recombination results in a molecule which both has an electron vacancy in the HOMO level through hole capture, and has an electron present in the ground state LUMO level from electron capture. This is the same electron population of energy levels, as in the excited state which results from photon absorption. According to the discussion above, the number of triplet states is three times of singlet ones when an excited molecule is formed by carrier combination. If emission comes from singlet excitons only, the maximum internal quantum efficiency (IQE), i.e. in terms of photons per injected electron–hole pair, is only 25%.

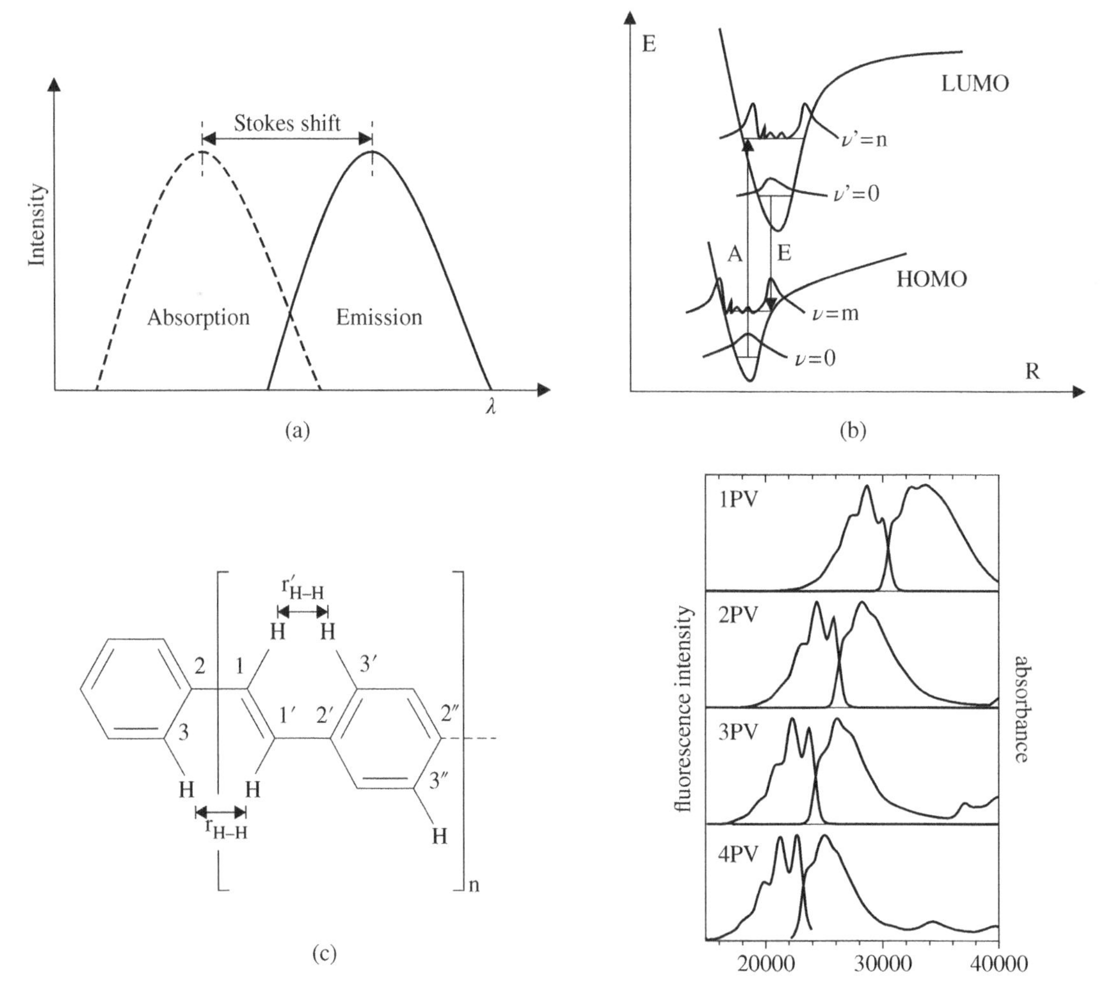

Figure 6.5 (a) Spectra of absorption and emission, (b) energy levels of Stokes shift, and (c) mirror image between absorption and emission spectra of nPVs [7].

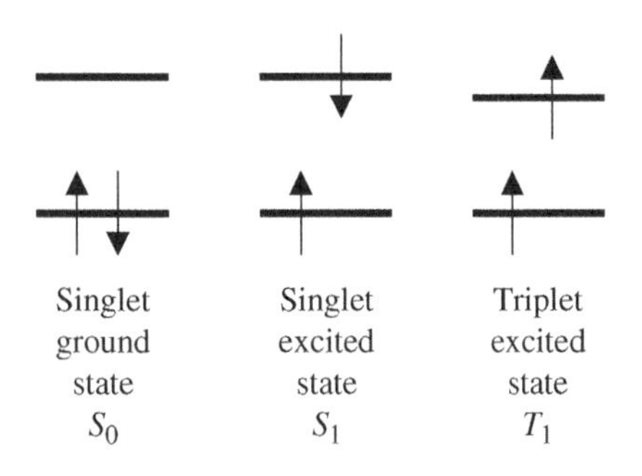

Figure 6.6 Singlet ground, singlet excited and triplet states.

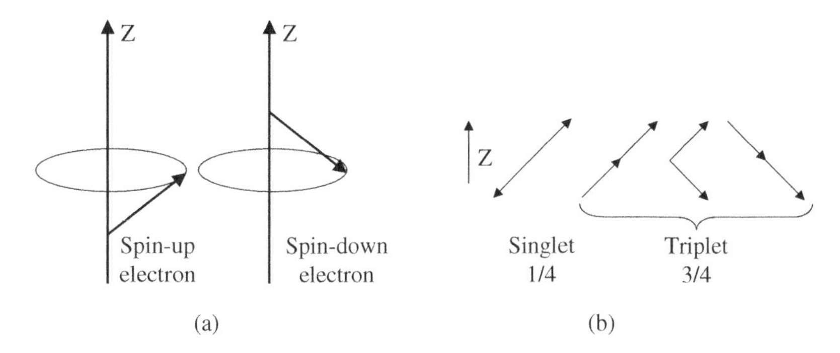

Figure 6.7 (a) Spin angular moment, and (b) coupling of spin angular momenta between two electrons at different orbitals [8].

However, the classification of "allowed" and "forbidden" transitions is not absolute in real emitting materials since perturbation terms reflecting weak electronic interactions are added into the total Hamiltonian which leads to a mixing of spin multiplicity. Spin–orbit al coupling, i.e. interaction between the spin and orbital angular momentum of an electron, is one of the most important terms which results in the mixing of singlet and triplet states. After considering the spin–orbital interaction, there are no pure singlet and triplet states. However, remember that spin–orbital coupling is only a perturbation term. That means the transition from $T_1 \rightarrow S_0$ is non-zero but still a small contribution, compared with $S_1 \rightarrow S_0$. In phosphorescence, the carrier lifetime is longer, because a spin inversion is involved, and it typically exhibits low efficiency. The strength of spin–orbit coupling is proportional to the atomic number. If an atom with a large atomic number is present in the molecule, spin–orbit coupling can be enhanced and the phosphorescence emission efficiency can be improved. This is called the "heavy atom effect."

6.3.3 Jablonski Diagram

Figure 6.8 shows the Jablonski diagram which is used to describe the energy transitions between electronic states. Figure 6.8 only shows the singlet ground state (S_0), two singlet excited states (S_1 and S_2), and two triplet excited states (T_1 and T_2). More excited states can be also included in the diagram.

As discussed above, absorption takes place from the singlet ground state (S_0) to singlet excited states (S_1 and S_2). The photon energy determines the final vibrational state in the excited state and the Franck–Condon principle governs the transition probability. Then, electrons in the singlet excited state may relax to the vibrational ground state by releasing the excess energy in the form of vibrational and rotational motion. It is also possible for non-emissive transitions to occur from a vibrational mode of S_2 to vibrational levels of S_1. Energy relaxation

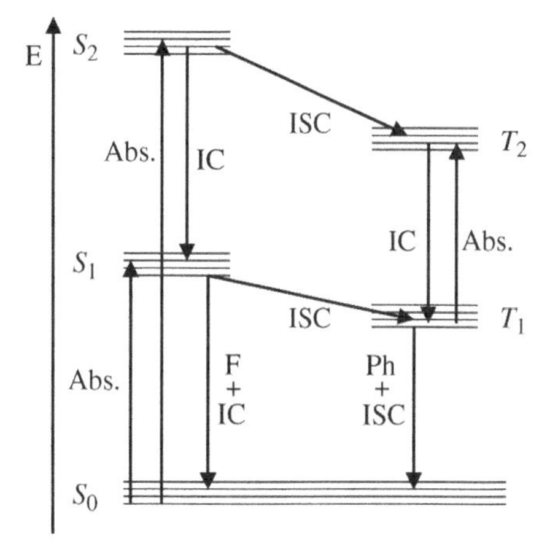

Figure 6.8 Jablonski diagram. Abs., absorption; IC, internal conversion; ISC, intersystem crossing; F, fluorescence; and Ph, phosphorescence.

(especially by thermal dissipation) without any spin inversion is called "internal conversion." After electrons relax their energy to the ground vibrational and rotational state of the singlet excited states S_1, fluorescent emission can take place from S_1 to S_0 through photon emission, which also has to obey the Franck–Condon principle. Obviously, the emission wavelength must be longer than the absorption one, the difference being the Stokes shift. It is also possible for molecules in the S_1 and S_2 states to relax to S_0 by shedding vibrational and rotational energy which contributes to non-radiative recombination. Electrons in a singlet excited state can transfer to the triplet state (T_1 and T_2) via intersystem crossing. Internal conversion may occur from T_2 to T_1. Optical absorption is possible, resulting in a transition from T_1 to T_2 since the two states exhibit the same spin angular momentum; this is called "triplet–triplet absorption." Relaxation from T_1 and T_2 may be radiative or non-radiative. Photon emission from $T_1 \rightarrow S_0$ is called phosphorescence, and exhibits a longer wavelength than fluorescent emission.

6.3.4 Intermolecular Processes

In addition to the energy relaxation routes which occur via intramolecular processes, as described by the Jablonski diagram, energy transfer and relaxation happen through mechanisms involving more than one molecule, which are called "intermolecular processes." One of the most important processes is energy transfer from donor to acceptor materials which is most useful in OLEDs for adjusting the emission wavelength, improve the efficiency, and to prolong the operational lifetime. When two molecules come into close proximity, new orbitals form only in the excited state. Such excited molecular aggregates are called "excimer" and "exciplex," consisting of the same or different molecules, respectively. Molecule at the excited state may transfer the energy to another ground-state molecule (or other species such as metallic material) which relaxes the energy non-radiatively, and it is called the quenching process. For example, impurity acts as a "quencher" in an OLED, which reduces the quantum yield and will be discussed in Sections 6.3.4.3 and 6.3.5.

6.3.4.1 Energy Transfer Processes

A molecule in an excited electronic state, which is called a "donor," may transfer its energy to another molecule, which is called the "acceptor." More precisely, these are "energy" donors and acceptors, respectively, rather than "electron" donor and acceptor molecules, which case will be discussed in the next section. After this energy transfer process, the donor molecule returns to its ground electronic state and the acceptor is promoted to a higher energy state [9], which can be represented as:

$$D^* + A \rightarrow D + A^* \tag{6.1}$$

where "D" and "A" are the energy donor and acceptor, respectively. The "*" indicates the excited state. Energy transfer can be a two-step process without any direct interaction between the donor and acceptor molecules as follows:

$$D^* \rightarrow D + h\upsilon \tag{6.2}$$

$$h\upsilon + A \rightarrow A^* \tag{6.3}$$

Here, photons from the radiative relaxation of the donor are absorbed by the acceptor and promote the acceptor to an excited state. This process is called "radiative energy transfer" since photons are involved. The strength of this energy transfer only depends on the emission efficiency of the donor and the absorption efficiency of the acceptor at this wavelength. A single-step process described by Eq. (6.1), without the intermediacy of photons, is also possible when the donor and acceptor molecules are close (less than 10 nm) and an energetic resonance occurs between the two molecules. This is called "non-radiative energy transfer." As shown in Figure 6.9, this process is isoenergetic and the transfer probability is proportional to the spectral overlap (J) between donor emission ($I_D(\nu)$) and acceptor absorption ($\varepsilon_A(\nu)$) spectra, as follows:

$$J = \int_0^\infty I_D(\nu)\varepsilon_A(\nu)d\nu \tag{6.4}$$

Figure 6.9 Donor emission and acceptor absorption spectra and their spectral integral.

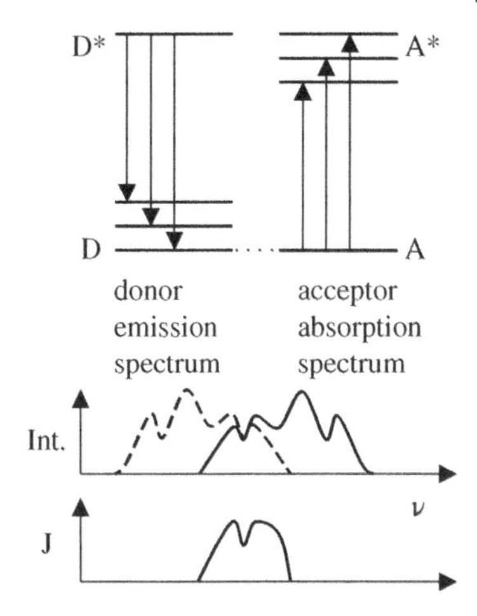

Two mechanisms can account for non-radiative energy transfer, based on dipole–dipole interactions or electron exchange, and called "Förster" and "Dexter" energy transfer respectively. The rate constant ($k_{ET(Coulomb)}$) of Förster energy transfer, which originates from dipole resonance from donor to acceptor molecules, can be represented as:

$$k_{ET}(\text{Coulomb}) \sim \frac{f_D f_A}{R^6{}_{DA} \nu^2} J k_{ET}(\text{Cou lom b}) \sim \frac{f_D f_A}{R^6{}_{DA} \nu^2} J \tag{6.5}$$

where f_D and f_A are the transition probabilities for donor emission and acceptor absorption, following the Franck–Condon principle. R_{DA} is the distance between donor and acceptor molecules. We can also see that the interaction strength decreases rapidly with increasing donor-acceptor separation. Typically, Förster energy transfer is efficient within an intermolecular distance up to 10 nm. Considering the spin angular momentum of the species involved, the following energy transfer is allowed:

$$^1D^* + {}^1A \rightarrow {}^1D + {}^1A^* \tag{6.6}$$

However, due to the characteristics of a dipole–dipole interaction, spin inversion is not permitted due to a selection rule. So, the following energy transfer is forbidden:

$$^3D^* + {}^1A \rightarrow {}^1D + {}^3A^* \tag{6.7}$$

The rate constant ($k_{ET(exchange)}$) of Dexter energy transfer can be written as:

$$k_{ET}(\text{exchange}) \sim e^{-2RDA/L} J kET(\text{exchange}) \sim e^{-2RDA/L} J \tag{6.8}$$

which decreases exponentially with an increasing donor-acceptor distance. Since it involves two electrons being exchanged between the donor and acceptor molecules, Dexter energy transfer is a short-range process which is only effective within a distance of 1 nm. Through the Dexter mechanism, the two energy transfers (6.9) and (6.10) are both allowed since the total spins are conserved [10]:

$$^1D^* + {}^1A \rightarrow {}^1D + {}^1A^* \tag{6.9}$$

$$^3D^* + {}^1A \rightarrow {}^1D + {}^3A^* \tag{6.10}$$

Singlet energy transfer as shown in Eqs. (6.6) and (6.9) is normally dominated by the Forster mechanism because it is effective at longer range. The triplet transfer represented in Eq. (6.10) is only permitted by the Dexter mechanism, which therefore determines the efficiency of this deactivation route.

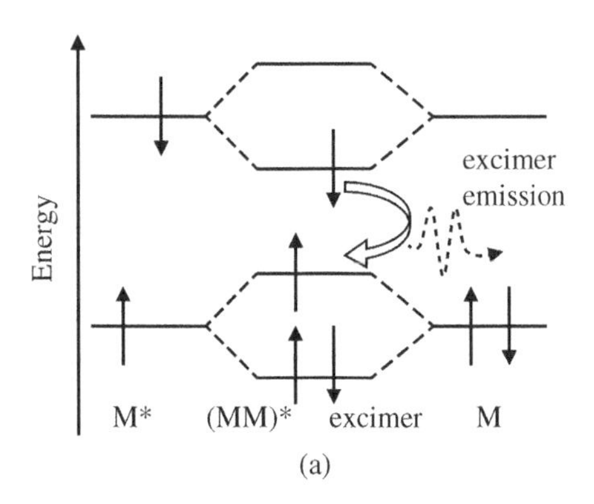

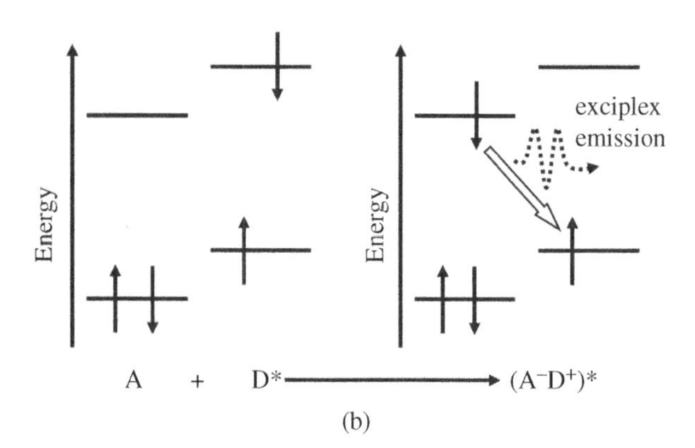

Figure 6.10 (a) Excimer and (b) exciplex formation.

6.3.4.2 Excimer and Exciplex Formation

When two identical molecules (denoted as "M" in Figure 6.10a) – one in its ground electronic state and the other in its excited state – approach one another closely, new electronic states with a lower energy may form which can bind the constituent molecules loosely together; the result can be viewed as a complex, as shown in Figure 6.10a. It is called an "excimer" (excited dimer). Two different molecules can also form a complex by a charge transfer process, as shown in Figure 6.10b, where "D" and "A" denote two different molecules, called "electron donor" and "electron acceptor" materials, respectively, different from the "energy" donor and acceptor in the last section [11]. In this configuration, the electron donor material is excited, and then donates an electron to the electron acceptor material thereby lowering the potential energy of the overall system. Then, the electron in the LUMO level of the electron acceptor material relaxes the energy and back to the HOMO level of the electron donor material. Note that the "electron" donor and "electron" acceptor functions in this section relate to the transfer of a single electron between two molecules. The donor and acceptor functions in the previous section relate to energy transfer between two molecules. Although the Dexter mechanism also involves electron transfer, that process is an electron-exchange involving two electrons moving between the two molecules in opposite directions and quite different from the single electron transition described in this section. Excimers and exciplexes may also relax their energy radiatively, back to the ground state. Note that excimers and exciplexes are bound complexes formed only in the excited state and they are dissociated in the ground state.

It is clear from Figure 6.10 that the emission wavelength from this excited state complex is longer than that of the original molecules due to its smaller bandgap. Figure 6.11 shows an example of excimer fluorescence [12]. Film 1 is a neat 4,4′-N,N′-dicarbazole-biphenyl (CBP) film. The absorption and PL emission peaks occur at 350 and 390 nm, respectively. Films 1–4 are all based on CPB as a host, so the absorption spectra are nearly identical for these four films. In Film 2, which contains <1 wt.-% of platinum(II)(2-(4′,6′-difluorophenyl)pyridinato-N, C2′)(2,4-pentanedionato) (FPt1) in CBP, the PL spectral peak shifts to 470 and 500 nm due to dopant emission. A small hump remains at 380 nm due to incomplete energy transfer. In Film 3, the FPt1 content is increased to 7 wt.-%, and one can see a broad long wavelength PL peak around 570 nm which comes from excimer emission. By mixing two dopant materials into CBP, i.e. 6 wt.-% iridium-bis(4,6,-difluorophenyl-pyridinato-N, C2)-picolinate (FIrPic) and 6 wt-% FPt1, a white PL emission can be achieved as is demonstrated in Film 4.

6.3.4.3 Quenching Processes

Fluorescence quenching as a photophysical process can be described by:

$$M* \xrightarrow{Q} MM* \xrightarrow{Q} M \tag{6.11}$$

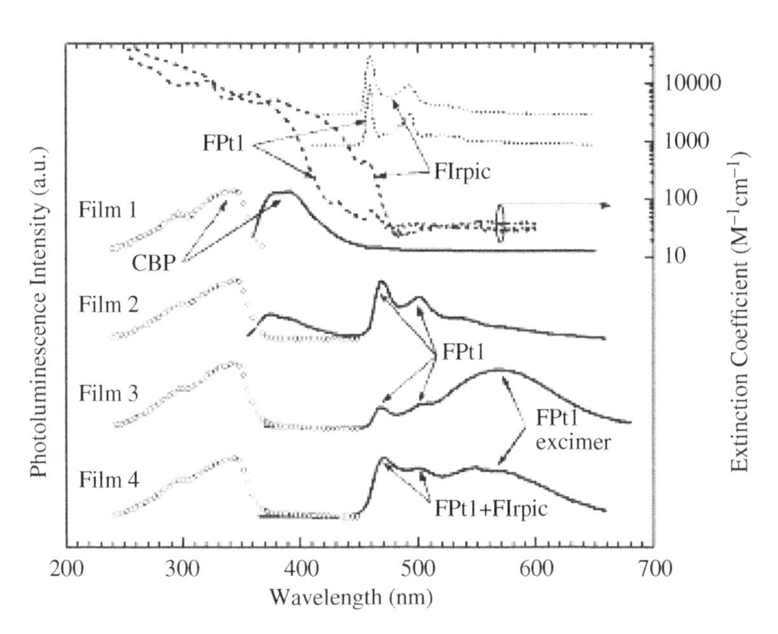

Figure 6.11 Absorption and fluorescence spectra of Films 1 to 4. Film1: neat CBP; Film 2: CBP + <1 wt.-% FPt1; Film 3: CBP + 7 wt.-% FPt1; and Film 4: CBP + 6 wt.-% FPt1 + 6 wt.-% FIrpic [12].

where Q is called the quencher. Quenching is an intermolecular process in which the excited molecules (M) relax their energy through the quencher. Some undesirable chemicals present in the organic OLED materials also provide an energy relaxation route which is called "impurity quenching." Excimer and exciplex relaxation may be non-radiative, providing their own quenching processes. In OLEDs, excitons may transfer their energy to a nearby metal electrode, which is called the "electrode quenching" process.

6.3.5 Quantum Yield Calculation

From a macroscopic viewpoint, light decreases in intensity as it passes through an absorptive medium. We can define the absorption coefficient (α) as:

$$dI = -\alpha I dx \tag{6.12}$$

where I and x represent the light intensity and the absorption thickness, respectively. In this equation, the intensity decrease is proportional to: (i) absorption coefficient, (ii) incident intensity, and (iii) absorption thickness. After integrating Eq. (6.12) over the sample thickness d, the resulting equation can be obtained:

$$I = I_0 e^{-\alpha d} \tag{6.13}$$

where I_0 is the initial light intensity. It is conventionally written in the modified form known as the Beer–Lambert law:

$$A = \log_{10}(I_0/I) = (1/2.303)\,\alpha d = \varepsilon c d \tag{6.14}$$

where A is the absorbance (or optical density) of the sample, c is the concentration (in moles per liter), the sample thickness d is given in centimeters, and ε (in terms of liters per mole per centimeter) is the molar absorption coefficient.

Example 6.1 Time-of-flight (TOF) is a technique which is commonly used to determine the carrier mobility in organic materials. The basic concept of TOF is to calculate the carrier-drift time through a material at a given electric field across it. Carriers are generated by optical excitation. Therefore, it is essential that the region of carrier generation is much thinner than the layer through which carrier transport occurs. Assume an organic material has absorption coefficient $\alpha = 5\times10^5\text{cm}^{-1}\alpha = 5\times10^5 cm^{-1}$ at the excitation wavelength of 355 nm. Find the penetration depth of light, which can be considered the width of carrier generation region in TOF measurement.

Answer
The penetration depth, L, of light impinging on a material is defined as the distance over which the incident intensity is decreased by a factor of e^{-1}. According to this definition, $L = \frac{1}{\alpha} = \frac{1}{5\times10^5\text{cm}^{-1}} = 20$ nm.

Therefore, the thickness of materials used in TOF measurements must be usually of the order of several μms to make sure that the carrier drift length approximates to the thickness of the deposited material.

In a system with many particles, when the carriers are promoted (optically or electrically) to the excited electronic state, typically a rate equation is used to describe the dynamics of their relaxation back to the ground state or to other excited electronic states, as follows:

$$\frac{\partial n}{\partial t} = G(t) - \frac{n}{\tau} \tag{6.15}$$

where $G(t)$ is the carrier generation rate which is used to describe the electrical or optical pumping, n is the excited state carrier density and τ_0 is the "carrier lifetime" of the excited state. The rate equation can alternatively be written as $\tau_0 = 1/k_0$, and k_0 is called the "rate constant." So the decay rate of the carriers in the excited state is proportional to the carrier density and the rate constant. Hence, the concentration of excited state carriers formed at $t <=0$, can be expressed as a function of time by an equation which holds for both radiative and non-radiative decay:

$$n(t) = n_0 \exp(-k_0 t) \tag{6.16}$$

The "quantum yield" of a relaxation process is defined as the fraction of the total excited state carriers which decay by this process. From the Jablonski diagram, carriers relax through several different pathways, such as fluorescence to the ground state, internal conversion, and intersystem crossing, with corresponding rate constants k_F, k_ic, and k_isc, respectively. The sum of the quantum yields for all the processes is equal to 1. The rate equation of the first singlet excited state can be written as:

$$\frac{\partial S_1}{\partial t} = G(t) - (k_\text{F} + k_\text{ic} + k_\text{isc})S_1 \tag{6.17}$$

And the quantum yield of fluorescence (ϕ_F) is denoted by:

$$\varphi_\text{F} = k_\text{F}/(k_\text{F} + k_\text{isc} + k_\text{isc}) \tag{6.18}$$

Example 6.2 Triplet–triplet exciton annihilation is a well-known phenomenon in phosphorescent OLEDs. The process is $T_1 + T_1 \xrightarrow{k_\text{TT}} T_n + S_0$ or $S_n + S_0$. Here, T (S) donates triplet (singlet) exciton concentration, k_TT is a rate constant and subscripts 0 and n denote the ground and nth excited state, respectively. This annihilation process is governed by $\frac{\partial T}{\partial t} = D\frac{\partial^2 T}{\partial x^2} - \frac{T}{\tau} - k_\text{TT}T^2$ where D is a diffusion coefficient and τ is the triplet exciton lifetime. Assume $k_\text{TT} = 1.8\times10^{-14}\text{cm}^3\cdot\text{s}^{-1}$; $\tau = 10\text{ms}$; $T = 5\times10^{16}\text{cm}^{-3}$. (i) Find the exciton lifetime in the annihilation process. (ii) Find the exciton lifetime in the entire system.

Answer

(1) Exciton lifetime is $\tau_{TT} = \frac{1}{k_{TT}T}$. Obviously, the higher the triplet exciton concentration, the faster the annihilation process. This leads to efficiency roll-off under high injection current, since the number of radiative excitons is reduced. $\tau_{TT} = \frac{1}{1.8\times10^{-14}\text{cm}^3\cdot\text{s}^{-1}\cdot5\times10^{16}\text{cm}^{-3}} = 1.1\text{ms}$.

(2) The exciton lifetime in the entire system $\tau_{total} = \frac{1}{\frac{1}{\tau}+\frac{1}{\tau_{TT}}} = \frac{1}{\frac{1}{10\text{ms}}+\frac{1}{1.1\text{ms}}} = 0.99\text{ms}$.

6.4 CARRIER INJECTION, TRANSPORT, AND RECOMBINATION

EL of organic solids comes from the recombination of injected carriers, which is similar to the emission mechanism of semiconductor LEDs, as described in Chapter 5. Typically, an OLED is fabricated on an ITO coated substrate. Thin (100–200 nm) amorphous organic layers with different functions such as a HIL, HTL, EML, ETL, and EIL are successively deposited on the ITO anode. Finally, a cathode metal is evaporated. In an OLED, charges are injected from electrodes and transported through the organic layers, where they recombine and form excitons which emit light through a transition from an excited state to the ground or an intermediate state [13].

Typically, Fowler–Nordheim (FN) tunneling and Richardson–Shottky [14] (RS) thermionic emission mechanisms are used to account for carrier injection in semiconductor devices. However, in an OLED, they can only partially describe the behavior at organic and organic/metal interfaces since: (i) chemical reactions may take place between the interfaces which result in the formation of interfacial layers, and (ii) the disordered nature of the organic materials.

Since OLED (or PLED) thin films are amorphous without a crystalline structure, the band theory developed for semiconductor or molecular crystals does not provide a good description of the charge transport. In organic thin films, the charge carriers hop between separate localized states, which explains the low carrier mobility of organic thin films (typically $<10^{-3}$ cm^2/Vs) [15]. Localization of the excited state also results in a large binding energy of the excitons. For an OLED display, emission wavelengths are within the visible range. One can understand that given the Stokes shift described in Section 6.3.1, the optical absorption must be in the UV region with a bandgap of ~2–4 eV, which implies low carrier concentrations in the organic materials (typically $<10^{10}$ cm^{-3}) since it is difficult to generate "free" carriers thermally. From a simple calculation:

$$\sigma = nq\mu \tag{6.19}$$

where σ is the conductivity (in terms of 1/Ω cm), n is the carrier concentration, q is the electron charge in coulombs and μ is the carrier mobility, one can estimate that the conductivity of the organic thin film is of the order of 10^{-12} 1/Ω cm. The conductivity of a semiconductor is between 10^{-8} and 10^2 1/Ω cm. Higher and lower conductivities correspond to conductors and insulators, respectively. For example, the conductivity of glass is between 10^{-11} and 10^{-10} 1/Ω cm, higher than many organic thin films. From this viewpoint, organic materials for OLED application can be regarded more as an insulator, rather than a semiconductor. For practical use, the organic layer thicknesses must be very thin (hundreds of nm) to achieve a driving voltage which is not too high (<10 V). The carriers in organic thin films are collectively described as a "space charge" since the free carrier are negligible in number, and all the carriers are injected from the electrodes. When carriers are injected, molecules become positively or negatively charged, and are termed cations or anions, respectively. The carriers "hop" among different molecules under electric field [16]. So, it is easy to understand that the charge mobility in organic thin films increases with increasing electric field, which is described by the "Poole–Frenkel" [17] model, which is outlined in Section 6.4.2. To describe carrier transport in organic thin-films, time dependent continuity equations, the drift-diffusion current equation, and Poisson's equation are still valid, as described in Chapter 5. However, some modifications are needed for the organic thin-films. Under certain approximations, the current–voltage relationship can be described by SCLC and TCLC. Figure 6.12 shows a typical current density (J) – voltage (V) characteristic of an OLED on a double-logarithmic scale. Ohmic current, $J_{Ohmic} \propto V$,

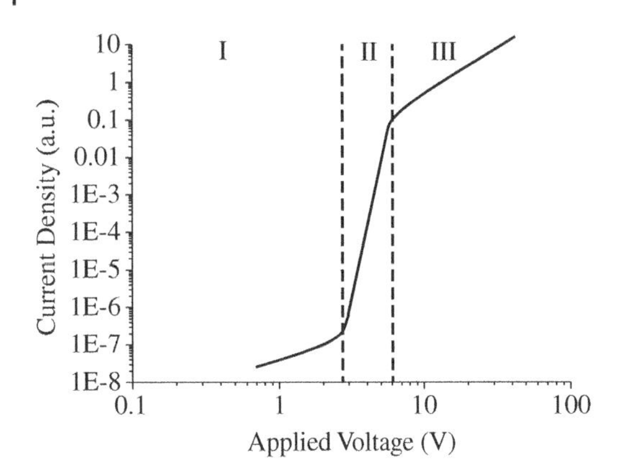

Figure 6.12 Typical J-V characteristics of an OLED.

dominates in Region I which arises from drift of the limited number of free carriers in the organic materials. As the driving voltage is increased, carriers are injected from the electrodes and their transport inside the organic material which include traps obeys TCLC, $J_{\mathrm{TCLC}} \propto V^{l+1}$. This is the case in Region II where the line slope is $l+1$. Once the traps are filled up, charge transport obeys SCLC, $J_{\mathrm{SCLC}} \propto V^2$. Region III corresponds to this phenomenon, and the slope of the J–V curve is then exactly two on a double-logarithmic scale.

The recombination rate in an OLED depends on the spatial overlap of the hole and electron densities. The low mobility values of the organic materials result in carrier transit times which are long, compared to the recombination time, so electrons and holes recombine and form excitons as soon as they meet in the space domain, following so-called "Langevin" recombination, as in the LED case described in Chapter 5. Once the excitons relax their energy through radiative decay and generate light, the electromagnetic wave propagates through the thin film structure which has a thickness comparable to the emitted wavelength. Strong interference effects occur which have a great influence on the optical characteristics of the device such as output intensity, emitted spectrum and device efficiency.

6.4.1 Richardson–Schottky Thermionic Emission

In a one-dimensional model, one can consider an organic material with a thickness L with the anode and the cathode at $x=0$ and $x=L$, respectively. The hole–current injected from the anode ($x=0$) to the organic layer can be represented by:

$$J_{\mathrm{p}}(0) = J_{\mathrm{th}} - J_{\mathrm{ir}} + J_{\mathrm{ptu}} \tag{6.20}$$

where J_{th} is the thermionic emission current, J_{ir} is the backward flowing interface recombination current, and J_{ptu} is the tunneling current.

The thermionic emission current has the well-known form [18]

$$J_{\mathrm{th}} = A^{*}T^{2}\exp(-E_{\mathrm{b}}/\eta kT) \tag{6.21}$$

where A^{*} is Richardson's constant and E_{b} is the interfacial energy barrier, η is the ideality factor, the value of which is typically between 1 and 2 in semiconductors [1]. Because of the image force, E_{b} depends on the electric field at the interface

$$E_{\mathrm{b}} = \phi_{\mathrm{B}} - q\sqrt{\frac{q|F(0)|}{\varepsilon}} \tag{6.22}$$

where φ_{B} is the Schottky energy barrier at zero field. In Figure 6.13a, the image charge and the electric field lines at the metal/organic interface are shown, and the change in barrier energy level due to the image force is also shown in Figure 6.13b.

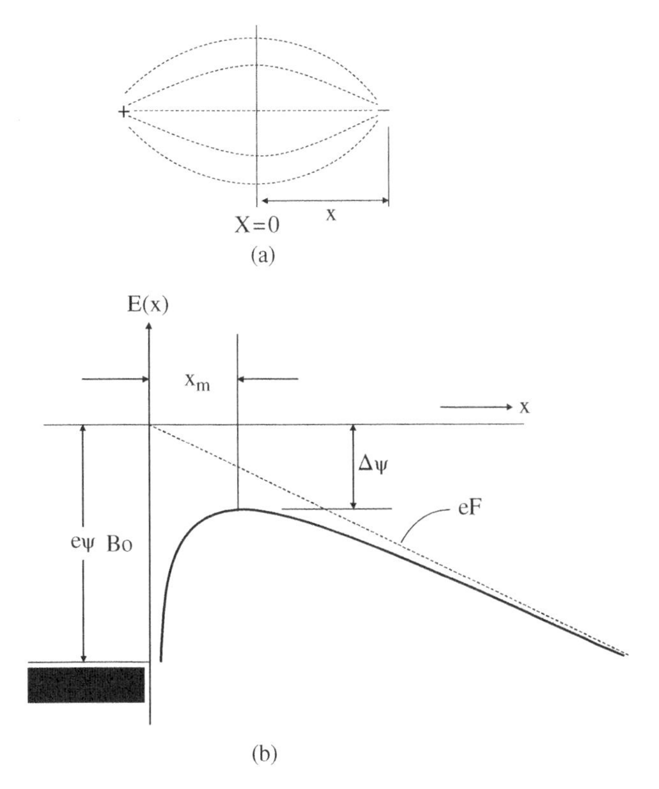

Figure 6.13 (a) Image charge and the electric field lines at the interface [19]; (b) Energy level change due to image force.

The interface recombination current is proportional to the hole density at the interface [18]

$$J_{\text{ir}} = qvp(0) = \frac{A^*T^2}{n_o}p(0) \tag{6.23}$$

n_o is the density of states [18, 20]. The kinetic coefficient v is determined by the detailed balance between thermionic and interface recombination.

$$v = \sqrt{\frac{kT}{2\pi m^*}} \tag{6.24}$$

The Fowler–Nordheim (FN) injection current at a tunneling contact [21] is:

$$J_{\text{ptu}} = \frac{q^3}{8\pi h\varphi_{\text{Bp}}}F^2 \exp\left(-\frac{8\pi\sqrt{2m^*\varphi_{\text{Bp}}^3}}{3qh\text{F}}\right) \tag{6.25}$$

where m^* is the effective mass of the carrier and h is Planck's constant. Equations similar to (6.24) and (6.25) can be written for the electron current density. Typically, in an OLED, thermionic emission and F–N tunneling (field emission) [22] dominate the injection mechanism under high field and low field conditions, respectively.

6.4.2 SCLC, TCLC, and P–F Mobility

The transport of electrons and holes in an OLED is described by time dependent continuity equations, with a drift-diffusion current, coupled to Poisson's Equation [18]. Considering electron transport alone in a unipolar organic material without generation or recombination, the continuity equation can be written as:

$$\frac{\partial n}{\partial t} = \frac{1}{q}\frac{\partial J_n}{\partial x} \tag{6.26}$$

where J_n is the electron current density, n is the electron density, as described in Chapter 5. Under steady state conditions, $\frac{\partial}{\partial t} = 0$, which implies:

$$J_n = \text{const.} \tag{6.27}$$

As described above, the carrier density in organic materials is negligible under thermal equilibrium, which means the diffusion current can be neglected. So, only the drift term appears in the current density equation:

$$J_n = \text{nq}\mu_n E \tag{6.28}$$

or

$$n = J_n / q\mu_n E \tag{6.29}$$

where E is the electric field, μ_n is the electron mobility, q is the electron charge. This may be substituted into Poisson's equation:

$$\frac{\partial E}{\partial x} = \frac{-q}{\varepsilon} n \tag{6.30}$$

to give:

$$\frac{\partial E}{\partial x} = \frac{-q}{\varepsilon}\frac{J_n}{q\mu_n E} \tag{6.31}$$

$V = \int E\,dx$ where ε is the static dielectric constant. Similar derivation can be done for hole transport alone in a unipolar organic material. And hence μ_n can be written as μ. Assuming μ is a constant, one can derive the well-known Mott–Gurney equation to describe the "space charge limited current" (SCLC) as a function of applied voltage:

$$J_{\text{SCLC}} = \frac{9}{8}\varepsilon\varepsilon_0\mu\frac{V^2}{d^3} \tag{6.32}$$

The significance of the SCLC is that it represents the maximum possible unipolar current that a trap-free dielectric layer, such as an organic material, can sustain at a given potential difference. In the presence of traps, the current will increase more rapidly than a quadratic dependency, until all the traps are filled. In this regime, bulk carrier transport forms a TCLC, which is governed by $J_{\text{TCLC}} \propto \frac{V^{l+1}}{d^{2l+1}}$23. The parameter l represents the effects of both the trap depth and density. In general, the TCLC model applies in the low electric field condition, whereas the SCLC is accurate under a high electric field. This observation implies that at lower fields, charge carrier traps are involved, which are filled up at higher field. Because of the nature of localization of π-electrons in organic molecules, carrier transport in solid-state organic layer can be viewed as "hopping" between molecular sites. The "hopping" process is essentially a one-electron oxidation–reduction process between neutral molecules and their charged derivatives. The charge hopping rate is known to be influenced not only by the effects of disorder, but is also strongly dependent on the applied electric field. This phenomenon directly leads to a field-dependent carrier mobility, which is described by a well-known model called Poole–Frenkel (PF) behavior [23]. The PF mobility, which is very frequently observed in amorphous molecular materials, is given by:

$$\mu(E) = \mu_0 \exp(\beta\sqrt{E}). \tag{6.33}$$

6.4.3 Charge Recombination

Recombination of opposite charged carriers creates excitons, which undergo the recombination of opposite charged carriers, excitons are created, which undergo 1D diffusion toward electrodes. The recombination process in an OLED is typically described as a bimolecular process, following the Langevin theory, which is governed by:

$$R = \gamma np, \tag{6.34}$$

where R is the carrier recombination rate, n and p the electron and hole densities, and γ is the recombination coefficient of the organic materials. The exciton is a neutral species, so its migration is not affected by the applied voltage. The exciton diffusion length (L) typically has a value of several nm, which roughly defines the width of recombination zone, and is related to the exciton diffusion coefficient (D) and its lifetime, τ [24] by

$$L = \sqrt{D\tau} \tag{6.35}$$

6.4.4 Electromagnetic Wave Radiation

As in case of the LEDs described in Chapter 5, from a classical viewpoint, the emission from the organic thin films has a radiation cone defined by Snell's law which only depends on the refractive index of the organic thin films and air [25]. Considering a unit sphere in Figure 6.14, the shaded region enclosed by the critical angle at the interface of the organic material and air indicates the exit through which the generated photons can escape.

For an organic layer with a refractive index of n_{EML}, the fraction of generated light escaping from the substrate can be calculated by [26]:

$$\begin{aligned} \eta_c &= \frac{2\cdot \int_0^{\theta_{\text{EML-air}}} 1\cdot d\theta \cdot 2\pi \cdot 1 \cdot \sin\theta}{4\pi \cdot 1^2} = \int_0^{\theta_{\text{EML-air}}} \sin\theta \, d\theta = 1 - \cos\theta_{\text{EML-air}} \\ &= 1 - \frac{\sqrt{n_{\text{EML}}^2 - 1}}{n_{\text{EML}}} = 1 - \sqrt{1 - \frac{1}{n_{\text{EML}}^2}} \cong \frac{1}{2n_{\text{EML}}^2} \end{aligned} \tag{6.36}$$

where η_c is the extraction efficiency, $\cos\theta_{\text{EML-air}}$ is the organic-air critical angle. For example, n_{EML} is typically 1.6, which means the extraction efficiency is only about 20% and 80% of the photons remain trapped inside the OLED. However, a multilayer OLED cannot be accurately described by the ray optics model since the layers have thicknesses less than the wavelength of the emitted light. For example, the dependence of the far-field emission pattern on the thickness of the organic layer, cannot be explained by the classical theory.

By using a transfer matrix method, optical interference effects in the thin-film structures can be described correctly. Typically, the OLED consists of a glass substrate carrying a transparent anode composed of ITO,

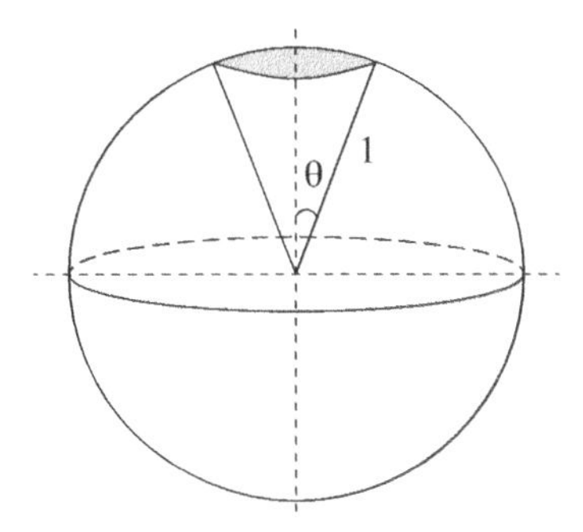

Figure 6.14 Escape cone from an OLED.

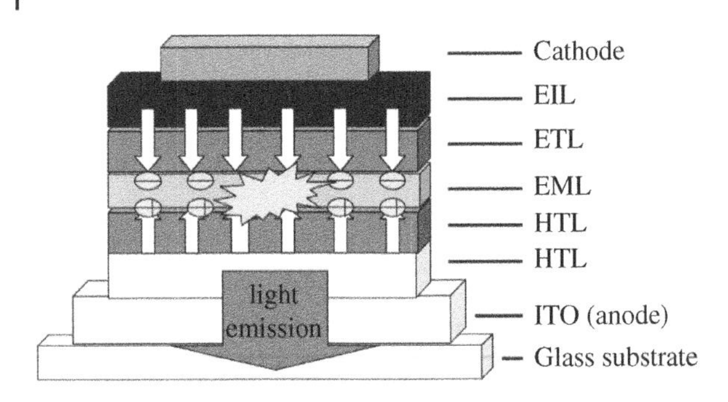

Figure 6.15 A typical OLED structure.

organic layers, and a reflective cathode, as shown in Figure 6.15. Photons generated in the organic layers propagate out of the OLED through the ITO anode and the glass substrate.

Taking into account that the photon emission from the organic materials originates from dipole oscillations, such a structure can be treated as a Fabry–Perot cavity and the emission spectrum in the normal direction can be described by the following Equation [27]:

$$|E_{\text{out}}^{\text{up}}(\lambda)|^2 = |E_{\text{in}}(\lambda)|^2 \times \frac{T_2 \times \left[1 + R_1 + 2\sqrt{R_1}\cos\left(\frac{4\pi x}{\lambda}\right)\right]}{1 + R_1R_2 - 2\sqrt{R_1R_2}\cos\left(\frac{4\pi L}{\lambda}\right)}. \tag{6.37}$$

Here, $|E_{\text{out}}^{\text{up}}(\lambda)|^2$ is the output intensity, $|E_{\text{in}}(\lambda)|^2$ is the free-space EL intensity, x is the optical distance of a dipole from the reflective cathode, and R_1 is the reflectivity of the cathode. Also, R_2 and T_2 are the reflectivity and transmittivity of the ITO anode side, respectively. L is the total optical thickness of the cavity and is given by:

$$L = \left|\frac{\phi_{\text{an}}\lambda}{4\pi}\right| + \sum_j n_jL_j + \left|\frac{\phi_{\text{ca}}\lambda}{4\pi}\right|. \tag{6.38}$$

Here, n_j and L_j are the refractive index and thickness of the jth layer between the two electrodes, and φ_{ca} and φ_{an} are the phase shifts which occur on reflection of light at the organic/cathode and organic/anode interfaces respectively.

Equation (6.37) can also be written as:

$$|E_{\text{out}}^{\text{up}}(\lambda)|^2 = |E_{\text{in}}(\lambda)|^2 \times \text{Tr} \times \left[1 + R_1 + 2\sqrt{R_1}\cos\left(\frac{4\pi x}{\lambda}\right)\right] \tag{6.39}$$

with

$$\text{Tr} = \frac{T_2}{1 + R_1R_2 - 2\sqrt{R_1R_2}\cos\left(\frac{4\pi L}{\lambda}\right)}. \tag{6.40}$$

The term $\left[1 + R_1 + 2\sqrt{R_1}\cos\left(\frac{4\pi x}{\lambda}\right)\right]$ represents the antinode enhancement factor, as shown in Figure 6.16a [28]. It has a maximum value when the emitting dipoles are located exactly at the antinode of the standing wave within the microcavity. On the other hand, the term, Tr, accounts for the effect of the multi-beam reflective interference, as shown in Figure 6.16b [29]. We can see that the effect of the antinode interference is determined only by the dipole position and the cathode reflectivity, while that of the multi-beam reflective interference is controlled by the cavity optical length, anode and cathode reflectivities.

Figure 6.17 shows a mode analysis of devices fabricated with different ETL thicknesses in a typical OLED structure, as shown in Figure 6.15 [30]. In this case, multiple beam interference is less significant than antinode interference because the reflectivity at the anode interface is small. The emissive dipole is located at the center of the EML as shown in Figure 6.15. The intensity of light emitted into the air oscillates as a function of ETL

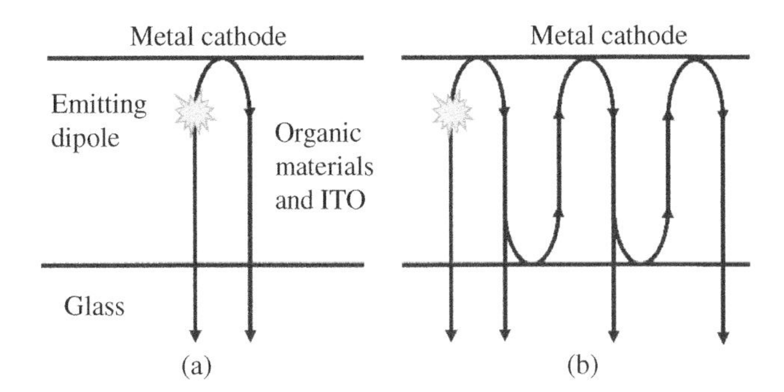

Figure 6.16 Schematic diagram of (a) antinode and (b) multiple-beam interference.

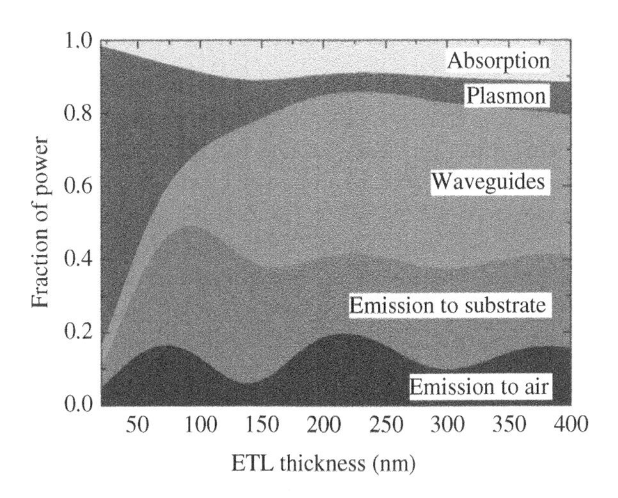

Figure 6.17 Power coupled to different modes of an OLED with different ETL thickness [30].

thickness, as expected from antinode interference with different x-values described by Eq. (6.39). Two total internal reflection (TIR) components can be identified as a waveguiding mode in which light is trapped in the ITO anode/organic layers (with $n = 1.6–1.8$) and a substrate mode, comprising light propagating in the glass substrate ($n = 1.5$), respectively. When an oscillating dipole is close to the metal surface, it excites a surface plasmon. The dipole and its mirror image polarization induced in the metal, effectively constitute a quadrupole which reduces the far-field intensity. This effect decreases as the distance increases, between the emitting dipole and the metal cathode, as shown in Figure 6.17. However, the waveguiding mode losses increase because more guided modes can be supported in a thicker slab waveguide.

6.5 STRUCTURE, FABRICATION AND CHARACTERIZATION

Considerations for the device design in a small molecule or polymer OLED include both electrical and optical optimization. In order to reduce the driving voltage, carrier injection should be efficient, which means careful energy level alignment is needed. The mobility value should be as high as possible, and the free carrier density should be raised. To optimize the recombination rate, suitable energy barriers are needed in the device to confine the carriers in the emissive region, which resemble the heterojunction structure in a semiconductor LED. Sometimes, it is difficult to find an organic material with good electrical properties and a high light emission efficiency. Organic mixtures, or a guest/host (dopant/matrix) system, are used to improve the IQE. Since the planar geometry of an OLED with a thickness of hundreds of nm is a micro-cavity structure, optical interference effects have to be taken into consideration to maximize the external quantum efficiency. In a small molecule OLED, precise control of the thin-film deposition technique makes possible a multi-layer structure. However, in a polymer OLED, the multilayer structure is difficult to fabricate by a solution process. The patterning processes and the encapsulation process are nearly identical for small molecule and polymer OLEDs. To drive a display, both passive matrix (PM) and active matrix (AM) techniques are typically used, as for LCDs. However, there are some important differences since the OLED is a current driven device, rather than a capacitive load like an LCD pixel. For example, at least two transistors and one capacitor are needed in an OLED pixel for AM driving. Lifetime is one of the main issues for OLED applications, and is determined by processes which can be classified as extrinsic and intrinsic degradation. Given proper encapsulation (or passivation) processes and careful environmental control during fabrication, extrinsic degradation can be greatly reduced. However, intrinsic degradation from organic material degradation dominates the ultimate operation lifetime of an OLED.

6.5.1 Device Structure of Organic Light-Emitting Device

The basic structure of an OLED is a multi-layer device in which the organic layer(s) is (are) sandwiched by two electrodes. Typically, ITO with high transparency and work-function is used as the anode. Then, one or more organic thin films were formed (by evaporation, spin-coating, ink-jet printing processes etc. which will be discussed in Sec. 6.5.3.1) followed by metal cathodes. The simplest structure is a single organic layer between the electrodes, as shown in Figure 6.18 [31]. In such a structure, the work functions of the anode and cathode should be matched to the HOMO and LUMO of the organic material, respectively, for better carrier injection. The organic material should exhibit ambipolar transport characteristics and emission capability.

Typically, it is difficult to find a bipolar charge transporting organic material with high emission efficiency. Even if an organic material which exhibits such good electrical and optical characteristics can be identified, the recombination zone needed for highly efficient emission is not easy to control in such a device. For example, if charge recombination takes place near an electrode, the dipole oscillation is at a node, rather than the antinode of a standing optical wave in the device (c.f. Section 6.4.4). This leads to destructive interference and decreases the device efficiency by "electrode quenching." To improve device performance, two-layer or multi-layer structures were introduced with separate layers performing functions of carrier injection, transport, and emission, respectively.

6.5.1.1 Two-Layer Organic Light-Emitting Device

Figure 6.19 shows the device structure of the two-layer OLED, first introduced by Tang et al. [4] The two organic layers are an aromatic diamine and the aluminum chelate tris(8-hydroxyquinoline) aluminum, (Alq_3) which function as HTL and ETL, respectively. In contrast to the bipolar conduction characteristics used in the single-layer device, HTL and ETL ideally only transport a single polarity of carrier-holes and electrons,

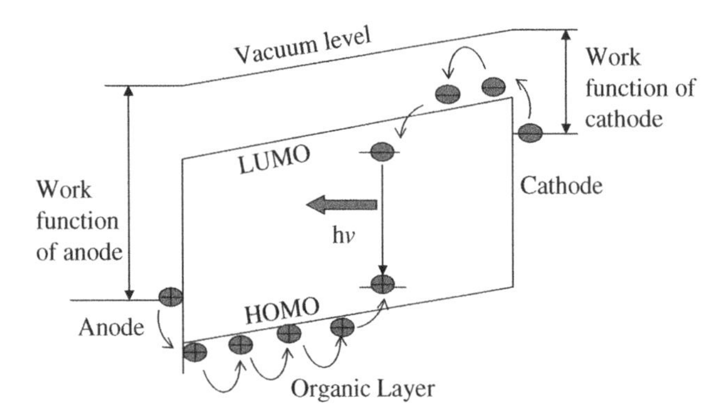

Figure 6.18 Device structure of a single-layer OLED.

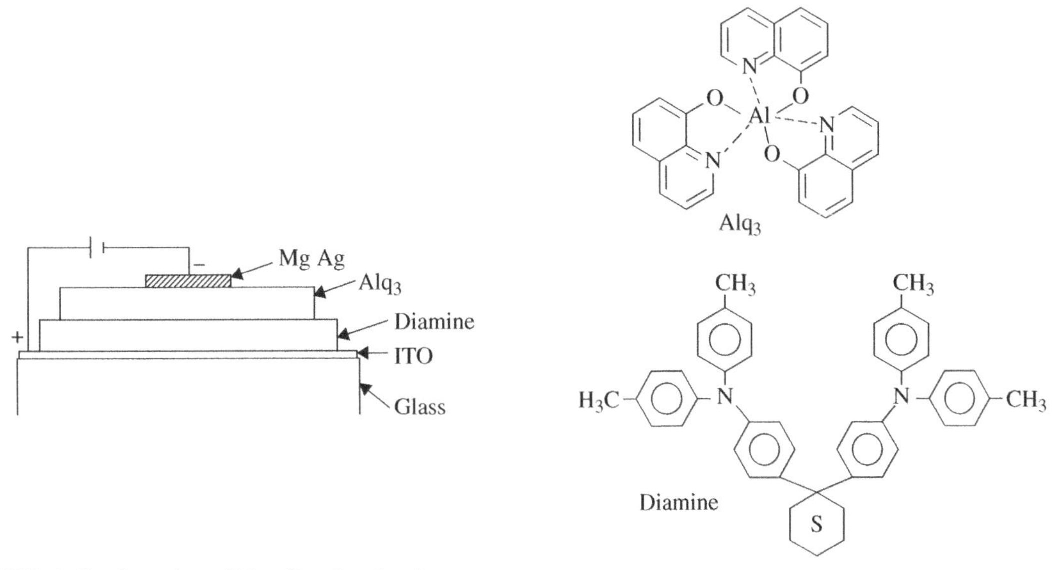

Figure 6.19 Layer structure and organic materials of the first two-layer OLED [4].

respectively. This provides some flexibility in selection of organic materials. In addition, this structure helps to improve the device efficiency due to better carrier confinement. In this two-layer structure, carrier recombination occurs near the HTL/ETL interface. Since the HTL and ETL are different organic materials, such a structure is similar to the heterojunction in semiconductor LEDs. Under forward bias, electrons from the cathode will drift through the ETL and be blocked at the HTL/ETL interface since the LUMO value of the HTL is typically lower than that of ETL. The HOMO value of ETL material is slightly higher than that of the HTL so that holes can readily enter into the ETL. The low hole-mobility in the ETL host results in a high hole-concentration inside the ETL and near the HTL/ETL interface. This enhances the collision capture process and recombination occurs in the ETL, so it acts as an EML at the same time. The recombination zone is in the ETL within 10 nm of the ETL/HTL interface [32].

Typically, the electron mobility in the ETL is at least one order of magnitude lower than the hole mobility in the HTL ($\mu_{e,\,ETL} \ll \mu_{h,\,HTL}$), which means that a high concentration of holes accumulates at the HTL/EML interface, ready for recombination. The recombination probability (P_R) for holes is given by:

$$P_R = (1 + \tau_{rec}/\tau_t)^{-1} \tag{6.41}$$

where τ_{rec} is the carrier recombination time, and τ_t is carrier transit time of the electrons from the cathode to the HTL/EML interface. Eq. (6.1) can be re-written as:

$$P_R = (1 + w/d_e)^{-1} \tag{6.42}$$

where w and d_e are width of the recombination zone and the ETL thickness, respectively [33]. From this equation, we can note that a narrow recombination width is preferred to prevent leakage of holes and electrons toward the electrodes, which leads to the low efficiency. Transparent ITO was used as the anode and the reflective Mg:Ag alloy was used as the cathode. Work functions of the anode and cathode should be high and low, respectively for effective hole and electron injection. Also, the emitted photons radiate out through the anode and glass substrate so the transparency of the ITO should be as high as possible.

Figure 6.20a shows the curves of current density and EL intensity versus voltage. The driving voltage is about 8 V with a current density of 100 mA/cm^2. By comparison, a conventional semiconductor LED, achieves a current density of 200 mA/cm^2 at only four V[4]. In this OLED, the HTL and ETL are 75 and 60 nm thick,

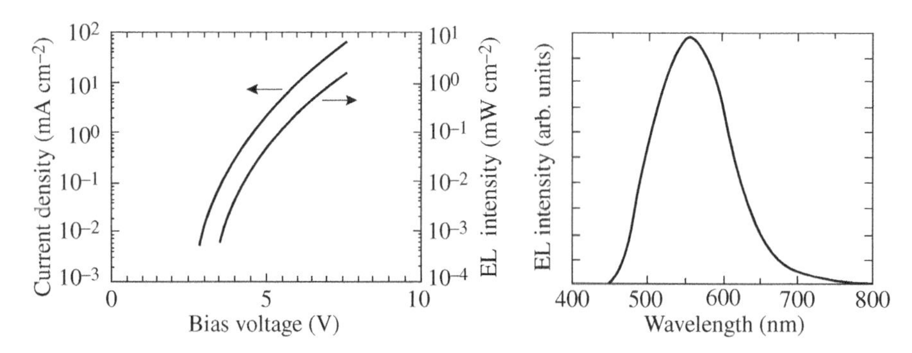

Figure 6.20 (a) J-L-V curves, and (b) EL spectrum of the two-layer OLED [4].

respectively, so are much thinner than the epitaxy layer (several μm) of a semiconductor LED. Although the device thickness is one order of magnitude less than the semiconductor LED, the driving voltage is more than two times higher due to the high resistivity of the organic materials. Figure 6.20b shows the EL spectrum of this device with the FWHM of about 100 nm which is twice as broad as that of the semiconductor LED due to the vibrational and rotational modes of the organic materials.

6.5.1.2 Matrix Doping in the EML

In the two-layer structure, recombination occurs in the ETL, which means the ETL is also the EML. So the functions of this layer include transport carrier and emission. However, although it is possible to find compounds which combine these properties, it is difficult to find such compounds which show high emission efficiency and high carrier mobility over the whole visible range. A common way to solve this problem is to introduce an additional layer composed of an EML doped a highly efficient luminescent material [9]. The criteria for the transporting material in the EML, which is usually called the matrix or the host, include: (i) exhibits good carrier transport characteristics; (ii) has suitable energy levels for carrier trapping and good overlap between the host photoluminescence and dopant absorption spectra for efficient energy transfer, as described in Section 6.3.4. For a dopant (or the guest) material, a high luminescence efficiency is a basic requirement. Not only does use of a guest–host system improve the electrical and optical characteristics of an OLED, it also effectively improves the stability of the device by transferring the recombined exciton to the emissive and stable dopant site thus minimizing non-radiative decay pathways which may degrade the ETL [34].

Figure 6.21a shows the device structure proposed by J. Shi et al. and the molecular structures of the layers [34]. Copper phthalocyanine (CuPC), N,N′-diphenyl-N,N′-bis(1-napthyl)-1,1′-biphenyl-4,4′-diamine (NPB), and Alq_3 are used in the HIL, HTL, and ETL, respectively. A detailed description of HILs will be presented in Section 6.5.1.3. In this device, the EML consists of a mixture of Alq_3 and N,N-dimethylquinacridone (DMQA), which exhibits high emission efficiency. The DMQA concentration should be low enough to keep the molecules apart from each other to prevent excimer quenching (c.f. Section 6.3.4). On the other hand, when the dopant concentration is not high enough, intermolecular energy transfer is not efficient since the average distance between the host exciton and the dopant is too great.

Table 6.1 shows the device performance with different DMQA concentrations. We can see that the luminance increases then decreases with increasing the dopant concentration. The initial increase results from the higher efficiency of DMQA than Alq_3. The decrease comes from excimer quenching, also called concentration quenching. We can also note that the operational lifetime increases with the incorporation of the dopant since higher radiative emission means less heat generation, which is beneficial to lifetime (c.f. Section 6.5.5). The emission wavelength and the Commission Internationale de l'Eclairage (CIE) coordinates of the doped and undoped devices are similar, and lie within the green spectral region.

Table 6.1 Device performance with different dopant concentrations [34]. The luminance outputs, efficiencies, color coordinates, and peak wavelengths were measured at 20 mA/cm^2. For lifetime measurement, OLED

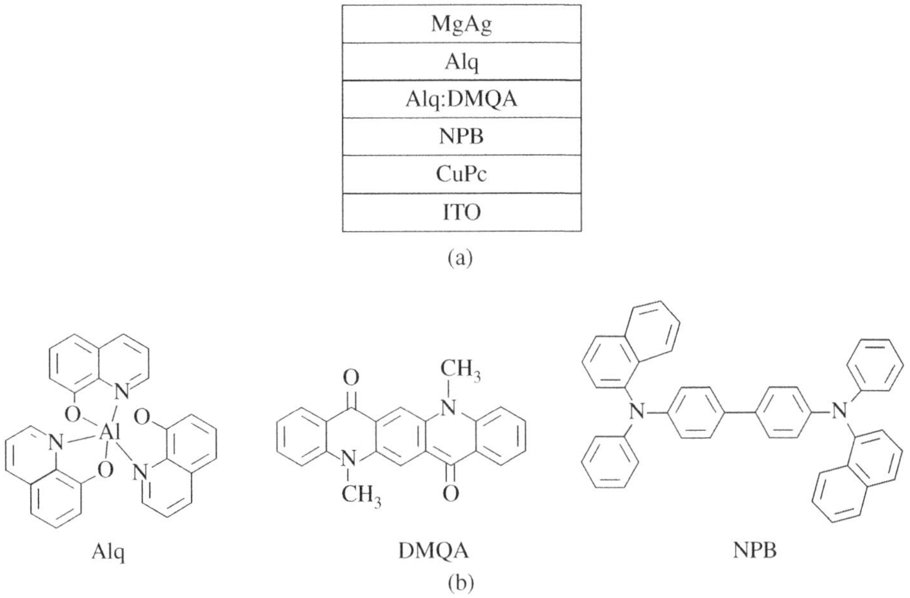

MgAg
Alq
Alq:DMQA
NPB
CuPc
ITO

(a)

Alq DMQA NPB

(b)

Figure 6.21 (a) Device structure and (b) molecular structures of a guest-host system [34].

Table 6.1 Luminance data of DMQA/Alq devices.

DMQA % in Alq	0.00	0.26	0.40	0.80	1.40	2.50
Lum. output (cd/m^{2})	518	1147	1322	1462	1287	1027
Efficiency (cd/A)	2.59	5.74	6.661	7.31	6.44	5.14
CIE_x	0.3872	0.3876	0.3785	0.3922	0.4046	0.4095
CIE_y	0.5469	0.5858	0.5995	0.5901	0.5799	0.5742
EL peak (nm)	544	540	540	544	544	544
$T_{1/2}$ (h)	4200	7335	7500	7340	5450	3650

was driven with 40 mA/cm^2 constant current in forward bias and − 14 V constant voltage in reverse bias. The waveform duty cycle and frequency were 50% and 1 kHz, respectively.

As well as improving the efficiency and lifetime, color-tuning is an important function of the guest–dopant system. By suitable molecular design, it is possible to emit different colors of light using different dopant materials even in the same host. For example, by doping the laser dye molecule DCM1 into Alq_3, a quantum efficiency of about 2.3% with an EL peak at 600 nm[9] is obtained. By suitable molecular modification, another red dopant material 4-(dicyanomethylene)-2-*t*-butyl-6-(1,1,7,7-tetramethyljulolidyl-9-enyl)-4H-pyran (DCJTB) was synthesized [35]. Efficiency of the red OLED based on DCJTB:Alq_3 EML can reach 2.0 cd/A at a luminance of 400 cd/m^2 under a drive current density of 20 mA/cm^2.

One of the criteria for efficient energy transfer from the host to the dopant is a good spectral overlap between the host PL and the dopant absorption, which implies that the dopant should have a larger bandgap than the dopant. That means it is not possible to use Alq_3 as the host of a blue OLED. As blue OLEDs have shorter emission wavelengths and hence higher photon energies than red and green devices, they require a wide bandgap material. The operational lifetime of blue devices is still lower than that of the other primary color devices due to the high photon energy. There are several published combinations of blue emitting guest/host systems. For example: distyrylarylene (DSA) derivatives may be used as a host material with the DSA amine styrylamine 4,4′-bis(2-(9-ethyl-9H-carbazol-3-yl)vinyl)biphenyl (BCzVBi) as the emitting dopant [36], or 2,5,8,11-tetra(t-butyl)-perylene (TBP) can be doped into 9,10-di(2-naphthyl)anthracene (ADN). Investigation of further combinations of host and dopant to achieve highly efficient blue emission and good device lifetime remains an active research field.

6.5.1.3 HIL, EIL, and p-i-n Structure

As described above, HTL and ETL materials have high carrier mobilities and can transport holes and electrons for recombination in the EML, respectively. Organic materials with a high quantum efficiency of fluorescence are used as the EML, which also determines the emission color. To further improve hole- and electron-injection from the anode and cathode into the organic layers, a HIL and EIL are included in many OLED structures. A HIL having its HOMO level between ITO work function and the HTL HOMO level is used to help injection of holes from the ITO anode into the HTL. Likewise, an EIL is used to help electron injection from the cathode into the ETL.

Figure 6.22 shows an example of the insertion of a HIL composed of 4,4′,4″-tris(3-methylphenylphenylaminotriphenylamine (m-MTDATA), between a 4,4′-bis(3-methylphenylphenylamino) biphenyl (TPD) HTL and the ITO anode [37]. From Eqs. (6.21, 6.22), we can see that the current density from thermionic emission decreases exponentially with increasing the barrier height. The HOMO value of m-MTDATA is 0.1 eV higher than the work function of ITO and 0.4 eV lower than the HOMO of TPD, and therefore forms a “ladder-like” sequence of energy levels for improved hole injection from the ITO to the HTL. Molecular structures and the energy diagram are shown in Fig. 6.22a and b, respectively. From Figure 6.22d, we can note that the luminance

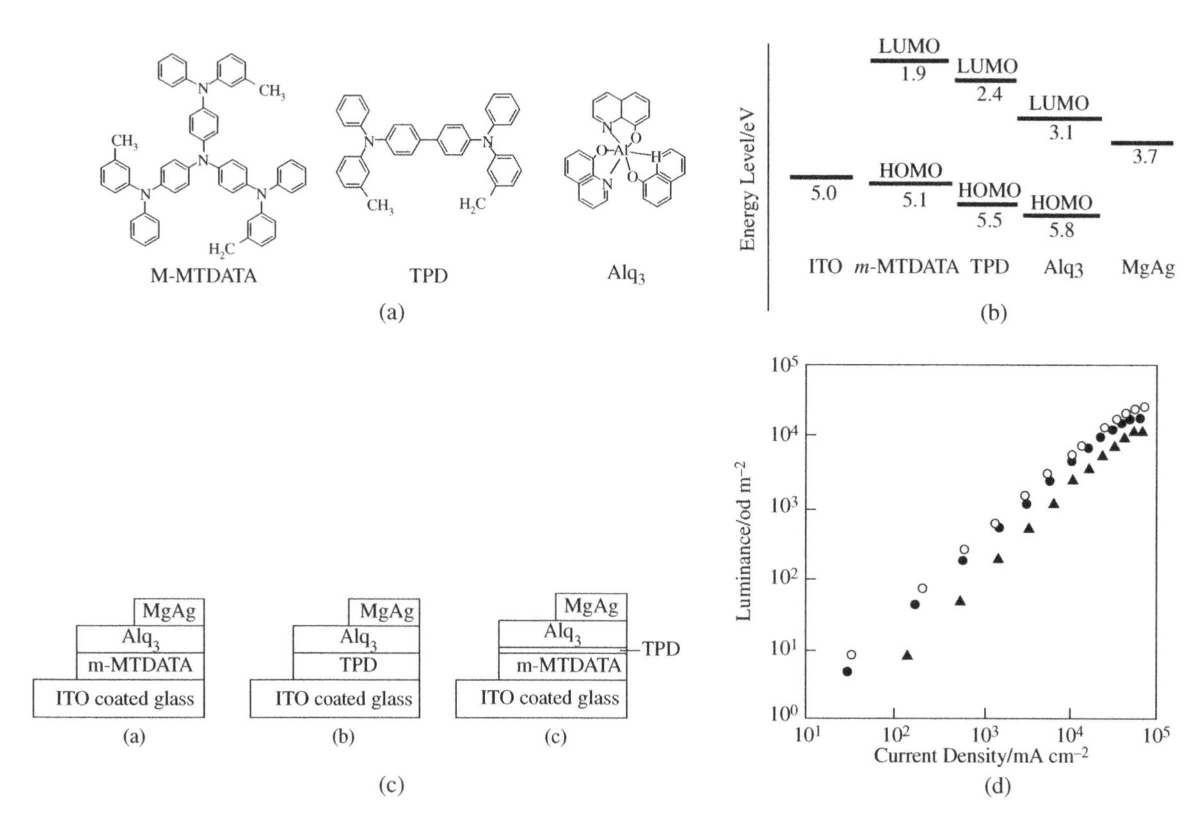

Figure 6.22 (a) Molecular structures, (b) energy level diagram, (c) device structures, and (d) luminance versus current density curves [37].

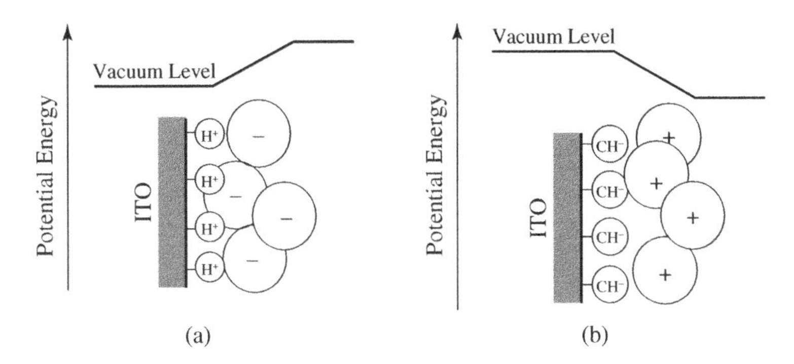

Figure 6.23 (a) Acid, and (b) base treatment at the ITO surface [38].

at a constant current density is highest in the three layer structure (HIL/HTL/ETL; device C), due to better hole-injection capability. When the conventional HTL is replaced by m-MTDATA (device A), the efficiency is lower since the barrier to hole-injection into the EML is highest. Another function of the HIL is to prevent oxygen from the ITO from attacking the organic layer (c.f. Section 6.5.5) which extends the lifetime. A copper phthalocyanine (CuPc) layer is used for this purpose. The layer also improves the charge balance in hole and electron recombination which is a key requirement to achieve high quantum efficiency in OLEDs.

From another viewpoint, the work function of the electrode could be also engineered to improve the injection efficiency. Figure 6.23 shows the effects of surface treatments on the ITO electrode. If the anode is treated with acid (Figure 6.23a), the protonated surface will attract negative charges which form a surface dipole directed toward the organic layer, which contributes to hole injection from the ITO into the organic layers. Typically, such a surface dipole is represented as a shift of the vacuum level (which is the energy outside any material in a free space) or an increase in the work function of the anode surface. Due to vacuum level misalignment, the effective barrier between the ITO and the organic layer is reduced. On the other hand, as shown in Figure 6.23b, by modification of the ITO with base, the effective hole-injection barrier is increased. Dipping the ITO substrate into the acid solution is therefore a method to improve the hole-injection. Another common process is to use oxygen plasma or a UV-ozone treatment before depositing organic layers [39, 40]. Oxygen plasma can not only increase the work function of ITO, but also effectively remove trace hydrocarbon contaminants on the as-grown ITO substrate.

On the other hand, the work function of the cathode should be as low as possible to achieve better electron injection. Figure 6.24 shows the current density versus applied voltage of OLEDs with different cathode structures. Devices with Al cathode exhibit the highest driving voltage (17 V) at 100 mA/cm^2. By using a $Mg_{0.9}Ag_{0.1}$ alloy as the cathode material, a 4 V reduction in driving voltage is achieved due to its lower work function. In this alloy, a small portion of Ag is added to improve its stability against the environment and to improve adhesion to the organic layers [4]. Although low work function materials, such as alkaline and alkaline earth metals, are preferred for the cathode, however, they are reactive under typical environments containing oxygen and water, which increases the difficulties in device process and material handling. An alternative approach is to deposit an ultrathin and insulating layer, such as LiF, between an air-stable metal such as Al, and the organic layers [41, 42]. The detailed physical mechanism for the success of this structure is not clear yet. One of the most popular explanations is that the LiF decomposes upon the Al evaporation. Then Li penetrates into the organic layer forming radical anions and effectively increasing the electron concentration in the ETL. In addition, dipolar layers are formed at the organic/cathode interface which facilitates electron injection. However, due to its insulating characteristics, deposition of a thicker LiF layer results in a voltage increase, as shown in Figure 6.24.

The high driving voltage of OLEDs arises only from energy level mismatch and low carrier mobility, but also from the low free carrier concentrations of organic materials. Both p- and n-type doping, in concept similar to that in the semiconductor LEDs, have been introduced. For p-type doping, a HTL material is usually doped

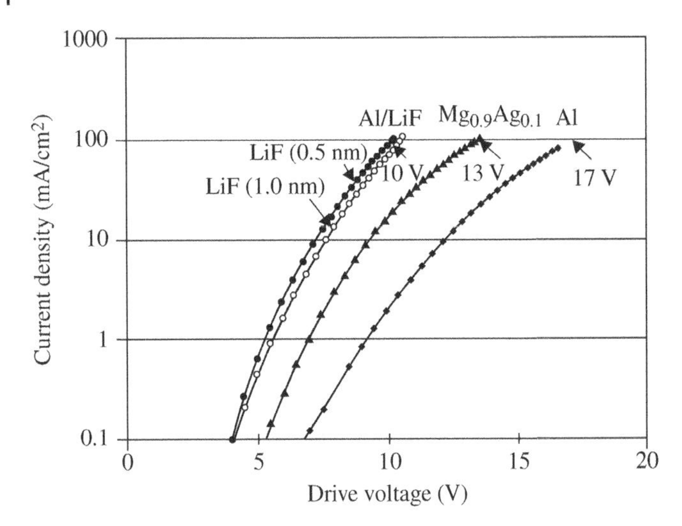

Figure 6.24 log J-V curves of OLEDs with different cathode materials [41].

with a Lewis acid, which binds to the electrons so that free holes are generated. It shifts the Fermi level in the organic materials from the center of the bandgap to close to the HOMO, which improves the layer's hole transport capability. Due to the increase in carrier concentration, Fermi level alignment between the electrode and the p-doped organic layer results in ohmic-like charge injection at the interface, which facilitates hole injection. Some effective material combinations providing guest/host systems for p-type organic materials include doping tetracyano-quinodimethane (TCNQ) derivatives [43], V_2O_5 or $FeCl_3$ into zinc phthalocyanine (ZnPC) or arylamine derivatives [44]. In contrast to p-type doping, n-type molecular doping is more difficult because the requirement that the HOMO value of the dopant be smaller than the LUMO of the host to generate "free electrons," makes such materials unstable against oxygen [45]. An alternative approach is the use of the alkali metals like lithium (Li), sodium (Na), or cesium (Cs) doped into the organic materials, which is called the metal dopant (MD) technique [46]. The alkali metals can readily release a free electron into the organic host. With such a MD technique, the carrier transport characteristics can be much improved by n-type doping, and the interface between the electrode and organic interface can be also modified from a Schottky barrier to an ohmic contact, which helps carrier injection and decreases the drive voltage. By applying p- and n-type doping techniques, a p–i–n OLED can be fabricated with a low driving voltage and high power efficiency [47]. However, the strong acid and base characteristics of the p- and n-type dopants make their fabrication difficult. The operational lifetime of MD technology is still an issue since it is easy for the dopant to diffuse into the EML during operation where it acts as an exciton quencher [48].

6.5.1.4 Top-Emission and Transparent OLEDs

Conventional OLED devices use a transparent ITO anode and a reflective cathode, which provide high and low work functions, respectively. Greater freedoms in designing the driving circuit (c.f. Section 6.5.3.3) and raising the external quantum efficiency (c.f. Section 6.8), are provided by top-emitting and transparent OLEDs. By replacing the reflective cathode with a transparent conductive layer (e.g. ITO), a see-through display can be achieved. However, the work function of ITO is high (~4.7 eV) and is therefore not suitable for electron injection, so a suitable buffer layer is needed between the ETL and ITO "cathode" to improve carrier injection. Such buffer layers are formed of opaque materials and must be thin enough to provide high transmittance. Mg:Ag or a lithium fluoride (LiF)/aluminum (Al) bi-layer several nm in thickness are suitable for this application and provide effective electron injection. Another function of the buffer layer is to protect the organic layers during the fabrication of the ITO "cathode," for which a sputter process is needed. Despite the buffer

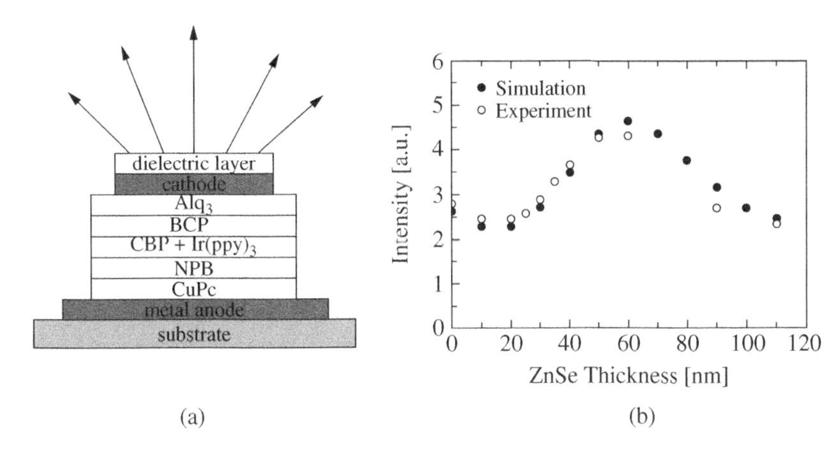

Figure 6.25 (a) Schmatic diagram of a top-emission OLED with dielectric layer (ZnSe), and (b) experimental and simulated intensity at normal direction [29].

layer, the radio-frequency power during the ITO sputtering stage has to be controlled very carefully to minimize damage to the underlying organic layer by energetic ions. Under these process conditions, the deposition rate is usually as low as several Å/s. The optical transmittance of the buffer layer is one of the major concerns in such an architecture. Thicker buffer layers absorb more light and hence decrease the OLED luminance and transmittance. On the other hand, thin buffer layers may not be sufficient to protect the organic layers from sputtering damage. A thick ITO cathode, about several hundreds of nanometers, is necessary to reducing ohmic losses, given the low conductivity of ITO, of around 10^4 S/cm. This bulk conductivity of ITO is one to two orders of magnitude lower than those of metals. Therefore, a thin, semi-transparent metal material, ten to one hundred times thinner than is required for ITO can be deposited by thermal evaporation and can be used as a transparent or semi-transparent cathode. Such a thin cathode can comprise 20–30 nm of samarium (Sm) [49], Ca/Mg [50], or LiF/Al/Ag [51].

In a top-emission OLED, where light is emitted through the top electrode rather than through the substrate, a reflective anode is also needed. It is straightforward to deposit a reflective metal such as Ag, underneath the ITO [52]. Alternatively, some high work function metals, such as Au and Ti, are good candidates as anodes, in terms of their injection capability. A high reflectivity over the whole visible range is also needed for better optical characteristics. Oxidation of Ag produces AgOx which has a high work function and forms an effective HIL, while the Ag itself provides good electrical characteristics [51]. In a top-emission OLED, the reflective anode and the semi-transparent cathode form an optical microcavity which exhibits strong multi-beam interference (c.f. Section 6.4.4). Riel and his coworkers fabricated top-emitting organic light-emitting devices (TOLEDs) with a dielectric layer of ZnSe capping the semi-transparent cathode, as shown in Figure 6.25a [29]. They demonstrated that the EL intensity and spectral characteristic is a function of the dielectric thickness, as shown in Figure 6.25b. The dielectric layer can enhance the outcoupled light intensity in the forward direction. In addition, angle-independent emission spectra can be achieved by tuning the dielectric thickness.

By engineering the transparent, semi-transparent, and reflective characteristics of the electrodes, some interesting displays can be fabricated. Based on a transparent OLED, the display can show information from both sides [53]. Moreover, a double-sided OLED display with different information content on each side, is also possible by engineering the optical characteristics of the electrode [54].

6.5.2 Polymer OLED

Two main kinds of OLEDs can be distinguished by the molecular weight of the organic materials. For molecular weights less than 1000 g/mol, devices are called small molecule OLEDs, which can usually be fabricated by thermal evaporation of the organics under high vacuum. On the other hand, devices using materials with a

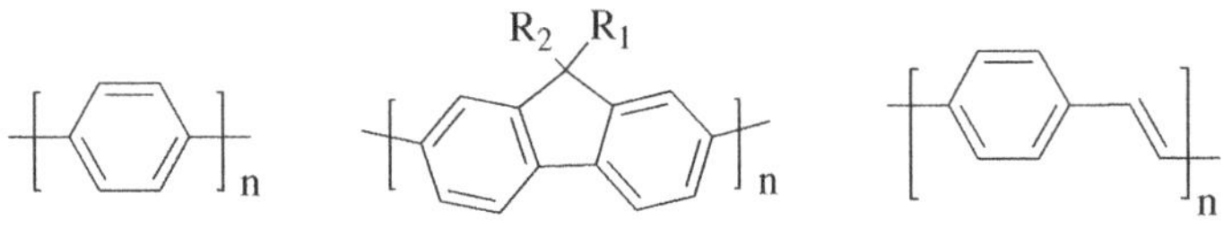

PPP PF PPV

Figure 6.26 Molecular structures of some polymer EL materials.

molecular weight over 10 000 g/mol is termed a polymer OLED (or PLED). Materials of intermediate molecular weight are more difficult to use and process, so are rarely used.

At Cambridge University in 1990, a polymer OLED was fabricated by a spin-coating technique devised by Burroughes et al. [5]. A yellowish-green emission was detected from a thin film of poly(p-phenylene vinylene) (PPV) with the device configuration of Al/$A1_2O_3$/PPV/Al. The emitted spectrum and EL efficiency of conjugated polymers can be tuned by suitable molecular design. Figure 6.26 shows some common repeating units of polymer EL materials, such as polyphenylenes (PPP), polyfluorene (PF), and poly(phenylenevinylene) (PPV). In 1991 D. Braun and A. J. Heeger reported a red polymer LED using poly[2-methoxy-5-(2-ethylhexyloxy)-1,4-phenylenevinylene] (MEH-PPV) [31]. Y. Ohmori et al. reported the blue-emitting polymer poly(9,9-dialkylfluorene) (PF) with an EL peak emission wavelength of 470 nm [55].

Since the polymer will thermally decompose at temperatures far lower than would be needed for evaporation, thin film formation for polymer OLEDs is by a solution process. The polymer is dissolved in a suitable solvent, and the resulting solution is coated onto the substrate. The solvent is removed by thermal or vacuum treatment leaving a uniform polymer layer. Residual solvent in the organic thin film is sometimes an impurity quencher which increases non-radiative recombination. It can also diffuse during device operation or storage, and shorten the device lifetime. To achieve low driving voltage, high recombination efficiency, and high emission efficiency, a multilayer structure is typically preferred to optimize the optical and electrical characteristics in an OLED. However, although possible it is not easy to fabricate a polymer OLED with more than two layers. Since typical organic non-polar solvents will dissolve most electroluminescent (EL) polymers, it is common to apply PEDOT:PSS, which can be dissolved in a polar, water-based solvent, as a HIL and HTL prior to EL polymer thin-film formation. Some methods have been proposed to fabricate a multi-layer polymer OLED. Self-assembly methods [56] or electropolymerization provide options to avoid the solubility problem [57]. Another approach is to use a series of functional polymers that can be spin-coated on the substrate from "orthogonal" solvents each of which does not affect the previously deposited layer. It requires polymers that can be dissolved in respectively polar and non-polar orthogonal solvent systems [58].

6.5.3 Device Fabrication

Since OLEDs and PLEDs are sensitive to moisture and oxygen, the thin film processing should be performed in a vacuum or an inert gas environment. After thin-film formation of organic and electrode layers, the sample should be encapsulated or passivated without exposure to the ambient environment. Hence, for an OLED display, the supporting circuitry which is fabricated by conventional semiconductor processes (e.g. lithography, etching and deposition) should be formed before the thin-film deposition.

Typically, conventional OLEDs are fabricated on glass substrates. Transparent conductive ITO is used as the anode. Light emitted from the organic layers propagates through the transparent ITO and glass substrate. However, since the resistivity of ITO is typically 10^{-4} Ohm/cm which is one to two orders of magnitude higher than that of the reflective metals, such as Al and Ag, the resulting high resistance of the ITO results in voltage being lost along the ITO transmission lines, which in turn induces: (i) non-uniformity of panel luminance, and (ii) delay and distortion of the control signals. Hence, low resistance lines made of metals such as Al or Cr, are used alongside the ITO but outside the emissive regions to improve the current conduction. Since the conductivities of the organic layers are quite low, they provide a significant current pathway only in the vertical direction (perpendicular to the glass substrate) and current will not propagate horizontally (parallel to the glass substrate). Light emission only occurs from the regions where the anode and cathode overlap in the vertical direction. A patterned insulating layer such as polyimide, is deposited on the ITO electrode below the organic layers to define the emissive regions of the pixels. Such a layer is critical since it provides a flat surface and planarizes the circuitry underneath which is vital for OLED film deposition due to the extreme thinness of the layers (100–200 nm). Thin-film transistors (TFTs) may be needed for active-matrix (AM) driving techniques. In an OLED display, at least two TFTs are needed at each pixel due to its diode junction characteristics. The mobility of the transistors should be as high as possible to provide sufficient current for light generation. Mobility decrease and threshold shift are two of the typical degradation effects observed in TFTs due to current stress.

Once the underlying structure of the OLED display has been fabricated, the organic thin-film and encapsulation processes should be accomplished without exposure to the ambient environment. For small molecule OLEDs, the organic materials can be deposited by thermal sublimation under high vacuum. By controlling the temperature, a deposition rate as low as 0.1 nm/sec, can be controlled precisely. For polymer materials, due to their large molecular weight, solution processes are needed. Two common fabrication methods applied to PLED materials are spin-coating and ink-jet printing. After film formation, the devices are transferred to the encapsulation process in an inert environment in order to achieve a longer lifetime. Depositing passivation layer(s) directly onto the OLED is also possible; this is beneficial in reducing the thickness of the display and has potential for flexible substrate applications.

6.5.3.1 Thin-film Formation

For small-molecule OLEDs, a common method to deposit the organic thin film is by thermal evaporation under high vacuum, since the molecular weight is low enough for sublimation. As illustrated in Figure 6.27, the organic layers and the cathode materials are deposited sequentially onto the patterned ITO glass substrate. Since no wet processing is permissible after thin-film deposition, the patterns of the organic layers and the metal are defined by shadow masks, which are located near to the glass substrate. Each mask is a metal plate with defined openings, through which the vapor from the thermal sources can pass and deposit on the substrate. Typically, the substrate is kept at room temperature, which means the deposited molecules cannot obtain enough thermal energy to move across the glass substrate and affect the intended patterning. It is important that the deposited thin film should remain amorphous in order to obtain high surface uniformity. The deposition rate is controlled by the evaporation source temperature. Higher temperature results in faster deposition rate, and higher throughput. Device performance parameters, such as efficiency and lifetime, are also affected by the deposition rate [59]. However, if the deposition temperature is too high, organic materials may decompose which results in premature device failure. To obtain a uniform thin film, the distance between

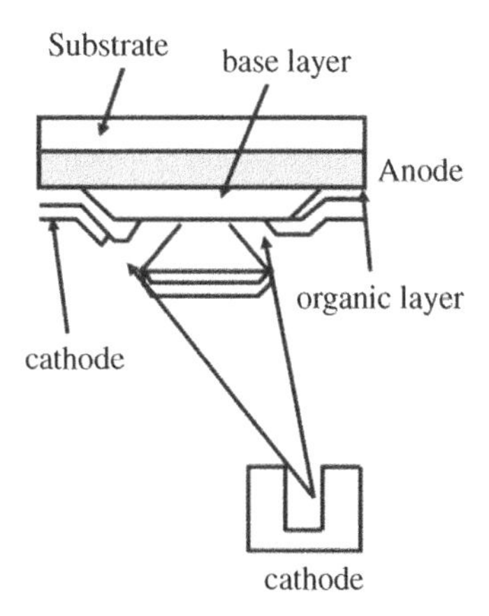

Figure 6.27 Illustration thermal evaporation.

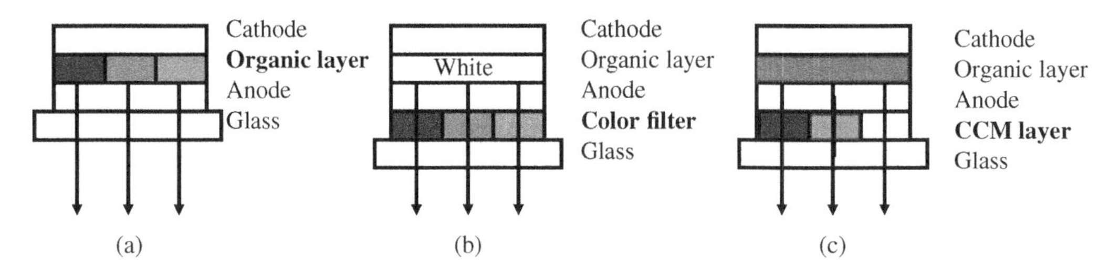

Figure 6.28 Fabrication methods for full color OLED: (a) lateral subpixelated, (b) white OLED with CF, and (c) CCM.

the substrate and the organic source should be as large as possible, which results in a low utilization rate of the organic material. Typically, using a conventional small area thermal evaporation source, only 10% or less is of the evaporated material is deposited on the glass substrate while the other >90% of materials are coated onto the wall of the chamber. This not only wastes the organic material, but also limits the panel size. A linear source, in which an array of many "point" thermal sources is assembled into a "linear" one, can effectively reduce the size of the chamber, increase the material usage rate, improve the throughput, and allow use of larger sized substrates [60].

In a full color display, red, green, and blue subpixels are needed. It is straightforward to use red, green, and blue OLEDs for this purpose. However, as described above, a conventional lithographic process is not available for OLED thin-film formation. However, there are three fabrication methods for full-color OLED displays, which are: (i) lateral subpixelated, (ii) white OLED with color filters, and (iii) blue OLED with color change materials (CCMs), and their structures are shown in Figure 6.28.

As shown in Figure 6.29, in the "lateral subpixelated" structure, the red, green, and blue OLEDs are fabricated in different locations though fine-pitch shadow masks. The device efficiency and operating lifetime are higher for a lateral emitter, compared with the other two structures, since the OLED for each primary color can be optimized independently. However, typically, the subpixel pitch is about 70 μm, which means the thickness

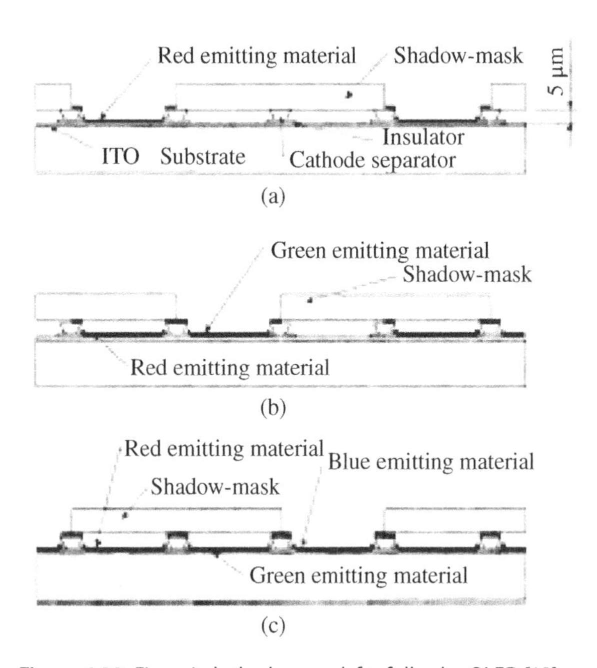

Figure 6.29 Fine-pitch shadow mask for full color OLED [13].

of the shadow mask is limited to about 100 μm. Since the glass substrate is large (370 × 470 mm even for Gen II glass), handling and processing with such a mask is not easy. When the shadow mask is used continuously for a long time (e.g. one week), an accumulation of evaporated organic materials around the edges of the gaps in a fine-pitch shadow mask may reduce the hole size, which in turn results in dimension shrinkage and misalignment of the deposited material. In addition, thermal expansion of the metal shadow mask also results in misalignment in a high-temperature deposition environment. The distance between the shadow mask and the substrate should be minimized for better alignment. However, the surface of the substrate is easily scratched by the shadow mask if they are too close, which results in more difficulty in device fabrication.

To simplify the process, the conventional color filter technique used in LCDs can be applied to full color OLEDs. By fabricating a white OLED without the need for a complex fine-pitch shadow mask, and providing color filters at each subpixel the fabrication yield can be improved. However, as in the LCD case, 2/3 of the light is absorbed by the color filter, which means the efficiency and lifetime of the display is reduced. The operating lifetime refers to the luminance decay over time, which is often different for different colors of OLEDs. In the lateral emitter structure, different degradation rates may result in a color shift with time, but the color stability is good when a white OLED and color filter are used. Another possible route to improve the efficiency and simplify fabrication is by using CCMs, which can absorb blue light, and re-emit red or green light. As shown in Figure 6.28, EL from the blue OLED itself is used for the blue subpixel. However, the material selection, fabrication process, and material stability still have some issues for this technique. By using a laser-assisted patterning technique, the lateral subpixelated arrangement can be fabricated with high resolution on a large-size substrate [61]. Laser induced thermal imaging (LITI) is one of these fabrication technologies and its operational principle is shown in Figure 6.30. An organic thin film is first deposited uniformly onto a donor film which includes a light-to-heat conversion layer (LTHC). The donor film is laminated onto the substrate, then under laser illumination on selected areas, the LTHC absorbs light and releases the organic thin film onto the substrate with high spatial resolution. Clearly the resolution of this technology depends on the beam size of the laser. The substrate size is limited to the scanning range of the laser head. By using multiple laser heads, the throughput can be increased [62].

Fabrication methods for polymer OLEDs must be solution based, since these materials decompose at temperatures below those needed for sublimation. Chloroform ($CHCl_3$), dichloroethane, toluene, and xylene are commonly used solvents. After dissolving the polymer materials, the solution is spin-coated or ink-jet printed upon the substrates. By driving off (e.g. under vacuum or by heating) the solvent, polymer thin-films can be obtained. Spin-coating is a common process in semiconductor fabrication. By controlling the rotation rate of the substrate and the viscosity of the solution, thin films with precise thickness (resolution about several nm) can be obtained. For fabrication of full color devices, different solutions must be deposited in different positions. Ink-jet printing, which is a well-established process technology, is used [63]. Seiko Epson Corporation demonstrated a 40 in. polymer OLED display by using multi-head ink-jet printing technology in 2004.

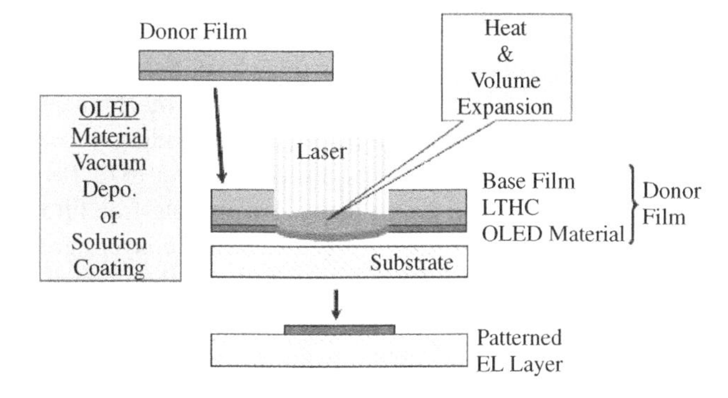

Figure 6.30 Illustrative diagram of the LITI process [61].

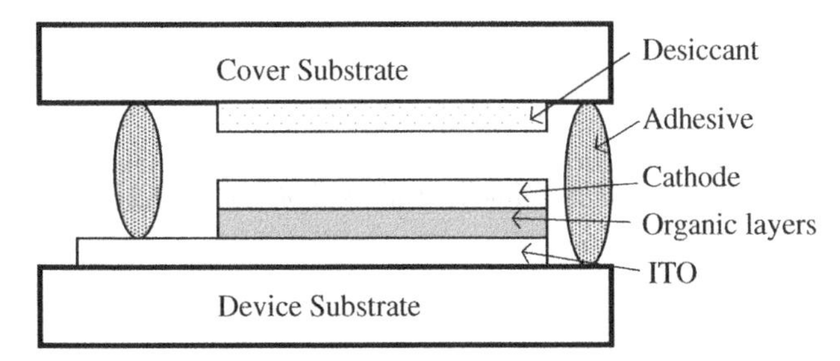

Figure 6.31 Illustration of an encapsulated OLED.

6.5.3.2 Encapsulation and Passivation

Many organic materials are sensitive to moisture and oxygen, and oxidation of the metal electrode may also degrade the OLED's properties [64]. Therefore, the exclusion of oxygen and water from the OLED package is necessary for a long lifetime. Typically, the OLED is hermetically sealed using a cover attached by an adhesive such as a UV-cured epoxy, as shown in Figure 6.31. The cover can be a glass or metal lid. Possible pathways for leakage of environmental moisture and oxygen are through the adhesive and at the interface of the adhesive with the glass substrate or the cover. To reduce the water and oxygen permeation into the OLED, the gap between the substrate and the cover should be as small as possible to reduce the "channel width." At the same time, the width of the epoxy should be as large as possible to increase the "channel length." A channel may form at the interface of the epoxy with the substrate or cover if the bond strength is not good enough. Even though the epoxy can entirely prevent attack from the environment, there may still be residual solvent trapped in the device during the fabrication steps prior to thin-film deposition, which will vaporize gradually and degrade the OLED. Hence, a desiccant such as CaO is incorporated to react with any water which is either residual or may diffuse through the epoxy seal, as shown in Figure 6.31 [65].

It is also possible to substitute the cover with a passivation layer deposited directly onto the OLED with a thickness of several μm. The thickness of the display can then be reduced to about that of one glass substrate (~0.55 mm). The process flow of OLED fabrication also becomes simpler. Since the organic materials cannot tolerate high temperatures, a process temperature lower than 100 °C is needed for the passivation layer. The passivation layer must also be free of cracks and pinholes which would provide channels for environmental attack. The thermal stress should be as small as possible. Hence, the organic and metal thin films underneath would not be damaged. Recent research has provided several concepts to form good passivation layers with low water vapor permeation rate (WVPR), such as the deposition of polymer by PECVD [66], single adhesive layer with inorganic cap [67], multiple layers consisting of alternating organic and inorganic materials [68] and an architecture based on a desiccant and barrier layer [65]. The main advantages of using a polymer as the passivation layer are its mechanical and chemical stability and its strong adhesion to the underlying layer [66]. The merits of inorganic materials are transparency, resistance to abrasion, and a permeability to water and oxygen which is much lower than polymers. However, low WVPR is only achieved by thick layer (~hundreds μm). It is not easy to obtain a defect-free film with such thickness, and it also results in a large mechanical stress which can crack the film beneath the passivation layer. To solve this problem while providing a low WVPR, an adhesive buffer layer is prepared before the deposition of an inorganic passivation layer [67]. The formation of pinholes is inevitable in a thick inorganic passivation layer. So, to provide a pinhole-free film which can also block water vapor ingress through grain boundaries, inorganic/organic multiple layers are adopted [64, 69, 70]. The organic films serve to isolate the inorganic layers from one another so that defects in different inorganic layers do not overlie one another, and oxygen and water vapor must follow an extended, convoluted path to pass through the entire layer.

As discussed in Section 3.6.6, such a passivation layer can be deposited, not only on the OLED, but also on the flexible substrate, for example: polyethylene terephthalate (PET) In this way the WVPR can be reduced from a typical figure of 10^{-1}~10^{1} g/m^2/day for an organic film, to <5×10^{-6} g/m^2/day which is the requirement for an OLED to reach a storage lifetime longer than 10 000 hours [66]. The buffer layer is also effective in reducing the surface roughness of the flexible substrate, which can cause dark spots after the fabrication process. The height of spikes on a PET surface is reduced from 150 Å to <10 Å by application of a multilayer structure. It is possible not only to block water vapor from outside, but also to absorb it before it can contact the device, by embedding getter particles into the polymer substrate or polymer lid.

6.5.3.3 Device Structures for AM Driving

As described in Chapter 3, silicon-based TFTs used in the AM backplane are sensitive to visible light and should be shielded by a black matrix. That means, that a proportion of the area of each pixel is devoted to the TFT layout and is not available for light emission. The aperture ratio of an AM-OLED is typically quite low, as shown in Figure 6.32a, which results in a low display luminance. This will be even worse if more than two transistors are used. The top-emission OLED (TE-OLED) is one of the most promising technologies for achieving a high aperture ratio AM-OLED, as shown in Figure 6.32b [71]. Typically, a TE-OLED has a highly reflective anode and a semi-transparent or transparent cathode; light is emitted from the cathode side. The TFTs can be hidden underneath the reflective anode and hence the aperture ratio is increased. As shown in Eqs. (3.19) and (3.20), by increasing the *W/L* value of the TFT channel, the current density will rise. Since the TFT and OLED can be stacked one above the other without affecting the performance of either, the TFTs can occupy the whole pixel area. Hence, it is possible to use a-Si TFTs to drive a TE-OLED even though the mobility value is low. However, the fabrication process and the optical design of such a device is quite different from the conventional bottom-emission OLED. Degradation of the a-Si TFT under current stress is another issue which needs to be considered. Configurations of TFTs and OLEDs for AM-OLED displays are still being developed.

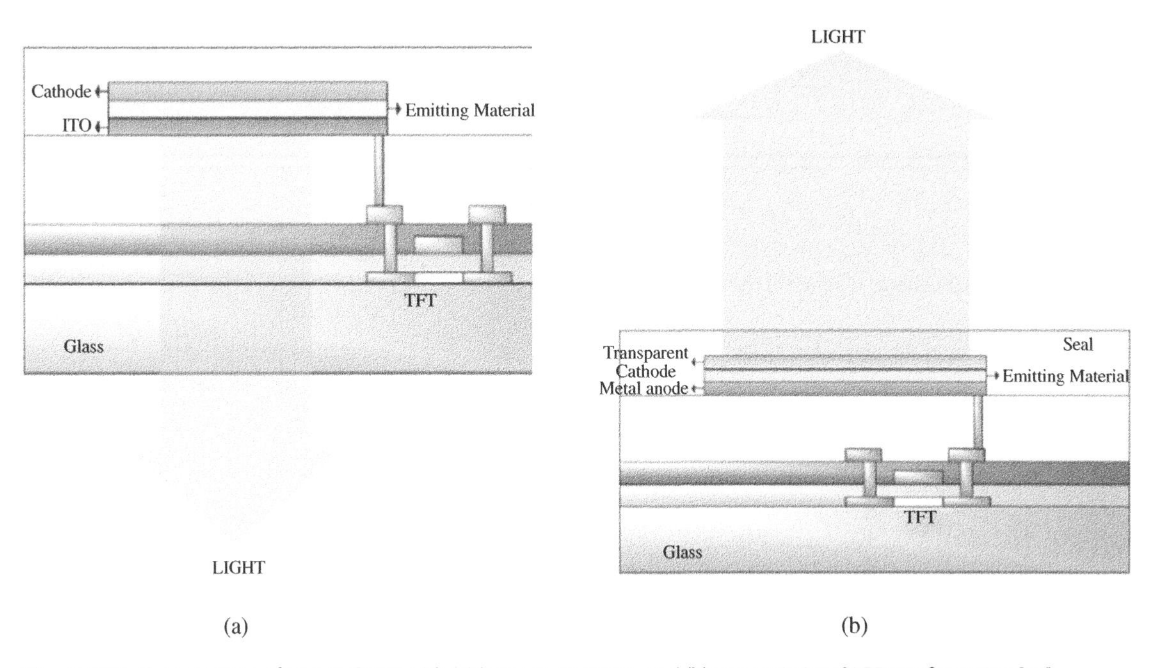

Figure 6.32 Cross section of an AM-OLED with (a) bottom-emission, and (b) top-emission OLED configuration [71].

6.5.4 Electrical and Optical Characteristics

In this section, the J-V characteristics and the optical performance of a two-layer OLED are demonstrated and analyzed, experimentally and numerically. The HTL and ETL (EML) materials are N,N -Bis(naphthalen-1-yl)-N,N -bis(phenyl) benzidine (NPB) and Alq_3, respectively. Their chemical structures and device band diagram are shown in Figure 6.33.

Five two-layer OLEDs were fabricated with varying HTL and ETL thicknesses. The HTL thickness was varied from 200 to 1000 Å, in steps of 200 Å, while the ETL thickness was varied from 1000 to 200 Å in order to keep the total organic layer thickness constant at 1200 Å. Table 6.2 shows the device structures.

Figure 6.34 shows experimental and simulated data on the J-V characteristics of the devices. Electrical simulation here was based on the Poisson, continuity, and drift/diffusion current equations described in Chapter 5, together with the thermionic emission and PF model for carrier injection and transport in Section 6.4, respectively. And optical simulation was based on the interference equations in Section 6.4.4. We can note that the current density increases sharply as the driving voltage is raised, which is typical behavior for charge transport by SCLC and TCLC. The driving voltage of device 1 is highest, and decreases as the HTL layer increases in thickness. The electron and hole mobilities of Alq_3 and NPB are between $10^{-6} - 10^{-3}$ and $10^{-4} - 10^{-3}$ cm^2/Vs at a field strength of 0.1–1 MV/cm, respectively. That means the hole transport is much more efficient than electron transport. It also implies that the voltage difference across the Alq_3 is lower than the hole mobility of NPB.

Figure 6.35 shows the distributions of charge density and recombination rate of devices 1, 3, and 5, respectively. We can see that holes and electrons are blocked at the HTL/ETL interface due to the energetic barriers formed by the differences in LUMO and HOMO value between NPB and Alq_3, respectively. Note that the barrier for holes (0.3 eV) is smaller than that for the electrons (0.6 eV) which means the holes can penetrate

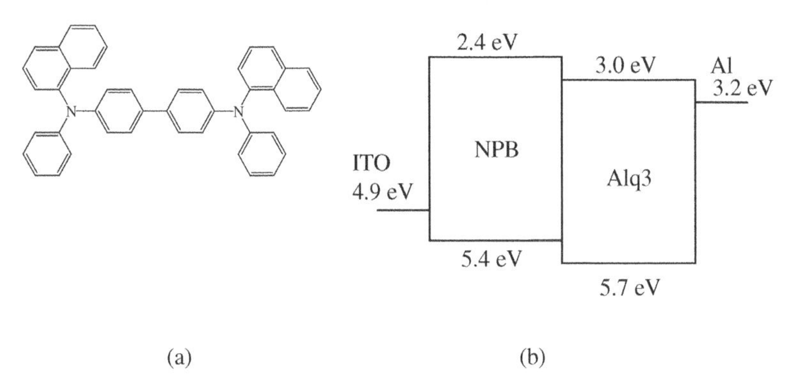

Figure 6.33 (a) NPB and (b) energy diagram of the two-layer OLED.

Table 6.2 Layer structures of the two-layer OLEDs.

(unit: Å)		HTL NPB	ETL Alq3	LiF	Al
No. 1		200	1000		
No. 2	ITO glass	400	800		
No. 3		600	600	12	1500
No. 4		800	400		
No. 5		1000	200		

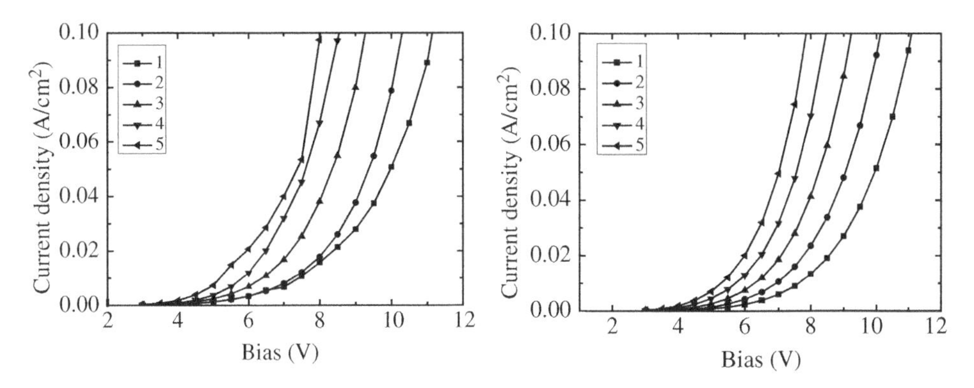

Figure 6.34 (a) Experimental and (b) simulated J-V characteristics of devices 1 to 5.

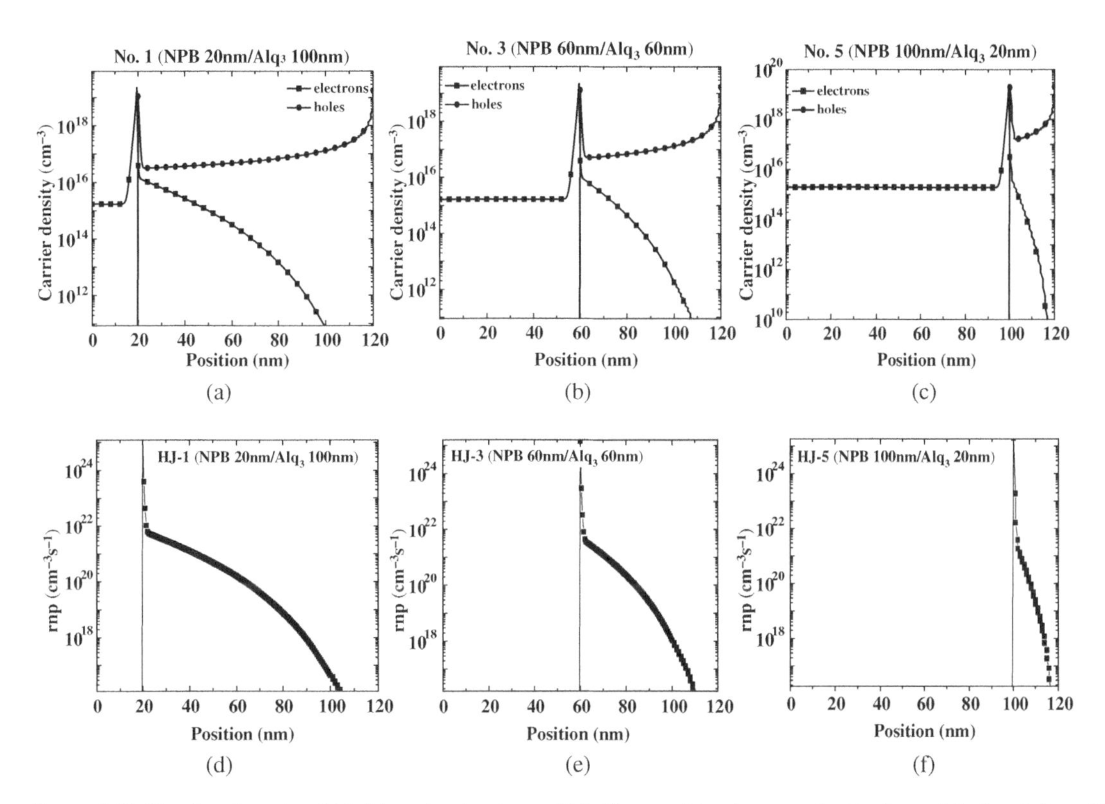

Figure 6.35 Simulated results of (a)–(c) carrier density and (d)–(f) recombination rate distributions of devices 1, 3, and 5.

into the ETL to recombine with the electrons, as shown in Figure 6.35a–c. Such a carrier accumulation near the HTL/ETL interface increases the recombination efficiency. Figure 6.35d–f shows the recombination rate (R), which follows the Langevin theory. We can also see that the recombination mainly takes place in the ETL and near the HTL/EML interface due to the high carrier density there.

Figure 6.36a shows the luminance versus current density for devices 1 to 5. We can see that the luminance increases and then decreases as the HTL thickness increases. As shown in Figure 6.35, maximum recombination occurs at the HTL/ETL interface. To have constructive wide angle optical interference, the recombination position must be located at a distance $\lambda/4n$ from the reflective cathode, according to Eq. (6.39). Considering

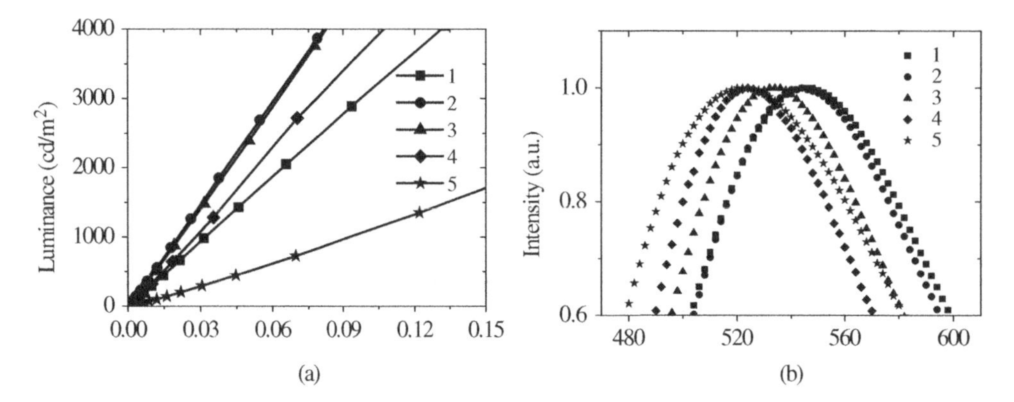

Figure 6.36 Measured (a) luminance versus current density and (b) EL spectra of devices 1 to 5.

the emission wavelength at 530 nm and that the refractive index of the organic material is about 1.6, the $\lambda/4n$ value is about 82 nm, which means that device 2 should be close to the optimization condition of the constructive wide angle interference. Experimental spectra of the devices are shown in Figure 6.36. The EL emission spectral peaks are at 558, 544, 536, 534, 526 nm for devices 1 to 5, respectively. Such a blue shift can be also understood from the antinode interference. The constructive interference condition is wavelength dependent, and blue shifts as the recombination zone is moved toward the cathode, i.e. when increasing the HTL thicknesses.

6.5.5 Degradation Mechanisms

Lifetime is one of the major obstacles to wider adoption of OLEDs in commercial applications. Although the operating principles of LEDs and OLEDs are similar, the lifetime of an OLED is much shorter (typically <1/10) than that of LEDs. The main reasons include: (i) OLEDs use thinner layer structures, (ii) organic materials are sensitive to moisture and oxygen, and (iii) organic materials are not as robust as the semiconductors due to larger vibration energy. In addition, inorganic semiconductor materials are maintained in their structure by multiple (generally four), strong covalent bonds, while the organic materials are Van der Waals solids in which even the constituent molecules are typically bonded by 1–2 covalent bonds. Degradation in OLED can be categorized according to three independent modes, which are: (i) dark-spot formation, (ii) catastrophic failure, and (iii) intrinsic degradation.

Dark-spot formation means the appearance of non-emissive regions which increase in size and number, with time. It develops even during OLED storage [72], and the growth rate may be accelerated by operation of the device. The formation mechanisms of the dark-spot include electrochemical reactions of the organic (or electrode) materials with environmental reactive contaminants. To slow down dark-spot formation, encapsulation or passivation as described in Section 6.5.3.2, is needed to prevent oxygen and moisture from attacking OLED. There are many possible underlying causes which contribute to the dark-spot phenomenon. Since the dark-spot is non-emissive in a certain area rather than causing a luminance decrease over the whole emissive device, the dark spots must arise from structural defects in OLED. The root cause of the structural defects includes: (i) loss of adhesion ("bubbles") at the interface between the cathode and organic layers which impedes electron injection and breaks contact between the cathode and organic layer, which leads to a non-emissive zone [73]; (ii) pinholes in the cathode provide a pathway for water and oxygen diffusion which oxidize the cathode metal and react with the organic materials [74]; (iii) ITO spikes which cause a local temperature rise. This in turn causes decomposition of the ITO and organic film and releases volatile species which locally delaminate and oxidize the cathode [75]. Figure 6.37a,b shows OLED pictures during operation taken under

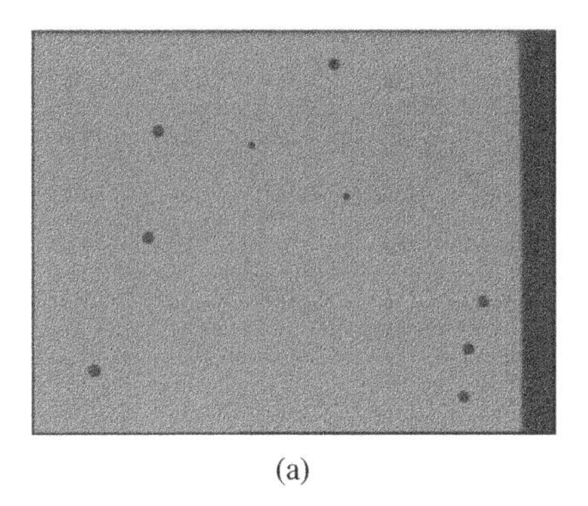
(a)

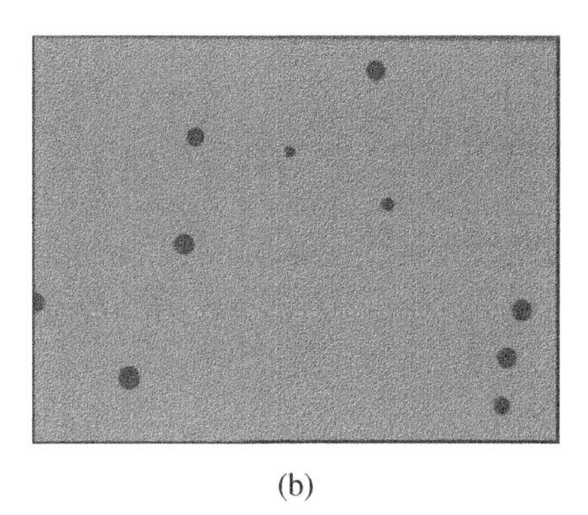
(b)

Figure 6.37 Pictures of EL from an OLED showing dark spots (a) when the device is newly fabricated, and (b) after storage at 85 °C/100% RH for three hours.

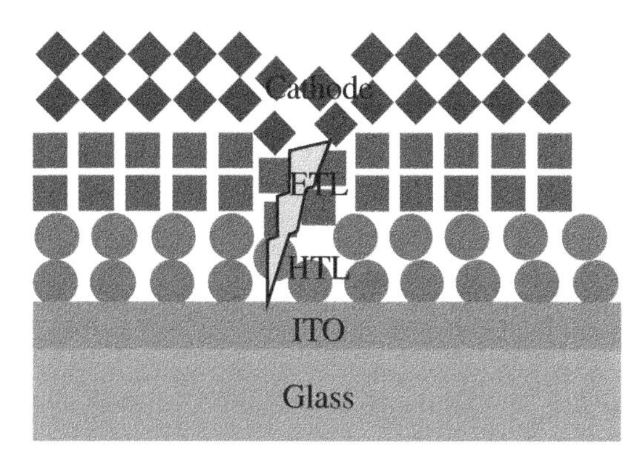

Figure 6.38 ITO spike in an OLED which results in catastrophic failure.

an optical microscope, just after device fabrication and after storage at 85 °C/100% relative humidity (RH) for three hours. It is clear that dark spots exist when the device is fabricated and develop further with time.

Typically, control of substrate cleanliness and smoothness, and careful control of the device fabrication process reduces the density of structural defects, and the development of dark-spots is also reduced. Using hermetic sealing of the OLED into an inert gas environment can also effectively control dark-spot degradation [76]. Heating the substrate in between the deposition of different organic layers [77] can also reduce the dark-spot area, but it changes some characteristics of the organic layer at the same time.

Catastrophic failure means that the luminance drops alongside a huge current increase because of electrical shorts which develop at defects in the organic layers [78]. Its mechanism also originates from morphological defects, such as ITO spikes. Since the organic layer thickness is only 100 nm, substantial roughness of the ITO results in a topographic change of the overlying organic thin-films. As shown in Figure 6.38, if there is an ITO spike, the organic thin films will be thinner there. When the device is operated under high electric field or after a long time, a progressive electric short appears and causes a collapse of the structure at the defect position. To improve device stability and lifetime, it is straightforward that uniform layers throughout the device structure will help keep the morphology of the OLED. After device fabrication, it is also possible to burn out localized conduction filaments by applying a reverse bias [79].

Even if the morphology and encapsulation of an OLED are completely perfect, under constant current driving a uniform luminance decay and increase in driving voltage with time can be still observed, which is called "intrinsic degradation." There are many reports on the mechanisms of OLED degradation, such as: (i) ions from the electrodes which diffuse to the light emitting zone: indium ions from the ITO anode and metal ions from the cathode (Mg:Ag or LiF/Al) will diffuse through into the EML [80] [81]. They form fluorescence quenchers and decrease the luminance under a fixed driving current. Meanwhile, the mobile ions also create

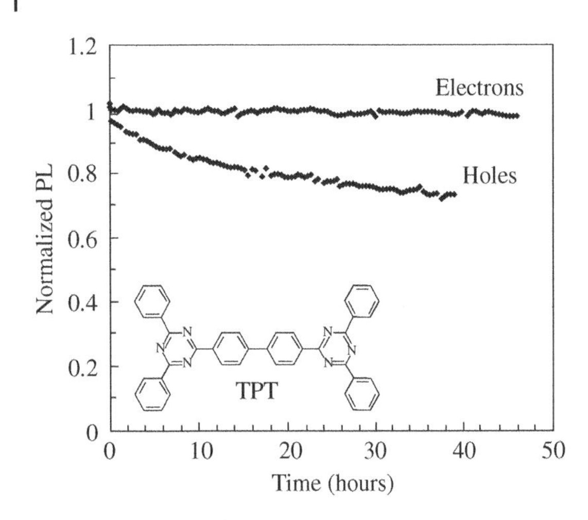

Figure 6.39 Photoluminescence as a function of time from the hole and electron only devices [86].

a built-in voltage that increases the driving voltage under the same electrical current; (ii) impurities that exist in the organic materials migrate in the devices. Zou et al. reported that ac driven devices may have longer lifetimes than those of dc driven devices because the impurities can be drawn back in the reverse bias phase of the ac waveform [82]; (iii) decomposition of light emitting materials: Cao et al. reported that loss of conjugation of poly (paraphenylenevinylene) (PPV) was the main decay mechanism and reported that the intrinsic lifetime of PPV based PLEDs was limited by the amount of charge flowing through the devices rather than ac or dc drive modes [83]; (iv) loss of contact of the electrode/organic layers, this type of decay, though uniform is not really intrinsic and the device may be recovered by recoating a cathode [84].

Another major degradation mechanism which may limit the long-term stability is the "unstable cation" model. As mentioned in Section 6.5.2, the heterojunction in an OLED helps to confine the carrier which improves the current efficiency. However, at the same time, the carriers congregate at the HTL/EML interface. Since holes injected from the HTL to the EML form cations in the EML material which are chemically unstable, this will accelerate the formation of non-radiative trapping centers and result in luminance decay and voltage increase in the device [85].

H. Aziz et al. designed an experiment to observe this degradation mechanism [86]. Figure 6.39 shows the PL intensity versus time of devices comprising a 5 nm-thick Alq_3 layer, sandwiched between two thick NPB or tetraphenyltriazine (TPT). NPB and TPT are HTL and ETL materials respectively, and provide only holes or electrons to be transported through the Alq_3 thin layer under a bias. We can see that, after hole transport through the Alq_3 layer, a gradual decrease in the photoluminescence (PL) intensity was observed, reflecting a continuous decrease in the PL efficiency of the Alq_3 on prolonged current flow. The decrease in the PL intensity of the Alq_3 points to its degradation as a result of hole transport, thus demonstrating that cationic Alq_3 species are unstable and that their degradation products are fluorescence quenchers. On the other hand, while significant PL decrease is observed in the case of transporting holes through the Alq_3, the PL stays remarkably constant in the case of transporting electrons. These results, therefore, provided evidence that injection of holes into the Alq_3 layers of the OLEDs was one of the main factors in device degradation.

When an OLED ages, not only does the luminance decay but also the voltage increases under constant current driving. J. H. Lee et al. have verified that the major degradation mechanism in an OLED is by cation formation by conducting a series of experiments with the same organic materials and different device structures [87]. If metal ion diffusion was the major degradation mechanism, a thicker HTL or ETL should result in a longer lifetime. However, the operational lifetime in the thicker device was not longer. On the other hand, the lifetime does correlate with the power efficiency as shown in Figure 6.40. As mentioned before, higher power efficiency indicates more effective energy transfer from electrical to optical power. Hence, the organic degradation comes from chemical reactions which proceed faster at higher temperatures, i.e. in devices with lower power efficiency. Given constant power efficiency, we can see that the lifetime is longer with a thicker HTL,

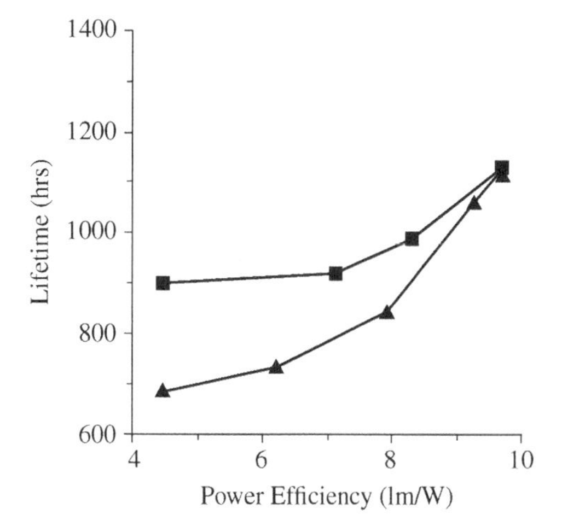

Figure 6.40 Dependency of OLED lifetimes on power efficiencies. Dashed line and square symbol correspond to electron-rich devices. Solid line and triangular symbol represent hole-rich devices [87].

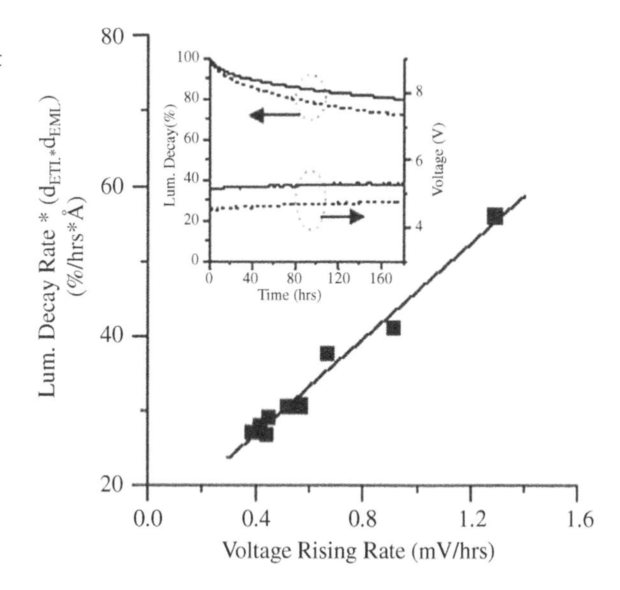

Figure 6.41 Voltage increase rate versus the term (luminance decay rate times the total thickness of EML and ETL). The inset shows typical curves of luminance decay and voltage as a function of time [87].

i.e. in an electron rich device. The thicker HTL means that the hole density is lower near the recombination interface and results in less cation formation.

A linear correlation between the trap formation rate and the quencher formation rate at the HTL/ETL interface was also observed by D. Y. Kondakov using a different measurement system, shown in Figure 6.41. Since the voltage increase and the luminance decay directly correlate to the trap and non-radiative center formation respectively, the linear relationship which was found means that a thermally-assisted chemical reaction induces degradation of the organic material near the recombination interface, which in turn quenches the photons and traps carriers.

Since the lifetime of an OLED is typically >10 000 hours under normal operation, accelerated measurement and extrapolation methods are needed to estimate the lifetime value systematically in a shorter period [88, 89]. However, accelerated aging might introduce other degradation mechanisms, making the estimation of the lifetime a complex issue. Under high luminance aging, a relationship which is commonly used is:

$$L_0{}^n t_{1/2} = \text{constant} \tag{6.43}$$

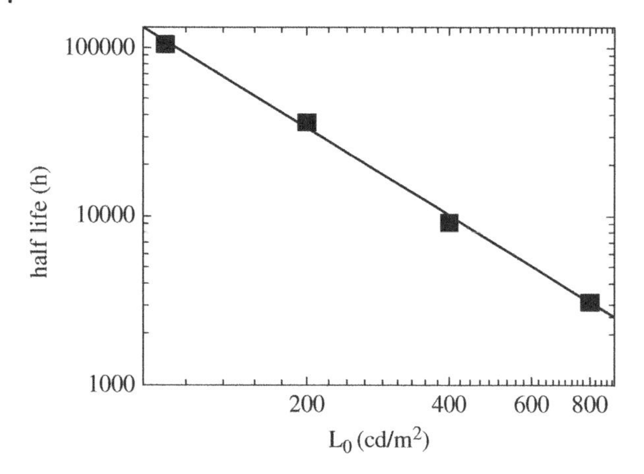

Figure 6.42 Half life versus the initial luminance on a double log scale [90].

where n is the acceleration coefficient, and $t_{1/2}$ is the half-life. As shown in Figure 6.42, the best fit has been obtained with $n = 1.7$ [90]. From this figure we can see that the half-life of this particular device is 40 000 hours when operated at 200 cd/m^2. If the luminance is increased to 10 000 cd/m^2 for acceleration test, we can obtain a half-life of 51 hours.

Another problem is that the luminance may not decrease monotonically, meaning that more than one mechanism influences the OLED degradation during the aging test [91]. This increases the complexity of analyzing the results. Figure 6.43 shows the luminance versus time at different environmental temperatures. We can

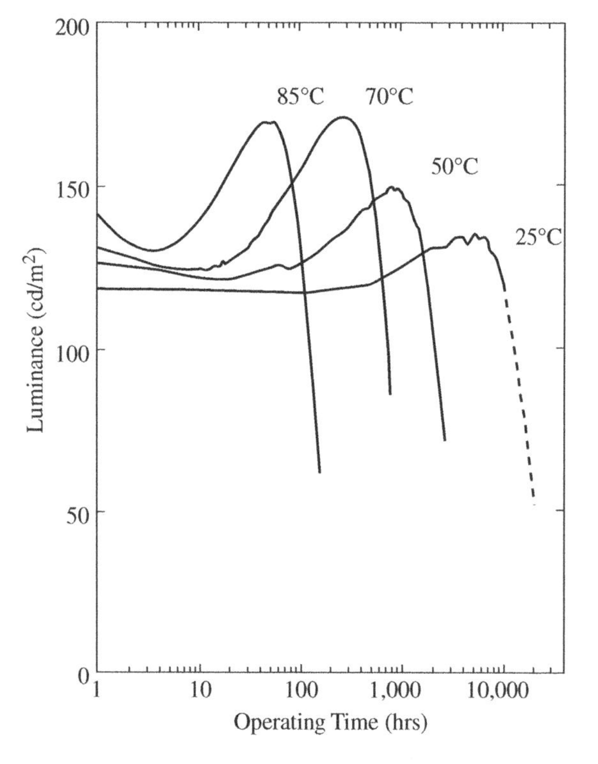

Figure 6.43 Luminance versus time with different environment temperatures on semi-log scale [91].

see that at 25 °C, the luminance at first remains constant, increases, and then drops rapidly. There are three distinct mechanisms to account for the degradation behaviors. The lifetime also changes rapidly at different temperatures.

6.6 TRIPLET EXCITON UTILIZATION

To optimize the efficiency of OLEDs, utilization of triplet excitons is necessary. This can potentially increase the OLED luminance by a factor of four, as described in Section 6.3.2. By introducing a heavy metal into the organic materials, the spin–orbital coupling efficiency can be increased and efficient phosphorescence due to the emission from the triplet excited state is observed, which makes a 100% IQE possible. Although red and green phosphorescent dopants have achieved reasonably high efficiency and lifetime, the wide bandgap required of a blue dopant still limits their performance [92]. The emission wavelength of phosphorescent blue OLEDs is still not short enough to generate pure blue emission, and the operational lifetime is much shorter than for red and green ones. An alternative to using triplet excitons for blue light emission is the triplet–triplet annihilation (TTA) mechanism [93]. In TTA materials, two triplets may combine to yield an excited singlet state which emits light, and hence no heavy metal is needed in the molecule. However, because two triplets can generate only one singlet, the maximum attainable IQE is lowered to 62.5%. Yet another approach is to reduce the energy gap between the singlet and triplet states by molecular engineering, again without a heavy metal. Hence, triplets generated by electrical excitation may be excited into the singlet state by ambient thermal energy, and then emit light, a process called thermally activated delayed fluorescence (TADF) [94]. Typically, TADF molecules consist of electron donor and acceptor moieties which reduces the wavefunction overlap between the HOMO and LUMO states, this in turn being the key to a small energy difference between the singlet and triplet states. An alternative approach to generating TADF is exciplex emission, as discussed in Section 6.3.4.2. By mixing electron donor and acceptor materials, exciplex emitters can be generated which have a small singlet/triplet energy difference [95].

6.6.1 Phosphorescent OLEDs

Typically, phosphorescent emission is a slower and less efficient process than that in a conventional fluorescent material since the decay of a triplet state in a conventional organic material is generally not allowed due to the conservation of spin symmetry [96]. It sets an upper limit on the IQE of a fluorescent OLED of 25% and 75% of the energy generates triplet excited states and is wasted. However, by introducing a heavy metal atom into the molecule, radiative decay of a triplet dopant becomes possible. There are two mechanisms accounting for the electroluminescent emission from phosphorescent dopants. The first is through energy transfer and the second is by carrier trapping. As shown in Figure 6.44, there are two mechanisms of energy transfer in OLED. First of all, Förster energy transfer is a long-range process (10 nm) where donor (D) transfers to acceptor (A) by dipole–dipole coupling. However, Förster transfer requires that the transition from excited state to ground state of both D and A must be allowed according to conservation of spin symmetry. Thus, it only transfers energy to the singlet state of acceptor. On the other hand, Dexter energy transfer, a short-range process, transfers energy by electron exchange between the donor and acceptor. Unlike Förster energy transfer, Dexter energy transfer requires only the total spin to be conserved. It allows both singlet–singlet and triplet–triplet energy transfer. The short range of Dexter transfer dictates that the dopant concentration in the EML of a phosphorescent OLED is typically high (e.g. >6%). Direct trapping of carriers is also important for the emission of phosphorescent OLEDs. Since the energy gap of a triplet dopant is usually smaller than that of the host, the triplet dopant may trap carriers leading to the direct formation of excitons at dopant sites.

As described above, with the strong spin–orbital coupling of heavy-metal complexes, phosphorescent materials can break the spin-forbidden selection rule controlling radiative relaxation of triplet states. Therefore, highly efficient phosphorescent OLEDs can be achieved. One of the most successful green phosphorescent

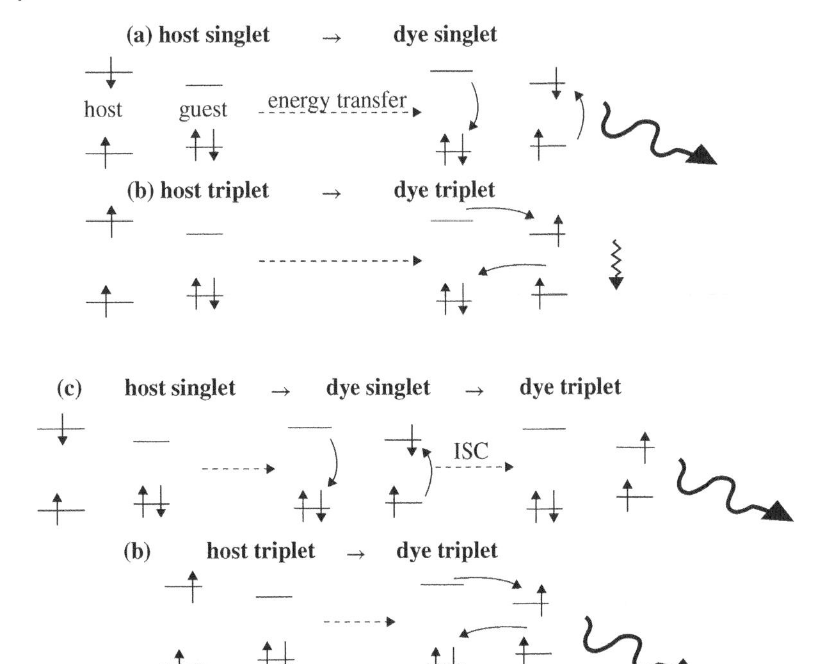

Figure 6.44 Energy transfer mechanisms in a guest-host system of an OLED. (a) and (b) are the energy transfer of the fluorescent dopant; (c) and (d) are the energy transfer of the phosphorescent dopant [96].

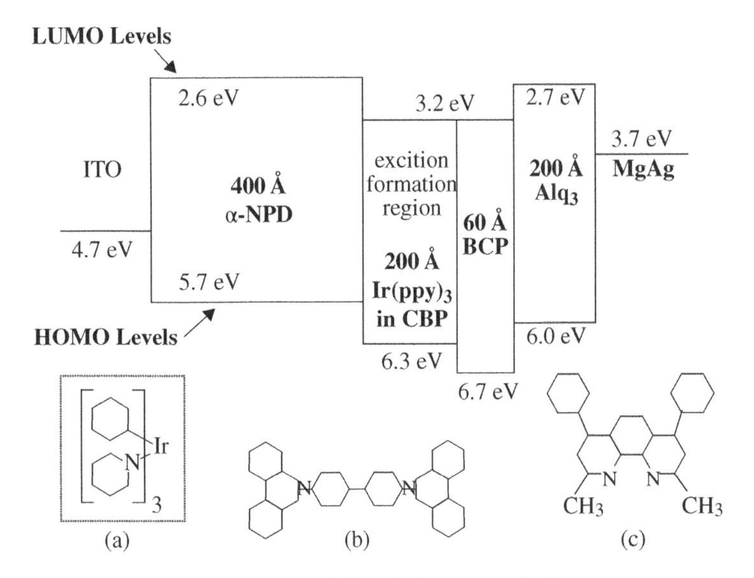

Figure 6.45 Device structure and chemical structures [97].

materials is fac-tris(2-phenylpyridine)iridium ($Ir(ppy)_3$) with 4,4′-N,N′-dicarbazole-biphenyl (CBP) as the host, which can demonstrates peak current efficiency and power efficiency of 26 cd/A and 19 lm/W at a luminance of 100 cd/m^2 [97]. Figure 6.45 shows the device structure and the organic structures. Since CBP is a bipolar transport material, i.e. is capable of transporting holes and electrons, an additional layer, called a HBL, is inserted to confine the excitons in the EML. 2,9-dimethyl-4,7-diphenyl-l,10-phenanthroline (BCP) is widely used as a HBL due to its wide bandgap and large HOMO value, which can hinder the diffusion of excitons formed in the EML and prolong their residence time, thereby increasing the probability of energy transfer from the host to the phosphorescent dopant. However, the low glass transition temperature (T_g) of BCP (83 °C) may lead to device instability. Thus, an alternative, aluminum(III) bis(2-methyl-8-quinolinato)4-phenylphenolate (BAlq) with a higher T_g value can be used which provides higher stability [98]. By using a p–i–n structure to improve the injection of carriers, green phosphorescent OLEDs with a power efficiency of 62 lm/W and current efficiency of 61 cd/A at 1000 cd/m^2 have been demonstrated [47] [99]. By further optimizing the carrier transport layers, power efficiency of 107 lm/W can be achieved at 1000 cd/m^2, corresponding to 26% external quantum efficiency (EQE), using an EML consisting of Ir(ppy)$_3$ and CBP [100].

By using 2,3,7,8,12,13,17,18-octaethyl-12H,23H-porphyrin platinum(II) (PtOEP) [101] as a red emissive dye in a CBP host, an OLED can achieve an external quantum yield of 5.6% [102]. However, due to its relatively long phosphorescent lifetime (~50 μs), PtOET tends to lead to TTA at high current density. Following this work, several red phosphorescent materials with shorter excited state lifetimes have been proposed, such as bis(2-(20-benzo[4,5-a]thienyl)pyridinato-N,C30)iridium(acetylactonate), and a red phosphorescent OLED achieved the highest current efficiency of 23.94 cd/A under a voltage of 8.29 V [103–106]. Compared to the great success of green and red phosphorescent materials, it is still not easy to get a "pure" blue phosphor due to the wide bandgap. The most common "sky-blue" phosphorescent emitter is iridium(III) bis[(4,6-Difluorophenyl)-pyridinato-*N*,*C*28]picolinate (FIrpic) which exhibits a current efficiency and power efficiency of 57.5 cd/A and 48.9 lm/W [107].

At the present time, the external quantum efficiencies of red, green, and (sky) blue phosphorescent OLEDs can reach >20%, which means that the IQE is almost 100% remembering that the extraction efficiency is ~20%. (This issue was mentioned in Section 6.4.4, and will be further discussed in Section 6.8). Typically, high efficiency means that the electrical power is efficiently converted to optical power and less heat is generated, which a crucial factor for OLED degradation, as discussed in Section 6.5.5. It results in the great success of red and green phosphorescent devices used in OLED displays for achieving both high efficiency and long lifetime. Although the efficiency of blue phosphorescent OLEDs is high, the operational lifetime is much shorter (~1/20) than the red and green ones. In a blue phosphorescent OLED, the triplet energy is high (~3 eV) and the triplet exciton lifetime is long (~μs). It may interact with a polaron (which also has an energy ~3 eV) which is called exciton–polaron interaction [108]. When a triplet transfers its energy to a polaron, it creates a hot polaron whose energy (~6 eV) is higher than the dissociation energy of most chemical bonds (for example: bonding energies of C–N and C–C are 3.04–3.64 eV, respectively). There are three key factors in the degradation mechanism. The first is the bandgap. In phosphorescent OLEDs, lower bandgap materials typically exhibit longer operational lifetimes (i.e. red> green> blue). The second is the triplet exciton lifetime. Shorter triplet exciton lifetimes typically increase not only the efficiency, but also the lifetime. And the third is the polaron. By controlling the polaron distribution in a blue phosphorescent OLED by making some device engineering, polaron–exciton interaction can be greatly reduced which increases the operating lifetime [92]. However, as mentioned above, the two bottlenecks of the blue phosphorescent OLED (sky-blue rather than deep blue emission wavelength and short lifetime) still hinder the use of blue phosphorescent OLEDs for mass production in display applications.

6.6.2 Triplet-Triplet Annihilation OLED

In the case of blue OLEDs, there seems to be a tradeoff between device efficiency and operational lifetime. Phosphorescent blue OLEDs exhibit high device efficiency but short lifetime due to polaron-exciton quenching with high energy (~6 eV) and long interaction time (~μs). The operating lifetime of fluorescent blue OLEDs is acceptable, but their efficiency is low because the triplet excitons relax their energy non-radiatively. A compromise between the operational lifetime and device efficiency is offered in the TTA-OLED. As mentioned in Section 6.3.3, the triplet state (T_1) energy is lower than that of the singlet state (S_1). In some organic molecules, two triplets may combine to form one singlet by what is called the TTA process. Under electric current drive,

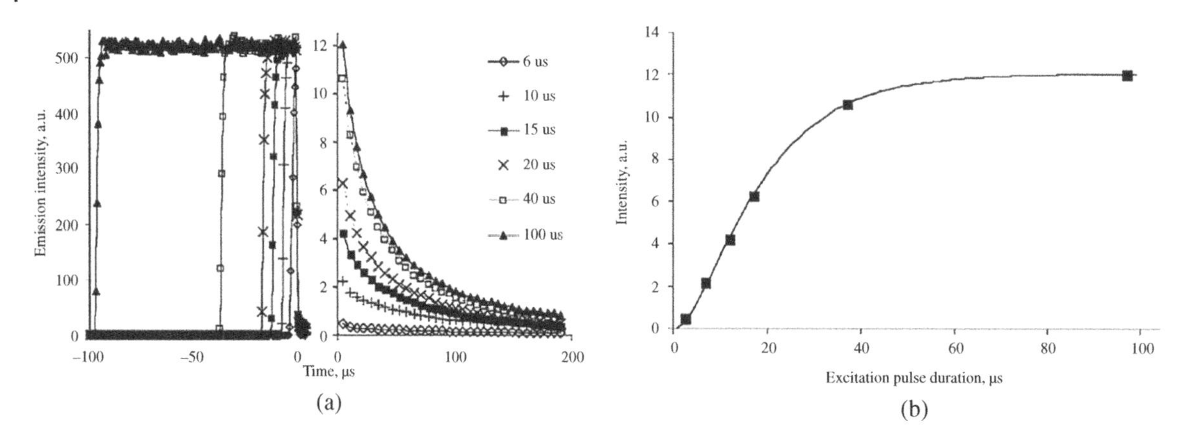

Figure 6.46 (a) TREL response, and (b) delayed fluorescence intensity with different pulse duration [93].

the 25% of excitons in the singlet state emit light, which is called (prompt) fluorescent emission. For the other 75% triplet excitons, it is possible for two triplets to annihilate forming one singlet which emits light. Due to the long exciton lifetime of the triplets, the emission is a slow process. As the emission comes from the singlet state, it is fluorescence, not phosphorescence, and is called "delayed fluorescence." Therefore, emission from the blue TTA-OLED consists of two components, prompt and delayed fluorescence from the contributions of singlet and triplet excitons, respectively. The theoretical limit of the IQE in TTA-OLED is: $25\% + 75\%/2 = 62.5\%$, which is between those of fluorescent (25%) and phosphorescent (100%) OLEDs. Despite the light extraction efficiency being limited to ~20%, a blue TTA-OLED with 13.7% EQE and 35 000 hours half-lifetime (from an initial brightness of 1000 nits) was demonstrated. Figure 6.46 shows the turn-off dynamics of the TTA-OLED, which shows two-step decrease representing the prompt and delayed fluorescence, respectively. Besides, delayed fluorescent intensity increases with increasing the electrical pulse width [93]. Because TTA is a bimolecular reaction, higher triplet density results in stronger delayed fluorescence. Figure 6.46 shows the transient electroluminescence (TrEL) of a blue TTA-OLED with different pulse width driving. With increasing the pulse width, more triplets were generated and hence higher delayed fluorescence intensity.

6.6.3 Thermally Activated Delayed Fluorescence

It is also possible to design organic molecules in order to raise the triplet state (T_1) energy. When the singlet-triplet splitting is sufficiently small (~100 meV), thermal energy (~25 meV for room temperature) can be sufficient to excite a triplet exciton back to the singlet state, which is called reverse intersystem crossing (RISC). Hence, the subsequent emission is called TADF, as shown in Figure 6.47. In TADF-OLED, the theoretical limit on the IQE is also 100%, the same as for the phosphorescent OLED, but no heavy metal (such as Pt and Ir) is required, which is advantageous as these metals are often rare and expensive and may cause problems during waste disposal. Hence, the TADF-OLED has potential advantages including better environmental acceptability and lower cost. Note that although both phosphorescent and TADF materials

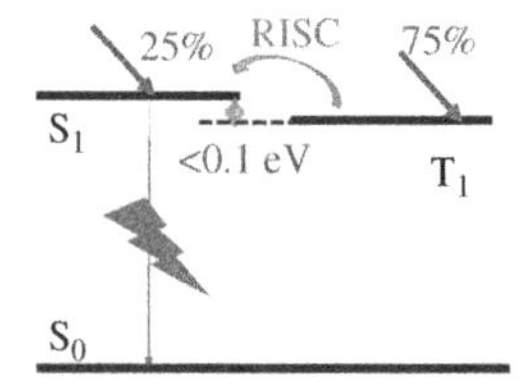

Figure 6.47 Schematic diagram of TADF emission.

can achieve 100% IQE, the physical mechanism is quite different. For the phosphorescent material, strong spin–orbital coupling makes the transition from T_1 back to S_0 possible, and hence it is a phosphorescent emission, as shown in the Jablonski diagram (Figure 6.8). On the other hand, in a TADF material, a triplet exciton transfers to the singlet state driven by thermal energy, and the emission is from the singlet state, which is a (delayed) fluorescent emission. By exploiting the contribution of triplet excitons to light emission, the maximum EQEs of the orange, green, and sky-blue TADF-OLED achieved to date are 11.2%, 19.3%, and 8.0%, respectively, which are all higher than the theoretical limit of the fluorescent OLED [94].

To utilize the triplet excitons for light emission in a TADF-OLED, the triplet energy (E_T) must be raised to a level close to the singlet energy (E_S) by engineering the organic molecular structure. E_S and E_T can be represented by:

$$E_S = E + K + J \tag{6.44}$$

$$E_T = E + K - J \tag{6.45}$$

where E is the orbital energy, K is the electron repulsion energy, and J is the exchange energy. Hence, the energy difference between singlet and triplet (ΔE_{ST}) is,

$$\Delta E_{ST} = 2J \tag{6.46}$$

Here, J is determined by the spatial overlap of the HOMO and LUMO wavefunctions. To reduce the J value, a successful approach is to connect an electron donor moiety and an electron acceptor moiety in one molecule. In the HOMO and LUMO levels, electron density accumulates at the electron donor and acceptor moiety, respectively, with a resultant small overlap of the wavefunctions. A spacer unit is sometimes needed to further reduce the spatial integral. An orthogonal (or twisted) structure between the electron donor and acceptor groups is even better. Hence, TADF molecules typically exhibit a donor–spacer–acceptor (D–X–A) configuration. However, as shown in Section 6.3.1, based on Frank–Condon principle, the transition probability from the LUMO to the HOMO level is also determined by the spatial overlap of the wavefunction. To achieve high fluorescence efficiency, wavefunction integral of the HOMO and LUMO states should therefore be as high as possible, which is contrary to the design rule needed for achieving TADF emission. The concept of combining both a TADF and a fluorophore in the host has therefore been proposed. The energy levels (S_1 and T_1) of the TADF molecules are higher than the singlet energy S_1 of the fluorophore. Hence, both singlet and triplet excitons can transfer their energy from the TADF molecules to the singlets of the fluorophores with high quantum yield for light emission [109]. The dopant concentration of the fluorescent materials should be kept low (1%), which is sufficient for Förster energy transfer (with the radius of 10 nm), but not effective for Dexter energy transfer (with the radius of 1 nm), to avoid energy transfer from the triplet of TADF to the fluorophore triplet. Also, TADF molecules must be doped into a wide bandgap host material to prevent quenching. For blue, green, yellow and red OLEDs, 13.4%, 15.8%, 18.0%, and 17.5% in EQE can be obtained. However, in this configuration, the fluorescent dopants must exhibit the lowest bandgap. Those of the TADF molecules are higher, and those of the host materials must be the highest. That means, the driving voltage in such a device is high because the charge carriers must be injected into and transported on the wide bandgap host.

6.6.4 Exciplex-Based OLED

An alternative to the TADF-OLED is to use an exciplex host, which was mentioned in Section 6.3.4.2. In this case, the EML host is a physical mixture of an electron donor and an acceptor material. The electron donor and acceptor moieties now are in different molecules, which greatly reduces the wavefunction overlap, so that a small ΔE_{ST} is easily achieved. One can also choose hole- and electron-transporting materials which will serve as the electron donor and acceptor materials respectively, in the mixed-host configuration. Due to the bipolar transport characteristics, the current density is raised which decreases the driving voltage. By mixing different electron donor and acceptor materials, red, green, and blue OLEDs can be obtained using exciplex-based hosts. EQE of 8% was achieved from a blue exciplex-OLED without any dopant material [95]. Fluorescent and phosphorescent dopant can also be added into the exciplex host to modify light emission. The exciplex host with fluorescence dopant, is similar to the case in Section 6.6.3 in combining a host and dopant (TADF). But the driving voltage can be reduced which improves the power efficiency in terms of lm/W. EQE of 14.5% was achieved in a green exciplex-OLED doped with a fluorescent emitter, while keeping the turn-on voltage as low as 2.8 V. A low fluorophore concentration (0.2%) was used to avoid triplet energy transfer from exciplex to dopant molecules [110]. In the case of a phosphorescent emitter, the small ΔE_{ST} of the exciplex-host also helps to reduce the driving voltage. In a phosphorescent OLED, a triplet exciton on the host can transfer its energy to the phosphor only by Dexter energy transfer with small range (1 nm), which means the dopant concentration must be high in this case. However, due to the efficient RISC process in the exciplex-host, the triplet changes to a singlet state in the host and can transfer energy to the dopant via Förster transfer. In other words, the dopant concentration can be reduced which saves the usage of heavy metals (e.g. Ir) in the phosphor. It was reported that, at 1000 cd/m^2, operating voltage, EQE, and power efficiency of 3.0 V, 28.7%, and 112.5 lm/W can be achieved from an exciplex-OLED doped with green phosphor [111]. However, the HOMO and LUMO levels of the exciplex host are determined by the electron donor and acceptor materials respectively, so the mixed host exhibits a smaller bandgap than either of the individual molecules. This means that blue emission is more difficult to achieve than green or red, because it requires higher bandgap materials.

6.7 TANDEM STRUCTURE

As shown in Eq. (6.43), as the luminance of an OLED is increased, the operating lifetime dramatically decreases. However, if one can connect two or three OLEDs in series, the light emission will double or triple under the same current density. Alternatively, the light emission from each unit can decrease to half or one-third the total, and the operating lifetime will be more than double or triple. Following this principle, OLED tandem structures have been proposed. The operating principle of a tandem device is to inject one electron–hole pair into the device stack to produce multiple electron–hole pairs – one per EML – through use of a charge generating layer (CGL) connected between each OLED unit, as shown in Figure 6.48 [112–115].

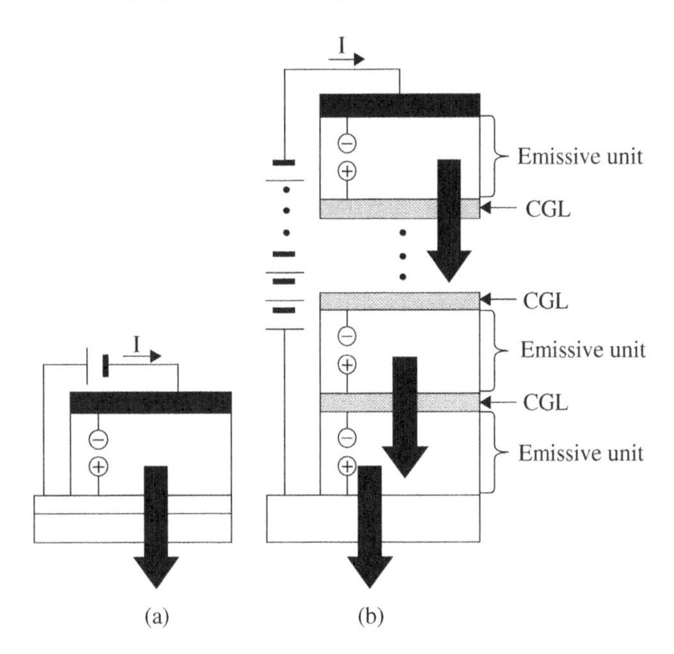

Figure 6.48 Structures of (a) a conventional OLED and (b) a tandem device [112].

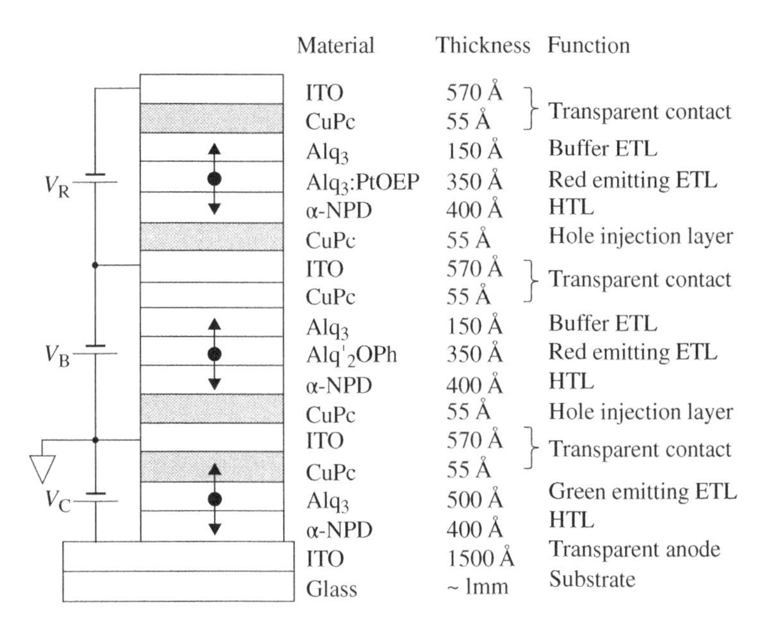

Figure 6.49 Schematic cross-section of a tandem OLED with conductive connecting electrodes [117].

In a tandem device, the current efficiency in terms of cd/A should scale linearly with the number of the single OLED units and exceed unity. Due to the high current efficiency, a tandem device can achieve the same luminance at lower driving current and voltage, which effectively improves the power efficiency and generates less heat, and it in turn increases the operating lifetime [116]. By connecting units emitting different colors, a high efficiency white OLED can be obtained. With suitable design of the layer structures, devices with a high external quantum efficiency and high color purity can be also achieved.

One of the most critical aspects of a tandem device is the design and the fabrication of the connecting unit, which can be either conductive or dielectric thin films. First we discuss the conductive case. With a proper layout design, the conductive connecting unit can be also used as the contact pads of vertically stacked OLEDs that emit three primary colors as shown in Figure 6.49. In this case, each OLED unit can be driven independently and so can be used for fabricating a full-color display with high aperture ratio. This structure was named a SOLED (stacked OLED) [117, 118].Typically, high work function materials, like ITO, gold, and nickel are used as the anode of an OLED device to effectively inject holes into the p-type organic materials. On the other hand, low work function materials, like alkali and alkaline metals, are good candidates for cathode materials [13]. However, in a tandem OLED, a common metal electrode is needed which can inject holes and electrons into the p- and n-type organic materials, respectively. The same techniques used for fabricating transparent or TOLEDs can be used here to fabricate the conductive electrode in a tandem device.

The concept of the insulating CGL, which operates by a tunneling junction is illustrated as follows. A reverse biased tunneling n–p junction is sandwiched between two OLED units. At the CGL, the electrons in the p-side valence band will tunnel into the n-side conduction band and generate electron and hole pair for the left and right device, respectively. Such a technique is widely used in semiconductor LEDs and laser diodes (LDs) to increase the IQE, improve thermal stability and decrease the threshold current of a LD as shown in Figure 6.50 [119, 120]. In semiconductor devices, such a tunneling junction is fabricated with n- and p-type materials of high carrier concentration by controlling the doping ratio during semiconductor epitaxy. However, intrinsic organic materials usually exhibit a large bandgap, of 2–3 eV. Therefore, the intrinsic concentration of thermally generated free carriers is generally negligible [121]. Similar to semiconductor materials, impurities are doped

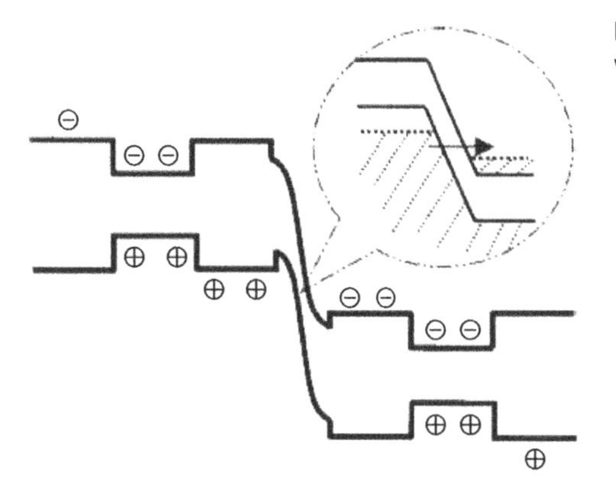

Figure 6.50 Energy band diagram of a tandem semiconductor LEDs with magnified reverse biased tunneling junction band energy diagram.

into the organic materials that can either transfer an electron to the LUMO states (n-type doping) or remove an electron from the HOMO states (p-type doping) to generate a free electron or hole, respectively (c.f. Section 6.5.1.3).

6.8 IMPROVEMENT OF EXTRACTION EFFICIENCY

Generally, the external quantum efficiency (EQE, η_{ex}) of an OLED device is determined by the internal quantum efficiency (IQE, η_{in}) and the light extraction efficiency η_{ext} [122]:

$$\eta_{ex} = \eta_{in}\eta_{ext} \tag{6.47}$$

where the η_{in} value is the product of the fluorescent quantum efficiency of the light emitting material η_F, the fraction of charge recombinations that result in emissive excitons χ, and the fraction of carriers which recombines in the EMLη_{re} [123].

$$\eta_{in} = \eta_F \chi \eta_{re} \tag{6.48}$$

For the commonly used ETL material Alq_3, η_F is only about 30%. It can be improved to nearly 100% by introducing high efficiency dopant materials into the matrix. According to spin statistics, the χ value in an OLED using typical fluorescent materials has a value of 1/4, but can be 1 for phosphorescent and TADF materials. A η_{re} approaching 100% can be achieved under charge-balanced conditions by adjusting the thicknesses of the organic layers. From the above discussions, we can see that the upper limit of the η_{in} value is about 25% for fluorescent OLEDs, 62.5% for TTA-OLEDs, and 100% for phosphorescent and TADF OLEDs [124].

Considering the TIR between the glass substrate of an OLED and the air according to the classical model shown in Eq. (6.36), the light extraction efficiency from the device is ~20%. Hence the upper limits on EQE from fluorescent, TTA, phosphorescent, and TADF OLEDs are 5%, 12.5%, 20%, and 20%, respectively. However, EQE values reported in the literature and in Section 6.6 are often higher than these values (e.g. EQE = 28.7% for exciplex-OLED with phosphorescent dopant), which shows that the extraction efficiency is underestimated by this simplified model. As mentioned in Section 6.4.4, electromagnetic wave propagation in organic materials can be treated as oscillating dipoles. Note the "dipole" here is a transition dipole in the emissive molecules, rather than any permanent dipoles in the molecule. Due to the anisotropic shape of the organic materials, some molecular orientational ordering can be achieved with certain organic materials under controlled fabrication conditions, in a film which nevertheless remains essentially amorphous. Formally, the dipole orientation can be resolved into components in the plane of ("horizontal"), and perpendicular ("vertical") to the substrate, as shown in Figure 6.51. In a perfectly amorphous film, 2/3 of the dipole component lies in the horizontal plane, and 1/3 is vertical. A horizontal dipole gives a

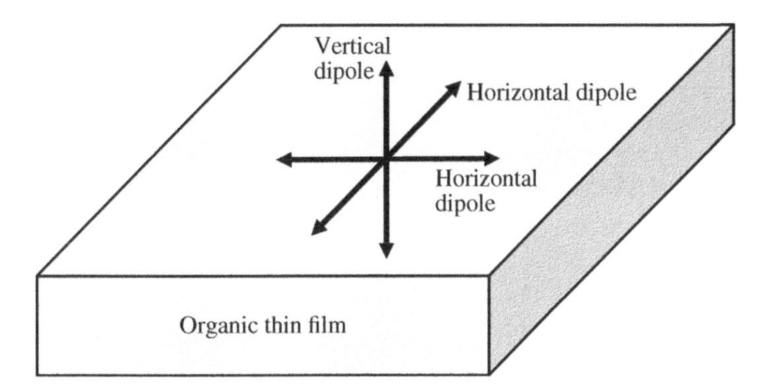

Figure 6.51 Horizontal and vertical dipoles in the organic thin film.

greater emission intensity, normal to the substrate, than its average in the plane of the substrate which results in higher extraction efficiency. On the other hand, light emitted from a vertical dipole tends to propagates along the substrate direction which results in higher TIR and other non-radiative loss. The first criterion for increasing the proportion of the horizontal dipole component is to choose the emitter which should be physically asymmetric. Then, the transition dipole vector (TDV) of the molecule should be parallel to the substrate direction, which is dependent on the guest molecule, host molecule, and fabrication technique. For example, the ligand acetylacetonate (acac) is effective in Ir-based phosphorescent dopants for promoting horizontal transition dipole alignment [125]. When one dopant material is processed under different conditions, the alignment may be different. Lampe et al. demonstrated horizontal alignment of the transition moments in layers formed by thermal evaporation, but isotropic orientation in solution processed layers using the same host and dopant, which implies that the interaction between the molecule and the surface during the evaporation process played an important role for the alignment [126]. It was reported that when co-depositing the same dopant with different hosts, alignment of the guest was better in a host with higher glass transition temperature (T_g), because the host with lower T_g allowed molecular migration which randomized the molecular orientation [127]. An EQE of 38.8% has been demonstrated experimentally in a phosphorescent OLED with 93% horizontal dipole [128] .

In a conventional bottom emission OLED with an ITO anode, photons propagate from the organic stacks (with a refractive index of 1.6–2.0) through the glass substrate (n = 1.5) to the air (n = 1) [129]. Two loss modes related to TIR are waveguiding (in the ITO and organic stacks) and substrate modes (confined in the glass) (Figure 6.52). Considering the interface between the air and the glass substrate, a non-planar structure

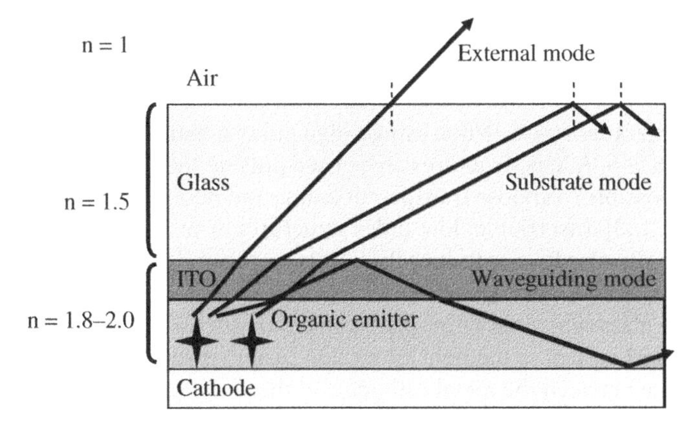

Figure 6.52 External, substrate, and waveguiding modes in a bottom-emission OLED.

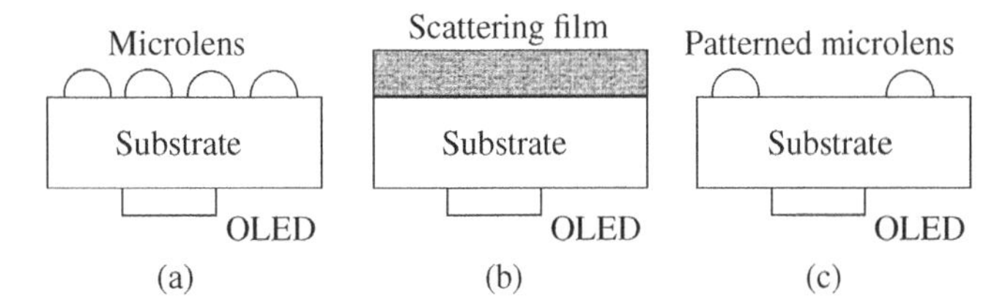

Figure 6.53 Methods for extracting substrate mode in an OLED.

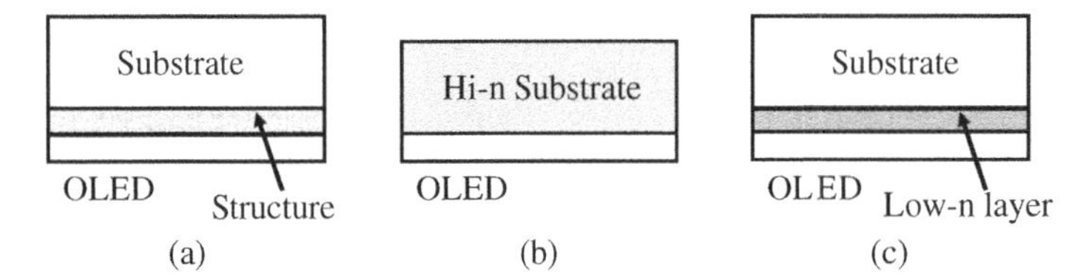

Figure 6.54 Methods for extracting waveguiding mode in an OLED.

such as a microlens array (similar to the structure of the LED package) at this interface effectively reduces the TIR and enhances the light extraction, as shown in Figure 6.53a [130–132]. Sometimes, a planar surface is preferred for display applications. A scattering film consisting of particles (or air voids) with refractive index different from the matrix can be used for this purpose, as shown in Figure 6.53b [133, 134]. However, there are several problems for display applications when applying such a microlens array or scattering film. First, the image from the display becomes blurred, and the image quality deteriorates [135]. Careful design of the microlens layout provides a patterned microlens array which can reduce the image blur effect, as shown in Figure 6.53c. However, such a film not only redirects the light from the OLED, but also from the ambient. Hence, the film appears white under ambient light when the OLED is turned off which reduces the effective contrast ratio of the display. Shaping the glass substrate on the OLED device side can help to redirect the light from the substrate, similar to the concept of the die shaping method in LEDs (Figure 5.34d). Emission can be redirected out of the device, similar to the effect of the brightness enhancement film discussed for LCDs in Chapter 4. However, for this method, shaping the glass substrate on the OLED device side is needed, which decreases the emitting area, and the fabrication is more complex [136].

To extract waveguiding modes, structures have been applied to the interface between the glass substrate and the ITO, as shown in Figure 6.54a, such as a nano-corrugated substrate which results in Bragg-scattering [137]. Anodic aluminum oxide (AAO) nanoporous film and a two-dimensional photonic crystal (PC) pattern have also been introduced between the substrate and the ITO to extract waveguided modes [138, 139]. Using patterned aluminum zinc oxide (AZO) embedded in the etched ITO, waveguided light was scattered due to the refractive index difference between the oxides [140]. Alternative methods exist, to manipulate the refractive index and reduce the waveguiding in an OLED. The origin of the waveguided mode in an OLED is the refractive index difference between the ITO/organic and the glass substrate. When using a high index substrate ($n = 1.8$), there is no waveguiding mode, as shown in Figure 6.54b. A lens structure can be used outside the substrate to outcouple all the substrate modes [141]. It is also possible to choose transparent anode and organic materials with lower refractive index to reduce the TIR [142, 143]. Insertion of low-index structures or materials at the interface between the glass substrate and the anode can also help to reduce the TIR at this boundary, as shown in Figure 6.54c [144, 145].

Considering the interface between the air and the glass substrate, a non-planar structure such as a microlens array at this interface effectively reduces the TIR and enhances the light extraction. Another loss channel is the plasmonic mode, which is related to the distance between the metal cathode and the emission dipole. For reducing the plasmonic loss, it is straightforward to increase the thickness of the organic layer, as shown in Figure 6.55a, because plasmonic mode decreases with increasing the distance between the emission dipole

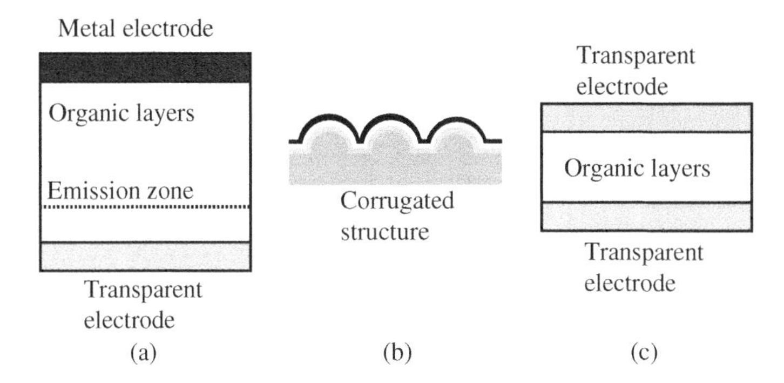

Figure 6.55 Methods for extracting plasmonic mode in an OLED.

and the metal cathode, and hence a high conductivity thin film is needed. Combining (i) thick organic film to reduce the plasmonic loss, (ii) high index substrate to decrease the waveguiding mode, and (iii) high-index macrolens to extract the substrate mode, white OLED with 124 lm/W in power efficiency, corresponding to 46% EQE, at 1000 cd/m^2 can be achieved [146]. With the introduction of the nanostructure, it is possible to couple the plasmonic mode back to the radiative mode, as shown in Figure 6.55b. Besides, as mentioned above, the nanostructure can also help to extract the light trapped inside the waveguiding mode. That means that plasmonic and waveguiding modes can be coupled simultaneously with a suitable design of the nanostructure. A green phosphorescent OLED with 63% in EQE and 225 cd/A was demonstrated fabricated on a patterned sapphire ($n = 1.8$) substrate with macrolens attachment [147]. However, the nanostructure scatters the light with different optical intensity into different angles. That means the emission from this display has a serious color shift at different viewing angles. A quasi-periodic structure may solve this problem [148]. The origin of the plasmonic mode comes from free electron oscillation at the metal electrode surface. When using metal oxide (such as ITO and indium zinc oxide (IZO)) both for anode and cathode, a plasmonic mode cannot be excited, as shown in Figure 6.55c. A transparent OLED with 62.9% in EQE (summation of emission from both sides) was demonstrated experimentally, combined with lens structures [149].

6.9 WHITE OLEDS

A white OLED can be used as the basis of a full color display, as shown in Figure 6.27b of Section 6.5.3.1. It can also be used as a lighting source. To generate a white light, the simplest approach is to use blue and yellow emissions, based on the colorimetric principles set out in Chapter 2. Typically, the full-width at half maximum of the EL spectra of the OLED is as broad as ~100 nm and hence multiple emitters can cover the whole visible range (~400–700 nm) which is suitable for a high CRI (>90) lighting source [150]. As discussed in Chapter 5, white LEDs can be used for general lighting. White OLEDs can be used for this application, too. Although the efficiency and lifetime of the white OLED is lower than those of white LEDs, there are several unique advantages to the white OLED. For example, the white OLED is a planar light source, unlike the point source LED which requires device packing and optical components to control the light emission. Hence, white OLEDs are more suitable for large-sized and planar light sources, which can also be fabricated on a flexible substrate and which exhibits the advantage of having a lower weight. However, compared to white LEDs, the power efficiency and operational lifetimes of white OLEDs are still far worse.

There are some differences in the device design between semiconductor LEDs and organic OLEDs. In a semiconductor LED, the emission wavelength is determined by the bandgap, which depends on the choice of material system. For example, phosphide and nitride materials are used for long (red) and short (green and blue) wavelength photon generation. However, in organic materials, the atomic compositions of different emitters are similar (C, H, O, N...), and hence it is possible to integrate different emitters (such as red,

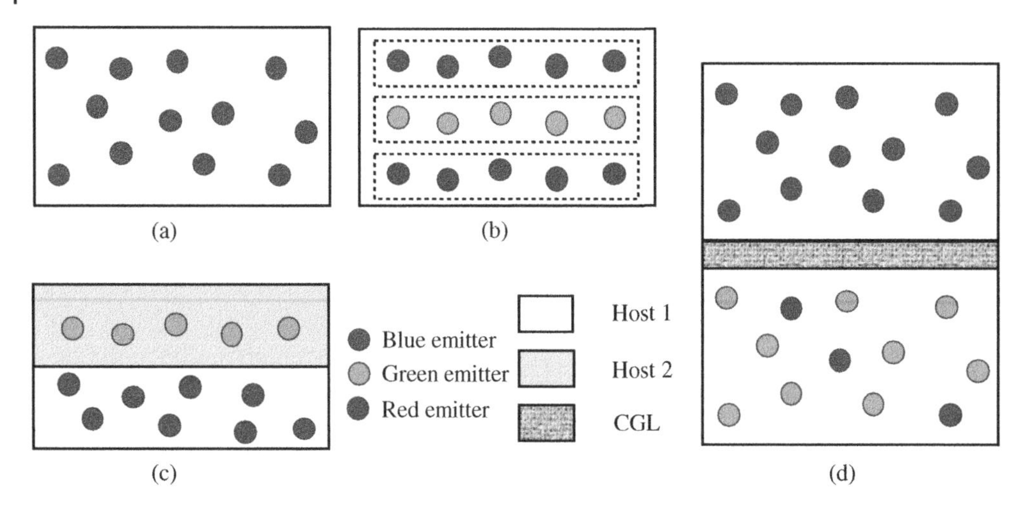

Figure 6.56 Device architectures of white OLED.

yellow, green, and blue) in one device between the anode and the cathode [151]. In the case of a guest–host system, that means that two (yellow and blue) or three (red, green, and blue) dopants are doped into a common host, as shown in Figure 6.56a. In this device architecture, one needs to consider not only energy transfer from the host to the dopants, but also from the high-energy dopant (blue) to low energy dopants (such as red and yellow). Hence, to achieve a white OLED, the dopant concentrations of the long-wavelength dopants should be decreased. For example, dopant concentrations for the sky-blue and orange emitters can be 6.5% and 0.75%, respectively, in the EML [152]. Dopant–dopant energy transfer may cause difficulties in fabrication for a three dopant system although it is still possible. In another approach, different emitters can be doped into separate regions of the host in the EML to achieve different wavelengths of emission, as shown in Figure 6.56b. However, one should note that the carrier mobility of the organic material obeys Poole–Frenkel model (Eq. (6.33)). The electron and hole mobilities change under a varying electric field, which results in a shift of the recombination zone in the EML of the white OLED. That means that the EL spectrum (and the efficiency) of the white OLED may change under different drive voltages. Although it is possible to control the carrier and exciton distribution by including carrier confinement layers in the EML, this increases the fabrication difficulties [153]. A double EML structure provides one solution to this problem. If a hole-transporting and an electron-transporting material are stacked in the EML, holes and electrons are blocked at the interface of these two layers, as shown in Figure 6.56c. Blue and red emitters are doped into the hole-transporting EML, and a green emitter is doped into the electron-transporting EML [154]. Excitons are transferred from the blue to red emitters to achieve long wavelength emission. In such a configuration, the recombination zone is fixed, which improves the color stability of the white OLED under different drive voltages. For example, a CIE shift of (0.01, 0.00) can be achieved over a luminance range from 100 to 10 000 cd/m^2. Considering now the microcavity effect in an OLED, the optimized optical thickness is different for different emission colors. The tandem device structure provides an additional dimension for achieving a white OLED, as shown in Figure 6.56d. By serially connecting two OLEDs with different emission wavelengths, white light can be generated independently without the color shift under different driving voltage [155]. For different applications, the requirements on the white OLED are different. For lighting applications, the CRI is one of the most important parameters and hence a broad EL spectrum is preferred. However, for displays (white OLED with color filter), distinct and saturated red, green, and blue emission peaks are better for achieving a wider color gamut. For lighting applications, comparing a white LED and an OLED, the white LED consists of a blue GaN diode with a yellow phosphor. Typically, the blue emission is most intense which results in a bluish white light with a high color temperature. Although the phosphor can be adjusted to obtain a lower color temperature, this sometimes results in a lower efficiency. On the other hand, the long wavelength (red and green) OLEDs

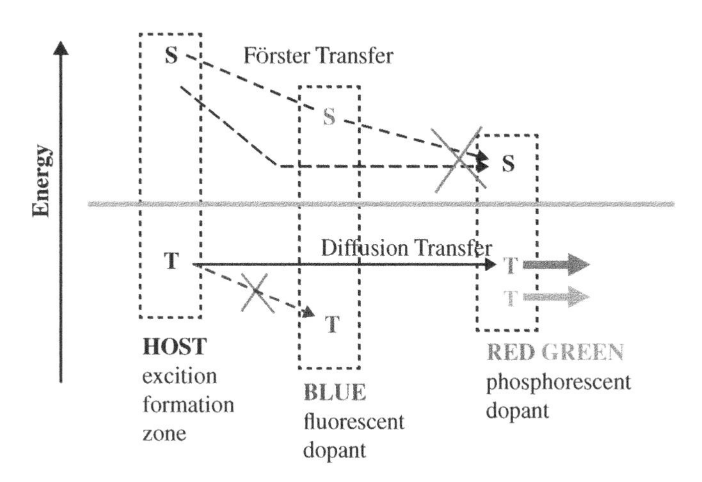

Figure 6.57 Energy transfer for exciton harvesting WOLED, which utilizes both singlet and triplet excitons [156].

have better performance parameters (efficiency and lifetime) than the short wavelength (blue) one, and are suitable to use in a low color temperature light source.

Regarding the organic materials used in a white OLED, obviously fluorescent materials are not good candidates due to their low efficiency. An all-phosphorescent white OLED is possible. For lighting applications, a deep-blue emitter is not necessary and a sky-blue phosphorescent emitter is good enough. However, lifetime is still an issue, hence hybrid (fluorescent blue and phosphorescent green and red OLED) structures have been proposed. In this case, the triplet excitons from the blue fluorescent material are wasted. Therefore, "exciton harvesting" was proposed. In such a device, electrons and holes recombine in the host material. By a suitable choice of organic materials and design of the layer structure, singlet and triplet excitons transfer their energy to fluorescent blue and phosphorescent green and red emitters, respectively, and fully utilize both the singlet and triplet excitons [156]. The principle by which singlet and triplet excitons are managed is illustrated in Figure 6.57. First, excitons are generated in the host molecules. Singlet excitons are then sequentially transferred to blue fluorescent dopants and triplet excitons are transferred to red and green phosphorescent dopants. This method not only eases the sharp decline of efficiency due to TTA, but also reduces the color-shift problem.

6.10 QUANTUM-DOT LIGHT-EMITTING DIODE

Compared to organic emitters, a semiconductor LED exhibits a narrower emission bandwidth which provides a more saturated color. However, LEDs must be epitaxially grown on a crystalline substrate, as discussed in Chapter 5. However, photoluminescent semiconductor quantum dots (QDs) have a narrow emission spectrum and can be synthesized by bulk chemical processing. They have been used in LCD backlight modules and provide a wide color gamut (as shown in Section 4.12). If QDs can be excited electrically rather than optically, in a quantum-dot light-emitting diode, (QLED), it may combine and exceed the technical advantages of LEDs and OLEDs [157]. For example, the color saturation is better than an OLED. The singlet and triplet energies are quite close in a semiconductor QD, and hence it can be regarded as an "intrinsic" TADF material. It is also compatible with large-size device fabrication and flexible substrates, which are difficult targets for LEDs (but it is possible for microLED). Also, compared to the QD-PL used in LCD backlights, direct driving of QLEDs improves the efficiency because there is no optical loss in color filters and polarizers as there is in a LCD.

A common QD structure for the QLED is shown in Figure 6.58a [158]. The size of the core (e.g. CdSe) determines the emission wavelength, as discussed in Section 4.12. The shell surrounding the core has a higher

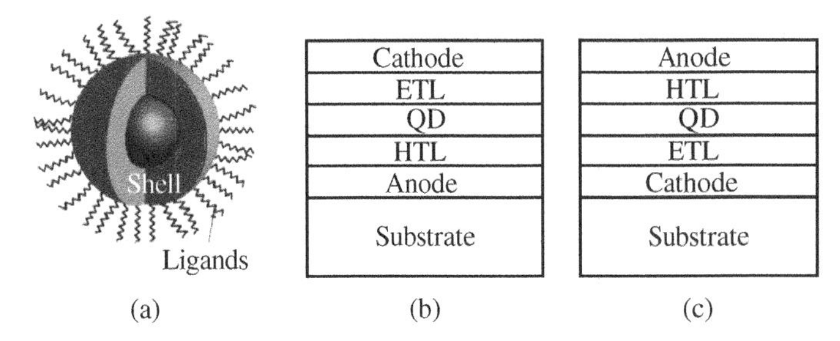

Figure 6.58 (a) Structure of QD for QLED. (b) normal and (c) inverted QLED structure.

bandgap than the core and prevents Förster resonant energy transfer (FRET) between the QDs which otherwise would reduce the quantum efficiency. Typically, CdS and ZnS can be used as shell material with on a CdSe core [159, 160]. QDs are formed of crystalline semiconductors and hence the lattice matching condition should be satisfied between core and shell materials. The energy offset of the conduction and valence bands between these two materials also affects the emission wavelength and the wavefunction distribution. For example, the lattice mismatch between CdS and CdSe is small and hence the crystallinity is better in this material system, compared to ZnS/CdSe. However, the shallow band offset limits the shortest wavelength which can be generated from this system. That means, that QDs with shorter wavelength (for example, for green emission) should use the CdSe/ZnS material system. Sometimes, more complex core shell structures (CdSe/CsS/ZnS) or ternary compounds (such as ZnSeS) have been used to achieve high crystallinity, suitable emission wavelength, and high quantum efficiency [161, 162]. Another issue is, that when the energy offset is small, the wavefunction extends from the core into the shell region. This increases the probability of energy transfer among the QDs and reduces their efficiency. To avoid FRET among the QDs, the shell thickness should be relatively thick. Ligands attached to the outside of the shell have many functions. Because QDs are too heavy for thermal evaporation, wet processing (such as spin coating, ink-jet printing, etc.) is needed. Ligands prevent QD aggregation in solution, which is essential to achieve a uniform QD film in the QLED. The ligand also passivates dangling bonds on the surface of the QD which would reduce the quantum efficiency. Finally, in an EL device, where electrons and holes are injected from the ligand into the QD, functional groups on the organic ligands change the surface potential which helps control the carrier injection characteristics.

The device structure of a QLED is similar to that of an OLED. A QD layer is sandwiched between an HTL and an ETL, which are connected to the anode and cathode, respectively. The HTL and ETL can be organic or inorganic materials. NiOx and ZnO are commonly used as an inorganic HTL and ETL, respectively. There are several organic HTLs with high hole mobility, suitable for QLED applications. The device structure can be normal or inverted, as shown in Figure 6.58b,c. When the anode (or cathode) is fabricated first, it is referred to as a normal (or inverted) structure, respectively. The structure chosen affects both the fabrication process and the device performance. For example, the QD layer must be fabricated via a solution process. In the normal structure, the solvent used to disperse the QDs may dissolve the organic HTL underneath. Conversely, in the inverted structure the QD layer may be damaged by the sputtering process used to deposit inorganic NiOx, which can serve as the HTL. An all-solution process for QLED fabrication has been proposed whereby the QDs and the charge transport layers (both organic and inorganic materials) are deposited from a series of orthogonal solvents to avoid etching the lower layers. Insertion of charge injection, transporting and blocking layers may improve the charge balance in the device and hence raise efficiency. For example, by inserting PMMA between the QDs and a ZnO ETL, and using a double HTL structure to achieve charge balance, EQE >20% can be achieved with a turn-on voltage of 1.7 V in a red QLED having its emission peak at 640 nm [163].

6.11 APPLICATIONS

The operating principle of the OLED is similar to that of the semiconductor LED, as shown in Chapter 5. When voltage is applied to the anode and cathode, EL from the organic materials gives light. However, for OLED fabrication, it is not necessary to use specific substrates (such as GaN on the sapphire substrate, and AlGaInP on the GaAs or GaP substrates). Hence, there are several advantages: (i) it is possible to fabricate OLEDs with different color emission on the same substrate; (ii) OLEDs can be fabricated on large substrates (e.g. >100 × 100 cm) without the limitation of semiconductor wafer sizes (e.g. <12 in. wafer); and (iii) they can be fabricated on flexible substrates. Compared to LCDs (especially transmissive LCDs), a backlight is not necessary for the OLED, and hence thinner (and lighter) OLED modules can be obtained which is attractive for mobile applications. Some novel display types (such as transparent displays) are possible by using transparent (or semitransparent) electrodes. However, compared to semiconductor LEDs, the operating lifetime of the OLED is much shorter, and the power efficiency of the OLED is much lower. Lifetime is an important parameter for some applications, such as TV, and efficiency is important for mobile applications. Compared to LCD displays, the pixel resolution of the OLED display is lower when it is fabricated by means of a fine-pitch shadow mask, and high pixel density is essential for high definition displays. The wide FWHM (~100 nm) of OLEDs is not satisfactory for achieving high quality color displays. Narrow EL emission is possible by selecting suitable organic materials, as well as by optimization of the optical cavity design. For different applications, different OLED designs are needed. In the following section, we will introduce the OLED technologies used for different applications, such as mobile and TV displays, lighting, flexible displays, and some novel displays.

6.11.1 Mobile OLED Display

The total thickness of the OLED (the anode, the organic thin films, and the cathode) is typically less than 1 μm. The substrate thickness can be as thin as 50 μm for flexibility. The thin film passivation process can provide a barrier with a thickness less than 10 μm. Hence, the complete OLED display can be very thin, which is very attractive for mobile applications. The bendable or foldable characteristics of such displays make possible many interesting designs beyond "flat" panel displays. For example, an OLED display can cover the side of a mobile device. The mobile device can even be folded to make it easier to carry, and it can be opened out when a large display is needed. In a mobile device, batteries must be used to provide all the electrical power, and hence one of the most important targets for OLED panels is to reduce the power consumption. As the viewing distance for mobile devices is <30 cm from the eyes, the pixel size should be small enough not to be resolved in normal use. Sometimes, the viewing angle performance can be sacrificed a little because mobile displays are typically used by a single viewer. The requirement for the lifetime of a mobile device is about two years, which is not too difficult for OLED materials and devices. The number of rows addressed in a passive-matrix (PM) OLED is typically limited to less than 240, otherwise the high peak luminance which appears in line-at-a-time addressing reduces the operation lifetime severely. That means that the size of a PM-OLED is limited to 2 in. with sufficient resolution for mobile use, and can be used for a watch or smart watch display. For larger mobile devices (>4 in.), an active-matrix (AM) OLED is needed. In designing an AM-OLED for a mobile device, TFTs can be hidden under the OLED in a top-emission configuration which increases the aperture ratio (and luminance) and decreases the power consumption. To achieve full-color, the lateral subpixelated configuration is typically used, fabricated with a fine-pitch shadow mask. The resolution remains one of the major problems for OLED mass production. A strong optical cavity effect generated between reflective and semi-transparent electrodes improves the color saturation. However, it also decreases the viewing angle. Optical thin films on the top-emission OLED help to broaden the viewing cone and improve the light-extraction efficiency. Note that in a mobile device, a touch panel is commonly used to serve as the input interface. Besides, although thin film encapsulation can effectively protect the OLED from attack by ambient oxygen and moisture, mechanical protection is needed for the panel. Hence, an "on-cell" touch panel can be used for this purpose (touch panel technologies will be introduced in Chapter 10). Between the display substrate and the touch panel substrate, an optical adhesive can be used which improves light extraction and the viewing angle performance. It can also help to reduce the water vapor permeation rate and prolong the panel lifetime.

For mobile applications, the ambient contrast ratio is an important characteristic, because these devices may be used under both low and high ambient light conditions. For the emissive OLED, sunlight readability is a serious issue. To improve it, the surface reflection of the display must be reduced. Typically, a circular polarizer is needed, which consists of a linear polarizer and a broadband quarter-wave plate. Ambient light incident on the device passes through the linear polarizer first. Then, the broadband quarter-wave plate changes polarization of the light from linear to circular. When it is reflected from the metal electrode of the OLED, the phase change results in reversal of the sense of circular polarization (i.e. from right circular polarization to left-handed or vice versa), which is converted to an orthogonal linear polarization by the phase plate, and blocked by the linear polarizer, therefore, the ambient reflection can be greatly reduced. However, compared to the black background in an LCD defined by a crossed-polarizer configuration, the cost of the circular polarizer is higher because of the need for a broadband quarter-wave plate. The birefringent film is also thicker, which is a disadvantage for flexible applications. And the reflection is higher which limits the ambient contrast ratio. Besides, due to the emissive characteristics of the OLED, ~50% of the light emission will be absorbed by the polarizer which decreases the luminance and increases power consumption.

6.11.2 OLED TV

In contrast to mobile devices, the size of the panel for TV displays is much larger (i.e. >50 in.), thehe viewing distance is longer (i.e. >2 m), and typically, TV is static in position and is operated indoors close to the wall. Multiple viewers share a single TV and watch from a wide range of angles. Excellent display performance is essential. For example: (i) the color gamut should be as large as possible to match standards such as BT 2020, (ii) the dynamic range of gray levels is important to achieve the "home theater" class performance in low ambient lighting, and (iii) the response time should be fast enough to avoid artifacts in moving images. In other words, TV displays should create an immersive experience for viewers. For TV application, a longer lifetime (both storage lifetime and operating lifetime) is expected (e.g. five years) compared to a mobile device. Power consumption is not the most important requirement because of the wall-plug power supply. However, certain criteria should be met (such as regulations defined by Energy Star).

Fabrication of such a "big" display is more challenging than mobile displays. Of course, the substrate should be big enough (e.g. Gen 8 or larger) to reduce the manufacturing cost. Remember that the total thickness of the OLED is <200 nm. For certain organic layers, the thickness is even as thin as 20 nm. A 5% difference in film thickness (1 nm) results in distinguishable non-uniformity in an OLED device. To achieve such precise control in film thickness and long operating lifetime, thermal evaporation is needed, which requires a huge vacuum chamber for such a big substrate. To provide a full-color OLED-TV, a white OLED with color filter is typically used rather than lateral emitters, due to the difficulty in fabricating such a big fine-pitch shadow mask. Pixel resolution is typically not an issue, due to the mature fabrication techniques which are established for color filters and the large pixel size of the TV. Obviously, the power consumption of the white OLED with color filter is higher because the color filter absorbs part of the emitted light. Different pixel arrangements such as red, green, blue, and white (RGBW) can be used, and can reduce power consumption. There is no color filter on the top of the white subpixel and hence the efficiency can be higher. A tandem structure is also sometimes employed, so that the luminance from each unit can be reduced thereby prolonging the operational lifetime. By combining blue emitting dopant with longer wavelength emitters (such as green, yellow, and red ones) the operating lifetime of the blue component in a white OLED can be increased, because energy transfers to the lower bandgap dopants and reduces the exciton–polaron interaction. Image sticking is an important issue for OLED-TV when a still-image is displayed for a long term. Due to the limited lifetime of the OLED device, the luminance of the brightest pixels decreases with time. When all the pixels are later turned on, the pixels which have operated for a longer time are dimmer. To reduce the image sticking issue, of course the device lifetime of the OLED should be as long as possible. Besides this, a TFT compensation circuit can effectively retain the uniformity of the display. For example, luminance decay of the OLED at constant current density is typically accompanied by a threshold voltage increase. Hence, by sensing the voltage drop across the OLED, it is possible to infer the luminance decay of the OLED and provide necessary compensation. The system surrounding the display is also important. For example, a screen saver function can provide unintentional still-images when the display is not in use. Not only the operating lifetime, but storage lifetime is also an issue for OLED-TVs. Thin-film passivation, adhesive, and encapsulation are all needed to achieve a low WVPR. The adhesive used here may not need to be transparent, provided that a bottom-emission OLED is used.

A weak optical cavity (reflective and transparent electrodes) is preferred for OLED-TV application because the viewing angle required is broader. For an immersive display, the OLED-TV should be bendable, but it is not necessary for it to be foldable. Ambient reflection is also not so critical for OLED-TVs because the indoor ambient light level is typically not very strong. Sometimes, a black matrix outside the emission zone is enough to reduce the ambient reflection without attachment of a circular polarizer. The thickness of the OLED-TV is not very critical, after all, OLED itself is ultra-thin ($<1\ \mu m$).

6.11.3 OLED Lighting

As discussed in Chapter 5, LEDs are very suitable for direct lighting, given careful optical design. On the other hand, OLED lighting provides a diffuse light source without any complex optical elements. OLEDs can be fabricated on large and flexible substrates and hence OLED lighting can provide a large area light. For example, OLED lighting can be embedded in wall paper, into a table, mirror, or even the window. It can be fabricated in any shape for various designs. In LEDs, the operating lifetime of the GaN-blue device is very long. With a suitable yellow phosphor, white light can be generated. On the other hand, for the OLED, the blue device exhibits the shortest lifetime. Hence, it is an advantage for OLED lighting to generate a "warm white" with low color temperature, which requires a much lower blue emission, compared to red and green. Given the wide FWHM spectra of the OLED ($\sim$100 nm) for a single emitter, a light source with a high color rendering index (CRI) can be realized easily by OLEDs. Compared to display applications, no TFTs are needed for OLED lighting, which greatly reduces the fabrication complexity. Color shift and luminance degradation are less serious for OLED lighting, compared to displays. Overall, it is easier to fabricate OLED lighting than displays, and, the cost of lighting devices is much lower than that of displays. To increase the throughput in lighting production, wet processing rather than physical vapor deposition may be a solution for organic layer deposition. Obviously, tandem devices and a weak cavity are suitable for lighting purposes.

Light extraction techniques are typically used for OLED lighting, rather than OLED display. In OLED display, a lot of light is trapped inside the glass substrate, ITO, and organic materials. Although we introduced several methods for extracting photons from the OLED, it is rare that those techniques can be applied to OLED displays, because the extracted light rays would result in image blurring and reduce the image resolution. However, this is not a problem for OLED lighting. An important parameter for the lighting technology is the power efficiency, in terms of lm/W. Power efficiency, EQE, and driving voltage of 149 lm/W, 50%, and 2.9 V of a white OLED lighting panel (9×9 cm) were reported at a brightness of 1000 cd/m^2 in 2016. The CCT was 2880 K, representing a warm white illumination and the CIE coordinates were (0.48, 0.47) [164].

6.11.4 Flexible OLEDs

The thin organic active layers and selected substrates used in OLEDs are flexible, and with care the complete device stack can be made suitable for use as a flexible display. To have a flexible OLED, one needs to include a flexible substrate, conductor, and TFT (Section 3.6.6). Suitable passivation and encapsulation are needed for long storage lifetime (Section 6.5.3.2). For mobile display applications, the requirement may be extended to have a display which is foldable, so that the device area is small enough to be put away in a bag or pocket, but can provide a large, attractive display when in use. Flexible OLED lighting provides many possibilities for curved designs. Flexible OLEDs are also important for wearable display applications, such as watches and clothing.

6.11.5 Novel Displays

Going back to the basics of the OLED device, it is organic stacks with a thickness of 100–200 nm sandwiched by anode and cathodes. By modifying the layout and the electrode materials, it is possible to fabricate reflective OLEDs, which look just like a domestic mirror when the OLED is not illuminated. By providing electrical current to the OLED, the device can be used for a display or for lighting. The transparent OLED is another example which looks like a window when the OLED is not powered on. Because the whole device (display or lighting) is transparent, it can be engineered to emit light from one or both sides, depending on the applications. In the case of a transparent OLED display with double-sided emission, the displayed image can be seen from both sides with mirror-like reversal from one direction. By dividing each pixel into two subpixels comprising respectively top- and bottom-emission OLEDs controlled by individual TFTs, it is possible to have an OLED display with double-sided emission built on a single glass substrate. The images displayed on such a double-sided device can be different, because they are controlled by individual TFTs and each side is optically separated from the other. Although only one substrate is needed for the OLED, sometimes a cover glass is added to provide mechanical protection and improve the storage lifetime. Hence, combining two glass substrates, each with a bottom-emitting configuration, provides another route to double-sided OLED displays.

OLEDs can be used as the display media for applications in virtual reality (VR) and augmented reality (AR) due to their fast response and light weight. These topics will be introduced in Chapter 9. For these purposes, the display is close to the eye and hence a super high resolution is needed. It is not easy to use such a fine-pitch shadow mask and therefore a white OLED with a color filter may be used. Because ambient light is excluded from VR displays, the luminance required is not high. A wide viewing angle is also not needed for such a single-user application. For AR applications, the situation is more challenging. Light emitted from the OLED has to compete with the ambient, and the transmittance of the OLED must be high to provide a see-through capability. Optimizing the performance of an AR display may need a more complex optical design.

Homework Problems

6.1 Is it possible to fabricate an OLED with a single active organic layer? Why? If it was possible, what were the criteria for the organic material and the electrodes?

6.2 Consider a system in which a host is doped with two guest materials. One guest is a blue singlet emitter, and the other is a yellow triplet emitter. Sketch the possible energy transfer routes from the host to the guests and the restrictions which allow 100% internal quantum efficiency to be achieved.

6.3 Derive the approximate SCLC equation as follows:
SCLC: $J = \mu\varepsilon\,(V^2/L^3)$;
J: current density, μ: carrier mobility, ε: static dielectric constant, V: driving voltage, L: device thickness.
Current conduction: $J = \rho v$; ρ: injected free charge concentration, v: drift velocity.
Capacitance: $Q = CV$; Q: total injected free charge, C: capacitance.
(You may need $Q = \rho AL$, $v = \mu\,[V/L]$, A: device cross section.)

6.4 Derive the approximate TCLC equation as follows:
TCLC: $J = \varepsilon\mu Nc\,(\varepsilon/\varepsilon N_0\,kT_t)^l\,(V^{(l+1)}/L^{(2l+1)})$;
J: current density, μ: carrier mobility, ε: static dielectric constant, V: driving voltage, L: device thickness, $l = T_t/T$.

(Assume the free and trap carrier concentration are defined as follows:
Free carrier concentration distribution: $n = Nc\, \exp[(F\text{-}Ec)\,/kT]$
Trap carrier concentration distribution: $n_t = kT_t\, N_0\, \exp[(F\text{-}Ec)/KT_t]$
Nc: effective density of states in the conduction band, *F*: quasi Fermi level, *Ec*: bottom edge of the conduction band energy, *k*: Boltzmann constant, *T*: temperature in degrees Kelvin, T_t: characteristic temperature.)
Current conduction: $J = \rho v$; ρ: injected free charge concentration, *v*: drift velocity.
Capacitance: $Qt = CV$; *Qt*: total trapped charge, *C*: capacitance.
(You may need $Qt = \rho_t AL$, $v = \mu[V/L]$, $n = \rho/\varepsilon$, $n_t = \rho_t/\varepsilon$, *A*: device cross section area.)

6.5 For a planar structure, does a top- or bottom-emission OLED exhibit higher EQE? Why?

References

1 Sze, S.M. (2001). *Semiconductor Devices – Physics and Technology*, 2e. Wiley.
2 Baldo, M.A. and Forrest, S.R. (2001). Interface-limited injection in amorphous organic semiconductors. *Phys. Rev. B* 64: 085201.
3 Pope, M., Kallmann, H., and Magnante, P. (1963). Electroluminescence in organic crystals. *J. Chem. Phys.* 38: 2042.
4 Tang, C.W. and Vanslyke, S.A. (1987). Organic electroluminescent diodes. *Appl. Phys. Lett.* 51: 913.
5 Burroughes, J.H., Bradley, D.D.C., Brown, A.R. et al. (1990). Light-emitting diodes based on conjugated polymers. *Nature* 347: 539.
6 Kagan, J. (1993). *Organic Photochemistry: Principles and Applications*. Academic Press Inc.
7 Gierschner, J., Mack, H.G., Luer, L., and Oelkrug, D. (2002). Fluorescence and absorption spectra of oligophenylenevinylenes: vibronic coupling, band shapes, and solvatochromism. *J. Chem. Phys.* 116: 8596.
8 Michl, J. and Bonacic-Koutecky, V. (1990). *Electronic Aspects of Organic Photochemistry*. New York: Wiley.
9 Tang, C.W., Vanslyke, S.A., and Chen, C.H. (1989). Electroluminescence of doped organic thin films. *J. Appl. Phys.* 65: 3610.
10 Klessinger, M. and Michl, J. (1995). *Excited States and Photochemistry of Organic Molecules*. New York: Wiley.
11 Yip, W.T. and Levy, D.H. (1996). Excimer/exciplex formation in van der Waals dimers of aromatic molecules. *J. Phys. Chem.* 100: 11539.
12 D'Andrade, B.W., Brooks, J., Adamovich, V. et al. (2002). White light emission using triplet excimers in electrophosporescent organic light-emitting devices. *Adv. Mater.* 14: 1032.
13 Hung, L.S. and Chen, C.H. (2002). Recent progress of molecular organic electroluminescent materials and devices. *Mater. Sci. Eng., R* 39: 143.
14 Petersson, G.P., Svensson, C.M., and Maserjian, J. (1975). Resonance effects observed at the onset of fowler-Nordheim tunneling in thin MOS structures. *Solid-State Electron.* 18: 449.
15 Pope, M. and Swenberg, C.E. (1999). *Electronic Processes in Organic Crystals and Polymers*, 2e. Oxford University Press.
16 Gill, W.D. (1972). Drift mobilities in amorphous charge-transfer complexes of trinitrofluorenone and poly-*n*-vinylcarbazole. *J. Appl. Phys.* 43: 5033.
17 Choi, W.K., Delima, J.J., and Owen, A.E. (1986). Model for the variations in the field-dependent behavior of the Poole-Frenkel effect. *Phys. Status Solidi B Basic Res.* 137: 345.
18 Davids, P.S., Campbell, I.H., and Smith, D.L. (1997). Device model for single carrier organic diodes. *J. Appl. Phys.* 82: 6319.

19 Rusu, G.I. (1993). On the current-voltage characteristics of some thin-film sandwich structures of the metal/organic semiconductor/metal type. *Appl. Surf. Sci.* 65-66: 381.

20 Staudigel, J., Stößel, M., Steuber, F., and Simmerer, J. (1999). A quantitative numerical model of multi-layer vapor-deposited organic light emitting diodes. *J. Appl. Phys.* 86: 3895.

21 Parker, I.D. (1994). Carrier tunneling and device characteristics in polymer light-emitting diodes. *J. Appl. Phys.* 75: 1656.

22 Kao, K.C. and Hwang, W. (1981). *Electrical Transport in Solid: With Particular Reference to Organic Semiconductors*. Pergamon Press.

23 Meyer, H., Haarrer, D., Naarmann, H., and Hohold, H.H. (1995). Trap distribution for charge carriers in poly (paraphenylene vinylene) (PPV) and its substituted derivative DPOP-PPV. *Phys. Rev. B* 52: 2587.

24 Markov, D.E., Hummelen, J.C., Blom, P.W.M., and Sieval, A.B. (2005). Dynamics of exciton diffusion in poly(*p*-phenylene vinylene)/fullerene heterostructures. *Phys. Rev. B* 72: 045216.

25 Greenham, N.C., Friend, R.H., and Bradley, D.D.C. (1994). Angular dependence of the emission from a conjugated polymer light-emitting diode: implications for efficiency calculations. *Adv. Mater.* 6: 491.

26 Saleh, B.E.A. and Teich, M.C. (1991). *Fundamentals of Photonics*. New York: Wiley.

27 Takada, N., Tsutsui, T., and Saito, S. (1993). Control of emission characteristics in organic thin-film electroluminescent diodes using an optical-microcavity structure. *Appl. Phys. Lett.* 63: 2032.

28 So, S.K., Choi, W.K., Leung, L.M., and Neyts, K. (2003). Interference effects in bilayer organic light-emitting diodes. *Appl. Phys. Lett.* 82: 466.

29 Riel, H., Karg, S., Beierlein, T., and Rieβ, W. (2003). Tuning the emission characteristics of top-emitting organic light-emitting devices by means of a dielectric capping layer: an experimental and theoretical study. *J. Appl. Phys.* 94: 5290.

30 Brutting, W., Frischeisen, J., Schmidt, T.D. et al. (2013). Device efficiency of organic light-emitting diodes: progress by improved light outcoupling. *Phys. Status Solidi A* 210: 44.

31 Braun, D. and Heeger, A.J. (1991). Visible light emission from semiconducting polymer diodes. *Appl. Phys. Lett.* 58: 1982.

32 Hsiao, C.H., Lee, J.H., and Tseng, C.A. (2006). Probing recombination-rate distribution in organic light-emitting devices with mixed-emitter structure. *Chem. Phys. Lett.* 427: 305.

33 Kalinowski, J., Palilis, L.C., Kim, W.H., and Kafafi, Z.H. (2003). Determination of the width of the carrier recombination zone in organic light-emitting diodes. *J. Appl. Phys.* 94: 7764.

34 Shi, J. and Tang, C.W. (1997). Doped organic electroluminescent devices with improved stability. *Appl. Phys. Lett.* 70: 1665.

35 C. H. Chen, K. P. Klubek, and J. Shi, "Red Organic Electroluminescent Materials", *U.S. Patent* No. 5908581.

36 Hosokawa, C., Higashi, H., Nakamura, H., and Kusumoto, T. (1995). Highly efficient blue electroluminescence from a distyrylarylene emitting layer with a new dopant. *Appl. Phys. Lett.* 67: 3853.

37 Shirota, Y., Kuwabara, Y., lnada, H. et al. (1994). Multilayered organic electroluminescent device using a novel starburst molecule, 4,4′,4″-tris(3-methylphenylphenylamino)tripheny~amine, as a hole transport material. *Appl. Phys. Lett.* 65: 807.

38 Nuesch, F., Rothberg, L.J., Forsythe, E.W. et al. (1999). A photoelectron spectroscopy study on the indium tin oxide treatment by acids and bases. *Appl. Phys. Lett.* 74: 880.

39 Wu, C.C., Wu, C.I., Sturm, J.C., and Kahn, A. (1997). Surface modification of indium tin oxide by plasma treatment: an effective method to improve the efficiency, brightness, and reliability of organic light emitting devices. *Appl. Phys. Lett.* 70: 1348.

40 Tadayyon, S.M., Grandin, H.M., Griffiths, K. et al. (2004). CuPc buffer layer role in OLED performance: a study of the interfacial band energies. *Org. Electron.* 5: 157.

41 Hung, L.S., Tang, C.W., and Mason, M.G. (1997). Enhanced electron injection in organic electroluminescence devices using an Al/LiF electrode. *Appl. Phys. Lett.* 70: 152.

42 Shaheen, S.E., Jabbour, G.E., Morrell, M.M. et al. (1998). Bright blue organic light-emitting diode with improved color purity using a LiF/Al cathode. *J. Appl. Phys.* 84: 2324.
43 Zhou, X., Pfeiffer, M., Blochwitz, J. et al. (2001). Very-low-operating-voltage organic light-emitting diodes using a p-doped amorphous hole injection layer. *Appl. Phys. Lett.* 78: 410.
44 Endo, J., Matsumoto, T., and Kido, J. (2002). Organic electroluminescent devices with a vacuum-deposited Lewis-acid-doped hole-injecting layer. *Jpn. J. Appl. Phys.* 41: L358.
45 Pfeiffer, M., Leo, K., Zhou, X. et al. (2003). Highly efficient organic light emitting diodes by doped transport layers. *Org. Electron.* 4: 89.
46 Lee, J.H., Wu, M.H., Chao, C.C. et al. (2005). High efficiency and long lifetime OLED based on a metal-doped electron transport layer. *Chem. Phys. Lett.* 416: 234.
47 He, G., Schneider, O., Qin, D. et al. (2004). Very high-efficiency and low voltage phosphorescent organic light-emitting diodes based on a p-i-n junction. *J. Appl. Phys.* 95: 5773.
48 D'Andrade, B.W., Forrest, S.R., and Chwang, A.B. (2003). Operational stability of electrophosphorescent devices containing *p* and *n* doped transport layers. *Appl. Phys. Lett.* 83: 3858.
49 Xie, Z., Hung, L.S., and Zhu, F. (2003). A flexible top-emitting organic light-emitting diode on steel foil. *Chem. Phys. Lett.* 381: 691.
50 Riel, H., Karg, S., Beierlein, T. et al. (2003). Phosphorescent top-emitting organic light-emitting devices with improved light outcoupling. *Appl. Phys. Lett.* 82: 466.
51 Chen, C.W., Hsieh, P.Y., Chiang, H.H. et al. (2003). Top-emitting organic light-emitting devices using surface-modified Ag anode. *Appl. Phys. Lett.* 83: 5127.
52 Han, S., Feng, X., and Lu, Z.H. (2003). Transparent-cathode for top-emission organic light-emitting diodes. *Appl. Phys. Lett.* 82: 2715.
53 Liu, K.C., Teng, C.W., Lu, Y.H. et al. (2007). Improving the performance of transparent PLEDs with LiF/Ag/ITO cathode. *Electrochem. Solid-State Lett.* 10: J120.
54 Ko, C.W., Hu, S.H., Li, S.H. et al. (2005). Development of 1.5-inch full color double sided active matrix OLED with novel arrays design. *SID Tech. Dig.* 36: 961.
55 Ohmori, Y., Uchida, M., Muro, K., and Yoshino, K. (1991). Blue electroluminescent diodes utilizing poly(alkylfluorene). *Jpn. J. Appl. Phys.* 30: L1941.
56 Ho, P.K.H., Kim, J.S., Burroughes, J.H. et al. (2000). Molecular-scale interface engineering for polymer light-emitting diodes. *Nature* 404: 481.
57 Chou, M.Y., Leung, M.K., Su, Y.O. et al. (2004). Electropolymerization of starburst triarylamines and their application to electrochromism and electroluminescence. *Chem. Mater.* 16: 654.
58 Ma, W., Iyer, P.K., Gong, X. et al. (2005). Water/methanol-soluble conjugated copolymer as an electron-transport layer in polymer light-emitting diodes. *Adv. Mater.* 17: 274.
59 Lee, C.B., Uddin, A., Hu, X., and Andersson, T.G. (2004). Study of Alq3 thermal evaporation rate effects on the OLED. *Mater. Sci. Eng., B* 112: 14.
60 Long, M., Grace, J.M., Freenman, D.R. et al. (2006). New capabilities in vacuum thermal evaporation sources for small molecule OLED manufacturing. *SID Tech. Dig.* 1474: 1474–1476.
61 Lee, S.T., Suh, M.C., Kang, T.M. et al. (1588). LITI (laser induced thermal imaging) technology for high-resolution and large-sized AMOLED. *SID Tech. Dig.* 38: 2007.
62 Hirano, T., Matsuo, K., Kohinata, K. et al. (1592). Novel laser transfer Technology for Manufacturing Large-Sized OLED displays. *SID Tech. Dig.* 38: 2007.
63 Iino, S. and Miyashita, S. (2006). Printable OLEDs promise for future TV market. *SID Tech. Dig.* 37: 1463.
64 Kwon, S.H., Paik, S.Y., Kwon, O.J., and Yoo, J.S. (2001). Triple-layer passivation for longevity of polymer light-emitting diodes. *Appl. Phys. Lett.* 79: 4450.
65 Lewis, J.S. and Weaver, M.S. (2004). Thin-film permeation-barrier technology for flexible organic light-emitting devices. *IEEE J. Sel. Top. Quantum Electron.* 10: 45.

66 Kho, S., Cho, D., and Jung, D. (2002). Passivation of organic light-emitting diodes by the plasma polymerized Para-xylene thin film. *Jpn. J. Appl. Phys.* 41: L1336.
67 Kim, G.H., Oh, J., Yang, Y.S. et al. (2004). Lamination process encapsulation for longevity of plastic-based organic light-emitting devices. *Thin Solid Films* 467: 1.
68 Lee, J.H., Kim, G.H., Kim, S.H. et al. (2004). Longevity enhancement of organic thin-film transistors by using a facile laminating passivation method. *Synth. Met.* 143: 21.
69 Weaver, M.S., Michalski, L.A., Rajan, K. et al. (2002). Organic light-emitting devices with extended operating lifetimes on plastic substrates. *Appl. Phys. Lett.* 81: 2929.
70 Burrows, P.E., Graff, G.L., Gross, M.E. et al. (2001). Ultra barrier flexible substrates for flat panel displays. *Displays* 22: 65.
71 Lee, C.J., Pode, R.B., Moon, D.G., and Han, J.I. (2004). Realization of an efficient top emission organic light-emitting device with novel electrodes. *Thin Solid Films* 467: 201.
72 Fung, M.K., Gao, Z.Q., Lee, C.S., and Lee, S.T. (2001). Inhibition of dark spots growth in organic electroluminescent devices. *Chem. Phys. Lett.* 333: 432.
73 Aziz, H., Popovic, Z., Tripp, C. et al. (1998). Degradation processes at the cathode/organic interface in organic light emitting devices with Mg:Ag cathodes. *Appl. Phys. Lett.* 72: 2642.
74 Lim, S.F., Ke, L., Wang, W., and Chua, S.J. (2001). Correlation between dark spot growth and pinhole size in organic light-emitting diodes. *Appl. Phys. Lett.* 78: 2116.
75 Gardonio, S., Gregoratti, L., Melpignano, P. et al. (2007). Degradation of organic light-emitting diodes under different environment at high drive conditions. *Org. Electron.* 8: 37.
76 Burrows, P.E., Bulovic, V., Forrest, S.R. et al. (1994). Reliability and degradation of organic light emitting devices. *Appl. Phys. Lett.* 65: 2922.
77 Chan, M.Y., Lai, S.L., Wong, F.L. et al. (2003). Efficiency enhancement and retarded dark-spots growth of organic light-emitting devices by high-temperature processing. *Chem. Phys. Lett.* 371: 700.
78 Kim, Y., Choi, D., Lim, H., and Ha, C.S. (2003). Accelerated pre-oxidation method for healing progressive electrical short in organic light-emitting devices. *Appl. Phys. Lett.* 82: 2200.
79 Luo, Y., Aziz, H., Popovic, Z.D., and Xu, G. (2006). Correlation between electroluminescence efficiency and stability in organic light-emitting devices under pulsed driving conditions. *J. Appl. Phys.* 054508: 99.
80 Kitamura, M., Imada, T., and Arakawa, Y. (2003). Organic light-emitting diodes driven by pentacene-based thin-film transistors. *Appl. Phys. Lett.* 83: 3410.
81 Hosokawa, C., Eida, M., Matsuura, M. et al. (1997). Organic multi-color electroluminescence display with fine pixels. *Synth. Met.* 91: 3.
82 Zou, D.C., Yahiro, M., and Tsutsui, T. (1998). Spontaneous and reverse-bias induced recovery behavior in organic electroluminescent diodes. *Appl. Phys. Lett.* 72: 2484.
83 Cao, Y., Yu, G., Parker, I.D., and Heeger, A.J. (2000). Ultrathin layer alkaline earth metals as stable electron-injecting electrodes for polymer light emitting diodes. *J. Appl. Phys.* 88: 3618.
84 Ni, S.Y., Wang, X.R., Wu, Y.Z. et al. (2004). Decay mechanisms of a blue organic light emitting diode. *Appl. Phys. Lett.* 85: 878.
85 Aziz, H., Popovic, Z.D., Hu, N.X. et al. (1999). Degradation mechanism of small molecule-based organic light-emitting devices. *Science* 283: 1900.
86 Popovic, Z.D. and Aziz, H. (2002). Reliability and degradation of small molecule-based organic light-emitting devices (OLEDs). *IEEE J. Sel. Top. Quantum Electron.* 8: 362.
87 Lee, J.H., Huang, J.J., Liao, C.C. et al. (2005). Operation lifetimes of organic light-emitting devices with different layer structures. *Chem. Phys. Lett.* 402: 335.
88 Tsai, C.H., Liao, C.H., Lee, M.T., and Chen, C.H. (2005). Highly stable organic light-emitting devices with a uniformly mixed hole transport layer. *Appl. Phys. Lett.* 87: 243505.
89 Mori, T., Mitsuoka, T., Ishii, M. et al. (2002). Improving the thermal stability of organic light-emitting diodes by using a modified phthalocyanine layer. *Appl. Phys. Lett.* 80: 3895.

90 Féry, C., Racine, B., Vaufrey, D. et al. (2005). Physical mechanism responsible for the stretched exponential decay behavior of aging organic light-emitting diodes. *Appl. Phys. Lett.* 87: 213502.

91 Parker, I.D., Cao, Y., and Yang, C.Y. (1999). Lifetime and degradation effects in polymer light-emitting diodes. *J. Appl. Phys.* 85: 2441.

92 Zhang, Y., Lee, J., and Forrest, S.R. (2014). Tenfold increase in the lifetime of blue phosphorescent organic light-emitting diodes. *Nat. Commun.* 5: 5008.

93 Kondakov, D.Y. (2007). Characterization of triplet-triplet annihilation in organic light-emitting diodes based on anthracene derivatives. *J. Appl. Phys.* 102: 114504.

94 Uoyama, H., Goushi, K., Shizu, K. et al. (2012). Highly efficient organic light-emitting diodes from delayed fluorescence. *Nature* 492: 234.

95 Hung, W.Y., Fang, G.C., Lin, S.W. et al. (2014). The first tandem, all-exciplex-based WOLED. *Sci. Rep.* 4: 5161.

96 Baldo, M.A., O'Brien, D.D., Thompson, M.E., and Forrest, S.R. (1999). Excitonic singlet-triplet ratio in a semiconducting organic film. *Phys. Rev. B.* 60: 14422.

97 Baldo, M.A., Lamansky, S., Burrow, P.E. et al. (1999). Very high-efficiency green organic light-emitting devices based on electrophosphorescence. *Appl. Phys. Lett.* 75: 4.

98 Kwong, R.C., Nugent, M.R., Michalski, L. et al. (2002). High operational stability of electrophosphorescent devices. *Appl. Phys. Lett.* 81: 162.

99 He, G., Pfeiffer, M., Leo, K. et al. (2004). High-efficiency and low-voltage p-i-n electrophosphorescent organic light-emitting diodes with double-emission layers. *Appl. Phys. Lett.* 85: 3911.

100 Tanaka, D., Sasabe, H., Li, Y.-J. et al. (2007). Ultra high efficiency green organic light-emitting devices. *Jpn. J. Appl. Phys.* 46: L10.

101 Baldo, M.A., O'Brien, D.F., You, Y. et al. (1998). Highly efficient phosphorescent emission from organic electroluminescent devices. *Nature* 395: 151.

102 O'Brien, D.F., Baldo, M.A., Thompson, M.E., and Forrest, S.R. (1999). Improved energy transfer in electrophosphorescent devices. *Appl. Phys. Lett.* 74: 442.

103 Adachi, C., Baldo, M.A., Thompson, M.E. et al. (2001). High-efficiency red electrophosphorescence devices. *Appl. Phys. Lett.* 78: 1622.

104 Jiang, X., Jen, A.K.-Y., Carlson, B., and Dalton, L.R. (2002). Red electrophosphorescence from osmium complexes. *Appl. Phys. Lett.* 80: 713.

105 Song, Y.H., Yeh, S.J., Chen, C.T. et al. (2004). Bright and efficient, non-doped, phosphorescent organic red-light-emitting diodes. *Adv. Funct. Mater.* 14: 1221.

106 Li, C.L., Su, Y.J., Tao, Y.T. et al. (2005). Yellow and red electrophosphors based on linkage isomers of phenylisoquinolinyliridium complexes : distinct difference in Photophysical and electroluminescence properties. *Adv. Funct. Mater.* 15: 387.

107 Huang, J.J., Hung, Y.H., Ting, P.L. et al. (2016). Orthogonally substituted Benzimidazole-Carbazole benzene as universal hosts for phosphorescent organic light-emitting diodes. *Org. Lett.* 18: 672.

108 Giebink, N.C., D'Andrade, B.W., Weaver, M.S. et al. (2008). Intrinsic luminance loss in phosphorescent small-molecule organic light emitting devices due to bimolecular annihilation reactions. *J. Appl. Phys.* 103: 044509.

109 Nakanotani, H., Higuchi, T., Furukawa, T. et al. (2014). *Nat. Commun.* 5: 4016.

110 Liu, X.K., Chen, Z., Zheng, C.J. et al. (2015). Nearly 100% triplet harvesting in conventional fluorescent dopant-based organic light-emitting devices through energy transfer from Exciplex. *Adv. Mater.* 27: 2025.

111 Park, Y.S., Lee, S., Kim, K.H. et al. (2013). Exciplex-forming co-host for organic light-emitting diodes with ultimate efficiency. *Adv. Funct. Mater.* 23: 4914.

112 Matsumoto, T., Nakada, T., Endo, J. et al. (2003). Multiphoton organic EL device having charge generation layer. *SID 03 Dig.*: 979.

113 Liao, L.S., Klubek, K.P., and Tang, C.W. (2004). High-efficiency tandem organic light-emitting diodes. *Appl. Phys. Lett.* 84: 167.

114 Yang, R.Q. and Qiu, Y. (2003). Bipolar cascade lasers with quantum well tunnel junctions. *J. Appl. Phys.* 94: 7370.

115 Guo, X., Shen, G.D., Wang, G.H. et al. (2001). Tunnel-regenerated multiple-active-region light-emitting diodes with high efficiency. *Appl. Phys. Lett.* 79: 2985.

116 Guo, X., Shen, G.D., Ji, Y. et al. (2003). Thermal property of tunnel-regenerated multiactive-region light-emitting diodes. *Appl. Phys. Lett.* 82: 4417.

117 Gu, G., Parthasarathy, G., Burrows, P.E. et al. (1999). Transparent stacked organic light emitting devices. I. Design principles and transparent compound electrodes. *J. Appl. Phys.* 86: 4067.

118 Gu, G., Parthasarathy, G., Tian, P. et al. (1999). Transparent stacked organic light emitting devices. II. Device performance and applications to displays. *J. Appl. Phys.* 86: 4076.

119 Kim, J.K., Hall, E., Sjölund, O., and Coldren, L.A. (1999). Epitaxially-stacked multiple-active-region 1.55 μm lasers for increased differential efficiency. *Appl. Phys. Lett.* 74: 3251.

120 Korshak, A.N., Gribnikov, Z.S., and Mitin, V.V. (1998). Tunnel-junction-connected distributed-feedback vertical-cavity surface-emitting laser. *Appl. Phys. Lett.* 73: 1475.

121 Brutting, W., Berleb, S., and Muckl, A.G. (2001). Device physics of organic light emitting diodes based on molecular materials. *Org. Electron.* 2: 1.

122 Do, Y.R., Kim, Y.-C., Song, Y.-W., and Lee, Y.-H. (2004). Enhanced light extraction efficiency from organic light emitting diodes by insertion of a two-dimensional photonic crystal structure. *J. Appl. Phys.* 96: 7629.

123 Gu, G. and Forrest, S.R. (1998). Design of flat-panel displays based on organic light-emitting devices. *IEEE J. Sel. Top. Quantum Electron.* 4: 83.

124 Moon, D.G., Pode, R.B., Lee, C.J., and Han, J.I. (2004). Transient electrophosphorescence in red top-emitting organic light-emitting devices. *Appl. Phys. Lett.* 85: 4771.

125 Jurow, M.J., Mayr, C., Schmidt, T.D. et al. (2016). Understanding and predicting the orientation of heteroleptic phosphors in organic light-emitting materials. *Nat. Mater.* 15: 85.

126 Lampe, T., Schmidt, T.D., Jurow, M.J. et al. (2016). Dependence of phosphorescent emitter orientation on deposition technique in doped organic films. *Chem. Mater.* 28: 712.

127 Mayr, C. and Brütting, W. (2015). Control of molecular dye orientation in organic luminescent films by the glass transition temperature of the host material. *Chem. Mater.* 27: 2759.

128 Kim, K.H., Liao, J.L., Lee, S.W. et al. (2016). Crystal organic light-emitting diodes with perfectly oriented non-doped Pt-based emitting layer. *Adv. Mater.* 28: 2526.

129 Nakamura, T., Tsutsumi, N., Juni, N., and Fujii, H. (2004). Improvement of coupling-out efficiency in organic electroluminescent devices by addition of a diffusive layer. *J. Appl. Phys.* 96: 6016.

130 Madigan, C.F., Lu, M.H., and Sturm, J.C. (2000). Improvement of output coupling efficiency of organic light-emitting diodes by backside substrate modification. *Appl. Phys. Lett.* 76: 1650.

131 Lin, L., Shia, T.K., and Chiu, C.J. (2000). Silicon-processed plastic micropyramids for brightness enhancement applications. *J. Micromech. Microeng.* 10: 395.

132 Yamasaki, T., Sumioka, K., and Tsutsui, T. (2000). Organic light-emitting device with an ordered monolayer of silica microspheres as a scattering medium. *Appl. Phys. Lett.* 76: 1243.

133 Shiang, J.J. and Duggal, A.R. (2004). Application of radiative transport theory to light extraction from organic light emitting diodes. *J. Appl. Phys.* 95: 2880.

134 Koh, T.W., Spechler, J.A., Lee, K.M. et al. (2015). Enhanced outcoupling in organic light-emitting diodes via a high- index contrast scattering layer. *ACS Photonics* 2: 1366.

135 Lin, H.Y., Chen, K.Y., Ho, Y.H. et al. (2010). Luminance and image quality analysis of an organic electroluminescent panel with a patterned microlens array attachment. *J. Opt.* 12: 085502.

136 Gu, G., Garbuzov, D.Z., Burrows, P.E. et al. (1997). High-external-quantum-efficiency organic light-emitting devices. *Opt. Lett.* 22: 396.

137 Matterson, B.J., Lupton, J.M., Safonov, A.F. et al. (2001). Increased efficiency and controlled light output from a microstructured light-emitting diode. *Adv. Mater.* 13: 123.
138 Peng, H.J., Ho, Y.L., Yu, X.J., and Kwok, H.S. (2004). Enhanced coupling of light from organic light emitting diodes using nanoporous films. *J. Appl. Phys.* 96: 1649.
139 Lee, Y.-J., Kim, S.-H., Huh, J. et al. (2003). A high-extraction-efficiency nanopatterned organic light-emitting diode. *Appl. Phys. Lett.* 82: 3779.
140 Hsu, C.M., Lin, B.T., Zeng, Y.X. et al. (2014). Light extraction enhancement of organic light-emitting diodes using aluminum zinc oxide embedded anodes. *Opt. Express* 22: A1695.
141 Liang, H., Luo, Z., Zhu, R. et al. (2016). High efficiency quantum dot and organic LEDs with a back-cavity and a high index substrate. *J. Phys. D: Appl. Phys.* 49: 145103.
142 Shin, H., Lee, J.H., Moon, C.K. et al. (2016). Sky-blue phosphorescent OLEDs with 34.1% external quantum efficiency using a low refractive index electron transporting layer. *Adv. Mater.* 28: 4920.
143 Wang, Z.B., Helander, M.G., Qiu, J. et al. (2011). Unlocking the full potential of organic light-emitting diodes on flexible plastic. *Nat. Photonics* 5: 753.
144 Tsutsui, T., Yahiro, M., Yokogawa, H. et al. (2001). Doubling coupling-out efficiency in organic light-emitting devices using a thin silica aerogel layer. *Adv. Mater.* 13: 1149.
145 Qu, Y., Slootsky, M., and Forrest, S.R. (2015). Enhanced light extraction from organic light-emitting devices using a sub-anode grid. *Nat. Photonics* 9: 758.
146 Reineke, S., Lindner, F., Schwartz, G. et al. (2009). White organic light-emitting diodes with fluorescent tube efficiency. *Nature* 459: 234.
147 Youn, W., Lee, J., Xu, M. et al. (2015). Corrugated sapphire substrates for organic light-emitting diode light extraction. *ACS Appl. Mater. Interfaces* 7: 8974.
148 Koo, W.H., Jeong, S.M., Araoka, F. et al. (2010). Light extraction from organic light-emitting diodes enhanced by spontaneously formed buckles. *Nat. Photonics* 4: 222.
149 Kim, J.B., Lee, J.H., Moon, C.K. et al. (2013). Highly enhanced light extraction from surface Plasmonic loss minimized organic light-emitting diodes. *Adv. Mater.* 25: 3571.
150 Jou, J.H., Chou, Y.C., Shen, S.M. et al. (2011). High-efficiency, very-high color rendering white organic light-emitting diode with a high triplet interlayer. *J. Mater. Chem.* 21: 18523.
151 Wu, Z. and Ma, D. (2016). Recent advances in white organic light-emitting diodes. *Mater. Sci. Eng., R* 107: 1.
152 Wang, Q., Ding, J.Q., Ma, D.G. et al. (2009). Harvesting Excitons via two parallel channels for efficient white organic LEDs with nearly 100% internal quantum efficiency: fabrication and emission-mechanism analysis. *Adv. Funct. Mater.* 19: 84.
153 Hsiao, C.H., Lan, Y.H., Lee, P.Y. et al. (2011). White organic light-emitting devices with ultra-high color stability over wide luminance range. *Org. Electron.* 12: 547.
154 Lan, Y.H., Hsiao, C.H., Lee, P.Y. et al. (2011). Dopant effects in phosphorescent white organic light-emitting device with double-emitting layer. *Org. Electron.* 12: 756.
155 Spindler, J., Kondakova, M., Boroson, M. et al. (2016). High brightness OLED lighting. *SID 16 Dig.*: 294.
156 Sun, Y., Giebink, N.C., Kanno, H. et al. (2006). Management of singlet and triplet excitons for efficient white organic light-emitting devices. *Nature* 440: 908.
157 Dai, X., Deng, Y., Peng, X., and Jin, Y. (2017). Quantum-dot light-emitting diodes for large-area displays: towards the Dawn of commercialization. *Adv. Mater.* 29: 1607022.
158 Shirasaki, Y., Supran, G.J., Bawendi, M.G., and Bulović, V. (2013). Emergence of colloidal quantum-dot light-emitting technologies. *Nat. Photonics* 7: 13.
159 Brovelli, S., Schaller, R.D., Crooker, S.A. et al. (2011). Nano-engineered electron–hole exchange interaction controls exciton dynamics in core–shell semiconductor nanocrystals. *Nat. Commun.* 2: 280.

160 Dabbousi, B.O., Rodriguez-Viejo, J., Mikulec, F.V. et al. (1997). (CdSe)ZnS Core–Shell quantum dots: synthesis and characterization of a size series of highly luminescent Nanocrystallites. *J. Phys. Chem. B* 101: 9463.
161 Yang, Y., Zheng, Y., Cao, W. et al. (2015). High-efficiency light-emitting devices based on quantum dots with tailored nanostructures. *Nat. Photonics* 9: 259.
162 Lim, J., Park, M., Bae, W.K. et al. (2013). Highly efficient cadmium-free quantum dot light-emitting diodes enabled by the direct formation of Excitons within InP@ZnSeS quantum dots. *ACS Nano* 7: 9019.
163 Dai, X., Zhang, Z., Jin, Y. et al. (2014). Solution-processed, high-performance light-emitting diodes based on quantum dots. *Nature* 515: 96.
164 Yamada, Y., Inoue, H., Mitsumori, S. et al. (2016). Achievement of blue phosphorescent organic light-emitting diode with high efficiency, low driving voltage, and Long lifetime by Exciplex–triplet energy transfer technology. *SID 16 Dig.* 47: 711.

7

Reflective Displays

7.1 INTRODUCTION

In contrast to transmissive and emissive display technologies, reflective displays show their images, just like paper, without the need for an internal light source. Reflective display technologies possess the attractive features of causing little eye strain, low power consumption, and superior optical contrast under sufficiently bright ambient light. They are attractive for portable reading applications and outdoor usage. There is a large variety of reflective display technologies. To display their images, some reflective displays need to be constantly addressed, referred as continuous refresh type, while some belong to the class of bistable devices. Bistable displays are capable of retaining the image last shown without any power consumption, and energy is only needed during the switching operation. In addition, some have a video rate switching capability, while some are more suitable for displaying still images. In this chapter, we will focus on discussing the basic working principles of the reflective display technologies based on electrophoresis (Section 7.2), reflective liquid crystal (Section 7.3), optical interference (Section 7.4), and electrowetting (Section 7.5), and the differences between them (Section 7.6).

7.2 ELECTROPHORETIC DISPLAYS

Electrophoretic images are created, based on the migration of charged colloidal particles within a dielectric fluid under the influence of an externally applied electric field. The particles are usually density matched closely to the dielectric fluid; therefore, the images are persistent after the removal of the voltage. In early electrophoretic displays, the film consisted of a single type of charged particle suspended in a dyed fluid sandwiched between the front and back electrode surfaces [1], as shown in Figure 7.1. When an external voltage, V, is applied to an electrode segment, depending on the charge polarity of the particles, they can be driven to the front or back surface to show the color of either the pigment or the dye. In this vertical configuration, the time, t, required for particles to move across the cell gap, h, can be approximated by [2]

$$t \approx \frac{h}{v} = \frac{h}{\mu E} = \frac{h^2}{\mu V}$$

where v, μ, and E denote the velocity of the particle relative to the surrounding fluid, electrophoretic mobility and the applied electrical field, respectively. The electrophoretic mobility is defined as the proportionality constant between an applied electrical field and the resulting velocity of the charged particle, and can be determined by [1]

$$\mu = \frac{\zeta\varepsilon}{6\pi\eta} = \frac{q\varepsilon}{12\pi r\eta}$$

where ε and η are the dielectric constant and viscosity of the surrounding fluid. ζ is the zeta potential of the suspension, which is proportional to the charge per particle, q, and inversely proportional to the particle

Introduction to Flat Panel Display, Second Edition. Jiun-Haw Lee, I-Chun Cheng, Hong Hua, and Shin-Tson Wu.

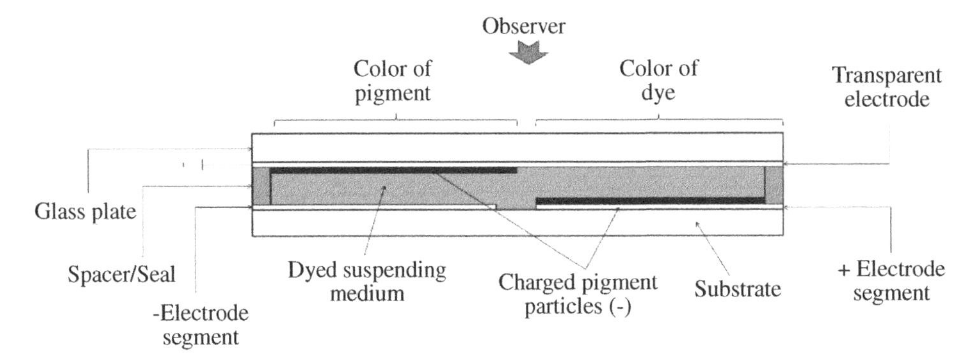

Figure 7.1 Schematic of a vertical electrophoretic display film containing negatively charged particles in a dyed fluid.

radius, r. Within the constraints of a fixed cell gap and applied voltage, fast switching can be achieved by using particles with a high charge and a fluid with low viscosity.

The cell of the Microcup® electrophoretic displays developed by SiPix Imaging, Inc., which was acquired by Eink Corporation in 2012, is formed by an embossed polymer rib pattern, as shown in Figure 7.2a, filled with a particle-dye electrophoretic fluid. It can be fabricated using a roll-to-roll process, as shown in

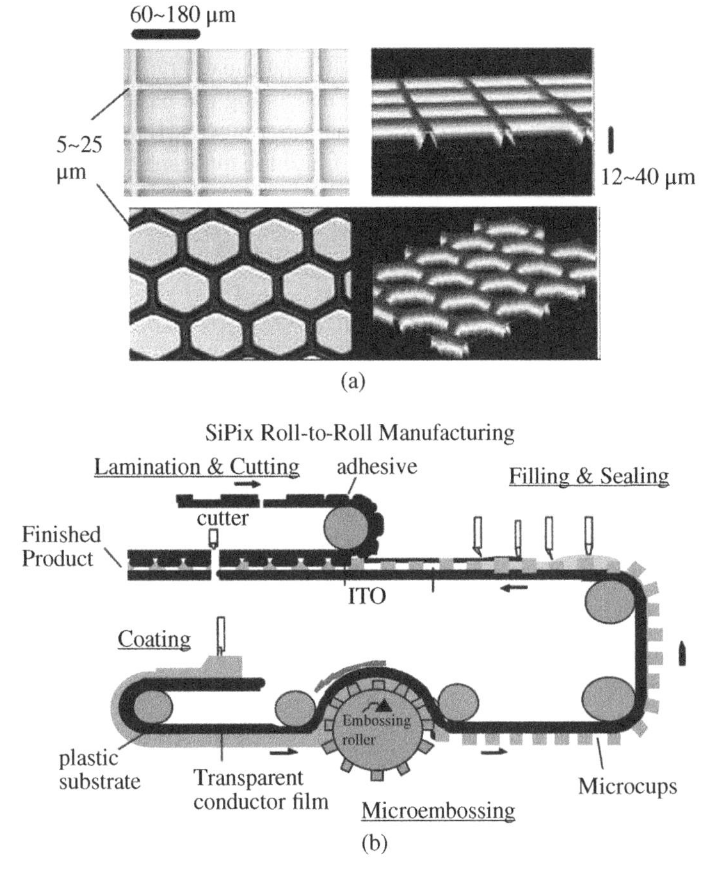

Figure 7.2 (a) Surface profiles of some typical Microcup arrays. (b) Schematic process flow of the SiPix roll-to-roll manufacturing process [3].

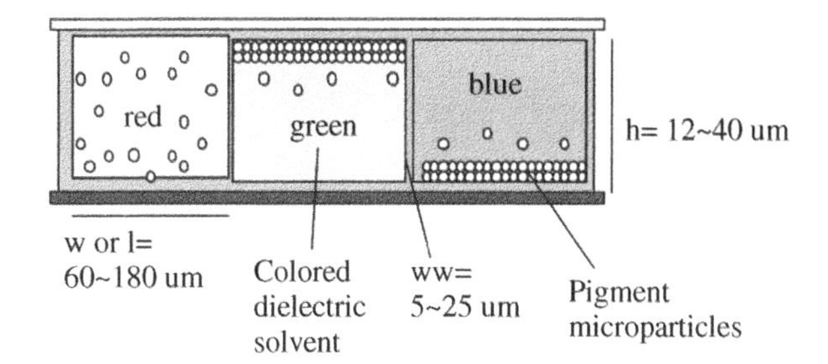

Figure 7.3 Schematic cross-section of a color Microcup EPD. Each Microcup is isolated and seamlessly top-sealed [3].

Figure 7.2b. The process comprises steps of (i) coating a radiation curable resin composition onto a film carrying conductive electrodes, (ii) embossing and hardening the resin composition, (iii) filling the Microcup with an electrophoretic fluid, (iv) seamlessly top-sealing the filled Microcup, and (v) laminating the top-sealed Microcups with a release liner or a second electrode-bearing film [3]. The white pigment particles are formed of polymer encapsulated TiO_2, and are dispersed in a dyed fluid. The rib structure serves to restrict the lateral migration of the particles within a very short length scale so that a non-uniform pigment distribution on the front surface is prevented. A gray-scale response can be achieved in Microcup electrophoretic displays by modulating the pulse amplitude, pulse width or pulse count of the applied voltage. A colored display can be implemented in a side-by-side architecture using subpixels with different dyed fluids, as shown in Figure 7.3.

Monochromatic electrophoretic display cells can be also realized using two types of charged particles with opposite polarities suspended in a clear fluid. The E-ink Corporation use a dual-particle electrophoretic fluid formulation in its microencapsulated black/white switching electrophoretic film to improve the image contrast, as shown in Figure 7.4. Positively charged TiO_2 and negatively charged carbon black are often chosen as the white and black particles, respectively. To obtain the white state, a positive voltage is applied to the bottom electrode, the black particles are attracted to the bottom and the white particles gather at the top. In the opposite case, a negatively-biased bottom electrode pushes the black particles to the top electrode, resulting in a black state. Gray tones in the dual-particle electrophoretic display can be achieved via states of intermediate switching between the black and white end states.

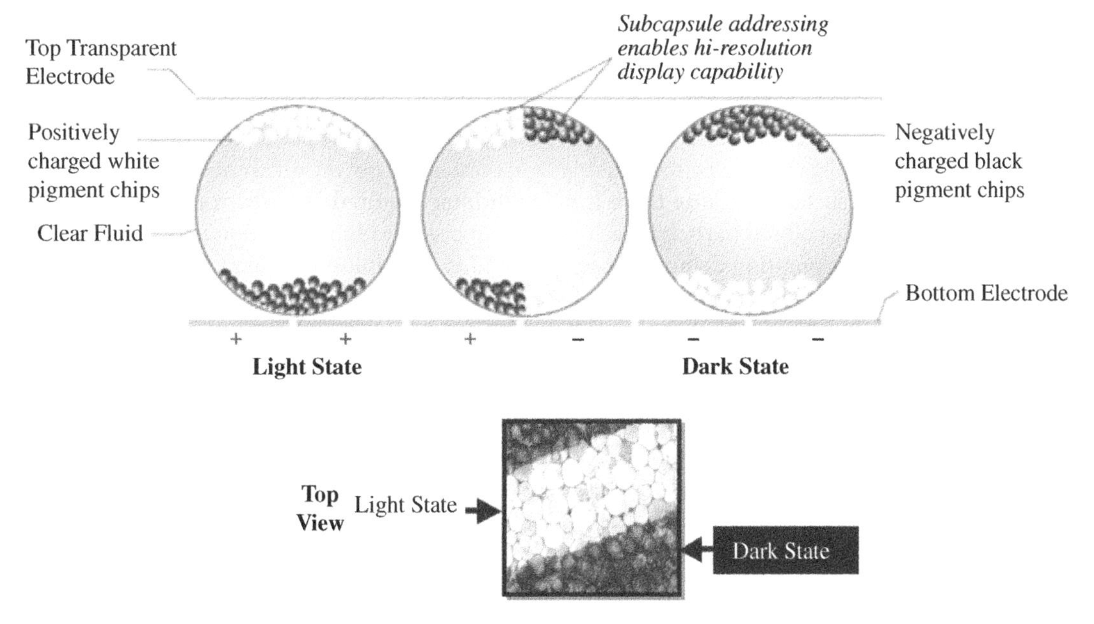

Figure 7.4 Schematic illustration of microcapsule electrophoretic image display [4].

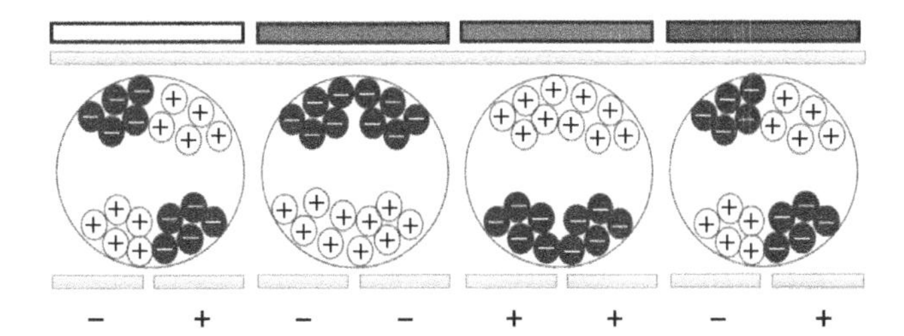

Figure 7.5 Schematic diagram of a full-color microencapsulated electrophoretic display using a black and white panel with a RGBW color filter.

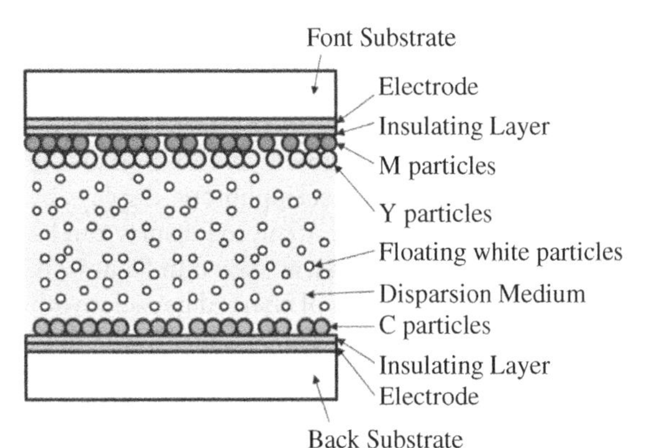

Figure 7.6 Cross section of an electrophoretic display based on "independently movable colored particles" technology [6].

The implementation of a color electrophoretic display is often based on the combination of a black and white panel with a RGBW color filter, as shown in Figure 7.5 [5]. A white subpixel is used to achieve a brighter white state. To improve the image quality of the colored states, Fuji Xerox proposed a color electrophoretic display based on substractive color mixing. Three primary colored particles, cyan, magenta, and yellow, which have different threshold electric fields, were used, as shown in Figure 7.6 [6]. The particles are designed to cohere on a substrate until a field larger than the threshold is applied. Colloidal white particles are mixed in the dispersion to hide the colored particles when they are moved toward the back substrate. The white particles are small enough to remain suspended under gravity and are designed to be largely insensitive to the electric field. The colored particles can move through the suspended white particles by applying a proper electric field. The color as seen from the front side of the display, results from mixing light scattered from the different colored particles which have been moved to that surface. Because there is no color filter required, the electrophoretic display based on "independently movable colored particles" technology can provide high saturation and bright white state. More recently, the E-ink Corporation demonstrated a color filter-free full-color electrophoretic display [7, 8]. The electrophoretic fluid contains four different types of pigment particles: yellow, cyan, magenta, and white, as shown in Figure 7.7. In addition to their color, they also differ in size, polarity and the strength of the charge they carry; therefore, they respond differently to the applied voltage. The switching combinations of these moving particles can display up to 32000 colors.

Image patterns can also be created based on the lateral movement of the charged particles in and out of the viewing area. This type of display is called an in-plane electrophoretic display. Various electrode layouts have been developed. Basically, two electrodes, a collector electrode and a large electrode covering the viewing area, are sufficient for the operation, as shown in Figure 7.8a. In one of the Philips's design, four electrodes, a collector, a gate and two view electrodes are used, as shown in Figure 7.8b [9]. The gate electrode is located between the viewing area and the collector area. When a repulsive voltage is applied to the gate electrode, no particles are allowed to move in or out of the viewing area. Without voltage applied to the gate electrode, the particles can move across freely, in a direction depending on the potential difference between the collector

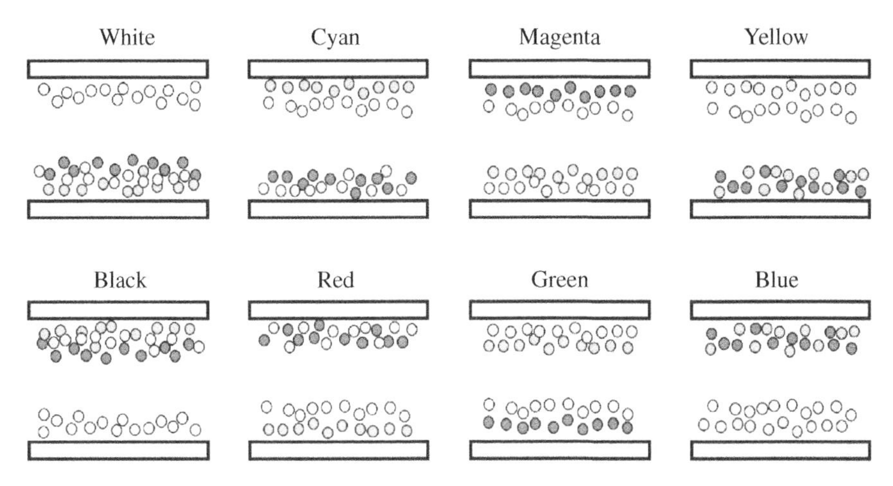

Figure 7.7 Changes in electrical voltage move different combinations of pigment particles to the surface of the color filter-free electrophoretic display (E-ink Advanced Color ePaper).

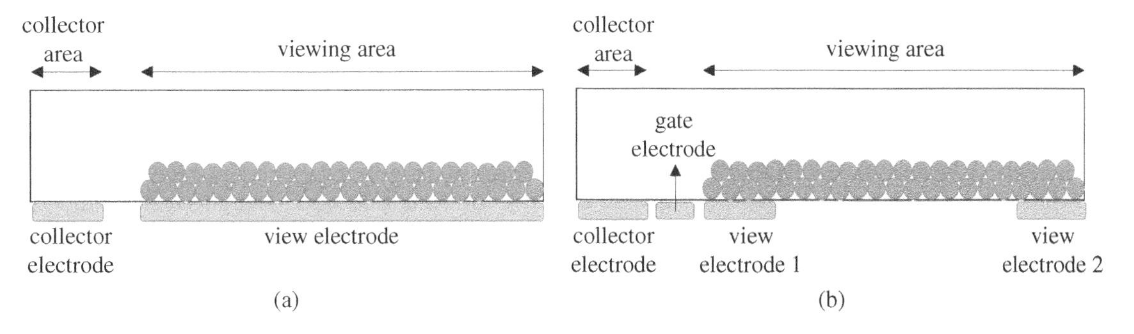

Figure 7.8 Schematic diagrams of in-plane electrophoretic pixels with (a) two-electrode and (b) four-electrode configurations.

and viewing area. A voltage gradient can be created between the two view electrodes to assist the particle spreading in the viewing area.

7.3 REFLECTIVE LIQUID CRYSTAL DISPLAYS

Reflective liquid crystal displays can be divided into two sub-groups: those which use polarizers and those which are polarizer-free. Polarizer-based liquid crystal displays may use one or two polarizers, a liquid crystal material and a reflective substrate at the rear. In a manner similar to transmissive liquid crystal displays, the liquid crystal material modulates the polarization of the incident light. Therefore, most polarizer-based liquid crystal displays need to be refreshed constantly. Another inevitable disadvantage of polarizer-based liquid crystal displays is their low reflectance because the polarizer may absorb more than 50% of the incident light intensity. In addition, it is difficult to fabricate polarizer-based liquid crystal displays on flexible plastic substrates since many types of plastic film have a non-uniform birefringence, particularly under mechanical deformation [10]. Nevertheless, some polarizer-based liquid crystal displays can offer features of low-power consumption, such as the Zenithal Bistable Displays (ZBD®). The ZBD has been used for signage in the retail sector. It has a similar structure to the twisted-nematic liquid crystal display (TN-LCD). The bistability is achieved by patterning one of the alignment surfaces of ZBD into a relief grating, which offers latching with a defined threshold. Depending on the opposing surface to the grating, several types of device geometries

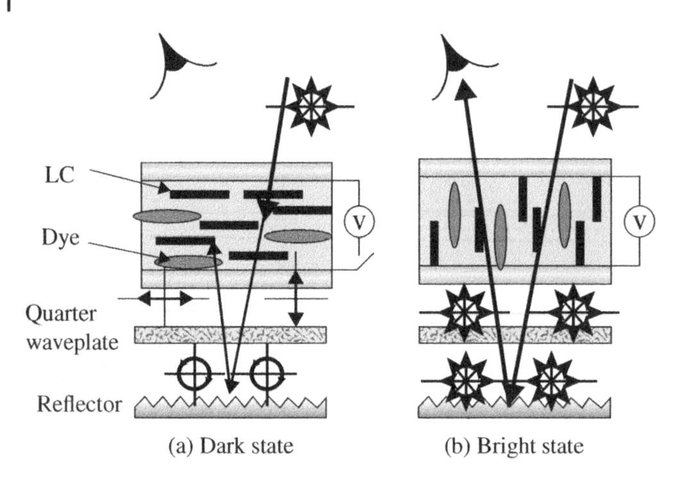

Figure 7.9 Schematic diagram of Cole–Kashnow reflective display [10].

are available. In twisted-nematic-type, the grating is used opposite the rubbed polymer alignment to give a 90° twist at the grating surface (low-tilt state) or a hybrid aligned state without twist at the grating surface (high-tilt state). When a reflective rear polarizer is used, it can be operated as a reflective display [11, 12].

In a polarizer-free liquid crystal display, the liquid crystal material either absorbs, scatters or reflects the incident light. In this section, three types of polarizer-free liquid crystal display will be discussed, including guest–host liquid crystal displays (based on absorption), polymer-dispersed liquid crystal (PDLC) displays (based on scattering), and cholesteric liquid crystal displays (based on reflection).

In guest–host liquid crystals, the liquid crystals are blended with dichroic dyes. The guest dichroic dye molecules are usually elongated. The average direction of the long molecular axis of the dye molecules aligns parallel to that of the host liquid crystals. For dyes which have a positive dichroism, incident light is absorbed when its polarization is parallel to the long molecular axes of the dye molecules. One type of polarizer-free guest–host liquid crystal display is the Cole–Kahnow display [10, 13], as shown in Figure 7.9. This display uses a liquid crystal of positive dielectric anisotropy in a planar alignment. A quarter-wave plate is placed in between the nematic liquid crystal layer and the reflector at 45° to the liquid crystal alignment. In the unswitched state, the (initially unpolarized) incident light is linearly polarized by absorption in the aligned dyed liquid crystal. The component of light polarized parallel to the long molecular axis of liquid crystal is absorbed by the dye, while the component with polarization perpendicular to the long molecular axis of liquid crystal reaches the quarter-wave plate and is converted into circularly polarized light. On reflection, its sense of circular polarization is reversed. When it passes through the quarter-wave plate again, the light is converted into linearly polarized light with its polarization rotated by 90°, i.e. parallel to the long molecular axis of the liquid crystal. Thus with the aid of the quarter wave plate and reflector, the dyed liquid crystal can absorb all the incident light, 50% on its first pass through the liquid crystal layer, and the remainder on its second pass after the polarization axis has been changed. When a voltage is applied to the liquid crystal, the liquid crystal and the guest dichroic dye are oriented perpendicular to the plane of the cell in response to the electrical field, and the dye transition axis is now perpendicular to any polarization direction of normally incident light. The absorption of the incident light by the dichroic dyes is substantially reduced, resulting in a bright state of the display.

Another polarizer-free guest–host liquid crystal display is the White–Taylor display [10, 14]. Its schematic is shown in Figure 7.10. In a White–Taylor display, a cholesteric liquid crystal with a positive dielectric anisotropy is used. In the unswitched condition, the liquid crystal and dye have a helical structure with the helical axis perpendicular to the cell and the liquid crystal long axis lies in a plane parallel to the substrate. The value of $P\Delta n/\lambda$ in the device is chosen to be relatively small, so light guiding does not occur. Under these conditions, incident light of any polarization is absorbed by the dye, resulting in a dark state. When a voltage

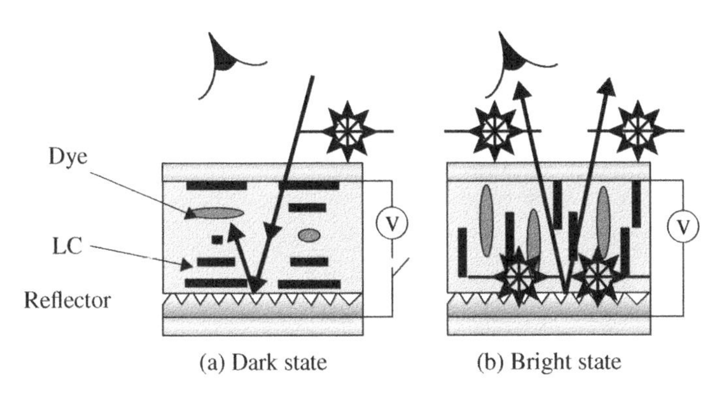

Figure 7.10 Schematic diagram of the White–Taylor reflective display [10].

is applied, the helical structure is untwisted and the liquid crystal undergoes a phase transition into a nematic state with homeotropic alignment in response to the vertical electrical field. In this state, the polarization of incident light is perpendicular to the long molecular axis of the dye molecules. Therefore, the light is not absorbed, leading to a bright state. Polarizer-free guest–host liquid crystal displays can provide full color performance using subtractive color mixing of yellow, magenta and cyan dye molecules in separately switched, stacked LC cells.

A PDLC consists of micron-sized liquid crystal droplets dispersed in a polymer binder. The droplet size, ranging from sub to tens of μm, and distribution can be controlled by phase separation techniques, such as thermally induced phase separation, solvent induced phase separation, and polymerization induced phase separation [15]. Among various techniques, polymerization induced phase separation is frequently used. In this technique, polymerization is carried out on a mixture of liquid crystal and a prepolymer containing initiator by either heat or ultraviolet radiation. Upon polymerization, the liquid crystal separates from the polymer matrix and forms droplets. In PDLC displays, reflection of light is produced by light scattering. The refractive index of the polymer is chosen to match the ordinary refractive index of the liquid crystal but is smaller than its extraordinary refractive index [10]. If no electric field is applied to the layer, the orientation of the liquid crystal is random and incident light is scattered due to the refractive-index mismatch between the dispersed liquid crystal and the polymer binder, as shown in Figure 7.11a. Light scattered in a backward direction generates a white appearance. When a voltage is applied, the liquid crystal is aligned parallel to the vertical electrical field and normally incident light sees a matched refractive index between the liquid crystal and the polymer binder. Therefore, light can propagate through to reach an absorption layer located underneath the PDLC and the display shows the appearance of the absorption layer, as shown in Figure 7.11b. The driving voltage ranges from a few to tens of V. The typical switching speed is 1–10 ms.

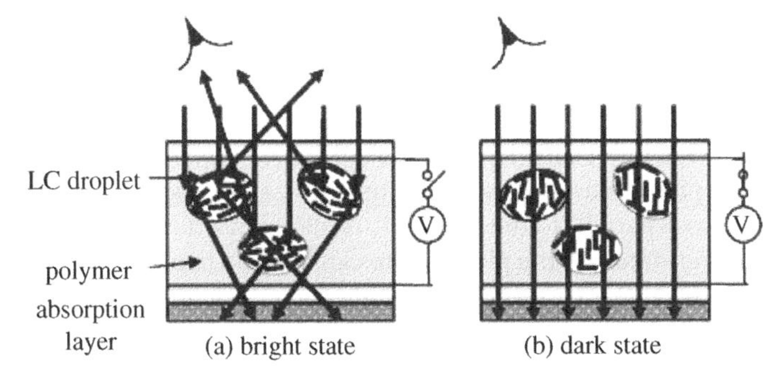

Figure 7.11 Schematic diagram of a polymer-dispersed liquid crystal display [10].

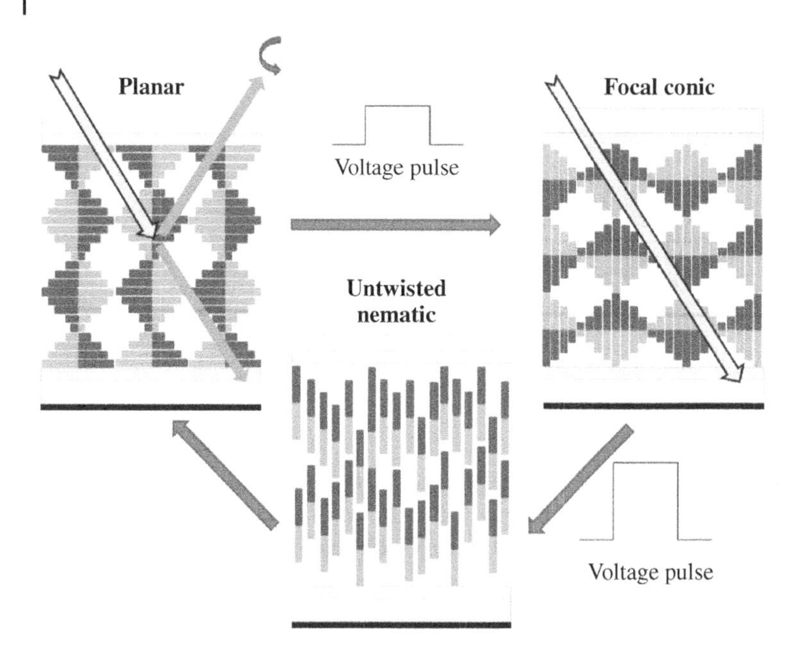

Figure 7.12 Schematic diagram of a bistable cholesteric liquid crystal display.

Cholesteric liquid crystals have the same degree of order as nematic liquid crystals. As in the nematic phase, the molecules are elongated spontaneously and order with their long axes parallel with one another. In the cholesteric phase, however, the director twists along a helical axis, as shown in Figure 7.12. The distance over which the direction of the molecules rotates through 2π radians is called the pitch, denoted by P. When the cholesteric liquid crystal is in a planar alignment state, the helical axis is perpendicular to the substrate. Owing to the periodic structure, it can strongly reflect a component of circular polarized light with the same handedness as the liquid crystal helical structure by a mechanism similar to Bragg reflection [10, 15]. The reflection band is centered at a wavelength of $\lambda_0 = 1/2[n_e + n_o]P \equiv \overline{n}P$ and has a bandwidth of $\Delta\lambda = (n_e - n_o)P \equiv \Delta nP$, where n_e and n_o are the extraordinary and ordinary refractive indices of the liquid crystal, and $\overline{n}$ and Δn are the average refractive index and birefringence respectively. For polarized incident light with the same handedness as the helical structure, the reflectance at the center of reflection band can be expressed by [16]

$$R = \left[\frac{\exp(2\Delta n\pi h/\overline{n}P) - 1}{\exp(2\Delta n\pi h/\overline{n}P) + 1}\right]^2$$

where h denotes the cell thickness. Given an average refractive index of about 1.6, more than 99% peak reflectance can be achieved when the cell thickness is $h = 2P/\Delta n$. For an unpolarized incident light, 100% reflectance within the reflection band can be realized by stacking a left-handed cholesteric liquid crystal layer and a right-handed cholesteric liquid crystal layer together.

When an electrical field is applied across the cell, i.e. perpendicular to the substrate, the liquid crystal molecules tend to align parallel to the field. If the voltage is not sufficiently large, the liquid crystal is switched to the focal conic state where the helical structure is preserved within a small domain size of a few micrometers with the preponderant orientation of the helical axis parallel to the substrate, resulting in a non-reflective state known as the focal conic texture. The focal conic state and the planar state can both be made stable by coating the cell with carefully chosen alignment layers [10]. The resulting bistable cholesteric liquid crystal cell requires no holding voltage either in the reflective planar state or in the non-reflective focal conic state.

If an applied voltage is sufficiently large, the helix is fully unwound and the liquid crystal is switched to the nematic-like homeotropic state where the molecules are aligned parallel to the electrical field. Depending on how quickly the voltage is removed, the liquid crystal relaxes to either the planar state (fast removal) or the focal conic state (slow removal) [17].

Because both the planar and focal conic states are poly-domain structures, gray scale can be achieved by switching some domains to the planar state and the others to the coal conic state [10]. Full color displays can be produced by stacking cholesteric liquid crystal layers which reflect red, green, and blue light, respectively. Due to the long relaxation time from homeotropic state to the planar state, it is difficult to display dynamic images using only vertical electric field switching in cholesteric liquid crystal displays. By applying a large in-plane electrical field, fast switching of $\sim$ 5 ms has been demonstrated between the reflecting planar state and non-reflecting in-plane-field-induced states [18].

7.4 REFLECTIVE DISPLAY BASED ON OPTICAL INTERFERENCE (MIRASOL DISPLAY)

The interferometric modulator display, trademarked Mirasol® by Qualcomm®, is a type of reflective display which uses a micro-electro-mechanical system (MEMS) to modulate light. The device operates as an optical resonant cavity, similar to a Fabry–Perot etalon, formed between a moveable fully-reflective mirror, a self-supporting deformable reflective membrane, and a fixed partially-reflective mirror, a thin-film stack deposited on a transparent substrate which usually forms the "top," viewing side of the device [19]. The thickness of the optical cavity is set by the vertical position of the moveable mirror, which is controlled by the balance between the electrostatic force arising from the voltage applied between mirrors and the restoring force due to the drumhead structure of the moveable mirror [20]. When light is incident on the structure, it is reflected off both the top thin-film stack and the reflective membrane behind it. Depending on the thickness of the optical cavity formed by these two surfaces, light reflected from each layer may interfere constructively or destructively according to its wavelength, as shown in Figure 7.13a. As a result, certain wavelengths can be selected for high reflectivity in comparison to others. Different sub-pixels have different cavity thicknesses and reflect different colors. When the reflective membrane is in the "collapsed" state, the reflection peak lies in the UV region of the spectrum so all colors of visible light are absorbed and the display shows an apparent black state, as shown in the middle subpixel of Figure 7.13b.

Mirasol displays are bistable; therefore, no refresh is required until the image is changed. The pixels are capable of switching speeds of 7 ms, which can enable video frame rate displays. Both spatial and temporal dithering can be used to generate grayscale. Spatial dithering divides each color subpixel into several smaller independently addressable elements, which are not resolved by the eye in ordinary viewing conditions. By switching some of these elements "on" and others "off," a spatially averaged greyscale is obtained. This approach offers lower power consumption compared to temporal dithering. Temporal dithering works by splitting each display refresh cycle into multiple subdivisions. Greyscale response is achieved by rapidly switching pixels on and off so that the time average reflectivity provides the desired lightness. The rapid addressing increased power consumption, but lower manufacturing tolerances can be used because the pixels are larger, and a higher fill factor can be obtained. To improve the number of gray levels, a combination of temporal and spatial dithering can be used [22]. A color display is fabricated using a side-by-side array of RGB sub-pixels; therefore, its maximum reflectance is limited to 33%. Compared to color filters and polarizers, color generation via interference is much more efficient in its use of light. However, a challenge for this technology is the dependence of color and contrast on the illumination condition and viewing angle, which is affected by the angular dependence of the effective optical cavity thickness which causes a change in peak reflectance wavelength for light away

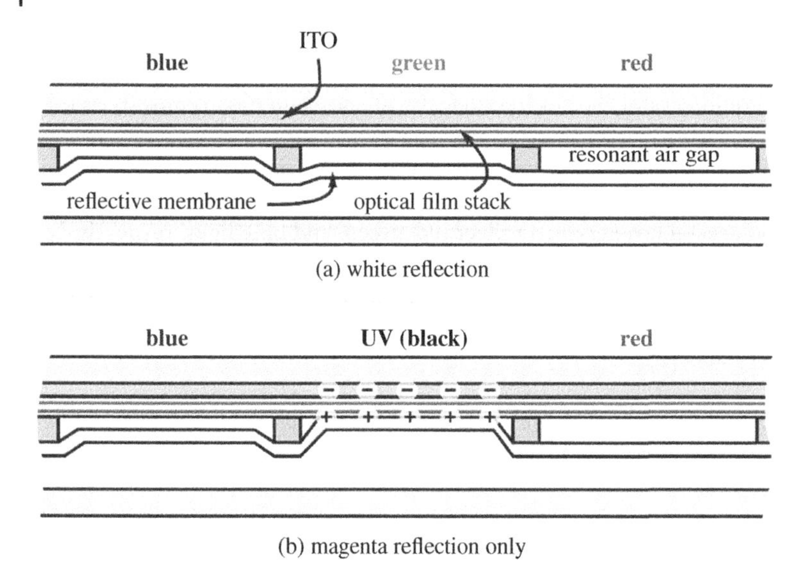

Figure 7.13 Interferometric modulator display structure showing light reflecting off the top thin-film stack and the bottom deformable reflective membrane to produce color [21].

from normal incidence [21]. The interferometric modulator display is mechanically robust, reliability over 12 billion cycles has been demonstrated. However, fabrication of flexible displays using this technology seems to be problematic because of the materials and processing route used which deliver the tight control of the gap in the MEMS structure which is required to ensure correct color performance.

A next generation of the Mirasol displays is called single mirror interferometric (SMI) display [23]. In the SMI display, the reflectance properties of individual pixels can tune continuously across the visible spectrum and also provide high contrast black and white states. The SMI architecture allows pure color states to overcome the 67% brightness loss, as imposed by the RGB sub-pixel structure. The SMI pixel is a three terminal device. The top electrode is fixed and semi-reflective, allowing partial transmission of the light. The middle electrode is movable and serves as a mirror. The bottom electrode is fixed and serves as an absorber. The SMI panel is driven using an indium gallium zinc oxide (IGZO) active matrix backplane, which enables a frame rate as high as 240 Hz [24].

7.5 ELECTROWETTING DISPLAY

Electrowetting is a microfluidic phenomenon where the wettability of a liquid on a solid surface changes as a voltage is applied between the liquid and the surface. A surface which is preferentially wetted by a dyed, insulating oil in the absence of an applied field, and displacement of the oil by a clear, conductive, polar liquid (typically not water in real devices) when a voltage is applied between the polar liquid and the electrode, as shown in Figure 7.14 [25]. The electrowetting display has a fast response of $\sim$ 1 ms, it is therefore capable of video-speed switching. A full color electrowetting display can be achieved by using side-by-side [25] or vertically stacked monochrome subpixels [26]. The electrowetting pixel is not inherently bistable: analog greyscale and color tones can be obtained by applying controlled intermediate voltage levels to the device. A bistable device can be obtained by a redesign of the pixel electrode structure to provide a single, common electrode on one substrate and two laterally separated electrodes on the other as shown in Figure 7.15. By applying an appropriate waveform to the common electrode and two in-plane control electrodes, the colored fluid can be moved from one bistable position to the other [27]. Based on this droplet driven technology, a flexible bistable electrowetting displays has been demonstrated on polyethylene terephthalate (PET) substrates [28].

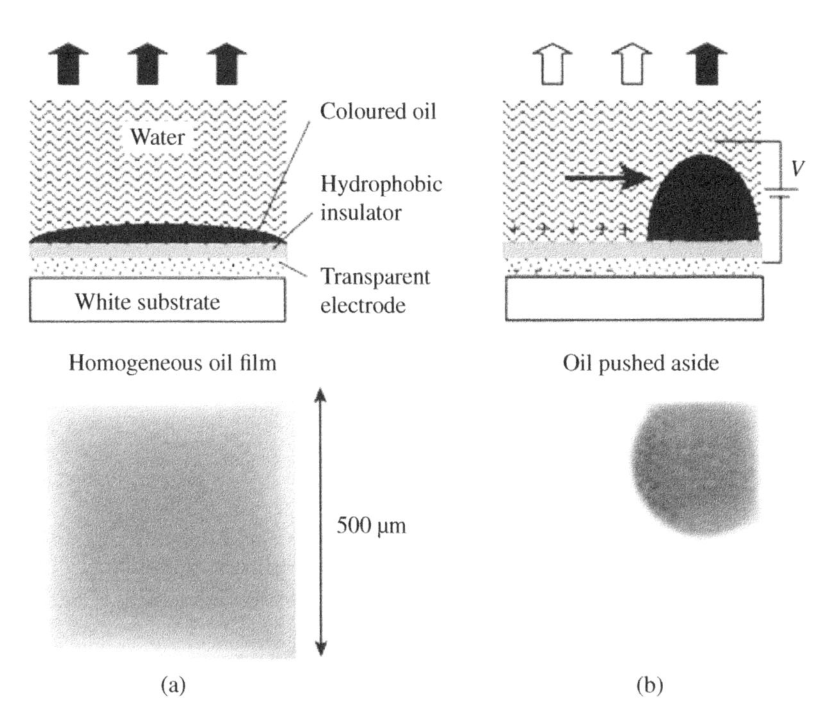

Figure 7.14 Electrowetting display principle. (a) A continuous film of dyed oil is covers the pixel area when no voltage is applied. (b) An applied voltage causes the polar liquid to wet the electrode, displacing the oil into a droplet which partially covers the surface [25].

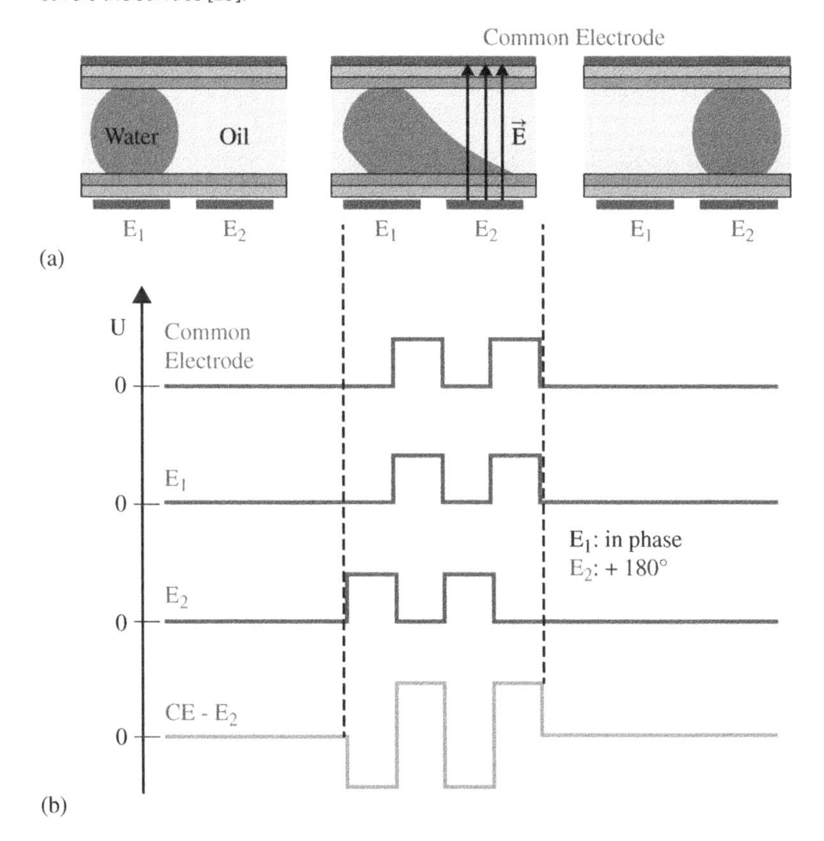

Figure 7.15 Droplet-driven electrowetting principle with common electrode CE and control electrodes E_1 and E_2. A droplet (a) is moved from one bistable position (left) to another one (right) by applying a waveform (b) [27].

7.6 COMPARISON OF DIFFERENT REFLECTIVE DISPLAY TECHNOLOGIES

A comparison of various monochrome reflective display technologies is summarized in Table 7.1 [10, 21, 29, 30]. Most of them can match the typical contrast ratio standard of 10 : 1 for printed images on paper, but their reflectance in the bright state is less than the typical value of 80% for a white paper. Regarding their power consumption, electrophoretic displays, cholesteric liquid crystal displays and interferometric modulator displays are bistable, so energy is only required during a switching operation. With inherent high switching speeds, interferometric modulator displays and electrowetting displays can easily support video frame rate. Electrophoretic displays, polarizer-free reflective liquid crystal displays, and electrowetting displays based on droplet driven technology can be implemented on flexible substrates, opening up their application opportunities as flexible or curved displays.

Homework Problems

7.1 In a chromatic electrophoretic display film based on the particle-dye formula, the image quality and response speed can be influenced by various parameters. Use the dye concentration and cell gap to explain the threefold reciprocity constraint between the quality of the bright state, the quality of dark state, and the response speed.

7.2 A right-handed circularly polarized light beam is incident on a right-handed cholesteric liquid crystal layer, which is in the planar state. If the reflection band is centered at 700 nm. Find (a) the bandwidth of the reflected light and (b) the cell thickness which can provide a peak reflectance of 99%. Assume the extraordinary and ordinary refractive indices of the liquid crystal are 1.7 and 1.5, respectively.

7.3 The images are formed based on the movement of colorants in both electrophoretic and electrowetting displays. However, they are quite different in many aspects, such as operation principles, image quality, and switching speed. Compare these two technologies.

Table 7.1 Comparison of various monochromic reflective display technologies.

	Electrophoretic	Polymer-dispersed liquid crystal	Cholesteric liquid crystal	Interferometric modulator display	Electrowetting
Contrast Ratio	15 : 1	10 : 1	8 : 1	15 : 1	15 : 1
Reflectivity for White or Color	40%	50%	40%	50%	60%
Lambertian	yes	no	no	no	Partial
Bistable	yes	no	yes	yes	no (yes with "droplet driven")
Voltage (V)	15	5	<4 V (lab) 25–40 V (product)	5–10	15–20
Switching speed (ms)	100's	100	300 (vertical field switch) 5 (in-plane field switch)	0.01's	10
Mechanical Flexibility	yes	yes	yes	no	yes

References

1 Dalisa, A.L. (1977). Electrophoretic display technology. *IEEE Trans. Electron Devices* 24: 827–834.

2 Amundson, K. (2005). Electrophoresis and elecophoretic imaging. In: *Flexible Flat Panel Displays* (ed. G.P. Crawford), 369–391. West Sussex, England: Wiley.

3 Liang, R.C., Hou, J., Zang, M. et al. (2003). Microcup® displays: electronic paper by roll-to-roll manufacturing processes. *J. SID* 11: 621–628.

4 Johnson, M.T., Zhou, G., Zehner, R. et al. (2006). High-quality images on electrophoretic displays. *J. SID* 14: 175–180.

5 Lu, Y.-H. and Tien, C.-H. (2013). Principle component analysis of multi-pigment scenario in full-color electrophoretic display. *J. Disp. Technol.* 9: 807–813.

6 Hiji, N., Machida, Y., Yamamoto, Y. et al. (2012). Novel color electrophoretic E-paper using independently movable colored particles. *SID Symp. Dig. Tech. Pap.* 43: 85–87.

7 L. Ulanoff, The future of ultra-low-powered displays is finally in living color, in http://mashable.com/2016/05/24/color-e/#ALQXu2qGXuqG, 2016.

8 D. Haynes, E Ink starting to show new filter-free full color displays aimed at digital signage market in, http://www.eink.com/press_releases/e_ink_announces_advanced_color_epaper_05-24-2016.html.

9 Lenssen, K.-M.H., Baesjou, P.J., Budzelaar, F.P.M. et al. (2009). Novel concept for full-color electronic paper. *J. SID* 17 (4): 383–388.

10 Yang, D.-K. (2008). Review of operating principle and performance of polarizer-free reflective liquid-crystal displays. *J. SID* 16: 117–124.

11 Jones, J.C. (2008). The zenithal bistable display: from concept to consumer. *J. Soc. Inf. Disp.* 16: 143–154.

12 E. T. Wood, G. P. Bryan-Brown, P. Brett, A. Graham, J. C. Jones, and J. R. Hughes, Zenithal bistable device suitable for portable applications, 2000 SID Symposium Digest of Technical Papers, (2000) Paper 11.2.

13 Cole, H.S. and Kashnow, R.A. (1977). A new reflective dichroic liquid-crystal display device. *Appl. Phys. Lett.* 30: 619–621.

14 White, D.L. and Taylor, G.N. (1974). New absorptive mode reflective liquid-crystal display device. *J. Appl. Phys.* 45: 4718–4723.

15 Crawford, G.P. (2005). Encapsulated liquid crystal materials for flexible display applications. In: *Flexible Flat Panel Displays* (ed. G.P. Crawford), 313–330. Wiley.

16 Yang, D.-K. and Wu, S.-T. (2014). *Fundamentals of Liquid Crystal Devices*, 2e. Wiley.

17 Yang, D.-K., Doane, J.W., Yaniv, Z., and Glasser, J. (1994). Cholesteric reflective display: drive scheme and contrast. *Appl. Phys. Lett.* 64: 1905–1907.

18 Kim, K.-H., Yu, B.-H., Choi, S.-W. et al. (2012). Dual mode switching of cholesteric liquid crystal device with three-terminal electrode structure. *Opt. Express* 20: 24376–24381.

19 Miles, M.W. (1997). A new reflective FPD technology using interferometric modulation. *J. SID* 5: 379–382.

20 R. A. Martin, A. Lewis, M. Mignard, N. Chuei, R. van Lier, A. Govil, M. Todorovich, K. Aflatooni, B. Gally, C. Chui, "Driving mirasol® displays: addressing methods and control electronics," 2011 SID Symposium Digest of Technical Papers, (2011) Paper 26.1.

21 Heikenfeld, J., Drzaic, P., Yeo, J.-S., and Koch, T. (2011). Review paper: a critical review of the present and future prospects for electronic paper. *J. SID* 19: 129–156.

22 Interferometric Modulator Technology Overview, in, https://www.qualcomm.com/media/documents/files/mirasol-imod-tech-overview.pdf, 2009.

23 J. Hong, E. Chan, T. Chang, R. Fung, C. Kim, J. Ma, Y. Pan, B. Wen, I. Reines, and C. Lee, Single mirror interferometric display – a new paradigm for reflective display technologies, 2014 SID Symposium Digest of Technical Papers, (2014) Paper 54.4L.

24 T. Chang, E. Chan, J. Hong, C. Kim, J. Ma, Y. Pan, R. van Lier, B. Wen, L. Zhou, P. Mulabagal, Single mirror IMOD display for practical wearable device, 2015 SID Symposium Digest of Technical Paper, (2015) Paper 3.3.

25 Hayes, R.A. and Feenstra, B.J. (2003). Video-speed electronic paper based on electrowetting. *Nature* 425: 383–385.

26 You, H. and Steckl, A.J. (2010). Three-color electrowetting display device for electronic paper. *Appl. Phys. Lett.* 97: 023514.

27 Blnkenbach, K., Schmoll, A., Bitman, A. et al. (2008). Novel highly reflective and bistable electrowetting displays. *J. SID* 16: 237–244.

28 Charipar, K.M., Charipar, N.A., Bellemare, J.V. et al. (2015). Electrowetting displays utilizing bistable, multi-color pixels via laser processing. *J. Disp. Technol.* 11: 175–182.

29 Bai, P.F., Hayes, R.A., Jin, M.L. et al. (2014). Review of paper-like display technologies. *Prog. Electromagnet. Res.* 147: 95–116.

30 Y. Itoh, K. Minoura, Y. Asaoka, I. Ihara, E. Satoh, S. Fujiwara, "Super reflective color LCD with PDLC Technology, 2007 SID Symposium Digest of Technical Papers, (2007) Paper 40.5.

8

Fundamentals of Head-Mounted Displays for Virtual and Augmented Reality

Head-mounted display (HMD) technology, often also known as head-worn display (HWD) or near-eye display (NED), refers to a display system that is typically attached in close proximity to the eyes and requires an optical system to couple light from a miniature image source into a user's eye [1–3]. Within military applications the acronym HMD has been used to refer to helmet-mounted displays where the display is attached to a military helmet. Although HMD technology has taken many decades of development, the high demand and rapid development of HMD solutions is a quite recent event which has been widely stimulated by several key modern technologies, including the ever-increasing bandwidth and accessibility of wireless networks, the explosion of digital information, the miniaturization of electronics and sensing technologies, and the ever-growing capability and affordability of computing. The rapidly growing interest in virtual reality (VR) and augmented reality (AR) technologies in recent years have further boosted the development and investment in HMD technologies.

HMD technology along with the emerging VR and AR technologies has been seen as a key enabler for a broad range of applications in many disciplines, spanning the fields of simulation, scientific visualization, medicine, engineering, education and training, wearable computing, defense, and entertainment [4, 5]. For instance, a lightweight optical see-through head-mounted display (OST-HMD), which enables optical superposition of two-dimensional (2D) or three-dimensional (3D) digital information onto a user's direct view of the physical world and maintains see-through vision to the real world, is viewed as a transformative technology in the digital age, enabling new ways of accessing and perceiving digital information essential to our daily life.

This chapter will focus on the fundamentals of HMD technology along with a review of recent developments. More specifically, the chapter is divided into five main sections. Section 8.1 provides a brief introduction which introduces the optical principles, defines key terminologies, and overviews historical developments. Section 8.2 provides a concise summary of the human visual system (HVS) parameters relevant to the design of an HMD system. Section 8.3 focuses on the fundamentals of HMD technology, including paraxial optical specifications, miniature display sources, optical principles and architectures, and optical combiner technologies. Section 8.4 reviews optical design methods and optical performance specifications. Section 8.5 reviews a few emerging HMD technologies with advanced capabilities, such as eyetracking, addressable focus cues, occlusion capability, high dynamic range (HDR), and light field rendering.

8.1 INTRODUCTION

Figure 8.1 shows a conceptual illustration of a monocular HMD. Per eye, an HMD is minimally composed of a modulated light source with drive electronics, such as a microdisplay device, which supplies the images to be displayed, an optical viewer for coupling the modulated light into the eye, an opto-mechanical unit for housing the optics and electronics and securing the device to a user' head, and a computer unit for image generation and data processing. A fully-developed HMD system for modern VR and AR applications also requires a suite of sensors necessary for sensing the various posture, gesture, and surrounding environments, and a suite of communication channels for exchanging data between the head-worn unit and remote date sources such as

Introduction to Flat Panel Display, Second Edition. Jiun-Haw Lee, I-Chun Cheng, Hong Hua, and Shin-Tson Wu.

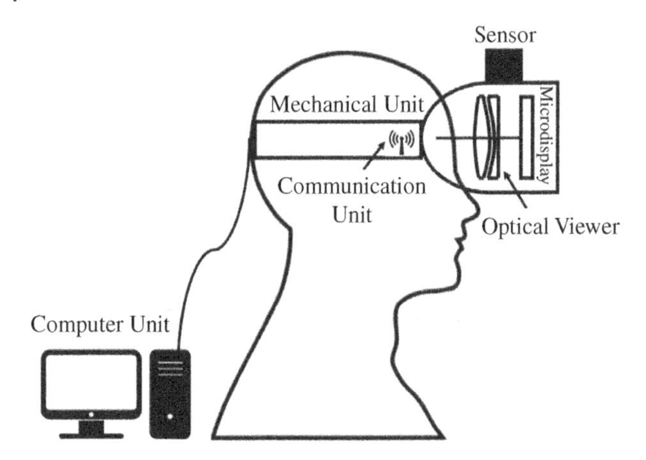

Figure 8.1 Conceptual illustration of a monocular HMD system.

cloud servers and other nearby devices. The sensor suite minimally requires a visual coupling system to display appropriate images based on a user's location, head position/orientation, and possibly eye gaze direction. Examples of visual coupling systems include a head tracker for tracking a user's head motion and an eyetracker for tracking a user's eye gaze. 3D depth sensors and built-in cameras have also become a necessity in the sensor suite of modern AR devices to create depth maps of the 3D scene surrounding a user and to capture images of the surrounding environment. Besides interfaces with the HVS, multimodal interfaces, such as 3D sound, haptics, and olfactory sensors that interact with the auditory, tactile, and olfactory senses of the user, may be integrated with an HMD system. The communication suite typically includes Wi-Fi, Bluetooth, and wireless connectivity. The computing unit may be a fully-integrated part of the device in some HMD systems, or built as a separate wearable unit such as a pocket unit, or the HMD may be tethered or wirelessly connected to a stand-alone computer system.

An HMD system may be monocular, biocular, or binocular. A monocular HMD system consists of a single microdisplay and optical viewer and only displays an image to one of the two eyes. Google Glass® is an example of a monocular HMD product (https://www.google.com/glass/start). Both biocular and binocular HMDs consist of two sets of typically identical optical assemblies, one for each eye. A biocular HMD displays the same image to both eyes, while a binocular HMD system displays two different images with slight disparity for stereoscopic viewing.

HMDs may be categorized into immersive and see-through displays [6]. An immersive display blocks a user's view of the real world and immerses a user in a solely computer-generated virtual environment. In an immersive HMD such as products commercially available by Oculus Inc. (https://www.oculus.com), the microdisplays and the optical viewers are placed directly in front of the viewer's eye, blocking the light path from the real world. A see-through display, on the other hand, blends real, and computer-generated scenes. See-through displays may be further classified into two types – video see-through and optical see-through. A video see-through display blocks the direct view of the physical world in a similar fashion to an immersive display, but it captures a view of the real world using one or more miniature video cameras mounted on the headgear and electronically fuses the video-captured view of the world with that of a computer-generated virtual environment (https://www.trivisio.com/hmd-nte). The core display optics of video see-through displays are similar to those of immersive displays. The performance parameters such as resolution, field coverage, fidelity, and latency of the real-world view are limited by the capability of the video cameras used for capture and post-processing. Digital mixing of the real and virtual scenes, means that video see-through HMDs can handle mutual occlusion of digital and physical objects in a more straightforward manner than an optical see-through display. An optical see-through display inserts an optical combiner between the optical viewer and the eye to maintain a user's direct view of the real world [7]. Through the optical combiner, images rendered by the display path are mixed with the view of the real-world scene. The optical see-through approach allows a user to see the real world with full resolution and is less intrusive into the user's view of the real scene than the video see-through approach. Generally speaking, it is a preferred method for tasks where hand-eye coordination or an unblocked view of the real-world is critical. Both types of see-through displays have been applied in various AR applications, from medical training to entertainment [4]. A thorough comparison between the video see-through and optical through methods can be found in the work by Rolland and Fuchs [6].

Interest in creating devices presenting visual experiences of different worlds can be dated back as early as around 1838 when Sir Charles Wheatstone proposed the concept of the stereoscope [8]. Through a stereoscope, a viewer is able to experience a sense of depth provided by a pair of static photographs with slight disparities. The concept of stereoscopes has evolved through many generations, leading to the development of View-Master by William Gruber and Harold Graves in 1938 [9], the invention of the Sensorama machine by Morton Heilig in the 1960s [10], the first machine rendering 4-dimensional immersive experiences, the Virtual Interface Environment Workstation (VIEW. system) by NASA in partnership with VPL Research Inc. in the 1980s [11], and the Nintendo Virtual Boy portable video game console in the 1990s (https://nintendo.fandom.com/wiki/Virtual_Boy). These technological developments over several decades have led to the more recent rapid rise of VR systems and applications. The visual experiences have evolved from static stereoscopic photographs in the 1830s to fully immersive, interactive, and very compelling experiences in high-fidelity digital worlds. VR headsets have become affordable, reached mass consumer markets, and broadened their applications into many new fields.

Parallel to the development of immersive VR headsets, a different visual experience in which views of digital objects are blended with that of our surrounding physical world has also been pursued. Dating back to around 1968, the pioneering work by Ivan Sutherland demonstrated the first OST-HMD [12], namely the "Sword of Damocles," through which computer-rendered wire-frame objects were superimposed on a viewer's real-world view. It became the foundation for decades of research efforts in the field of mixed reality (MR) and AR; the term Augmented Reality was formally introduced in the early 1990s by researchers at Boeing Inc. [13]. The pursuit of MR-AR experiences has evolved from early demonstration-of-use cases for knowledge-based AR systems [14, 15], mobile AR system [16] and wearable computing [17], tele-collaboration [18], Magic-Book [19], the development of AR Toolkits [20], to the massive success of the popular game Pokemon Go® (https://www.pokemongo.com/en-us), the recent releases of AR Toolkits and hardware and billions of dollars of investment into the development of fully integrated spatial computing platforms. The visual experiences presented by MR-AR technology have significantly evolved from wireframe graphical objects to compelling and complex scenes that truly blur the boundaries between bits and atoms.

In addition to the efforts of developing HMDs for VR and AR applications, there is a long history and a substantial amount of work in the area of developing HMD technologies for military defense applications. In fact, one of the earliest patents on military HMD can be dated back to the 1900s [21]. Some well-known examples of more recent HMDs for defense use include the Integrated Helmet and Display Sighting System (IHADSS) integral to the AH-64 Apache attack helicopter built by US Army in the 1970s [22].

Despite the rapid advances of HMD technology, a light-weight, non-intrusive, and comfortable HMD is one of the most critical and perhaps most challenging enablers for building compelling VR and AR applications. This chapter will therefore serve as a tutorial to students and engineers interested in the HMD technology. Though the fundamentals reviewed in this chapter are generally applicable and relevant to the engineering of HMD technology for defense applications, an excellent book by Rash provides a specialized account of the domain-specific design considerations and requirements unique to military and defense applications [22].

8.2 HUMAN VISUAL SYSTEM

The HVS is the final core element in an HMD system. Conventional design of HMDs typically focuses on optimizing the optics to deliver a high-quality 2D image to be viewed by the eye and do not attempt to compensate for aberrations of the eye optics or to optimize the perceived retinal image accounting for eye aberrations or photoreceptor sampling characteristics. However, several of the emerging HMD displays, such as light field HMDs (Section 8.5.4), go beyond simply rendering 2D images and require accurate modeling of visual optics to complete the image formation process. In this case, system optimization needs to be done based on a retinal image and the eye optics becomes an integral part of the overall optical system. Therefore, understanding some of the anatomical, optical, and visual properties of the HVS is not only critical for defining the requirements for new HMD systems, but also for evaluating the performance and usability of a conventional HMD system and for designing and optimizing some of the emerging HMD systems. This section will briefly summarize HVS properties and discuss how they are related to the HMD–HVS interface.

The optical system of the human eye consists of the cornea, pupil, crystalline lens, and retina [23]. Aqueous humor is the liquid filling the space between the cornea and the eye lens, and vitreous humor fills the chamber between the eye lens and the retina. Figure 8.2 shows the schematic layout of the eye optics. The mean optical power of the eye optics is 59.63 diopters, with a range of 54–65 diopters, approximately two-thirds of which is produced by the cornea and one-third by the crystalline lens. Due to the refractive index of the vitreous humor material of 1.336, the front focal length of the eye optics is about 17 mm and the rear focal length is about 23 mm. The eye optics is able to adjust its optical power to allow focusing of objects on the retina from far to near distances through a mechanism known as accommodation. Eye accommodation is achieved by modifying the shape of the crystalline lens by relaxing or contracting the ciliary muscles attached to the lens. The curvatures, the thickness, and the position of the crystalline lens change in the process of accommodation. Depending on age, the crystalline lens can contribute approximately a 10-diopter change in optical power to adapt its focus from far to near objects.

A variety of schematic eye models exist that are used to model the optical properties of the eye to various levels of accuracy and complexity. Some of the models simply describe the paraxial imaging properties of an unaccommodated schematic eye and do not accurately model aberrations and neglect accommodation. For instance, the Gullstrand–LeGrand schematic eye is a common paraxial model which models the eye optics with four spherical surfaces and approximates the gradient index of the crystalline lens with a uniform effective

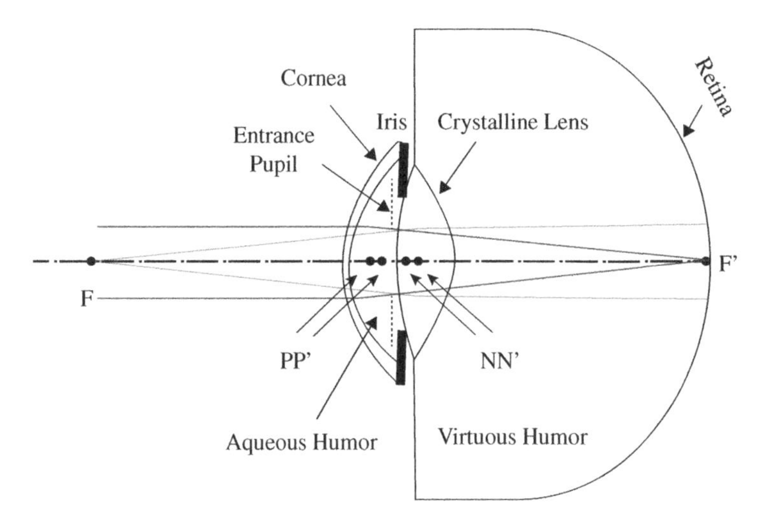

Figure 8.2 Schematic layout of the eye optics.

index of 1.42. The details of the model are summarized in Table 8.1. This type of paraxial eye model should only be used to examine cardinal points, pupils, magnifications, and other first-order paraxial imaging effects.

On the other hand, several more sophisticated models using aspheric surfaces have been developed to more accurately model aberrations and accommodation [24, 25]. These aspheric eye models are better suited for examining clinical levels of eye aberrations, both on- and off-axis, and simulating the retinal image formation of display systems. Among these models, the Arizona Eye Model is designed to model the eye accommodation process and to match clinical levels of on- and off-axis aberrations [24], and is one of the most commonly adopted models for simulating the retinal image formation process in various HMD designs [26–29]. Table 8.2 summarizes the optical properties of the Arizona Eye Model.

The optics of the eye has three main types of residual aberration, including spherical aberration, chromatic aberration, and astigmatism. Longitudinal spherical aberration (LSA) of the eye optics is measured by the optical power difference between marginal and paraxial rays and there is an LSA of about 1.25 diopters for a 6 mm pupil diameter. Longitudinal chromatic aberration (LCA) of the eye is measured by the optical power difference between a given wavelength and a reference wavelength such as 589 nm. There are roughly 2.5 diopters of LCA between 400 and 700 nm wavelengths. Astigmatism is calculated by the optical power

Table 8.1 Gullstrand–LeGrand schematic eye.

Name		Radius (mm)	Thickness (mm)	Index	Abbe
Cornea	Anterior	7.8	0.55	1.3771	57.1
	Posterior	6.5			
Aqueous humor			3.05	1.3374	61.3
Lens	Anterior	10.20	4.0	1.42	47.7
	Posterior	−6.00			
Vitreous humor			16.5966	1.336	61.2

Table 8.2 Arizona eye model.

Name		Radius	Conic	Thickness	Index	Abbe
Cornea	Anterior	7.8 mm	−0.25	0.55 mm	1.377	57.1
	Posterior	6.5 mm	−0.25			
Aqueous humor				t_{aq}	1.337	61.3
Lens	Anterior	R_{ant}	K_{ant}	t_{lens}	n_{lens}	51.9
	Posterior	R_{post}	K_{post}			
Vitreous humor				16.713 mm	1.336	61.1
Retina		−13.4 mm	0			

where

$$R_{ant} = 12.0 - 0.4A \qquad K_{ant} = -7.518749 + 1.285720A$$
$$R_{post} = -5.224557 + 0.2A \qquad K_{post} = -1.353971 - 0.431762A$$
$$t_{aq} = 2.97 - 0.04A \qquad t_{lens} = 3.767 + 0.04A$$
$$n_{lens} = 1.42 + 0.00256A - 0.00022A^2$$

where A is the accommodation level measured in diopters.

difference between the tangential plane and the sagittal plane for different field angles, and the eye can have as astigmatism as large as 12 diopters for field angles over 60°.

The pupil is the optical aperture of the human eye and its size changes via dilation and contraction of the muscles that control the iris to adapt to the brightness of the scene being viewed. The pupil adapts faster to bright light, typically in less than one second, but more slowly to dark, typically in several seconds or up to a minute. The diameter of the pupil typically varies from 2 mm under bright light condition such as sunlight to about 8 mm in the dark. Several models have been developed to model the relationship between pupil diameter and ambient lighting conditions. For example, the average pupil size of the general population can be modeled as

$$D = 4.9 - 3 * \tanh[0.4(\log_{10}(L) + 1)] \tag{8.1}$$

where D is the pupil diameter in mm and L is the luminance in cd/m^2. Based on this model, the mean pupil size is about 3 mm under a luminance level of 10 cd/m^2. The normal eye is considered to be near-diffraction limited for a 2 mm pupil diameter.

The pupil diameter plays a significant role in the depth of field (DOF) of the optics of the eye. As quantified in a study by Ogale and Schwatz [30], the DOF of the eye decreases by about 0.12 diopters as the pupil diameter increases by 1 mm. As illustrated in Figure 8.2, the entrance pupil of the eye is the image of the eye pupil through the cornea, and is located about 3.05 mm behind the cornea of the eye as measured from the vertex of the cornea. To efficiently couple light from an HMD system into the eye, the location of the exit pupil of an HMD system needs to match the location of the entrance pupil of the eye.

The retina has two types of light-sensitive photoreceptors: cones and rods. The cones are responsible for color, photopic vision where the luminance level generally is greater than 3 cd/m^2 and saturates the rods, the rods are considered as being color blind and responsible for scotopic vision where the luminance level is generally less than 0.03 cd/m^2 and is too low to activate the cones. Both the cones and rods are activated for the in-between, mesopic condition where the luminance level is generally between 0.03 and 3 cd/m^2. Spectrally, the eye optics transmit the wavelength range from approximately 380 nm up to about 1400 nm, but only a portion of the wavelength range, about 380 to 700 nm, are absorbed by the rods and cones and are thus considered as the visible spectrum. The photopic luminance response peaks at 550 nm and the scotopic luminance response peaks at 505 nm. In normal color vision, three types of cones, known as S, M, and L cones, are found in the human retina, with their peak sensitivities approximately corresponding to wavelengths of 440, 545, and 580 nm, respectively. The optical performance of the optical viewer of an HMD system needs to be optimized for the dominant wavelengths of the light sources of the chosen microdisplay with a spectral weighting scheme based on the relative photonic sensitivities of the corresponding wavelengths.

Two important features are found on the retina, the slightly oval-shaped fovea measuring about 1.25 mm in diameter on the retina, and the oval-shaped optic disc measuring about 1.5 mm by 2 mm. The cones dominate the fovea while rods are dominant outside the fovea. The area of the optic disc is free of photoreceptors and it is therefore commonly known as the blind spot of the eye. The foveal pit, measuring approximately 250 μm and covering a visual angle of 1.37°, is the central region of the fovea which has the highest cone density and is free of rods. The cone diameter is roughly 2.5 μm in the foveal pit, but increases rapidly outside the fovea to about 10 μm in the periphery. The rod density, on the other hand, increases rapidly from zero at the fovea and reaches its peak at roughly 18° from the fovea.

The instantaneous field of view (FOV) of a human eye without eye or head movements is approximately 160° horizontally (about 60° nasally and 100° temporally) and 130° vertically (60° superiorly and 70° inferiorly). Considering the eyes as a pair, the overall FOV measures approximately 130° vertically and 200° horizontally when the eyes are stationary. The binocular visual field, within which a target is visible to both eyes, measures about 114° when the eyes converge symmetrically, and less when they converge asymmetrically on an eccentric point. The eyeball is held in position by three pairs of muscles in its orbital cavity and the muscles rotate the eyeball in the orbits by as much as about ±35° horizontally and vertically, and about 2–10° about the visual axis. The center of rotation of the eye is about 13 mm behind the corneal vertex. Such eye movements allow the image of a scene to be focused at all times on the foveal pit of the central retina, and extend the overall FOV of the HVS to about 290° horizontally. It is worth noting that the eye pupils rotate along as the eyeballs rotate in their orbits, which makes it necessary to ensure the eyebox of an HMD design is larger than the range of the expected eye pupil motion when designing an HMD optics.

Visual acuity (VA) is a term which defines the ability of the human eye to see high-contrast fine details. Particularly, separable VA, also known as distance VA, is commonly used to measure the ability to resolve closely spaced details such as thin lines. The separable VA of a normal eye is typically one minute of arc; it can be 0.5 minutes of arc in the foveal region of about 5°, but degrades rapidly beyond 10–15° from the fovea (e.g. about 20–25% of the peak VA at 10° from the fovea). Various eye charts have been used to measure VA. The most commonly used is the Snellen eye chart from which Snellen acuity, also known as Snellen fraction, S, is defined as

$$S_{\mathrm{VA}} = \frac{D'}{D_{\mathrm{normal}}} \tag{8.2}$$

where D' is the standard viewing distance, typically 6 meters or 20 feet, at which a subject can just correctly read a given line on the chart, and D_{normal} is the furthest distance from which a "normal" observer can just read the same line. The relative VA of a human eye is normalized by the peak acuity at the foveal pit and is generally modeled as a function of eccentricity, given by

$$\mathrm{VA}(e_x, e_y) = \frac{e_2}{e_2 + \sqrt{{e_x}^2 + {e_y}^2}} \tag{8.3}$$

where e_2 is the retinal eccentricity where spatial resolution falls to half of what it is in the center of the fovea and $e_2 \cong 2.\ 3°$; e_x and e_y are the horizontal and vertical retinal eccentricities in degrees.

The contrast sensitivity function (CSF) evaluates the visual ability to detect targets of different spatial frequency with low contrasts. Typically, sinusoidal patterns of varying contrast are used to test the contrast sensitivity of the eye. At a given spatial frequency, the contrast sensitivity is defined as the reciprocal of the minimal detectable modulation M_t. The relationship between contrast modulation sensitivity and spatial frequency is referred to as contrast sensitivity function, or CSF. Several models exist for modeling the CSF of the HVS. The Barten CSF model is one of the most sophisticated models and has been commonly adopted by the display industry [31].

Besides these monocular properties of a single eye, several aspects of the binocular properties of the HVS are also very important for all operational modes of HMD systems, including monocular, biocular, and binocular. The interpupillary distance (IPD) defines the physical separation between the left and right eye. The mean of IPD values measured from 4000 flying personnel by the US Air Force anthropologists is 63.3 mm [32].

8.3 FUNDAMENTALS OF HEAD-MOUNTED DISPLAYS

This section will mainly focus on four fundamental aspects involved in the process of designing an HMD system, namely paraxial optical specifications, microdisplay sources, basic principles of HMD optics, optical combiner technology, as well as optical design and performance.

8.3.1 Paraxial Optical Specifications

Driven by the requirements of each specific application, the design process of an HMD system often starts with defining its system-level specifications such as FOV, resolution, microdisplay dimensions, see-through capability, luminance, spectral band of operation, and eye clearance. These specifications help to derive the paraxial parameters of the optical viewer (or eyepiece) such as its equivalent focal length (EFL), optical magnification, exit pupil diameter (EPD), eye relief, etc.

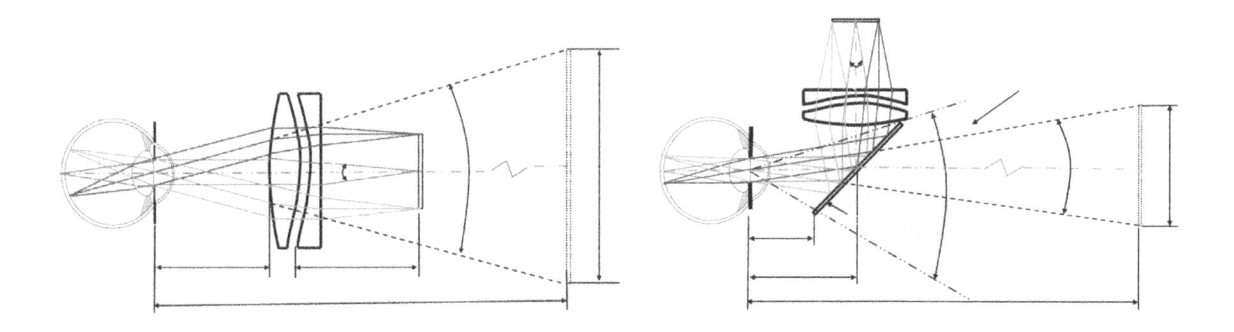

Figure 8.3 Schematic optical layouts of (a) an immersive and (b) optical see-through HMD.

Figure 8.3a and b show schematic optical layouts of an immersive and an optical see-through HMD, respectively. In both types of display, light rays emitted or modulated by a microdisplay device are collected and projected by an eyepiece toward its exit pupil. The projected rays form a virtual image of the display, namely a virtual display, which appears to be located at a sufficiently far distance from the exit pupil for viewing. By placing a viewer's eye so that its entrance pupil coincides with the exit pupil of the eyepiece, the eye perceives the virtual display and the rays projected by the eyepiece are then further focused by the eye optics to form a retinal image of the virtual display. In the case of an optical see-through HMD, an optical combiner such as a simple flat beamsplitter is inserted between the eyepiece and the exit pupil to combine the ray paths of the virtual display and real-world scene.

FOV is one of the most important specifications of an HMD system, defining the angular limits within which a viewer sees information. The FOV of an immersive or video see-through HMD is solely determined by the angular subtense of the virtual display, while the FOV of an OST-HMD has two different aspects, a virtual display FOV and a see-through FOV. The virtual display FOV of a monocular HMD is typically defined as the instantaneous angular extent, measured in degrees, which the displayed monocular virtual image subtends to the user's eye. It depends on the dimensions of the active area of a microdisplay and the effective optical magnification of the eyepiece. It is specified diagonally by the angle, θ_{D}, subtended by the diagonal dimension of the virtual display to the eye, and may be expressed as

$$\theta_{\mathrm{D}} = 2\mathrm{atan}\left(\frac{D_{\mathrm{VD}}}{2Z_{\mathrm{VD}}}\right) \tag{8.4}$$

where D_{VD} is the diagonal size of the virtual display and Z_{VD} is the apparent distance of the virtual display to the eye. Given the aspect ratio of the virtual display, the horizontal and vertical FOVs, θ_{H} and θ_{V}, respectively, of the virtual display can be conveniently obtained.

The apparent distance of the virtual display, Z_{VD}, can be set as close as 25 cm or 4 diopters, which is considered as the near point of accommodation limit for young adults, or as far as optical infinity or 0 diopter. In practice, it is set according to the nature of its targeted applications. For instance, for far field applications such as flight simulators it is preferred to set the virtual display distance to optical infinity (i.e. at least 6 m), while for near field applications such as surgical operations or arm-length tasks it is preferred to set the virtual display around 750 mm. For dual near-field applications such as driving navigation, vari-focal or multi-focal plane displays or light field displays capable of rendering correct focus cues will be necessary, which will be discussed in detail in Section 8.5.

Regardless of complexity, an eyepiece can always be characterized by its EFL, f_{EP}, along with its cardinal points, as schematically illustrated in Figure 8.4 where rays are traced from right to left. Given the axial distance, Z_{MD}, between the microdisplay and the front principal plane, P, of the eyepiece, the distance, Z_{ER}, between the exit pupil and the rear principal plane, P', of the eyepiece, and the diagonal size of the

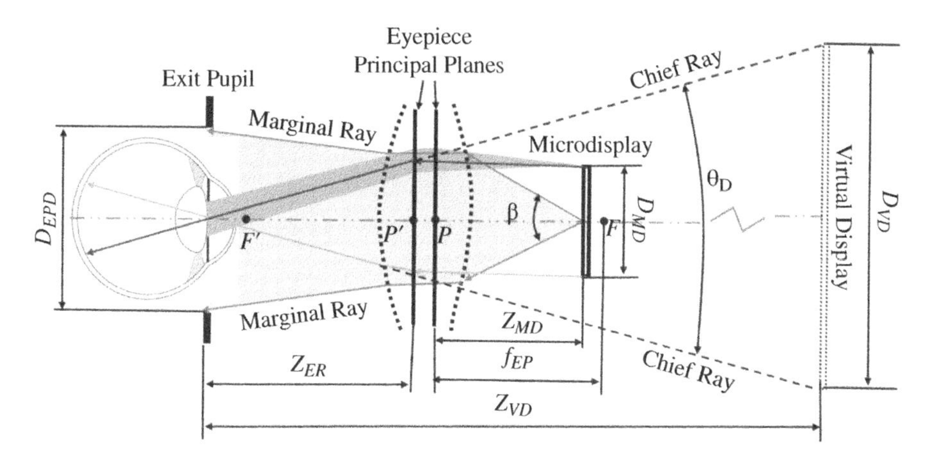

Figure 8.4 Illustration of cardinal points and paraxial properties of an eyepiece.

microdisplay, D_{MD}, the location and size of the virtual display can be determined by tracing the chief rays from the edges of the microdisplay and be expressed as

$$Z_{\mathrm{VD}} = \frac{f_{\mathrm{EP}} \cdot Z_{\mathrm{MD}}}{f_{\mathrm{EP}} - Z_{\mathrm{MD}}} + Z_{\mathrm{ER}} \tag{8.5}$$

and

$$D_{\mathrm{VD}} = \frac{f_{\mathrm{EP}} \cdot D_{\mathrm{MD}}}{f_{\mathrm{EP}} - Z_{\mathrm{MD}}} \tag{8.6}$$

When a microdisplay is placed to coincide with the front focal point, F, of the eyepiece, the projected rays of the virtual display are considered as being collimated and the virtual display appears to be at optical infinity from the viewer. In such circumstances, the monocular FOV of Eq. (8.4) can be simplified as

$$\theta_{\mathrm{D}} = 2\mathrm{atan}\left(\frac{D_{\mathrm{MD}}}{2f_{\mathrm{EP}}}\right) \tag{8.7}$$

In an OST-HMD, the see-through FOV, defining the angular range, α, in which a viewer can see the real-world, is equally, if not more important than the virtual display FOV. Depending on the type of optical combiner being used in a particular design, the see-through FOV is often limited by the dimensions of the combiner or its mechanical housing, as illustrated in Figure 8.3b. Although it is highly desirable that an HMD system deliver FOVs of both the virtual display and see-through paths to match the extremely wide FOV of the human eye, practical FOVs are not only driven by the requirements of specific applications but also constrained by several engineering factors. In general, immersive and video see-through HMDs require a very wide FOV for the virtual display as close to the FOV of the eye as possible to create the sense of immersion and presence. For an OST-HMD, on the other hand, the requirements for the virtual display FOV are very task-driven. For instance, a 15° virtual display FOV can be quite adequate for a simple information display used for reading emails or navigation guides, while a 50° or more virtual display FOV is needed for spatial computing applications. It is important to specify FOV requirements based on the main tasks to be performed through the display. A general recommendation by Patterson is that a display FOV of 50° can be adequate for tasks such as targeting and object recognition and a FOV greater than 60° appears to be necessary for tasks requiring peripheral stimulation. It is highly desirable for the see-through FOV of an OST-HMD, to be as large as possible, especially for applications where situational awareness of the real world is critical. As discussed in Section 8.3.4, not all optical combiners can allow a large see-through FOV.

The overall FOV of a binocular HMD is the overall angular range visible by the two eyes and the binocular overlap FOV is the region simultaneously visible to both eyes and capable of rendering stereoscopic scenes.

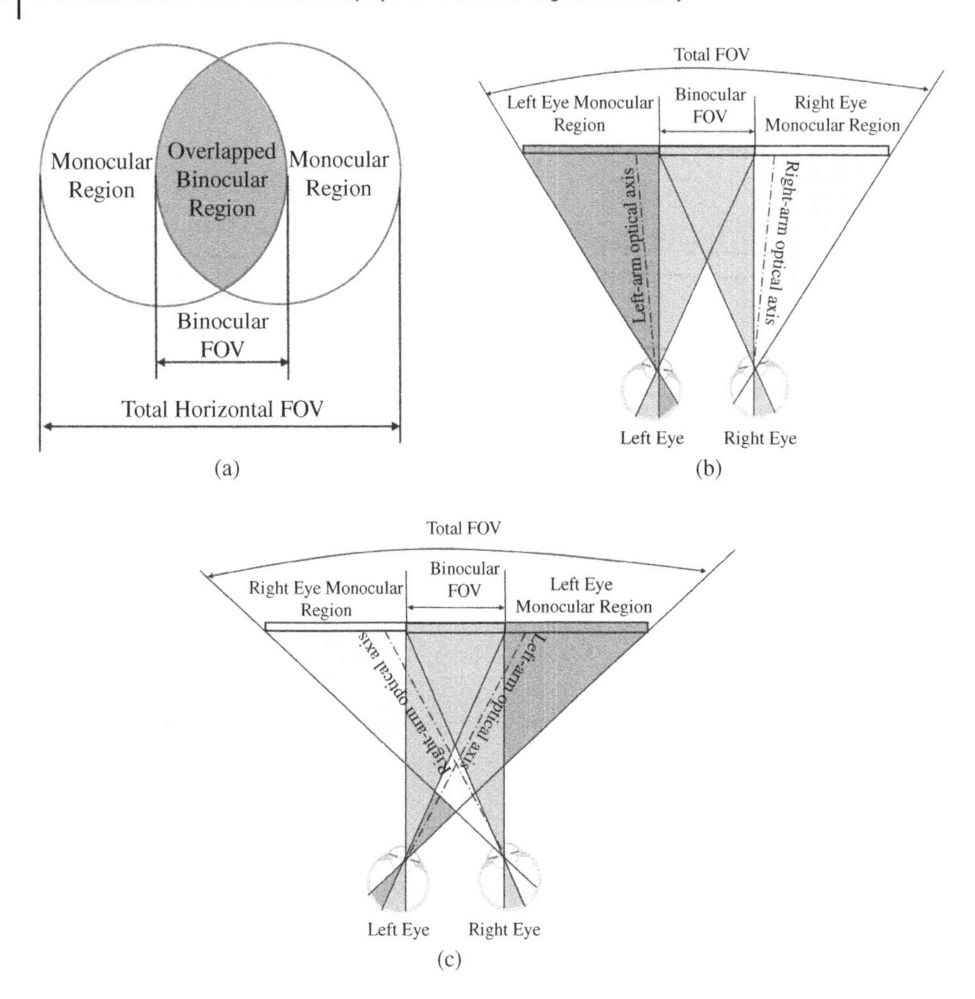

Figure 8.5 Illustration of partial overlap schemes in an HMD.

Both of these two measurements depend on the overlap schemes utilized in assembling the two arms of the binocular system. Full overlapping is the most commonly used scheme in which the optic axes of the left-arm and right-arm systems are arranged such that their monocular FOVs fully overlap. In this case, the total FOV and the binocular FOV of the system are identical to the monocular FOV. When the monocular FOV is not adequate for its targeted applications, partially overlapping schemes may be adopted in which the optical axes of the optics for each eye are tilted inwards or outwards to increase the total horizontal FOV at the cost of a reduced binocular overlap FOV. As illustrated in Figure 8.5a, the partially overlapped FOV consists of a central binocular region, and two monocular regions. Figure 8.5b and c further illustrate two different partially overlapping schemes using divergent overlap and convergent overlap respectively, where the optical axes of the two arms are tilted toward or away from each other. The appropriate amount of overlap for a given monocular display FOV remains an unresolved question. Based on visual considerations Grigsby and Tsou have argued for a binocular overlap of at least 40° while studies by scientists at Kaiser Electro-Optics suggest a user preference for partial overlap as a percentage of the overall FOV [33, 34].

Although they provide the benefit of an increased total FOV, there are a few drawbacks associated with partial overlapping schemes. First, while the full overlap FOV consists of one contiguous binocular region, the partial overlap FOV consists of three regions separated by the binocular overlap borders. This can result in visual fragmentation, which may cause incorrect visual perception and interpretations due to a fragmented

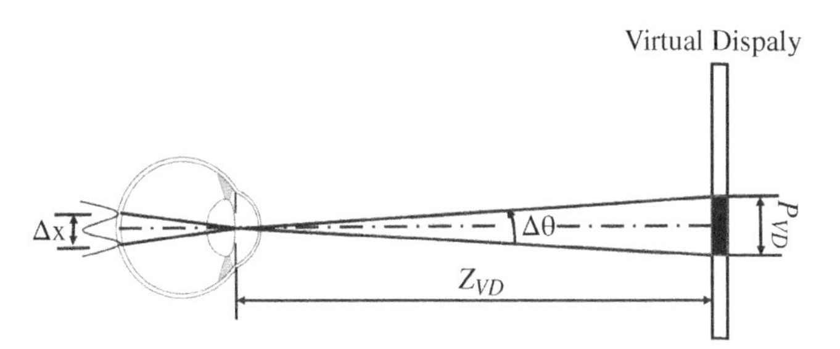

Figure 8.6 Illustration of angular resolution of the virtual display in an HMD.

rendering of a continuous world in reality. Secondly, luning, a temporally varying subjective darkening of the flanking monocular regions, may occur in the partial overlap displays near the binocular overlap borders. This phenomenon is due to the nature of the dichoptic stimulation of the monocular regions. This results in various forms of binocular rivalry where these inputs compete for awareness and the inputs of each eye alternate in suppressing the input of the other eye. Third, competing visual input can result in less detectable targets in the monocular regions of the partial overlap FOV. The fourth issue is the choice of divergent or convergent overlap. Klymenko et al. have found that there is less luning and fragmentation in convergent displays than in divergent displays [35].

Resolution is another specification important to an HMD system, defining the spatial detail limits what a viewer can see through the display. The resolution of an immersive or video see-through HMD is solely determined by the resolution limit of the virtual display, while the resolution of an OST-HMD has two different aspects, a virtual display resolution and a see-through resolution. Although the number of addressable pixels in horizontal and vertical directions or alternatively pixel pitch has been commonly used as a resolution metric for conventional 2D display panels, neither of them meaningfully reflects the apparent image quality a user sees in HMDs. Instead, similar to the term of VA defining the eye's ability to see spatial details, the virtual display resolution of an HMD is best defined by the visual angle, $\Delta\theta$, subtended by the pixel pitch, P_{VD}, of the virtual display to the eye, as illustrated by Figure 8.6. The visual angle per pixel, known as the angular resolution of an HMD, is a more accurate parameter for characterizing the apparent image quality at the eye independent of the virtual display distance or FOV, and may be expressed as

$$\Delta\theta = 2\text{atan}\left(\frac{P_{\mathrm{VD}}}{2Z_{\mathrm{VD}}}\right) \approx 2\text{atan}\left(\frac{P_{\mathrm{MD}}}{2f_{\mathrm{EP}}}\right) \tag{8.8}$$

where P_{VD} is the pixel pitch of the virtual display. To be intuitively comparable to the VA of the eye, the angular resolution in Eq. (8.8) is typically expressed in minutes of arc (or arc minutes) per pixel. The angular resolution of an HMD in a given direction can be estimated by dividing the FOV in the corresponding direction by the corresponding number of pixels along that direction and conversion to minutes of arcs (i.e. 1° equals 60 minutes of arc). For instance, for an HMD of 50° and 1000 pixels in a given direction, the corresponding angular resolution is approximately three minutes of arc per pixel.

The see-through resolution of an OST-HMD is also important, and is defined in a similar fashion to that of VA. Some optical combiners, such as flat beamsplitters, introduce nearly negligible degradation to the optical resolution of the see-through path, while some combiners, such as freeform combiners may introduce significant effects to the see-through path if not optimized properly.

Although wide FOV and high resolution are both desirable characteristics of high-performance HMDs, in practice a tradeoff is often made to balance factors such as FOV, resolution, as well as weight and volume to address the needs imposed by specific tasks. With a fixed number of available pixels in microdisplays, the angular resolution of the display decreases as the FOV increases. To demonstrate the tradeoff relationship, Figure 8.7 plots the display angular resolution in minutes of arc as a function of full horizontal or vertical FOV in degrees and the number of pixels in the corresponding direction. For a given number of available pixels, the

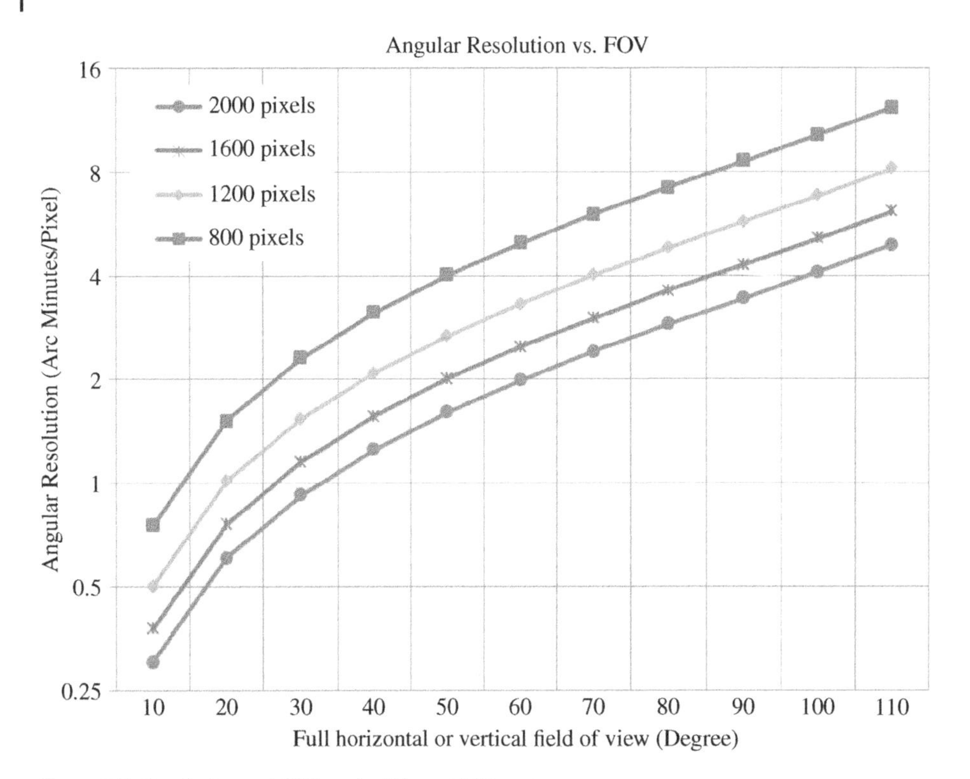

Figure 8.7 Resolution and FOV tradeoff in an HMD system.

angular resolution of the virtual display degrades as the FOV increases. It is highly desirable that the angular resolution of an HMD matches the spatial resolution of the HVS, that is, one minute of arc or better. However, state-of-the-art VR displays have prioritized FOV over resolution and typically deliver an angular resolution of about 5–10 minutes of arc.

Besides the function of creating a virtual display image at a comfortable viewing distance for the HVS, an eyepiece also forms an exit pupil through which the projected rays from the microdisplay are coupled into a viewer's eye, as illustrated in Figure 8.3. A nominal EPD is typically defined for the eyepiece optics to yield a non-vigenetted virtual image across its entire FOV at a specified nominal exit pupil location. Considering a typical eye pupil diameter and the nature of eye movements, it is suggested that the EPD of an HMD should typically be at least 10–12 mm. This range of EPD allows an eye swivel of about ±21° up to ±26.5° within the eye socket without causing vignetting or loss of image for a typical 3 mm eye pupil under the lighting conditions provided by typical HMDs. Furthermore, it allows a ±5 mm to ±6 mm IPD tolerance for different users without the need to mechanically adjust the IPD of the binocular optics.

The nominal exit pupil location may be specified by the axial distance between the nominal exit pupil and the closest approach point of the nearest surface in an eyepiece design to any part of the eye, referred to as the eye clearance, Z_{EC}. Viewing comfort for HMDs depends on adequate eye clearance so that the HMD does not contact the eye, eyelash or eyeglasses (if the viewer wears them). The location of the entrance pupil of the eye is inside the eyeball, about 3 mm from the vertex of the anterior cornea, so the recommended smallest value of eye clearance is 17 mm to allow the wearing of standard 2 mm-thick eyeglasses, which assumes a 15 mm distance from the eye entrance pupil to the inner surface of the eyeglasses. While 23 mm of eye clearance is preferable to accommodate most eyeglasses, extra specifications may vary depending on the application and user population. An alternative way to provide adequate eye clearance without considering eyeglasses is to provide refractive correction for each user, either via a focus adjustment or via an optical insert. An increase in eye clearance is often accompanied by an increase of viewing optics size for a given FOV, resulting in a loss of compactness, increased weight, and a degradation of optical performance because more of the outer periphery of the optics needs to be optimized as we increase eye clearance.

Another commonly used figure of merit to specify the nominal exist pupil location is the eye relief, Z_{ER}, which refers to the axis distance between the nominal exit pupil and the vertex of the last surface in the eyepiece design. As illustrated in Figure 8.3b, the two specifications may differ significantly in the case of an OST-HMD using an optical combiner which is tilted relative to the visual axis. Large values of eye relief do not necessarily provide sufficient clearance. It is therefore worth emphasizing that eye clearance is the most relevant figure of merit used to describe the HMD positioning with respect to the eyes.

As well as needing specification of the nominal EPD and eye clearance, an HMD must allow for a viewer's eye to move closer or further away to the optics or move laterally from the center of the exit pupil. For this reason, the eyebox of an HMD system may be specified, which defines a 3D volume centered at the nominal exit pupil location in which the eye can be placed to see the full FOV of the display without or with a certain level of vignetting. Figure 8.8a and b illustrate the eyebox in the lateral plane and the axial direction, respectively, in relation to the nominal exit pupil specifications and vignetting. For instance, given an eyepiece design with an EPD represented by the black solid circle in Figure 8.8a and eye pupil diameter of D represented by a small green circle, the blue dashed circle represents the eyebox limit of the lateral eye movement within the exit pupil plane where the entire eye pupil falls inside but just reaches the boundary of the EPD and no vignetting is induced to entire FOV, and the red dashed circle represents the eyebox limit where the entire eye pupil just falls outside the boundary of the eyepiece EPD and the viewer is just about to completely lose sight to the furthest edge of the FOV. In other words, the furthest edge point of the FOV just reaches a 100% vignetting. As the eye pupil continues to move further away from the EPD boundary, the eye will lose sight of an increasing portion of the FOV. Similarly, the eye pupil can move further away from its nominal exit pupil position axially as illustrated in Figure 8.8b within the lightly shaded area if no vignetting is allowed or within the darkly shaded area which marks the furthest axial positions where a viewer is about to completely lose sight to the furthest edge of the FOV or a 100% vignetting is reached. It is worth clarifying that vignetting leads to a proportional loss of perceived image brightness and the level of acceptable vignetting may differ for different applications. A 100% vignetting to a given field angle means the cut-off of visibility of that part of the scene and thus is used as an extreme boundary condition for defining an eyebox limit in Figure 8.8.

Although a wide FOV and a large eyebox are two highly desirable specifications in an HMD system, in practice, a design tradeoff between FOV and EPD is often made due to the Lagrange invariant of optical

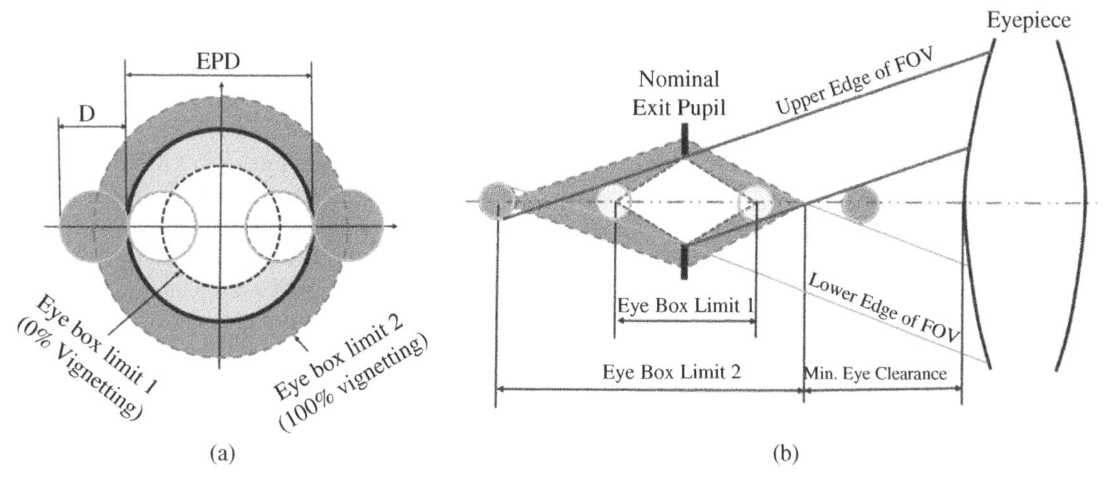

Figure 8.8 Illustration of eyebox in the (a) lateral plane of the exit pupil and (b) axial direction.

systems [36]. The Lagrange invariant, Ж, is formed by the paraxial marginal and chief rays through a system, and when applied to the exit pupil it can be expressed as

$$Ж = \frac{n \cdot \theta_D \cdot D_{EPD}}{4} \tag{8.9}$$

where n is refractive index in the visual space and it is typically 1 assuming (as is normal) an airgap between the eyepiece and the eye. As shown by Eq. (8.9), for a fixed value of the Lagrange invariant, the FOV of an HMD system in the visual space, θ_D, is inversely proportional to the diameter of the EPD.

The Lagrange invariant of an optical system is related to the throughput, or etendue of an optical system in radiative transfer, which determines the amount of light or radiant energy from an object that is collected by the optics and reaches the image. The throughput of an optical system, η, is related to the Lagrange invariant by

$$\eta = \pi^2 Ж^2 = n^2 A_{EPD} \Omega_{FOV} \tag{8.10}$$

where A_{EPD} can be considered as the area of the exit pupil and Ω_{FOV} as the solid angle subtended by the virtual display. The throughput of an optical system is conserved and the flux collected by the eyepiece from a microdisplay is transferred to the virtual display. Therefore, we can similarly apply the Lagrange invariant to the space of the microdisplay and express its throughput as

$$\eta = \pi^2 Ж^2 = n'^2 A_{MD} \Omega_{EP} \tag{8.11}$$

where n' is the refractive index in the microdisplay space, A_{MD} can be considered as the area of the microdisplay and Ω_{EP} as the solid angle subtended by the entrance pupil of the eyepiece. The solid angle of the entrance pupil, $\Omega_{EP,}$ is proportional to the square of the numerical aperture (NA), or equivalently the f/#, of the eyepiece in the microdisplay space.

Changing the Lagrange invariant of an optical system leads to a proportional scaling of the throughput and dimensions of an optical system. Therefore, tradeoff between FOV and EPD also goes hand-in-hand with a consideration of the overall size of the system. For instance, doubling the EPD specification of an HMD system while maintaining the same FOV will lead to a four times increase in the system throughput, requiring either double the size of microdisplay for the same eyepiece NA, or double the NA of the eyepiece for the same microdisplay size. To illustrate these tradeoffs, Figure 8.9 plots the maximum affordable non-vignetted EPD as a function of the full FOV for different choices of microdisplay panel size or eyepiece NA.

8.3.2 Microdisplay Sources

One of the core elements in HMD systems is a modulated light source for rendering the images to be displayed, and each HMD system minimally requires one or more such devices. The most commonly used modulated light source is a microdisplay device consisting of an array of pixels each of which either emits light or modulates another light source to render a displayed image. Alternatively, the modulated light source can be a point or line-type light source, such as a laser or laser-diode, which is modulated and scanned across the visual field to form a 2D or 3D image, which is generally referred to as a scanning-type display. Microdisplays can be further categorized into the self-emissive type which emits its own light and spatial-light modulating (SLM) types that modulate a light source such as a laser, laser diode, light-emitting diode (LED), or another lamp. Examples of self-emissive microdisplays include miniature cathode ray tubes (CRTs), organic light-emitting diode displays (OLED), and the more recent micro-LED devices, and examples of SLM-type displays include active-matrix liquid crystal displays (AM-LCDs), liquid-crystal-on-silicon (LCoS) panels, ferroelectric liquid-crystal-on-silicon (FLCoS) displays, and digital-mirror displays (DMD). Among the SLM types, AM-LCDs are transmissive and require a backlight source, while all others are reflective types that require front illumination.

Advances in microdisplay technologies have had a great impact on HMD development. A few decades ago miniature CRTs were the sole choice for HMD designs, while limited resolution AM-LCDs became the dominant technology for HMDs in the 1990s. More recently, a variety of new microdisplay technologies have

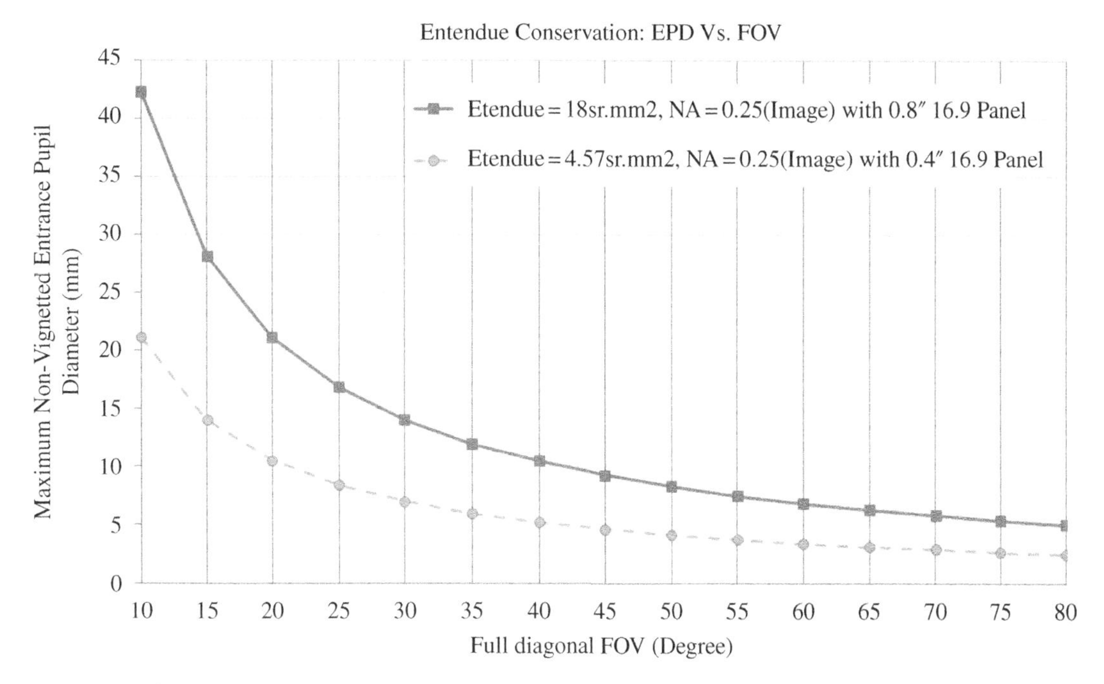

Figure 8.9 Effects of Lagrange invariant on the tradeoffs of system FOV and EPD for different choices of microdisplay size or eyepiece numerical aperture.

evolved to become mainstream choices, including high-resolution AM-LCDs, DMD panels, LCoS panels, FLCoS displays, and OLEDs. These microdisplay technologies readily offer pixel resolutions from VGA and SVGA to SXGA and full high-definition (HD) resolutions. Some of these technologies are already capable of offering 4K or higher resolution. Table 8.3 provides examples of these array-type microdisplays available from major manufacturers. The emerging micro-LED displays are under research and development and are expected to become another technology of choice. Additionally, laser diode-based image sources have been employed in scanning displays, which have demonstrated advantages in applications requiring high image brightness.

Choice of microdisplay has a critical impact on both the ergonomics and optical performance of an HMD system and must be one of the first design decisions to be made following basic system specifications. The important parameters to consider when choosing microdisplays include, but are not limited to, microdisplay type, physical dimensions, color schemes, pixel resolution, pixel pitch, luminance, power consumption, refresh rate, and pixel persistence. Considering the microdisplay type as an example, self-emissive and backlit displays tend to yield more compact systems, while reflective SLM displays tend to yield a more bulky system but with higher luminance.

The self-emissive nature of OLEDs and their compact packaging offer potentially the most compact system design among these technologies. Additionally, the wide emission angle of OLED displays does not require telecentricity for the eyepiece design, which often leads to a more compact lens structure. Existing OLED microdisplays have a good range of panel size from about 0.4″ to 1″ diagonally and the pixel pitch from about 6–15 μm by various manufacturers. However, OLEDs have several drawbacks which are worth noting. First of all, the luminances of existing OLEDs are relatively low compared to reflective SLMs such as FLCoS and LCoS microdisplays, although higher luminance OLED panels have emerged recently. The low luminance makes OLED a less viable choice for OST-HMDs designed for bright indoor and outdoor environments. Additionally, OLEDs suffer from a shorter operating life than other microdisplays, and the life span is further shortened when the display is operated at high luminance. Also, the luminance of various colors degrades non-uniformly

Table 8.3 A survey of array-type microdisplay technologies.

	OLED	Micro-LED	AM-LCD	LCoS (Color filter)	LCoS (Color sequential)	LCoS (Color sequential)	FLCoS	DMD
Manufacturer	Emagin	Plessey	Kopin	HimaxDisplay	Jasper Display	Holoeye	Forth Dimension	Texas Instrument
website	http://www.emagin.com	http://www.plesseysemiconductors.com	http://www.kopin.com	http://www.himaxdisplay.com	https://www.jasperdisplay.com	http://www.holoeye.com	http://www.forthdd.com	http://www.dlp.com
Model	EMA-101306-01	Prototype	SXGA LBC	HX7097	JD2552 SP55	HED-2200	M150 QXGA	DLP4710
Type of illumination	Self-emissive	Self-emissive	Backlight	Front illumination optics				
Size(diagonal) (inch)	1.07	0.7	0.97	0.294	0.7	0.26	0.83	0.47
Resolution (pixel)	2048 × 2048	1920 × 1080	1280 × 1024	800 × 3 × 480	1920 × 1080	1280 × 720	2048 × 1536	1920 × 1080
Pixel size (μm)	9.3 × 9.3	8 × 8	15 × 15	7.7 × 7.7	6.4 × 6.4	4.5 × 4.5	8.2 × 8.2	5.4 × 5.4
Max Luminance (cd/m2)	250	100 000		Depending on illumination source and optics				
Contrast Ratio	100 000 : 1	N/A	300 : 1	NA		1000 : 1	NA	>1000 : 1 depending on illumination
Color method	RGB sub-pixels		RGB sub-pixels	RGB sub-pixels	Color sequential	Color sequential	Color sequential	Color sequential
Color depth (bit)	24	10 (monochrome)	24	24	8-bit gray	8-bit gray	24	24
Frequency (Hz)	120		60	60	480 color field	300 color field	100	120
Fill factor	75%		N/A	N/A	>93%	92%	>94%	N/A
Optical efficiency	N/A		<10% (Estimated)	<10% (Estimated)	94%	70%	60–70%	N/A
Power Consumption (typical)	600 mW	5 mW	N/A	N/A	N/A	N/A	N/A	N/A

over the display life. The rapidly developing micro-LED displays are expected to significantly improve the prospect of self-emissive displays and produce microdisplays with much higher luminance and low power consumption.

Among SLM-type microdisplays, the AM-LCD offers a potentially more compact system than other SLMs because its transmissive nature only needs a backlit panel rather than a bulky illumination unit. Additionally, backlit AM-LCDs have a larger viewing angle than reflective types, thus imposing less constraint on the incident angle of the light cones for different field angles, which tends to yield more compact eyepiece design than microdisplays which impose telecentricity requirements on the eyepiece design. However, AM-LCDs tend to have low transmission efficiency, relatively low contrast ratio, and a low dynamic range (LDR). Furthermore, AM-LCDs require three subpixel cells for each pixel to form a color display, which leads to large pixel dimensions compared with other available microdisplay technologies.

LCoS, FLCoS, and DMD-based microdisplays generally have much higher fill factors and reflectance than transmissive-type LCDs. They can take advantage of high brightness LEDs or laser diodes and produce displays with higher brightness and longer life-time than OLEDs. Currently these microdisplays offer high pixel resolution (1080p is quite common), luminance output, optical efficiency, and image contrast, and they are the primary choices for OST-HMD systems for bright environments. On the other hand, due to their reflective nature, they require a carefully designed illumination unit to illuminate the microdisplay from its front, which makes the overall display system less compact than a system using AM-LCD or OLED microdisplays. The LCoS and FLCoS microdisplays work most efficiently when the illumination rays are normally incident upon the display surface. Under normal incidence, all the rays experience the same optical path through the liquid crystal material, while skewed incidence leads to a reduction in image contrast. To ensure high contrast in the output image, it is recommended to limit the incident angle to a range of $\pm 16^\circ$, which imposes a critical requirement on the design of both the light engine and the imaging lens. Similarly, DMD-based microdisplays require that the incident illuminating beam and the reflected output beam be separated by an angle which is limited by the tilting angle of the micro mirrors to achieve images of adequate contrast. Therefore, such ray angle requirements with these reflective SLMs essentially set limits on the effective NA or F/# of the eyepiece optics or the available throughput of an HMD system. Furthermore, the refresh rate of the liquid crystal used in some LCoS microdisplays may be inadequate to provide a full color display using a single panel with the color sequential method. In this case, either the color depth must be compromised, or multiple display panels must be employed with a bulky illumination system to achieve full color. Alternatively, the fast response speed of ferroelectric liquid crystals or DMD devices, makes it possible to achieve a full-color display using a single-panel with a color sequential scheme.

Choosing the right size of microdisplay plays a critical role in both the optical performance and ergonomics of HMD systems. For HMD designs, smaller panels have the advantage of enabling a more compact system, but often at the cost of a narrower FOV than is possible with a larger panel, because more optical magnification is required to achieve an equivalent FOV with a smaller panel. As given by Eqs. (8.4–8.7), a smaller panel requires a shorter focal length for the eyepiece than a larger panel to achieve the same FOV. Furthermore, as suggested by Eq. (8.11), a smaller panel affords smaller throughput when assuming the same eyepiece NA, or requires a higher NA for the eyepiece than a larger panel to maintain the same throughput. When the display panel is too small, it requires a shorter focal length and larger magnification to achieve a reasonable FOV, which leads to a challenging design of a low f/# system. To provide the same pixel count, smaller panels must use a smaller pixel pitch than larger ones. Microdisplays with small pixel pitch require more optical magnification to create images of adequately magnified pixels for viewing and impose daunting challenges on the optical design in terms of optical performance requirements.

8.3.3 HMD Optics Principles and Architectures

The microdisplays used for HMDs are too small and are placed too close to the eye to be viewed directly. Instead, an optical viewer, commonly referred to as an eyepiece, is required as a core element in both

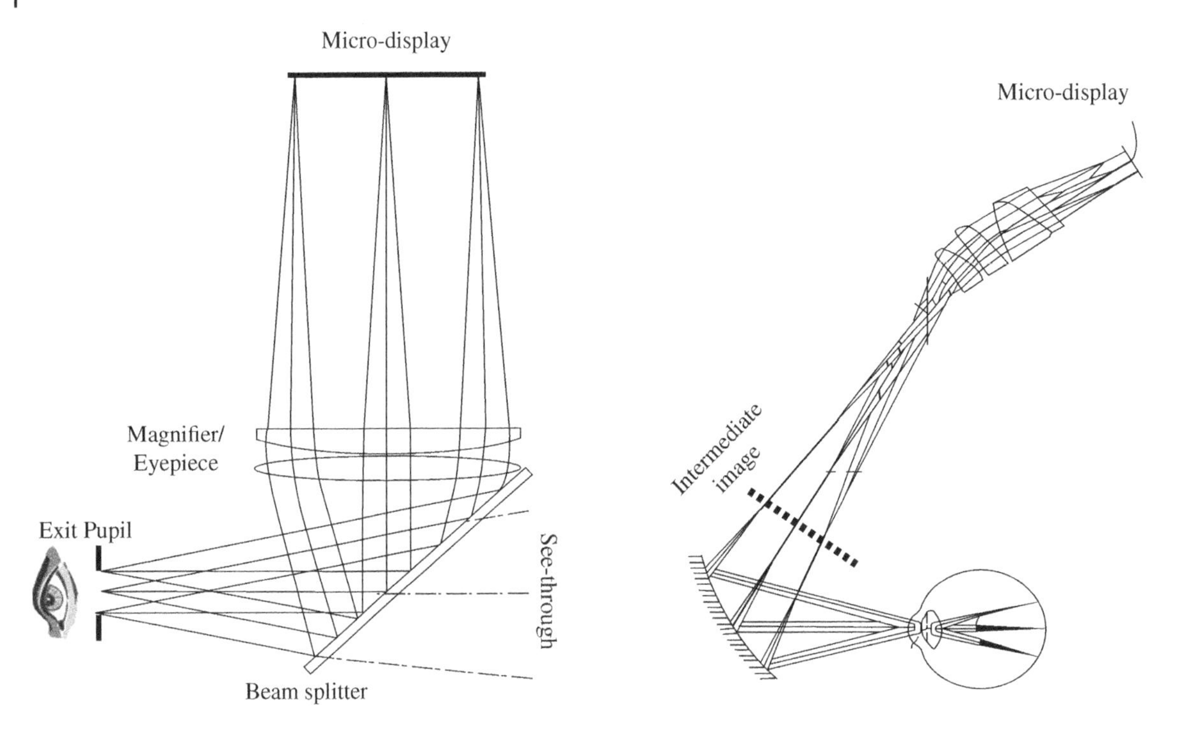

Figure 8.10 Examples of (a) non-pupil forming and (b) pupil-forming optical design for HMDs [37].

immersive and see-through HMDs. It collects light rays from the microdisplay source, forms a magnified image of the microdisplay apparently located at a distance for comfortable viewing, and properly couples the light rays into the eye. It can be thought of as an optical interface to the HVS.

Design of compact, lightweight, and high-performance eyepieces for HMDs has been researched over several decades, which has led to many different forms of optical architectures. The existing forms of optical designs for HMDs may be categorized into non-pupil forming and pupil forming types. In non-pupil forming designs, as illustrated by the example in Figure 8.10a, there is no other location inside the eyepiece optically conjugate to the exit pupil of the optics limiting the passing ray bundles. In this case, the entrance pupil of the viewer's eye serves as the stop aperture of the optical system. This type of system typically has a more compact form, and allows a large range of eye locations axially from the nominal exit pupil location, because a user will see the virtual image of a given point as long as a possible light path exists between the corresponding point on the microdisplay and the eye. The main disadvantage for non-pupil forming design is that the optics size and aberrations both scale with increasing eye pupil distance, which causes difficulty in folding the optical path with a flat beamsplitter to provide a see-through capability, without significantly reducing the FOV. On the other hand, as illustrated in Figure 8.10b, a pupil-forming viewer design forms an intermediary image conjugate to the microdisplay source besides the final virtual display and its optics provide an internal stop limiting the transmitted ray bundles. In this case, it is necessary that the optics form optical conjugates between the internal stop and the exit pupil. A large mismatch in the conjugates will cause part of or the entire virtual display image to disappear.

From the point of view of optical structure, the existing designs for HMDs fall into four broad architectures: magnifier type, objective-eyepiece compound type, projection type, and retinal scanning type. Figure 8.11a–d show examples for each type, respectively. As shown in Figure 8.11a, a magnifier type is the simplest form of optical viewer where the microdisplay is located at or inside the back focal point of the optics so that a magnified virtual image of the microdisplay is formed. The light rays emerging from the optics do not come to a real focus which can be cast on a screen, but appear to be divergent or collimated and not detectable with a

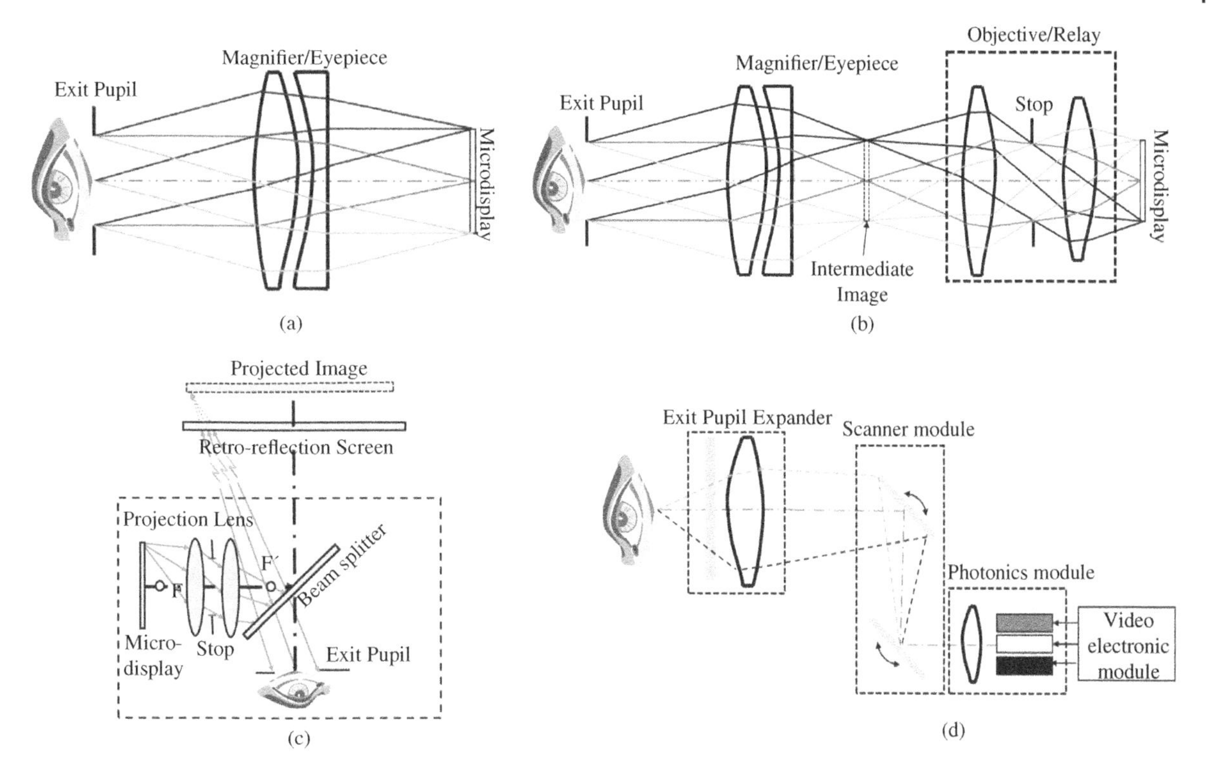

Figure 8.11 Examples of (a) magnifier, (b) objective-eyepiece compound, (c) projection, and (d) retinal scanning types of optical design forms for HMDs.

screen. Through such a simple magnifier, there is no intermediate image surface conjugate to the microdisplay source within the optics, and thus the simple magnifier type viewer is a non-pupil forming design.

An objective-eyepiece compound design, shown in Figure 8.11b, consists of an objective lens group and an eyepiece group. The objective group functions as a relay lens and forms an intermediary image of the microdisplay and typically has an internal stop as the system aperture limiting the passing ray bundles, while the eyepiece group magnifies the intermediate image to form a magnified virtual image for viewing and relays the internal stop to form a conjugate image at the exit pupil of the system. Therefore, this compound type of optical viewer is naturally a pupil-forming type of design and inherits the limitation of a more constrained eyebox. It is worth noting that this compound type of architecture is gaining popularity in many of the emerging HMD designs such as vari-focal or multi-focal plane HMDs (see Section 8.5.3) which require a vari-focal element (VFE) to allow the focal depth of the virtual display to be varied. In such systems it is preferred to place the VFE at an internal stop location optically conjugate to the exit pupil such that the apparent FOV angle of the virtual display is independent of the varying optical power of the system [38, 39]. More details about such HMD designs can be found in Section 8.5.3.

The projection type of viewing optics is an alternative optical design for HMDs, used in the head-mounted projection display (HMPD) technology pioneered by Kijima and Hirose [40], Fisher [41], Fergason [42], and extensively researched by Hua and Rolland's research groups [43–53]. As illustrated in Figure 8.11c, a monocular configuration of an HMPD consists of a miniature display, a projection lens, a beamsplitter combiner, and a retroreflective screen. In contrast to a conventional HMD, an HMPD replaces the magnifier-type or objective-eyepiece compound type optics with a projection lens. An image rendered on the miniature display, which is located beyond the focal point of the projection lens, is projected through the lens and thus forms a magnified real image instead of a virtual image. The combiner, such as a flat beamsplitter, is placed after the projection lens and reflects the light from the projection lens toward the retroreflective screen which replaces the diffusing screen typically used in a conventional projection system. The light rays projected onto the retroreflective screen will be redirected back toward the beamsplitter in the opposite direction to the incident rays. The retroreflected light forms a magnified virtual image for the eye to view at the exit pupil of the system which is located at the conjugate position of the aperture of the projection lens. An HMPD design is thus a pupil-forming type of design. Compared to a conventional projection system with a diffusing screen, the retroreflective screen only reflects the projected light rays back along the opposite direction, therefore a virtual image is formed at the same location as the real image projected by the projection lens, independent of the location of the retroreflective screen and it can only be viewed at the exit pupil. Due to this critical virtue, a binocular HMPD system can be configured by integrating a pair of head-mounted projectors and the unique combination of projection and retroreflection eliminates crosstalk between the projected images for the left and right eyes and thus enables stereoscopic and multi-viewing capabilities. Figure 8.12a–d demonstrate the optical design and prototype of an HMPD system prototype by Zhang and Hua [7], an augmented virtual environment built upon HMPD technology [46], and an application example captured through the prototype [51]. Typically, the components inside the dotted-line box in Figure 8.11c are integrated into a head-worn device, while the retroreflective screen is positioned remotely, away from the user. To facilitate mobility, work by Martins et al. [49] demonstrated a design that integrates a screen within the head-worn device. A detailed analysis of the imaging properties of retroreflective screen can be found in [53] and a more general review of HMPD technology and its development can be found in [52].

Another form of viewing optics found in HMDs uses scanning, and has mainly been used in retinal scanning displays (RSDs) which were pioneered in the 1990s [54, 55]. As shown in Figure 8.11d, a RSD consists of a photonics module, a video electronic module, a scanner module, and an exit pupil expander [56]. The photonics module generates light beams of the desired shape in different colors. Typically, LEDs or laser diodes of different colors and beam shaping optics are often needed to collimate and filter the beams. The video electronics module converts the input video signals to signals needed for modulating the intensity and the mix of the color beams and generates the timing control signals needed for the scanner module. For instance, the intensity of the LED or laser-based beams can be controlled directly by varying their driving current. Alternatively, an external acoustic-optic or electro-optic modulator may be used for intensity modulation of the beams [55]. The scanner module receives the beams of light from the photonics module and directs them to a desired location on an image plane. Examples of scanning techniques being explored for RSD include rotating polygons, galvanometers, piezoelectric deflectors, acousto-optic scanners, holographic scanners, and micro-electro-mechanical scanners (MEMS) [57]. A comprehensive review and comparison of some of these scanning techniques in terms of their resolution, speed, and cost can be found in [58]. Among these scanning techniques, MEMS-based biaxial scanners, in which a small mirror is suspended with two flexures in a gimbal and actuated by electrostatic plates underneath the mirror, are capable of operating at high enough scanning speed to support a high-resolution (e.g. 1280 × 1024 pixel definitions) display of a 2D or even 3D image by repeatedly scanning the beam onto the retina in a pixel-by-pixel fashion. They have been implemented in commercial products by Microvision Inc. (http://www.microvision.com). Due to the nature of beam scanning, the exit pupil natively created in a typical RSD system is very small, typically 1–2 mm, due to the effect of the Lagrange invariant discussed in Section 8.3.1. It is therefore often necessary to include an exit pupil expander to improve the usability and user friendliness of such displays [59].

Each of the four different optical architectures shown in Figure 8.11 has its own advantages and limitations, and has its intended applications. For instance, the magnifier type is the simplest of all and yields the shortest optical path between the display source and the eye. Owing to its non-pupil forming nature, this type of optical viewer is compact and allows a large range of eye locations axially from the nominal exit pupil location, because a user will see the virtual image of a given point as long as a possible light path exists between the corresponding point on the microdispay and the eye. On the other hand, the size of the magnifier as well as aberrations scales with an increase of eye clearance distance and EPD. The objective-eyepiece compound type has a much longer optical path than the simple magnifier type due to the formation of an intermediate image

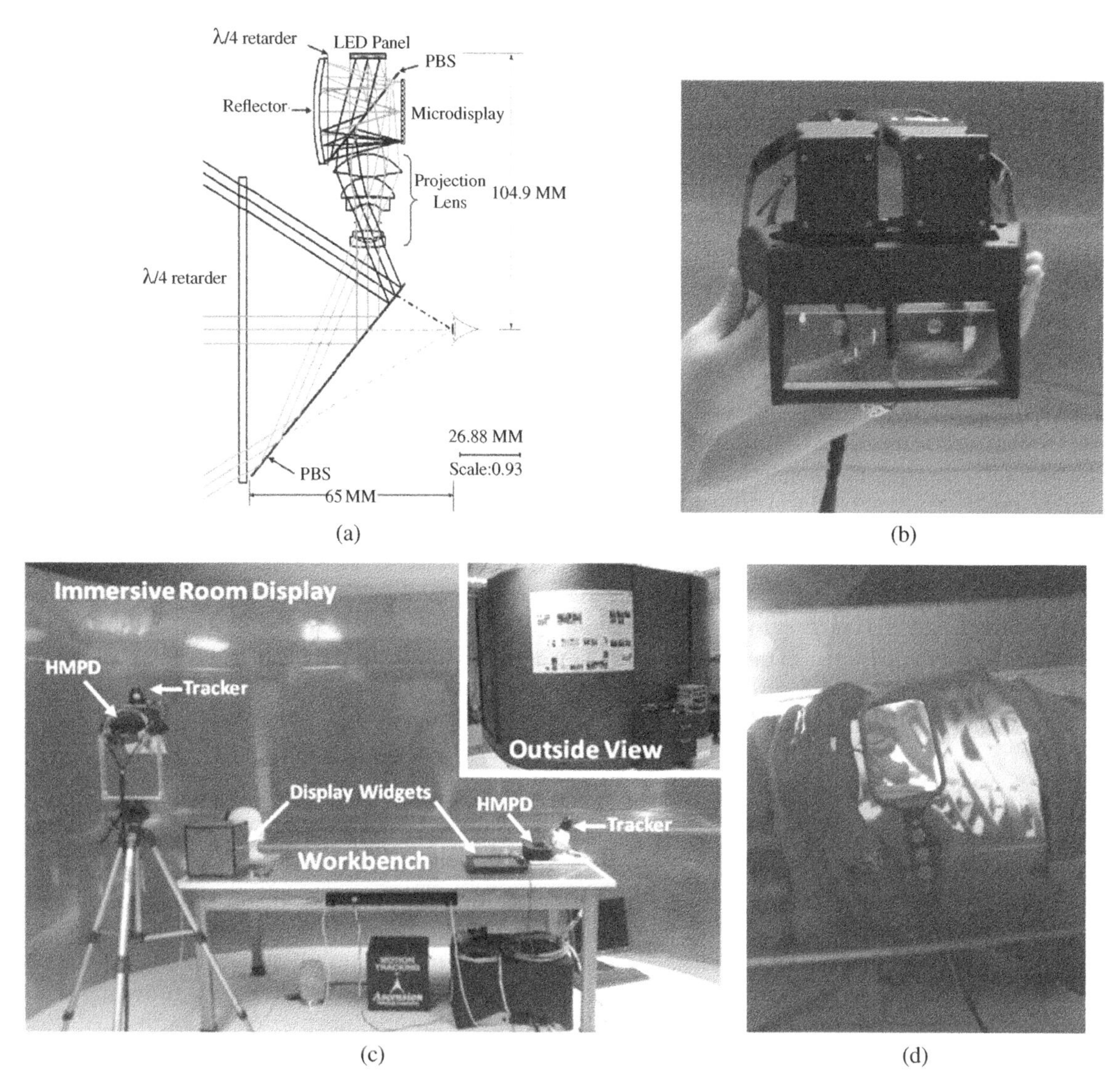

Figure 8.12 (a) Optical design and (b) prototype of an HMPD system; (c) SCAPE – an augmented virtual environment based on HMPDs; (d) an application example captured through the HMPD prototype where a physical hand-held device blocks part of virtual scene (Mars terrain) and is augmented by high-detail view (magnified crater).

and stop-pupil conjugate. It offers opportunities to fold the system and redistribute the weight and volume for improved weight balance and to enable new capabilities such as inserting a VFE for focal depth control (see Section 8.5.3). However, due to its pupil-forming nature, it has a more constrained eyebox. Additionally, the eyepiece size in the objective-eyepiece-compound type HMD designs scales in a similar fashion to the simple magnifier-type with an increase in eye clearance or EPD, thus yielding a system of increased volume, increased weight, and degradation of optical performance. In the case of the projection type design, the stop surface is located within the optics and thus compact and high-performance projection optics can be designed [7, 47, 48]. However, as shown in Figure 8.11c, a beamsplitter needs to be properly positioned and oriented so the user's eye pupil will be comfortably conjugated to the stop of the optics and a retro-reflective screen is required for proper viewing of the projected image. The beamsplitter size then becomes a limiting factor for compactness as the required eyebox and eye clearance increases. Additionally, requiring a retro-reflective screen may not be possible in many applications, which limits the utility of this type of architecture. More importantly, as demonstrated in [53], the image quality perceived via a retro-reflective screen can be severely degraded owing to the unique imaging properties of the screen, and the level of degradation depends on geometric parameters, construction of the screen and the distance between the projected image depth and the screen. Although the RSD design is regaining popularity in recent years with a promise of a more compact and brighter system especially when combined with fiber scanning techniques [60], its limited eyebox, which often leads to the requirement for an exit pupil expander, makes it a challenge to develop systems with wide FOV and large eyebox.

8.3.4 Optical Combiner

A key component in an OST-HMD is an optical combiner which plays a fundamental role in determining the optical performance and system compactness of the display and thus is a key component in distinguishing designs. Several different technologies for constructing optical combiners have been developed, including a flat beamsplitter, a curved or freeform surface combiner, a segmented contact lens, a diffractive or holographic waveguide, and a geometrical lightguide. Table 8.4 provides schematic illustrations as well as design examples of these optical combiners.

Regardless of the combiner types, two of the important parameters to generally characterize an optical combiner are its orientation with respect to the visual axis and its relative distance to the exit pupil of the display because typically the required size of the combiner scales with these parameters. Each of the schematics in the second column of Table 8.4 illustrates the simple geometric relationship between these parameters. Generally, the required size of a combiner in a given direction, $D_{combiner}$, can be simply expressed as

$$D_{combiner} = \frac{2L_{ECLR} \cdot \sin(\theta) + D_{EPD} \cdot \cos(\theta)}{\cos(\alpha + \theta)} \tag{8.12}$$

where L_{ECLR} is the eye clearance distance, D_{EPD} is EPD, α is the orientation of the optical combiner with respect to the direction parallel to the exit pupil plane, and θ is the maximum half field angle to be reflected or transmitted via the combiner in the corresponding direction. Equation (8.12) assumes that the top and bottom halves of the FOV are symmetric. Under the assumptions of an 8 mm EPD and 20 mm eye clearance distance, Figure 8.16 plots the beamsplitter dimension as a function of the full field angle to be reflected or transmitted for three different combiner orientations, 0°, 20°, and 45°, respectively. Clearly, the larger the tilting angle of the combiner with respect to the exit pupil plane, the more rapidly the combiner size scales with FOV. It is generally preferred to maintain an angle of zero, between the combiner and the exit pupil plane.

Among the different types of optical combiners, a flat beamsplitter is simple, low-cost, and readily available in many shapes and sizes. It is typically used in optical viewer designs with rotational symmetry and is oriented at a 45° angle with respect to the exit pupil plane to fold the optical path either vertically or horizontally. Therefore, the folded system generally remains srotationally symmetric. A single flat beamsplitter design is often extended to a dual combiner configuration, as illustrated in Table 8.4, in which a flat beamsplitter is combined with a spherical mirror [61, 67–70]. This configuration is often known as the "bird-bath" configuration. In the first example shown in Table 8.4, light from a display source is reflected off the beamsplitter plate followed by reflection from the spherical mirror before being transmitted through the plate toward the exit pupil [67]. Alternatively, light from a display source is transmitted through the beamsplitter plate followed by reflection off the spherical mirror before being reflected by the plate beamsplitter toward the exit pupil [68, 70]. With a flat beamsplitter, the combiner itself introduces a minimum of complication to the optical design process and smallest change to the image quality of the optical viewer design. However, due to its large tilt angle, the combiner size scales up rapidly with the transmissive or reflective FOV of a system as well as the eye clearance distance and can become impractical when the required FOV is large, as shown by the green curve in Figure 8.13. Figure 8.14 demonstrates an example of how a flat beamsplitter scales with FOV using the

Table 8.4 A survey of optical combiner technologies and design examples.

Combiner type	Schematic illustration	Design example
Flat beamsplitter	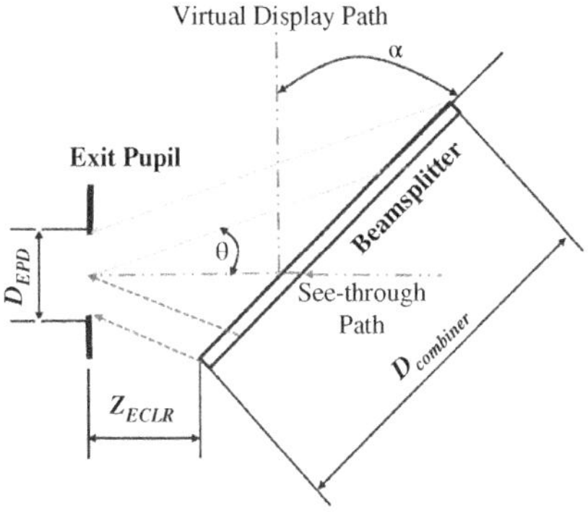 Virtual Display Path; α; Exit Pupil; Beamsplitter; θ; D_{EPD}; See-through Path; $D_{combiner}$; Z_{ECLR}	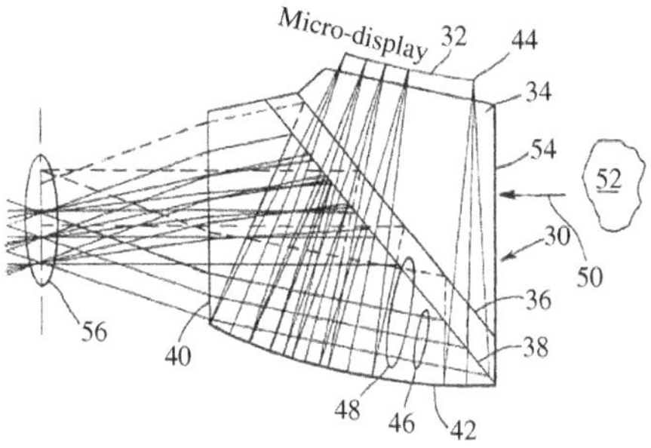 Micro-display 32 44 34 54 52 30 50 36 38 56 40 48 46 42 Ref: US Patent 5,506,718 [61]
Curved combiner	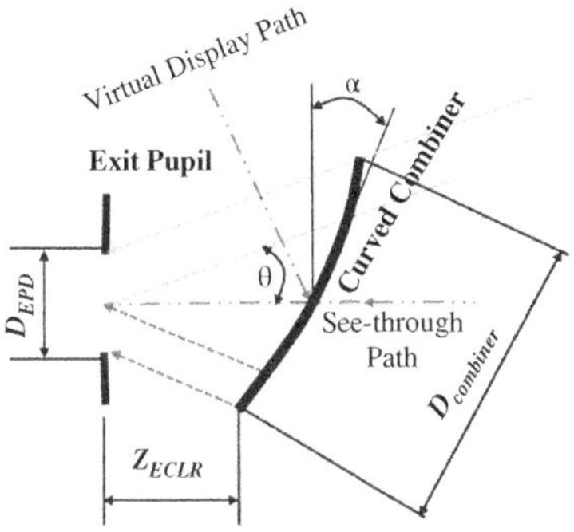 Virtual Display Path; α; Exit Pupil; Curved Combiner; θ; D_{EPD}; See-through Path; $D_{combiner}$; Z_{ECLR}	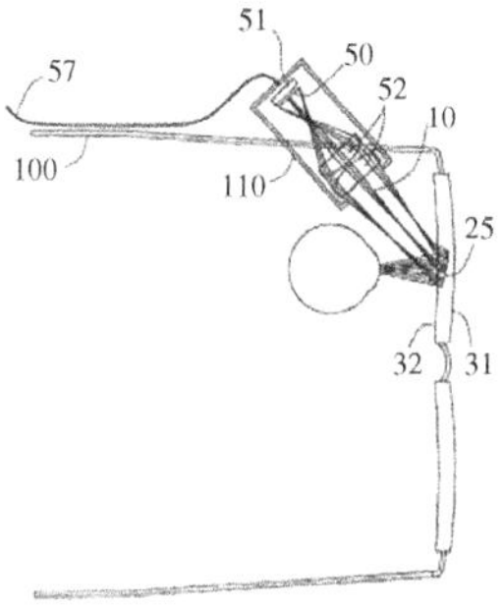 51 57 50 52 10 100 110 25 32 31 Ref: US Patent 6,353,503 [62]
		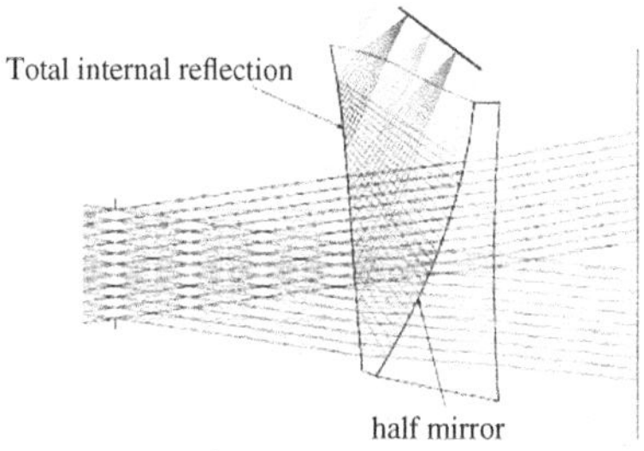 Total internal reflection; half mirror Ref: Cheng et al. [63]
Contact lens combiner	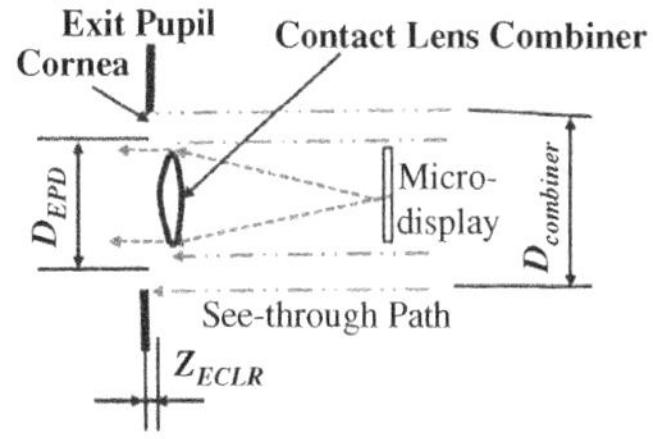 Exit Pupil; Cornea; Contact Lens Combiner; D_{EPD}; Micro-display; $D_{combiner}$; See-through Path; Z_{ECLR}	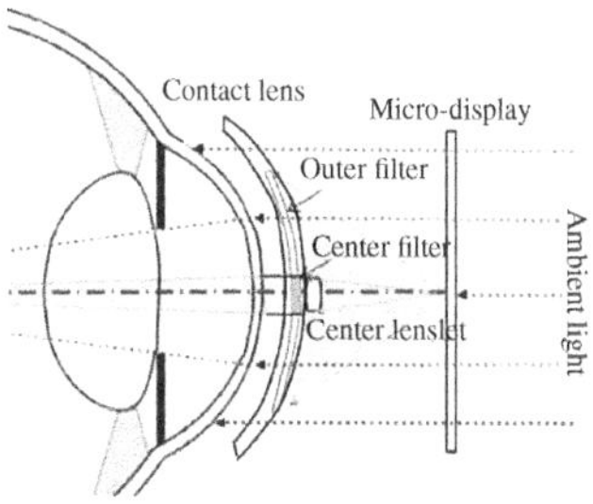 Contact lens; Micro-display; Outer filter; Center filter; Center lenslet; Ambient light Ref: US Patent 8,142,016 [64]

(continued)

Table 8.4 (Continued)

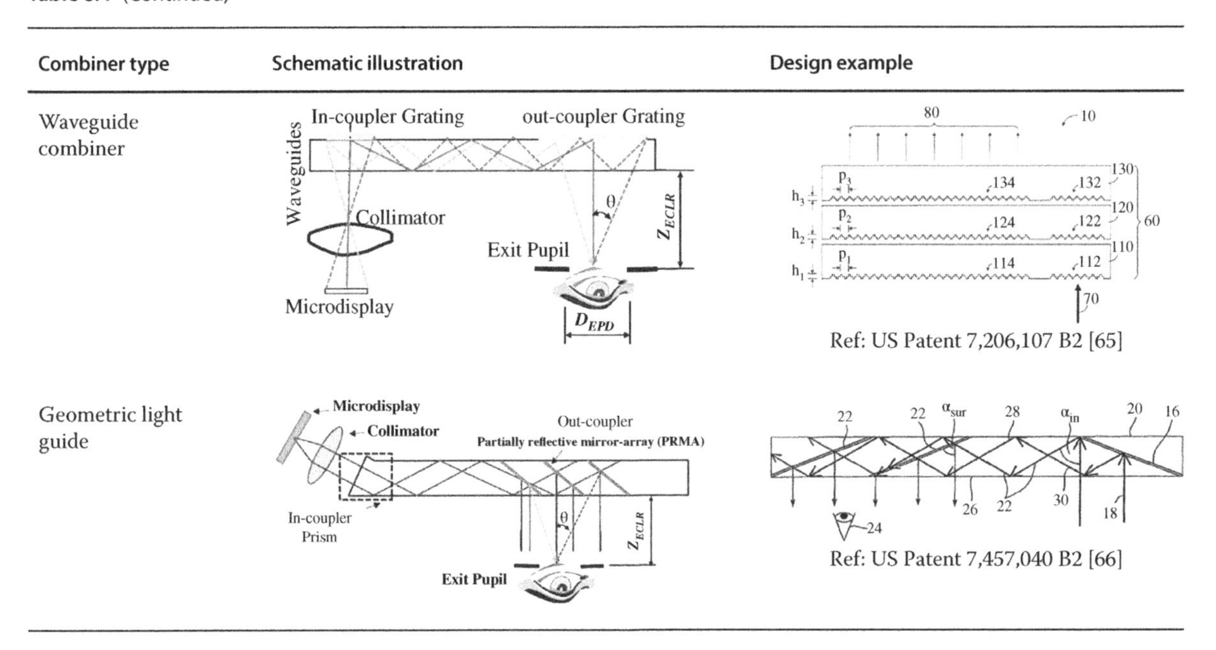

Combiner type	Schematic illustration	Design example
Waveguide combiner		Ref: US Patent 7,206,107 B2 [65]
Geometric light guide		Ref: US Patent 7,457,040 B2 [66]

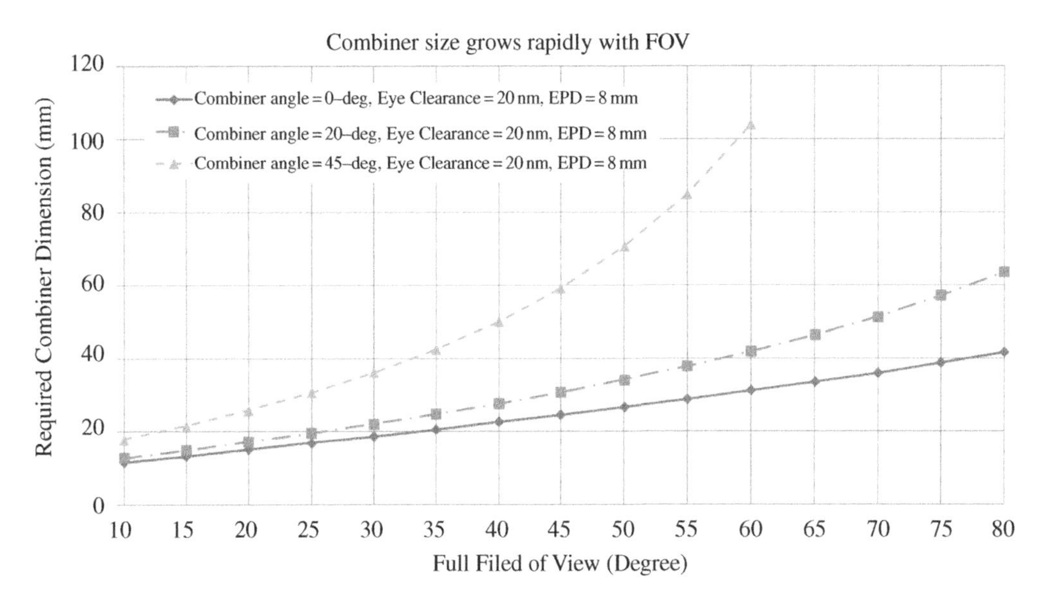

Figure 8.13 Beamsplitter dimension as a function of the full field angle.

dual-combiner configuration. The example assumed an EPD of 8 mm, eye clearance of 18 mm, and a refractive index of 1.5 for the optical parts. The overall thickness of the eyepiece increases from about 13 mm to about 140 mm as the total FOV of the virtual display increases from 12.5° to 70°, and the dimensions of the optics increase to a level that is no longer practically possible with FOV greater than 50°.

A curved or freeform surface combiner can be made of a tilted and curved partial reflector [62, 71, 72], or be a freeform prism with one of surfaces tilted and coated with a beamsplitting coating and cemented with a freeform lens compensator [63, 73–76]. The tilt angle of the beamsplitting surface in such a combiner is

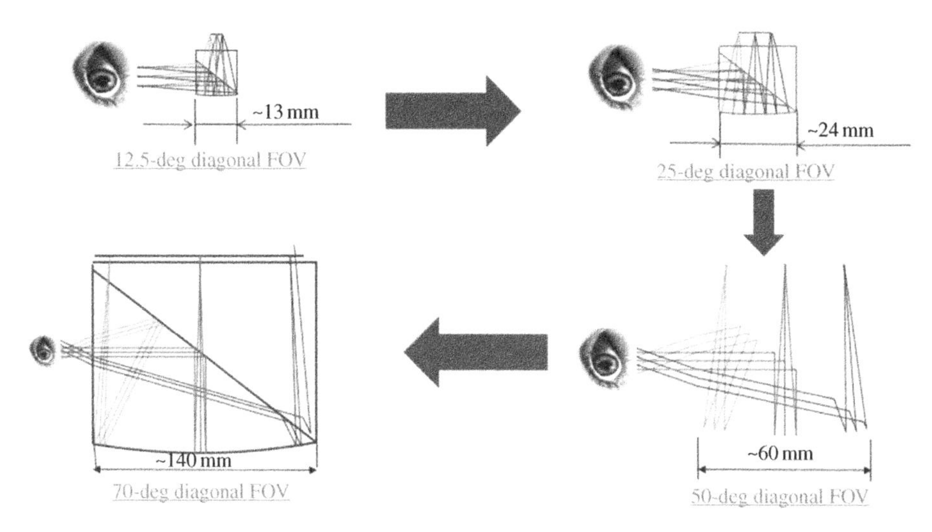

Figure 8.14 Dimensions of a flat beamsplitter increase rapidly with FOV.

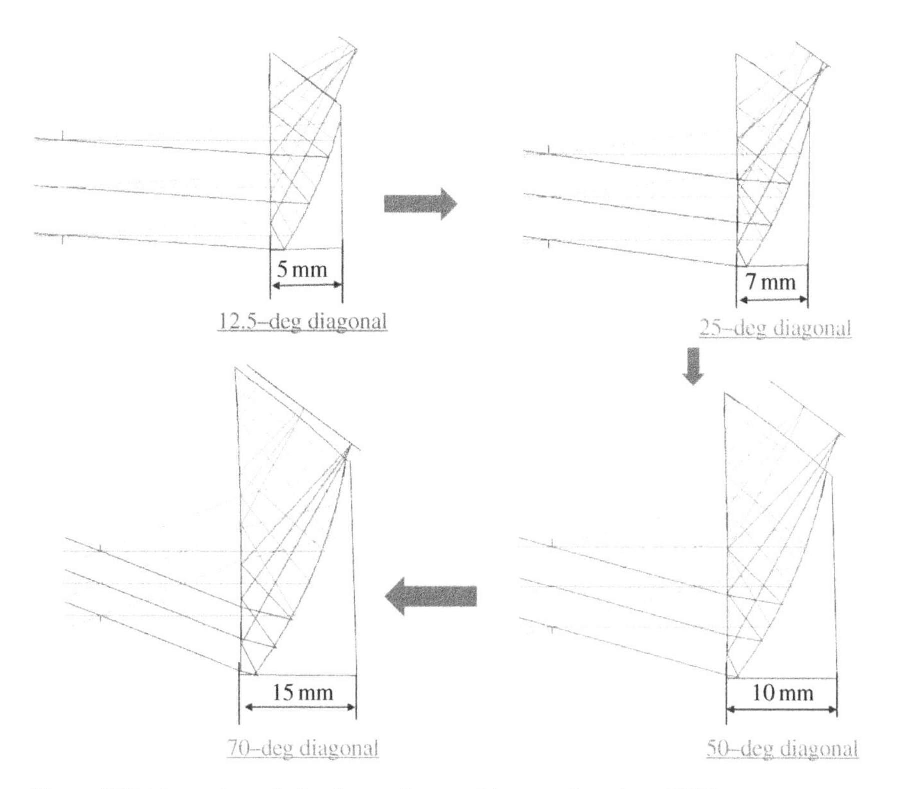

Figure 8.15 Dimensions of a freeform prism combiner as a function of FOV.

typically much smaller than that of a flat beamsplitter, and thus its size is less sensitive to the scaling effects of FOV and eye clearance. It can therefore support a much larger FOV without significantly increasing the size and volume. Figure 8.15 demonstrates an example of how a freeform combiner scales with FOV using the well-known freeform prism architecture. This example assumes the same conditions as those in Figure 8.14 in terms of EPD, eye clearance, and refractive index. The FOV of the virtual display was also scaled in the same range. The overall thickness of the freeform eyepiece increases from about 5 mm to about 15 mm as the total FOV of the virtual display increases from 12.5 to 70°, and the dimensions of the eyepiece remain practically acceptable with a FOV of 70°. A number of OST-HMD prototypes based on this freeform prism architecture or a modified structure have been designed recently for systems with FOVs from about 30° to as large as 100° [63, 74–76]. Although the capabilities and functionalities of these systems vary a lot, the overall thickness of the prism eyepiece remains approximately 20 mm or less and all of the systems achieved high optical performance and delivered optical resolution compatible with microdisplays with a pixel density as high as about 6000PPI, equivalent to a pixel size of about 4.5 μm. The main drawbacks of the freeform prism combiners are their complexity and cost for design, fabrication, and testing for prototyping purposes. Also, a freeform prism combiner is still considerably thicker than some of the other combiner solutions such as waveguide or lightguide combiners.

A segmented contact lens combiner, [64], as illustrated by the schematics in the Table 8.4, is attached to the cornea of the eye. It is divided into two concentric zones. The inner zone is an embedded lenslet providing the necessary optical power for magnifying the image rendered on a microdisplay mounted on an eyeglass frame and collimating the light rays from the display. It also has an embedded filter which blocks light rays from the real-world scene from entering the center lens and only transmits the rays from the display. The outer zone is a filter with its shape conformed to the shape of the cornea. It blocks rays from the display but transmits light from the real-world scene. The nearly negligible eye-clearance distance effectively eliminates the scaling effects of the FOV on combiner size and offers the potential of creating a wide FOV system. On the other hand, the diameter of the embedded lenslet is much smaller than the entrance pupil diameter of the eye and thus may potentially limit the image quality. Additionally, the requirement to wear a contact lens is more invasive and may not be broadly acceptable for consumer applications.

A waveguide combiner couples and guides light through a substrate using diffractive techniques. As illustrated by the schematics in the Table 8.4, these grating-based waveguide combiners typically consist of an in-coupler, a guiding substrate, and an out-coupler. The in-coupler couples light rays from a miniature projection system into the guiding substrate along which the in-coupled rays propagate via total internal reflection (TIR) when the incidence angles of the in-coupled rays are greater than the critical angle, as defined by the refractive indices of the substrate and its surrounding media. On reaching the out-coupler, the TIR-guided rays are extracted in the direction of the exit pupil where the viewer's eye is located. The in-coupler and out-coupler are core to a waveguide combiner and different types of diffractive techniques are used for couplers, including surface relief gratings (SRG) [65, 77], holographic volume gratings (HVG) [78–81], switchable Bragg gratings (SBGs) [82], and the more recently proposed polarization volume gratings (PVGs) [83, 84]. A SRG, illustrated schematically in Table 8.4, is a reflective or transmissive diffractive optical element (DOE) consisting of periodic deep-slanted gratings made via a mature nanolithography process and is one of the most widely used structures. It was initially developed by Nokia® [65, 77] and is currently adopted by existing waveguide-based AR displays such as the Microsoft Hololens® (https://www.microsoft.com/en-us/hololens) and Magic Leap One® (www.magicleap.com). Though a SRG can be fabricated at low cost, it is very common to observe rainbow effects through this type of grating combiner due to the angular and wavelength sensitivity of the grating. A HVG is a reflective or transmissive holographic optical element (HOE) consisting of a periodic phase or absorption variation through the entire volume of the element which is fabricated in a holographic recording process, though its basic operating principle is similar to that of a conventional grating. When an incident beam satisfies the Bragg phase matching condition, it is diffracted by the periodic variation of the element. A SBG, which is commercially developed by DigiLens Inc., is a Bragg hologram recorded in a liquid crystal/polymer material system with a high index modulation which can be switched due to the LC component. It therefore offers more flexibility to optimize its diffraction efficiency and angular bandwidth [82]. A PVG is a reflective or transmissive bulk periodic LC element which generates periodically slanted planes of refractive index perturbations by applying Pancharatnam–Berry deflectors for waveguide coupling [83, 84]. The fabrication of a PVG is adapted from the process of making the well-known diffractive waveplate (also called polarization grating or optical axis grating), which exhibits periodic spatial variations, by adding a chiral

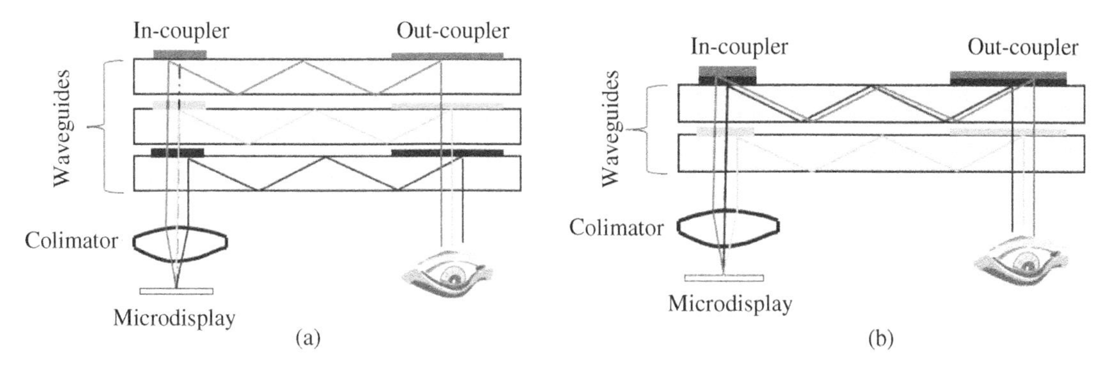

Figure 8.16 Schematic layout of (a) three-layer waveguide design and (b) two-layer waveguide design.

dopant to a nematic LC host to create a helical structure and periodicity perpendicular to the substrate surface [83, 84]. This modification generates periodically slanted refractive index planes in the direction perpendicular to the substrate and creates a volume grating structure. With an adequate number of periodic refractive index planes, Bragg diffraction condition can be established as a diffractive element [83, 84]. In general, these grating-based waveguide combiners have the advantage of being thin and compact, but it is very challenging to achieve a large FOV, high-quality uniform image, and low chromatic artifacts due to its high sensitivity to wavelength and angle of incidence. Due to the generally high wavelength sensitivity, waveguide multiplexing [78, 80] is required where multiple sets of waveguides are required to guide the different color channels separately through different waveguides. Figure 8.16a shows the schematic layout of a typical three-layer waveguide design where light rays from the red, green, and blue color channels are coupled and guided separately through three layers of waveguides. Figure 8.16b shows a layout of a two-layer waveguide design where the red and blue color channels share the same waveguide and the green channel is guided by its own dedicated channels. Waveguide multiplexing can partially suppress some of the chromatic artifacts. However, it is very common to observe rainbow effects in waveguide-based displays due to the variation of spectral reflectivity versus incident angle. Another typically observed artifact is luminance and color non-uniformity across the display's FOV as well as luminance and color variation observed at different positions within the eyebox. The levels of spatially resolvable detail are generally inferior to those achieved by more conventional combiners such as flat beamsplitter and curved or freeform combiners. Last but not least, some waveguide-based combiners have drawbacks such as stray light artifacts due to leakage through the substrates, high fabrication costs, and low yield.

A geometrical lightguide provides an alternative way of guiding light through a substrate and out-coupling light to the eye using arrays of either cascaded beamsplitters or microstructure mirrors. As schematically shown in Figure 8.17a, a beamsplitter-array lightguide uses a cascaded array of partially reflective mirrors (PRMA) as the out-coupling optics, in which the PRMA are coated transversely through the substrate and are placed in parallel at equal distances from each other [66]. Rays propagating inside the substrate are partially coupled out when they are reflected by the PRMA. A major problem of this structure is that complicated stray light paths are caused by unexpected reflections and degrade the outcoupled image performance [86, 87]. As shown in Figure 8.17b, a micro-mirror-array (MMA) geometrical lightguide contains an array of microstructure mirrors coated at the top of wedged grooves [88, 89]. These shallow micro-mirror grooves are spaced apart by uncoated flat regions through which the light rays from real world scenes, shown in red dashed lines, can propagate through the parallel-plate substrate. Compared with holographic waveguides which suffer from angular non-uniformity and color crosstalk issues due to their diffractive nature, geometrical lightguides are immune to chromatic aberration and angular sensitivity. Moreover, the shallow mirror grooves in an MMA are typically hundreds of microns deep and can be easily manufactured by diamond cutting or molding on a plastic substrate, which makes the fabrication process much simpler and lower cost. While a typical freeform prism combiner is at least tens of millimeters thick, a geometric lightguide can be as thin as 3–5 mm, and it

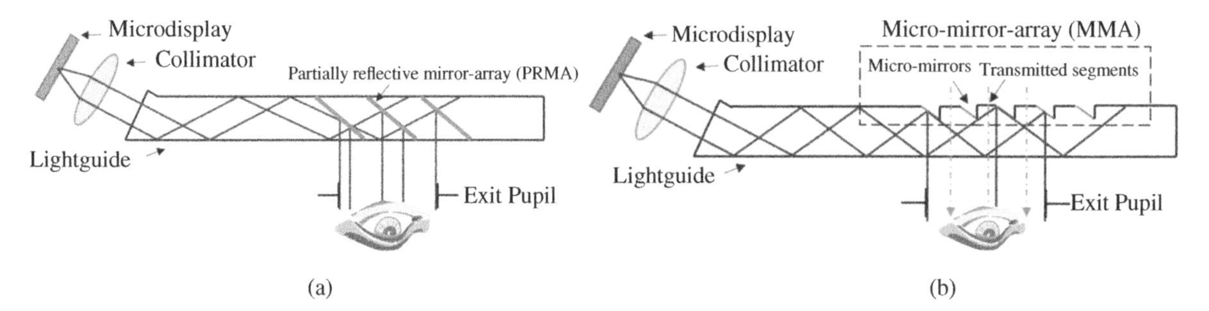

Figure 8.17 Schematic layout of (a) partially reflective mirror-array lightguide and (b) micro-mirror array lightguide [85].

has better photostability than holographic waveguides. Designing and optimizing a lightguide-based AR display, however, presents several major challenges mainly due to the nature of non-sequential ray propagation inside a geometric lightguide, which does not arise in the design of conventional imaging optics. While the optical performance of conventional imaging systems where sequential ray tracing is applicable can be readily evaluated using well-accepted metrics such as modulation transfer functions (MTFs), the non-sequential ray paths in a lightguide require new means of assessing the optical performance for imaging purposes where the toolbox available for non-imaging optics is insufficient. Moreover, the non-sequential ray paths lead to lightguide-based imaging systems being much more prone to artifacts such as non-uniform images and stray light than sequential imaging systems. A recent work by Xu and Hua proposed a framework, a set of quality metrics, and strategies for modeling, optimizing, and evaluating the resulted image quality and artifacts for geometric lightguide-based AR displays [85].

Viewed from the perspective of the Lagrange invariant or etendue/throughput of a system discussed in Section 8.3.1, the different types of optical combiners discussed above may be categorized into two groups, the etendue conserved type and the etendue non-conserved type. The etendue conserved type of combiners include the flat beamsplitter, curved or freeform surface combiner, and segmented contact lens combiner, where the etendue in the viewing space defined by the area of the nominal eyebox and the solid angle of the virtual display. FOV remains the same as the etendue in the microdisplay space defined by the solid angle of the eyepiece NA and the area of microdisplay, as given by Eqs. (8.10, 8.11), respectively. The etendue non-conserved type includes the waveguide and geometric lightguide combiners, where the etendue in the viewing space defined by its eyebox and FOV is typically significantly larger than that in the microdisplay space due to pupil replication effects. This type of combiner is therefore also referred to as an exit pupil expander. The optical viewer is generally designed with a small NA and thus low throughput to ensure its compactness and serves the function of creating a collimated and magnified image of a microdisplay. The collimated light rays are then coupled into and out of the combiner via gratings or cascaded beamsplitter or mirror arrays.

8.4 HMD OPTICAL DESIGNS AND PERFORMANCE SPECIFICATIONS

8.4.1 HMD Optical Designs

Although the four different forms of optical architectures summarized in Section 8.3.3 are well understood by optical designers, designing wide FOV, lightweight, compact, and high-performance HMD optical systems with adequate size of eyebox and eye clearance continues to be a major challenge and pursuit for HMD developers. Over the past decades, researchers have made great advances in HMD optical designs with the advancement of optical technology, improvements in microdisplays, and miniaturization of electronics. Different optical design methods have been applied to HMD systems to improve one or several aspects of the HMD system performance. These methods include applying a catadioptric technique (i.e. hybrid of reflective and refractive surfaces), introducing tilt and decenters to rotationally symmetric systems, and capitalizing on emerging optical technologies that are becoming more readily available such as aspheric surfaces, freeform technologies, DOEs, and HOE.

From an applications perspective, although the four different forms of optics architectures can be used for both immersive and see-through HMD designs, the preferred choice of optics architecture tends to be different. Table 8.5 provides a few optical design examples for immersive HMDs implementing some of the

Table 8.5 Optical design examples of immersive HMDs.

Number	Optical layout	Reference
1	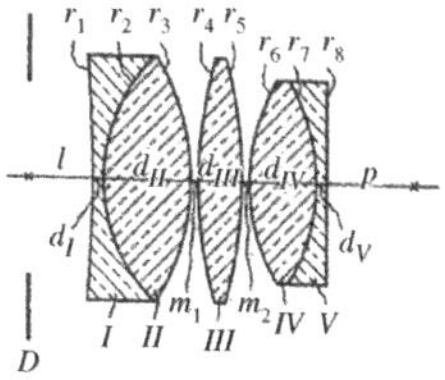	US Patent 1,478,704 Dec, 1923 [90]
2	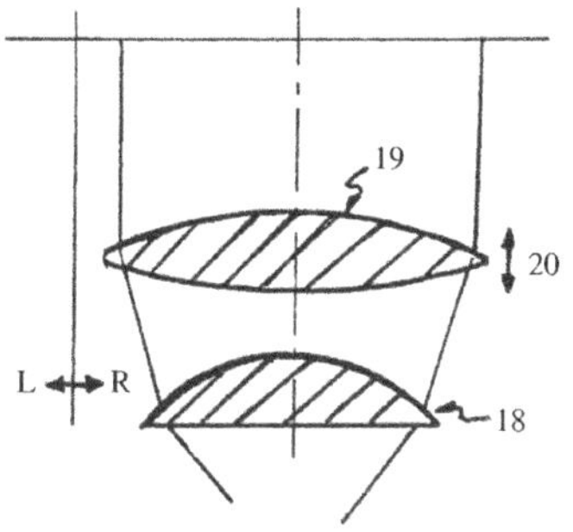	US Patent 4,406,532 1983 [91]
3	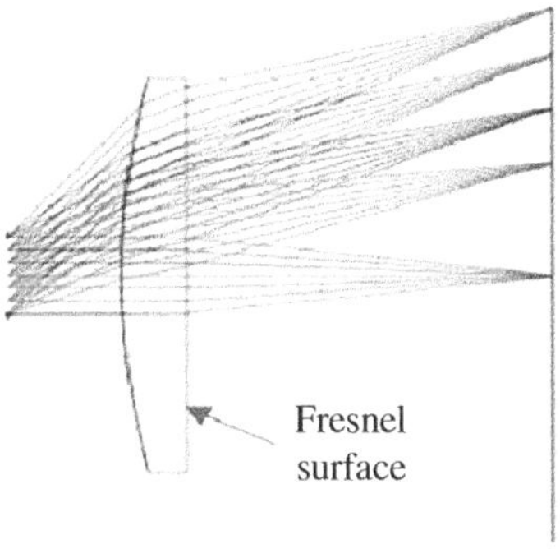	Geng et al. Viewing optics for immersive near-eye displays [92]
4	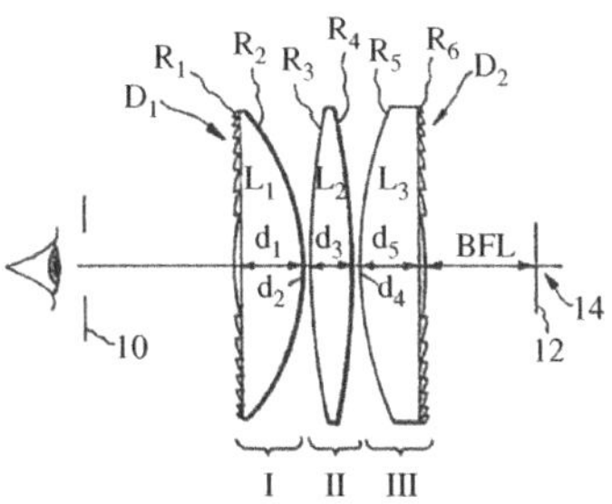	US 5,446,588 1995 [93]

(continued)

Table 8.5 (Continued)

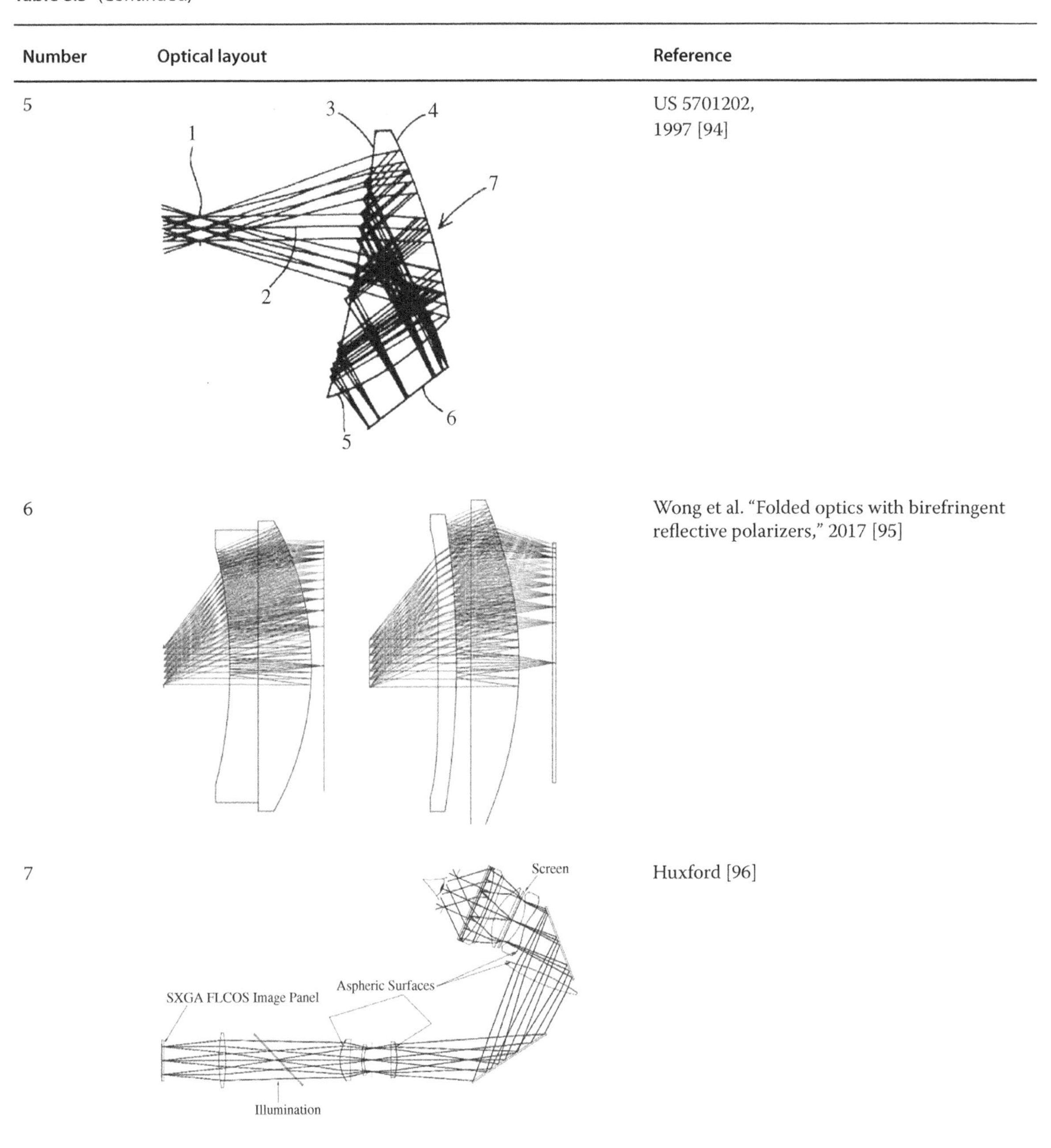

Number	Optical layout	Reference
5		US 5701202, 1997 [94]
6		Wong et al. "Folded optics with birefringent reflective polarizers," 2017 [95]
7		Huxford [96]

different optical design methods listed above. Most of the existing immersive HMDs for VR application adopt the non-pupil forming, simple eyepiece/magnifier type of architecture owing to its simplicity and compactness. The optical design can adopt the classical Erfle eyepiece [90] or other eyepiece designs as shown in the first two examples of the table, using a non-folded layout where multiple rotationally-symmetric refractive elements are stacked together to deliver the necessary magnifying power and correct aberrations. The LEEP optics design shown as example 2 in the table was the first large FOV non-pupil forming optics extensively used in the pioneering phase of VR HMD design [91]. Recent advances in fabricating and testing aspherical surfaces have led to designs of single-element aspheric eyepieces or the use of a Fresnel surface to further reduce volume and weight of the eyepiece in some low-cost VR HMDs systems [92]. Design methods such as DOEs, catadioptric technique, tilt, decenter, and freeform technologies have also been explored for designing compact and high-performance eyepieces for immersive HMDs. Example 4 shows a wide-angle eyepiece design employing a hybrid of refractive and DOEs [93], and Example 5 is a monolithic, freeform eyepiece formed by folding multiple refractive and reflective surfaces into a prism shape [94]. The light rays from a microdisplay are refracted at a surface and coupled into the prism and are then reflected consecutively by multiple reflective surfaces and finally refracted at an exit surface which couples the rays out of the prism into the exit pupil. The pancake design is another method used to reduce volume and weight of immersive eyepieces and achieve a wide FOV. The original pancake window design by La Russa consisted of a single, curved, spherical beam-splitting mirror along with polarization elements [97]. This original design had very low light throughput of about 1–2%. Berman and Melzer improved the light efficiency of a pancake window design by using a cholesteric LC [98]. As shown in example 6, recent advances yield pancake windows with up to 20% transmittance and wide FOV [95, 99]. Tiled configurations of pancake windows have also been reported, providing FOVs as large as about 150° by 50° with 4 arc minute resolution [34]. Besides these non-pupil forming examples, pupil-forming designs can also be found in immersive HMDs. The last example in the table shows a design adopting a pupil-forming architecture, where the first lens group functions as an objective, the second lens group serves as an eyepiece magnifying the intermediate image, and a mirror may be used to fold the optical path and move the center of mass back to avoid the system being front-heavy [96].

Optical see-through HMD designs require an optical combiner and often adopt folded architectures which allow the center of mass to be moved back and keep the system package more balanced in weight. Table 8.6 provides a few optical design examples for OST-HMDs demonstrating the utility of the four different optical architectures discussed above. This table mainly distinguishes the designs by their different optical architectures, while Section 8.3.4 discussed in more detail the different types of optical combiners and their different uses and properties. For instance, the example shown in Figure 8.10a adopts a simple eyepiece/magnifier structure by inserting a flat beamsplitter oriented at 45° between the eyepiece and the eye to provide a see-through path. As shown in Figure 8.13, the optics size as well as aberrations of the magnifier type eyepiece scale rapidly with an increase in FOV and eye clearance distance. Therefore, adopting the flat, tilted beamsplitter in this structure makes it challenging to achieve a wide FOV and compact design. Example 1 shows a catadioptric non-pupil forming design where a flat beamsplitter is cemented between two prisms and a curved reflector is attached to the end of the prism as a reflective magnifier [68]. Example 2 shows a freeform monolithic eyepiece cemented with a freeform see-through lens which corrects distortion which the freeform prism imposes on the see-through rays [73]. The reflective surface S2 is typically a tilted freeform surface coated with a beamsplitting coating. The tilt angle of the freeform beamsplitting surface is typically less than 30° with respect to the vertical direction and thus its size scales much more slowly than a simple flat beamsplitter would, which makes it possible to support a substantially wider FOV in a more compact form than a rotationally-symmetric magnifier. Example 3 shows a multi-channel OSTHMD formed by tiling several freeform eyepiece prisms together to provide a wide FOV and high-resolution system [74]. Examples 4–6 show designs adopting an objective–eyepiece compound architecture where an intermediate image is formed. Example 4 shows a design using a combination of a dual combiner and off-axis optics in a pupil forming structure [100]. Example 5 is a design with a tilted, curved combiner and off-axis objective optics with DOEs [101], while the design in example 6 is adapted from the non-pupil forming freeform prism eyepiece with an objective/relay lens group [102]. Examples 7–9 show designs using a waveguide combiner [78], beamsplitter-array lightguide combiner [86], and a mirror-array combiner [85], respectively. Example 10 shows a design adopting the projection optics architecture [48]. Example 11 is a design of the RSD type [103].

8.4.2 HMD Optical Performance Specifications

Regardless of the optics architecture for the eyepiece or the type of optical combiner (if any), the ultimate goal and also the most challenging task of optical design for HMDs is to achieve high optical performance with minimal aberrations below the levels of perception of the HVS and maintain high yield through dimensional tolerance values compatible with low-cost fabrication processes.

Table 8.6 Optical design examples of optical see-through HMDs.

Number	Optical layout	Reference
1	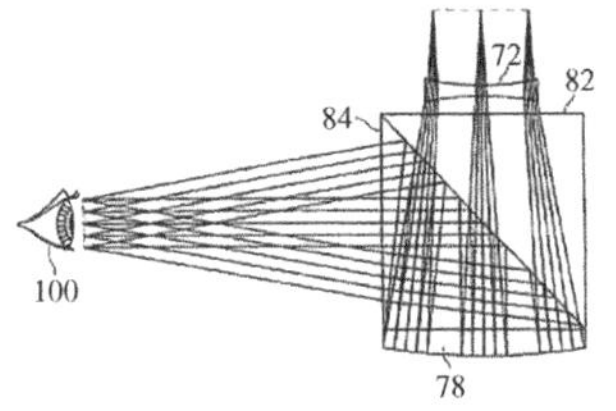	US Patent: 5,696,521 [68]
2	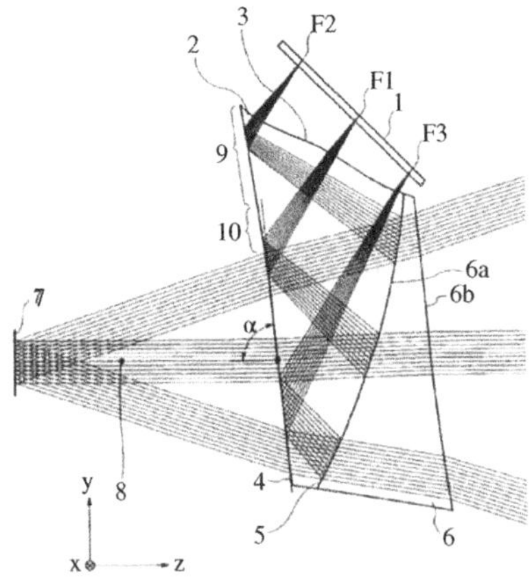	US Patent: 6,384,983, May 2002 [73].
3	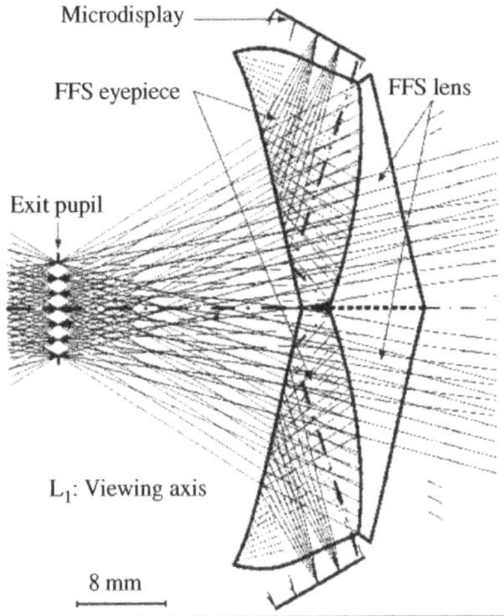	Cheng et al. "Design of a wide-angle, lightweight head-mounted display using freeform optics tiling," Optics Letters [74].

(continued)

Table 8.6 (Continued)

Number	Optical layout	Reference
4	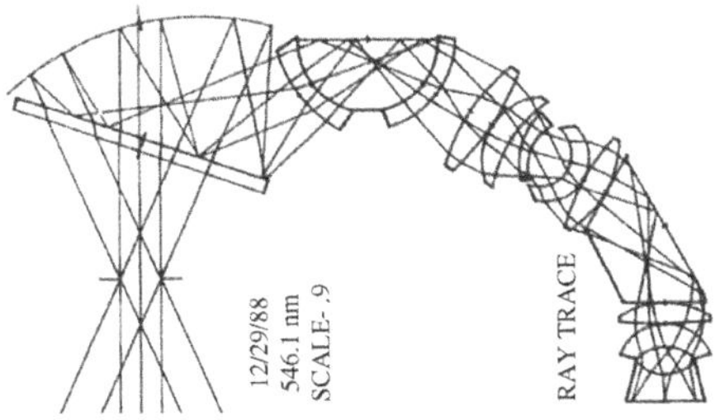	Droessler and Rotier, "Tilted cat helmet-mounted display," Optical Engineering, 29(8) 24–49, 1995 [100].
5	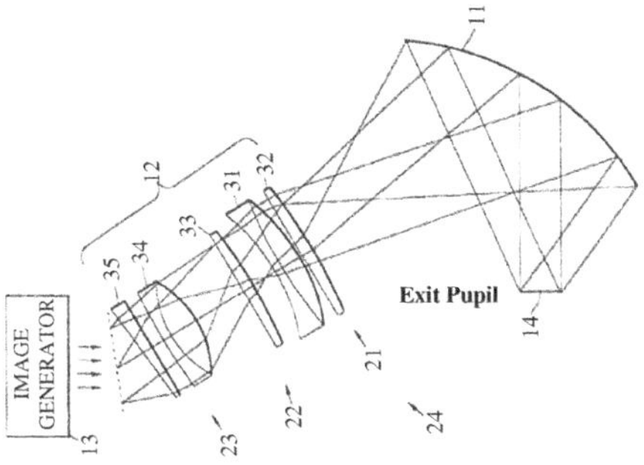	B. Chen, "Helmet visor display employing reflective, refractive, and diffractive optical elements," US Patent 5,526,183, 1996 [101].
6	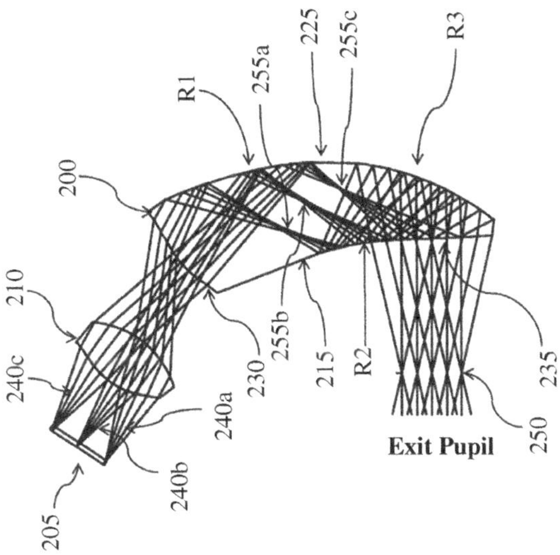	Gao, "Ergonomic head mounted display device and optical system," US Patent 9,740,006 B2 [102].
7	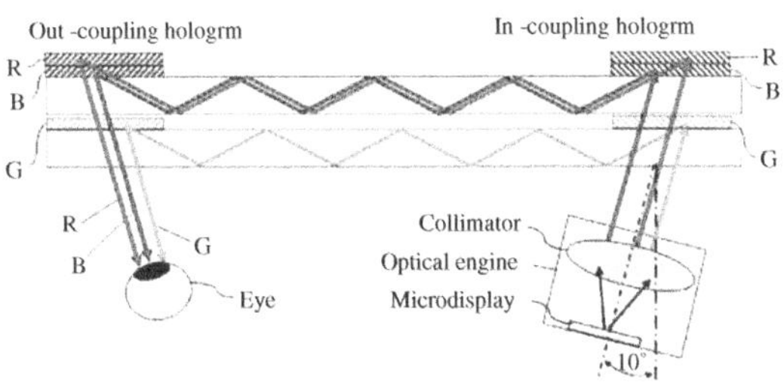	Mukawa et al. "A full-color eyewear display using planar waveguides with reflection volume holograms," JSID [78].

(continued)

Table 8.6 (Continued)

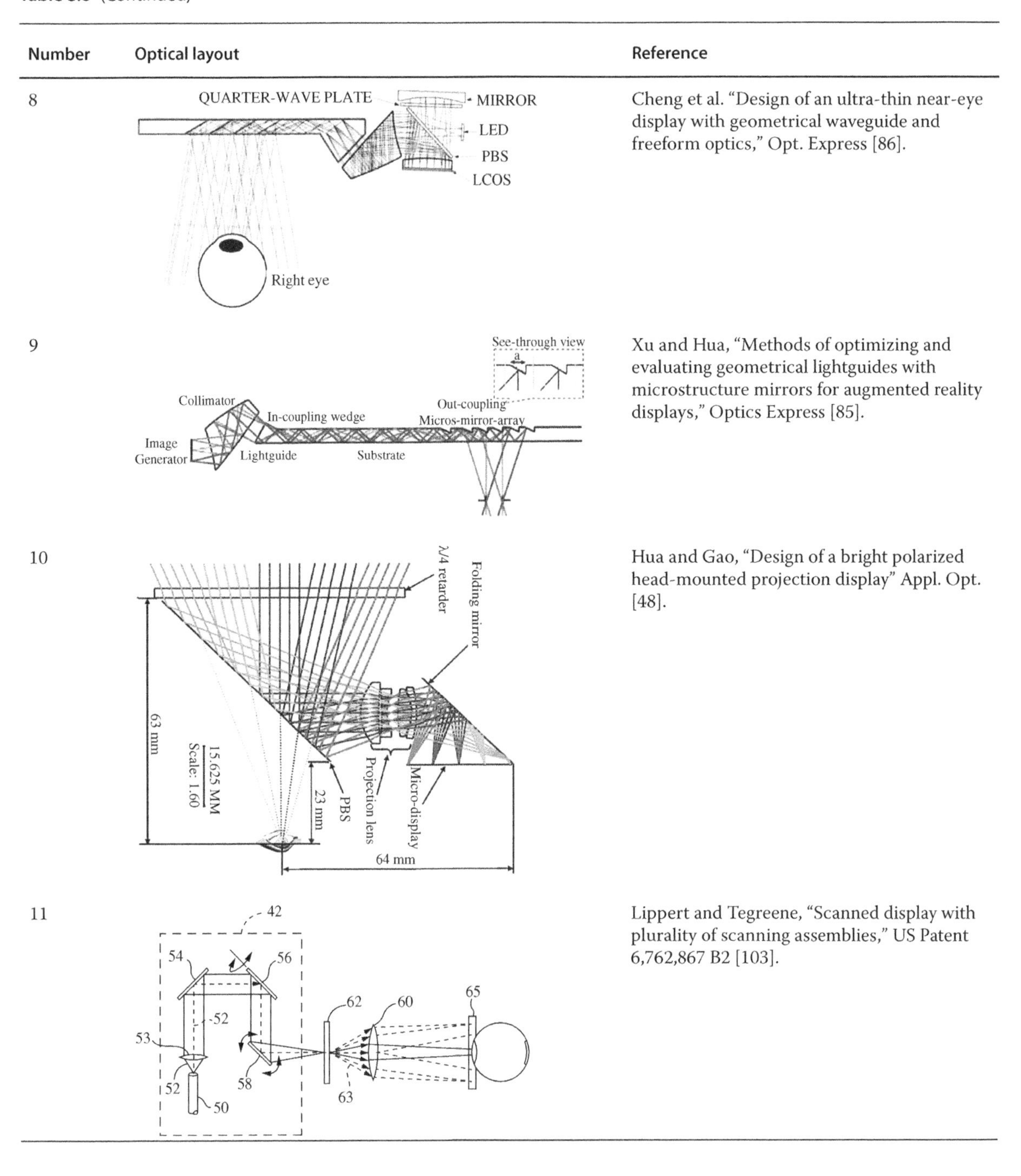

Number	Optical layout	Reference
8		Cheng et al. "Design of an ultra-thin near-eye display with geometrical waveguide and freeform optics," Opt. Express [86].
9		Xu and Hua, "Methods of optimizing and evaluating geometrical lightguides with microstructure mirrors for augmented reality displays," Optics Express [85].
10		Hua and Gao, "Design of a bright polarized head-mounted projection display" Appl. Opt. [48].
11		Lippert and Tegreene, "Scanned display with plurality of scanning assemblies," US Patent 6,762,867 B2 [103].

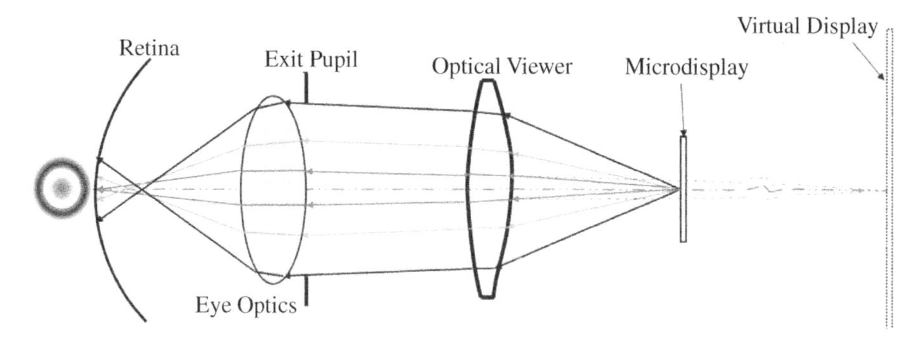

Figure 8.18 Schematic illustration of spherical aberration on image sharpness.

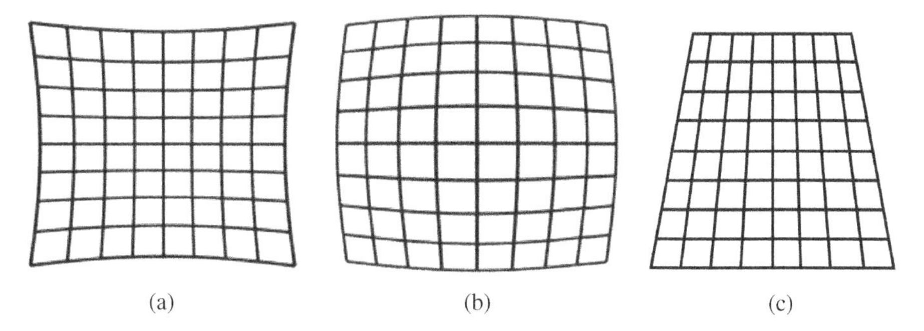

Figure 8.19 Illustration of (a) pincushion (b) barrel and (c) keystone types of distortion of a regular grid.

Optical aberrations are the main sources of errors causing optical performance degradation and low tolerance to fabrication errors. The various aberrations of interest in eyepiece designs may be classified as blurring-type and warping-type aberrations. The blurring-type aberrations cause image blur, degrade image resolution, and reduce image contrast. Examples of the blurring type include spherical aberration, coma, astigmatism, field curvature, and LCAs, and detailed descriptions of these aberrations can be found in [36]. Due to their adverse effects on image quality, blurring-type aberrations should be well-corrected to meet the resolution and contrast requirements of the display. For instance, Figure 8.18 illustrates an example of spherical aberration where the optical power of the system varies with the ray height on the exit pupil. As a result, assuming the eye accommodates at the depth of paraxial focus of the virtual display, rays of different pupil heights are imaged at different lateral positions on the retina and cause image blurring.

Warping-type aberrations, including distortion and lateral chromatic aberrations, do not degrade image resolution or contrast but cause some form of image distortion. For instance, Figure 8.19 shows the image of a regular, evenly spaced grid under three different types of distortions – pincushion, barrel, and keystone – where the optical magnification of the eyepiece varies with the relative height of the field angle from the optical axis in a non-linear fashion. The rays of different pupil heights are imaged onto the same position, but the image position is deviated from its paraxial image position. Figure 8.20 shows the image of the same regular grid through a system with lateral chromatic aberration, where the transverse optical magnification of different fields varies with wavelength. Rays of different wavelengths, such as the red, green, and blue-wavelength rays, appear to be laterally separated, though rays of the same wavelength are imaged sharply at the same position. In general, some amount of warping aberrations in the optics can be acceptable as both the distortion and lateral chromatic aberrations can be corrected by pre-warping the images rendered on the microdisplay in hardware or software. It is essential, however, that any correction is achieved in real time to avoid perceptible artifacts. The minimum required speed of correction depends on applications. For instance, less than 10 ms of delay per frame may be adequate for typical VR applications, but a faster frame rate, e.g.

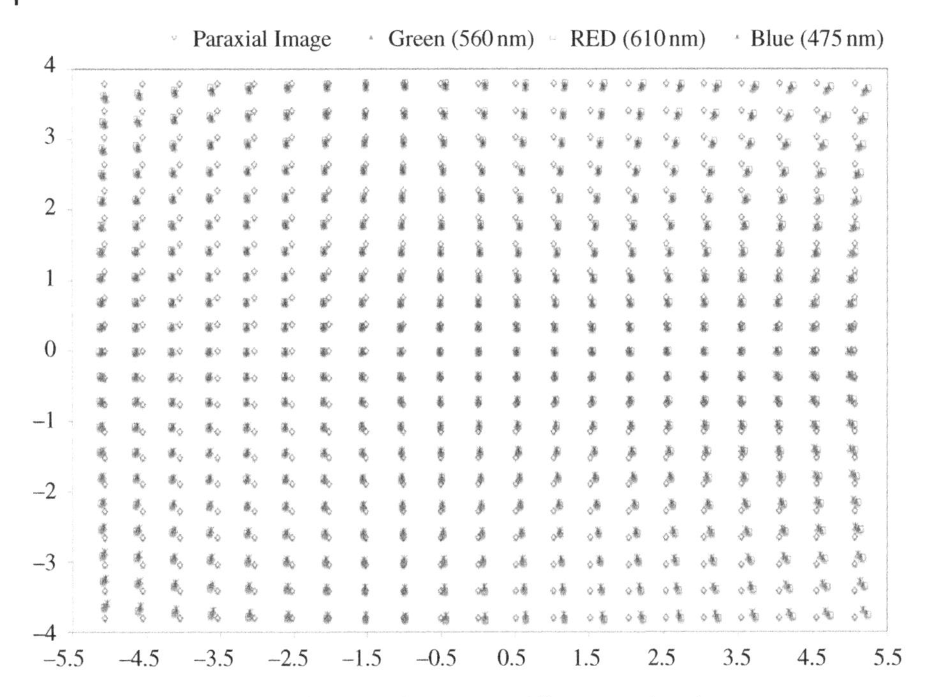

Figure 8.20 Effects of lateral chromatic aberration on different wavelengths.

~1 ms, is necessary in optical see-through AR applications where the real-world scene is seen in real time and delay of virtual-scene rendering leads to perception of mis-registration.

Optical resolution is one of the metrics that are often used to quantify the optical performance of an eyepiece, defining the limiting spatial detail that the eyepiece can resolve. The optical resolution of an eyepiece can be evaluated in the microdisplay space by quantifying the dimensions of equivalently resolvable pixels, or in the visual space by measuring the resolvable angular resolution. For the former, an eyepiece can be modeled by tracing rays from the visual space to the microdisplay space, assuming the virtual display is an ideal object to be imaged, as shown by the example in Figure 8.11a. By tracing a large number of rays of different wavelengths sampled at different field angles through the eyepiece, the intersections of these rays on the plane of microdisplay – known as spot diagrams – can be simulated to estimate the image blur produced by the eyepiece aberrations for a point object, as shown in Figure 8.21. For a given field angle, the centroid of a spot diagram, $(\overline{\varepsilon_x}, \overline{\varepsilon_y})$ is computed to characterize the deviation of the image location relative to its paraxial image by averaging the ray errors, expressed as

$$\begin{cases} \overline{\varepsilon_x} = \frac{1}{N} \sum_{i=1}^{N} \varepsilon_{xi} \\ \overline{\varepsilon_y} = \frac{1}{N} \sum_{i=1}^{N} \varepsilon_{yi} \end{cases} \tag{8.13}$$

where N is the number of rays being traced through the exit pupil for the given field angle, ε_{xi} and ε_{yi} are the ray deviations from its paraxial location along the X- and Y-directions of the image plane. The dimensions of the equivalently resolved pixels of the eyepiece for a given field angle can then be quantified by the root mean square (RMS) spot size and are computed by integrating the ray errors over the pupil by the formulae

$$\begin{cases} P_{\text{RMSX}} = \sqrt{\frac{1}{N} \sum_{i=1}^{N} (\varepsilon_{xi} - \overline{\varepsilon_x})^2} \\ P_{\text{RMSY}} = \sqrt{\frac{1}{N} \sum_{i=1}^{N} (\varepsilon_{yi} - \overline{\varepsilon_y})^2} \end{cases} \tag{8.14}$$

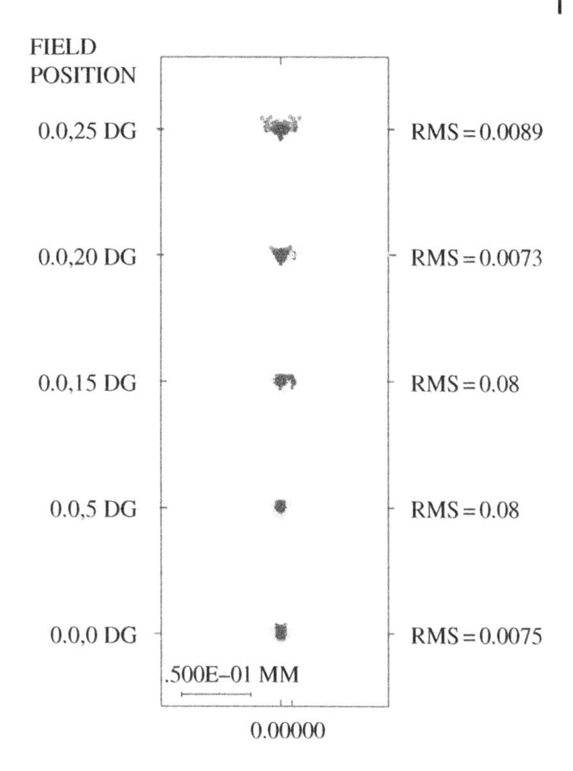

Figure 8.21 Example of spot diagram of an eyepiece design.

For instance, the spot diagram in Figure 8.21 suggests an equivalent pixel resolution of about 7.5 μm in the center and 8.9 μm at the 25° field angles, respectively. The resolvable angular resolution in the visual space can be calculated from the equivalent pixel dimensions by accounting for the eyepiece magnification and the distance of virtual display using Eq. (8.8). It is worth mentioning that the method of quantifying equivalent spatial resolution via a geometric spot diagram is valid only if the diffraction effects of the eyepiece are negligible compared to the effects of aberrations. Otherwise, wave optics propagation is necessary.

Regardless of the optical architecture, the resolution of the eyepiece must be designed to be compatible with the pixel resolution of the microdisplay being used. In other words, the eyepiece should neither under-perform (i.e. optical resolving power is significantly lower than the pixel resolution) nor over-perform (i.e. optical resolving power is significantly higher that the pixel resolution). In the case of under-performing, the valuable pixels rendered by the microdisplay are wasted and the resulting system does not deliver a high-quality image to the Nyquist limit of the microdisplay. In the case of over-performing, the eyepiece may reveal structural details of the display pixels and lead to perception of image artifacts, for instance, the well-understood screen-door effect. For instance, the spot diagram example in Figure 8.21 would be considered as being an under-performing design if the microdisplay pixels were less than 6.5 μm or being over-performing if the microdisplay pixels were 10 μm or larger.

Though equivalent pixel resolution is an intuitive measurement of optics resolution, it fails to convey the resulted image contrast resulting from the design and thus does not fully characterize the performance of an eyepiece. For this purpose, the MTF, defined as the ratio of the modulation in the image to the modulation in the object as a function of the frequency of a sine-wave pattern, is the most commonly used metric for quantifying the performance of an eyepiece. In conventional eyepiece designs, the optics is typically modeled with reversed rays propagating from the visual space to the microdisplay. Under such a raytracing configuration, the MTF performance of an eyepiece is commonly evaluated in the microdisplay space as shown by the polychromatic example in Figure 8.22a, where the spatial frequency is measured in cycles/mm. This type of MTF plot can provide direct assessment of the image resolution and contrast limits of the optics related to the

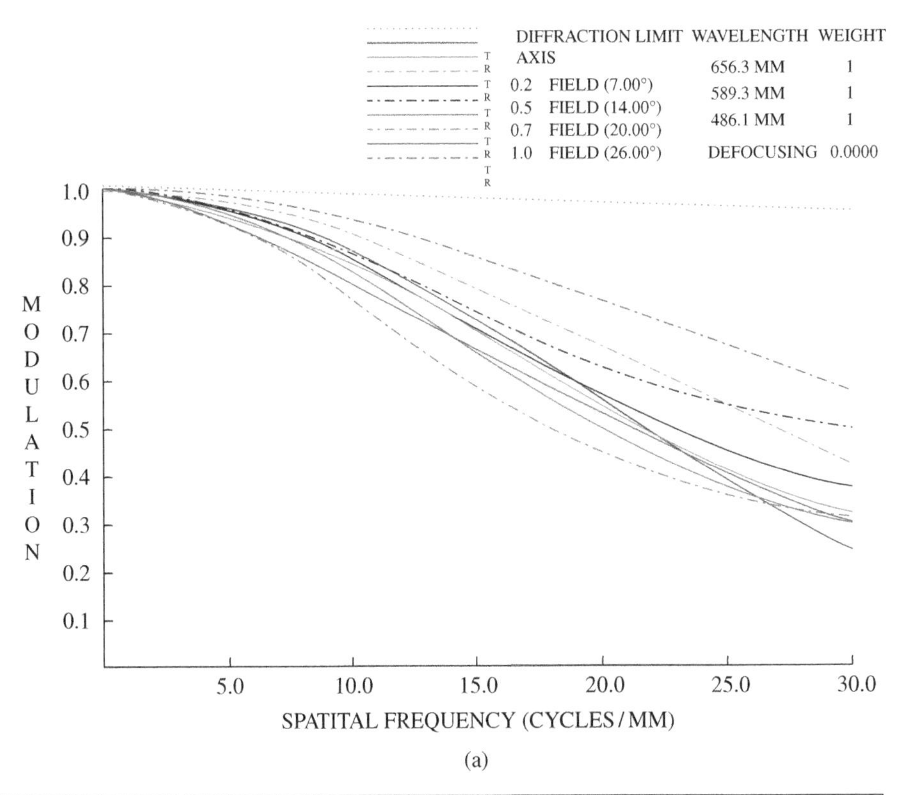

(a)

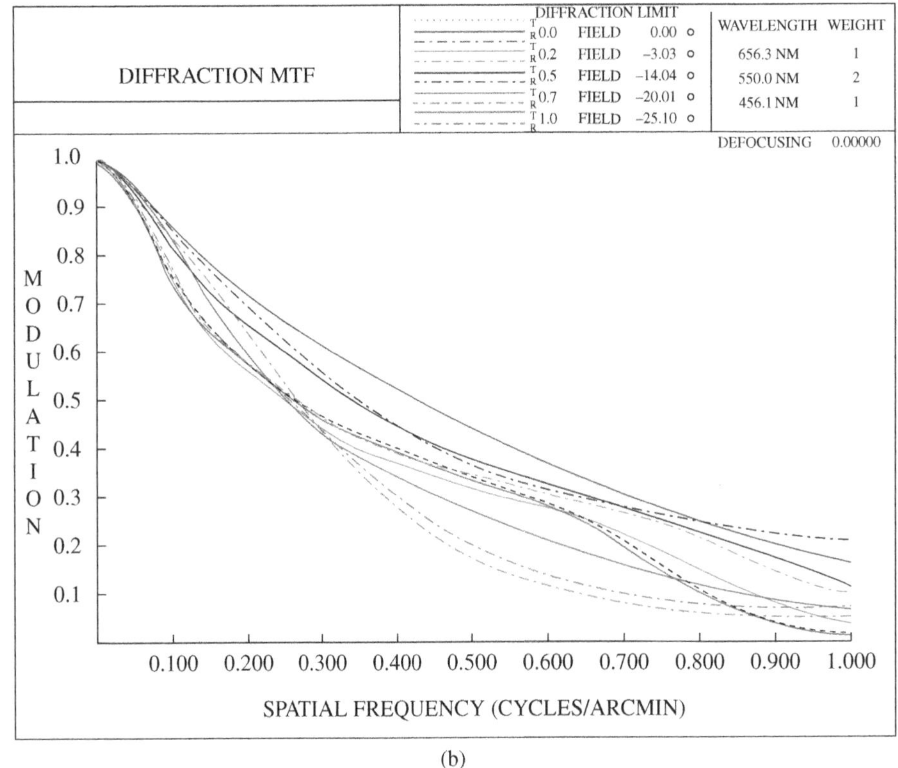

(b)

Figure 8.22 Examples of polychromatic MTF plots simulated (a) in microdisplay space measured in terms of linear frequency of cycles/mm and (b) in visual space measured in terms angular frequency of cycles/arc minute.

pixel dimensions of the microdisplay. For instance, this MTF example shows the performance of an eyepiece with a half FOV of 26°. It suggests that the eyepiece can achieve approximately 75% modulation contrast if used with a microdisplay pixel size of 30 μm and about 40% contrast if used with a microdisplay pixel size of 15 μm.

Evaluating the MTF performance in the microdisplay space is helpful for choosing the appropriate pixel resolution of the microdisplay for the eyepiece, but it fails to evaluate properly, the performance of the optics to the HVS. The MTF in Figure 8.22a does not indicate how the system image resolution and contrast would appear to normal vision as it does not account for the effects of optical magnification and virtual display distance. For instance, a 15 μm-pixel microdisplay being magnified 100 times with the virtual display located 1 m away from the eye would render a poor-resolution virtual image of about 5.2 arc minutes per pixel, while a 30 μm-pixel microdisplay magnified 10 times at the same virtual display distance would render a high-resolution virtual image at about 1 arc minute per pixel. Therefore, it is preferable to evaluate the MTF performance in the visual space in terms of cycles/° or cycles/arc minute by modeling the system with rays traced from the microdisplay to the visual space. A detailed description of the conversion method can be found in [104, 105]. Figure 8.22b shows an MTF example plotted in the visual space for the same eyepiece as the MTF plot for Figure 8.22a. It suggests that the systems can render a virtual image with 55% modulation contrast at 0.3 cycles/arc minute, or equivalent to 1.67 arc minutes of angular resolution and an image with about 40% contrast at 0.5 cycles/arc minute, equivalent to 1 arc minute of angular resolution. Such an assessment provides intuitive quantification of the optical performance of the display system to the HVS.

Both the resolution and MTF assessments above have been performed, evaluating the display optical system alone which is the method commonly adopted for conventional HMD designs. Such a method is generally adequate for evaluating the optical performance of conventional HMDs. However, in many newly emerging HMD architectures, such as multi-focal plane (MFP) displays, light field displays, computational displays, and displays with waveguide and lightguide combiners, the eye optics plays a critical role in the image formation process. In most of these displays, the eye optics cumulatively integrates more than one elemental view to form the perception of each object point and the perceived image of the point often depends on the accommodation status of the eye optics. In these cases, it is essential to model the eye optics together with the display viewing optics and to evaluate the optical performance by simulating the accumulated point spread function (PSF) or MTF of the whole system. Examples of modeling methods can be found in [26–29].

The performance metrics above mainly focus on the various aspects of optical quality for a monocular display. For a binocular display, besides these optical quality metrics, several additional metrics are needed to evaluate the binocular properties, including interocular vertical misalignment, interocular rotation difference, interocular magnification difference, interocular luminance and contrast difference, and binocular disparity. In a binocular or biocular system, the arms for the left- and right-eyes may be built with slight differences due to alignment errors. The eyes are particularly sensitive and intolerant to vertical misalignment. According to ISO 9241-303, the interocular vertical misalignment, illustrated in Figure 8.23a, must be less than 8.6 arc minutes, though a tighter tolerance such as 3.4 arc minutes was recommended in [22, 106]; the interocular rotation difference, as illustrated in Figure 8.23b, defines the relative orientation between the two images and must be less than 1°. An interocular magnification difference, as illustrated in Figure 8.23c leads to image size mismatching and misalignment. According to ISO 9241-303, the overall magnification difference should be less than 1% of the image size and at the edge of the overlapping FOV, it must produce a vertical misalignment of 8.6 arc minutes or less. The eye is less sensitive to any interocular luminance and contrast difference. It is preferable to maintain a luminance difference less than 10%, but can be acceptable with a difference less than 25%. Binocular disparity refers to the slight difference of retinal image positions when the eyes are converged inside (crossed disparity, e.g. point A Figure 8.24) or outside (uncrossed disparity, e.g. Point B in Figure 8.24) the horopter which is defined as a surface of zero disparity on the retinal images. An excessive amount of binocular disparity rendered by a binocular display system prevents the eyes from achieving binocular fusion and objects may be seen double or diplopic. Fusion limits depend on the disparity magnitude, stimulus size, and spatial frequency, and they are generally given to be 10 arc minutes for small disparity, 2° for medium

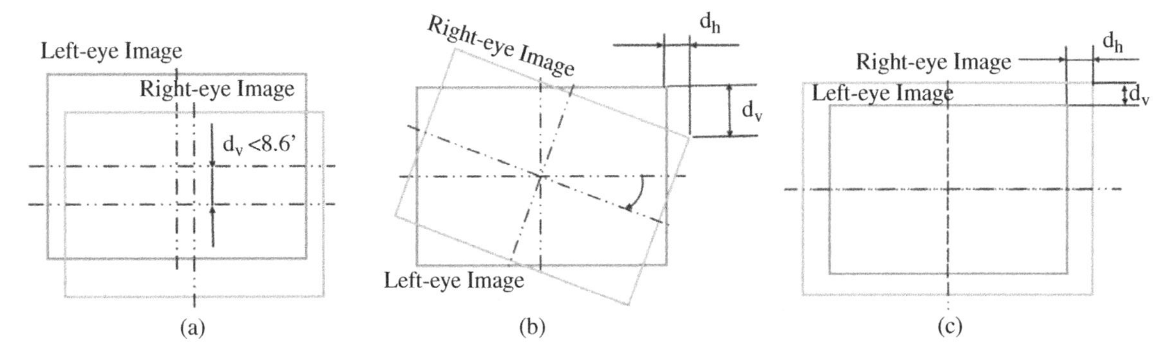

Figure 8.23 Illustration of (a) interocular vertical misalignment, (b) interocular rotation difference, and (c) interocular magnification difference.

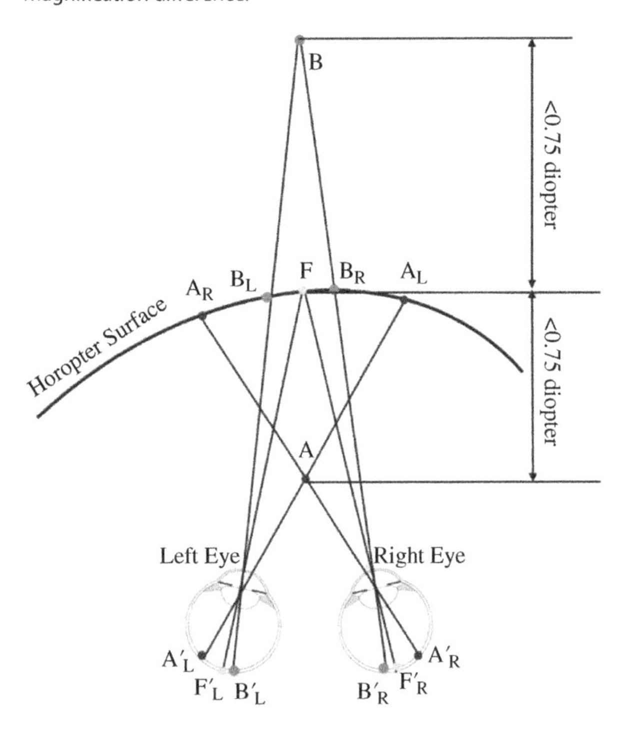

Figure 8.24 Illustration of crossed and uncrossed binocular disparities and the recommended limits in a binocular system.

disparity, and 8° for high disparity. As illustrated in Figure 8.24, a common way of setting disparity limits is by limiting the depth differences between the vergence depth and the virtual display depth. A rule of thumb suggested is that these depth differences should be less than ±0.75 diopters.

8.5 ADVANCED HMD TECHNOLOGIES

Besides the conventional optics architectures and designs reviewed in Sections 8.3 and 8.4, a range of advanced methods has emerged for developing HMD systems with new capabilities to address different persistent challenges in HMD development. This section will provide a review on these emerging technologies, including eyetracked and foveated displays, multi-channel displays, dynamic range enhancements, mutual occlusion capability, addressable focus cues, and light field displays.

8.5.1 Eyetracked and Fovea-Contingent HMDs

Eyetracking has long been considered a key technology and has been applied to several disciplines including vision research, human computer interfaces, tele-operation environments, and visual communication. The benefits of eyetracking for multi-modal human-computer interfaces and the technical benefits of data compression have been well-recognized and studied. Integrating eyetracking capability with HMDs (ET-HMDs), which allow systems to display virtual images as a classical HMD does and additionally track the gaze direction of the user, can potentially offer multiple benefits, not only to fundamental scientific research but also to emerging applications of such technology. In recent years, eyetracking has been viewed as an essential capability contributing toward the creation of advanced spatial computing platforms.

For instance, from the technology point of view, eyetracking capability integrated with HMD systems can enable new user interaction methods with VR/AR applications, enable multi-resolution gaze-contingent schemes for displays, image processing, graphics rendering, and data management, and enable solutions to the vergence-accommodation conflict (VAC) problem (Section 8.5.3). From the applications point of view, many new opportunities will emerge from the technology such as assistive communication.

The approaches to creating an integrated ET-HMD system can be classified into two categories, functionality-level integration and systematic integration. Functionality-level integration brings together standalone and separately developed HMD and eyetracking systems at a late stage of development, and limited opportunity exists to create optimal designs for compact and mobile ET-HMDs. This approach does not take advantage of low-level optimization and its products typically lack the attributes of compactness, accuracy, and robustness. Efforts toward functionality-level integration started as early as the pioneering work of developing high resolution inset displays by CAE Corporation [107]. Other examples include the bench prototype by Iwamoto et al. [108] where a commercial eyetracker was added to a prototype display providing feedback to a mechanical driver which moved a high resolution inset, the bench prototype by Liu and Hua where a binocular eyetracker was added to a vari-focal plane (VFP) display prototype for detection of eye convergence depth [109], and the work by Andy Duchowski [110] where a commercially available ISCAN eyetracker was integrated with a commercially available V8-HMD from Virtual Research Corporation to study a software-based fovea-contingent scheme.

In contrast to the functionality integration approach, a systematic approach seeks as a fundamental design principle, to conceive and optimize both display and eyetracker as one single system. This approach potentially has many advantages in creating a fully integrated ET-HMD device. The greatest benefits of this systematic approach include the ability to explore design constraints and requirements for both the display and eyetracker units, conceive new solutions, and optimize the designs for a compact and robust system. Several pioneering efforts explored the possibility of complete integration with low-level optimization [111, 112]. Following these early efforts, developers continued to pursue a fully integrated design approach and explored methods for modeling and optimizing illumination schemes in an ET-HMD system to gain optimally illuminated eye areas for capturing high-quality eye images, which are essential for eyetracking robustness and accuracy [113]. They also led the optimization and improving accuracy of eyetracking methods and algorithms which can be readily coupled with an ET-HMD system [114], and investigated new optics architectures for fully-integrated designs [115]. For instance, an iterative simulation-optimization approach was proposed based on non-sequential ray-tracing which not only systematically models the essential elements involved in an eye illumination-imaging system such as infrared (IR) source selection and placement, imaging optics, and IR detector selection but also models the necessary human factors such as eye-facial features, iris color differences and skin reflectance variations [113]. This approach allowed inclusion of human factors in the design loop and systematic analysis of the effects of these factors for obtaining optimal illumination methods that are applicable and robust over an entire population. It took a further step by utilizing simulated eye images to evaluate a set of quality criteria driven by image analysis algorithms.

Typically the systematic approach of integration attempts to combine as much as possible of the optical paths for the display and eyetracking sub-systems. Figure 8.25 shows a first-order layout of an ET-HMD optical system, in which the optical system is simplified with ideal lens modules to emphasize the concept and

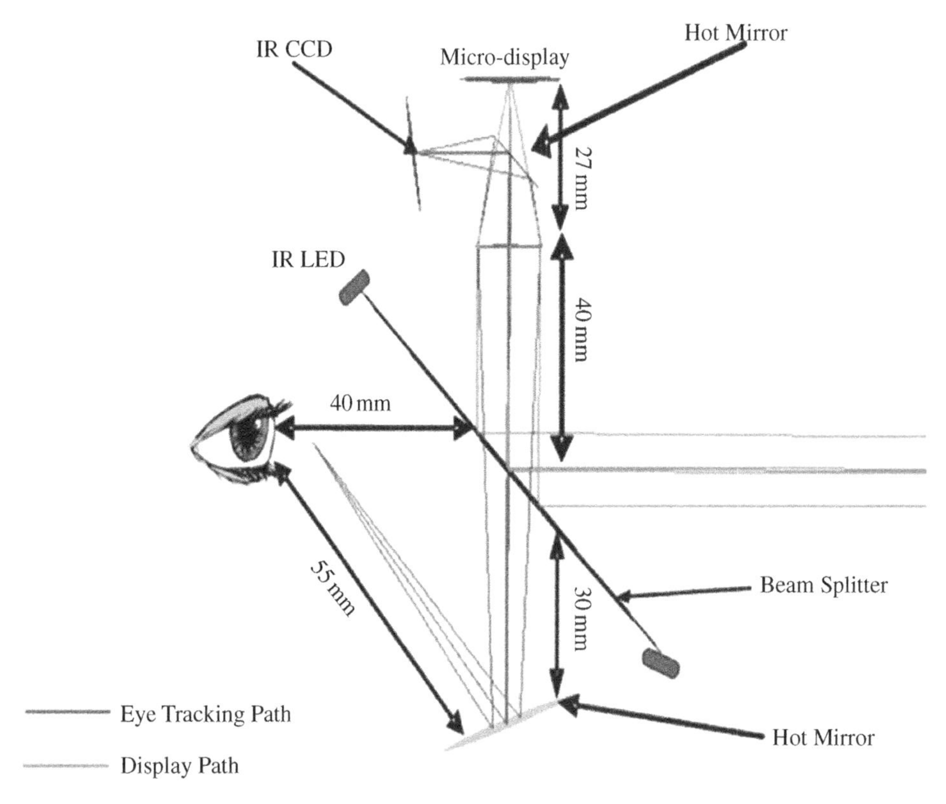

Figure 8.25 An ET-HMD system based on rotationally symmetric optical technology [115].

the scale [115]. The HMD design was based on the HMPD architecture discussed in Section 8.3.3. The design took a full integration approach and combines most of the optical paths for the display and eyetracking subsystems. The same projection optics was shared for both display and eye imaging function. The main limitation of this design, however, lies in the fact that conventional rotationally symmetric optical surfaces and design methods were adopted. As a result, the overall integrated ET-HMD system, although significantly improved compared to other designs, is still bulky and heavy. Baumgarten et al. [116] presented an ET-HMD prototype based on a bi-directional microdisplay which embeds photodiodes within the pixel array of an OLED panel to enable both functionalities of display and eye image capture in the same device. Another example is a novel design and implementation of a lightweight, high-resolution, optical see-through ET-HMD based on freeform optical technology. Figure 8.26a shows the optical system layout which consists of a freeform wedge-shaped prism, a freeform corrector, and a singlet imaging lens [75]. The freeform wedge-shaped prism was a core element shared by four optical paths used for eye illumination, eye imaging, virtual display, and real-world see-through. First, the prism serves as an illumination optics that collimates the light from one or multiple near infrared NIR-LEDs and uniformly illuminates the eye area and the critical features (e.g. glints and darkened pupil) that are to be imaged for eyetracking. Secondly, the same freeform element is the core element in the eye imaging subsystem that captures NIR-illuminated eye images of a user for tracking. The freeform prism folds the light path within a single element so that the image detector is placed at the side of the element. Thirdly, the prism serves as a single-element HMD eyepiece for viewing images on a microdisplay. Finally, the prism, when cemented with a freeform corrective lens to correct the viewing axis deviation and undesirable aberrations introduced by the prism, enables see-through capability of the system which offers low peripheral obscurations and negligible distortions to the real-world view. Overall, this unique optical scheme achieves the capabilities of eyetracking and display without additional hardware cost. Figure 8.26b shows a photograph

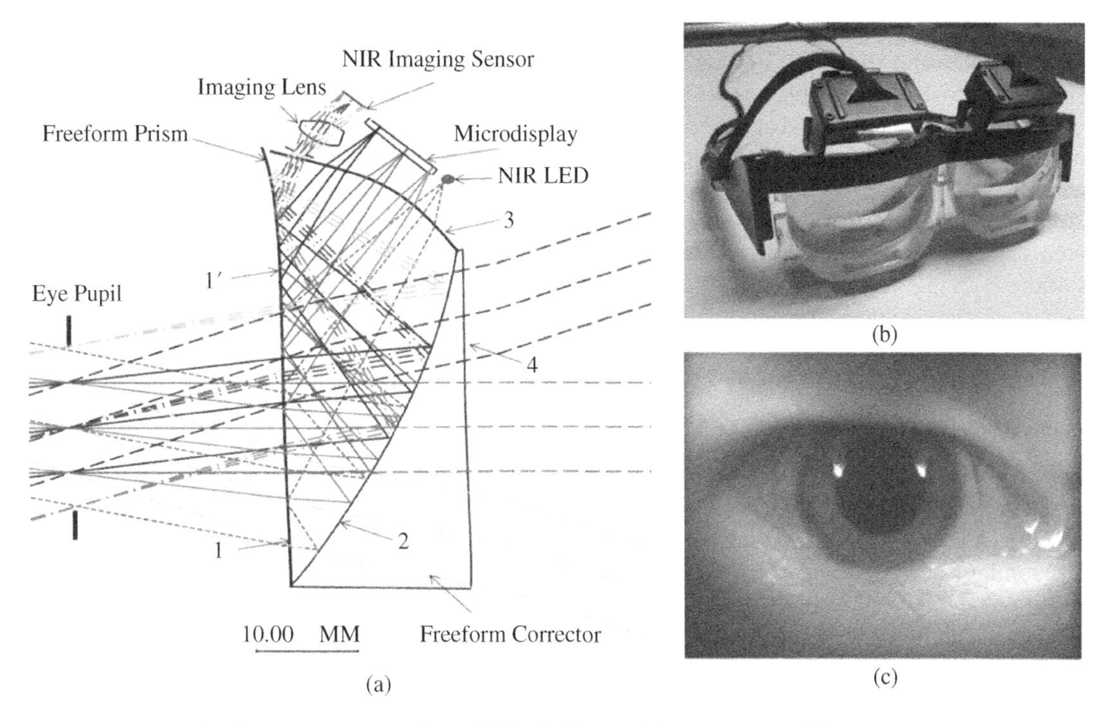

Figure 8.26 Example of compact ET-HMD design (a) Optical layout (b) prototype, and (c) captured eye image under IR illumination for eye-gaze tracking [75].

of a prototype based on optical layout in Figure 8.26a. Figure 8.26c shows an image of the eye under NIR illumination captured by the integrated IR camera. The dark pupil and the glint reflections of the IR sources are the main features to be exploited for computing eye gaze direction.

As discussed in Section 8.3.1, achieving wide FOV while maintaining high resolution is very challenging. An eyetracking capability is essential for an emerging HMD design architecture known as a gaze-contingent display, or foveated display, which can overcome the FOV-resolution tradeoff. It is inspired by the foveation properties of the HVS where only a narrow region around the fovea offers exceptional VA, contrast, and color sensitivities: these properties fall off rapidly with increasing retinal eccentricity, and eye movements can dynamically track the fovea onto the region of interest in the scene. A foveated display renders a small area of a high-resolution image referred to as an inset, in the area of foveal vision, over a wide FOV but lower resolution background for peripheral view. The position of the high-resolution inset can be dynamically controlled to follow the region of interest corresponding to the gaze point.

Existing foveated display techniques that are applicable to HMD systems can be categorized into two types. The first type is an algorithmic approach in which the display hardware has a uniform high resolution, but the resolution of the rendered image is reduced away from the gaze point. For example, much of the work in gaze-contingent display research produces simulated images to experimentally investigate visual processing and perceptual artifacts such as perceptible blur and image motion in foveated multi-resolution displays [117, 118], or applies perceptually-driven foveation treatments to 3D graphics rendering such as level-of-detail management and polygon reduction in 3D models based on the view eccentricity [119, 120].

The second type of foveated display takes a hardware approach in which displays with spatially varying resolution are developed to reduce requirements on high-resolution displays or high-quality and complex optical systems. For instance, one of the pioneer works is the CAE Fiber Optic HMD [107] which provided a binocular FOV of 127° horizontally and 55° vertically with about 38° overlap and angular resolution about 4 arc minutes/pixel. Iwamoto et al. [108] demonstrated a bench prototype of a foveated display which used the eyetracked gaze point to control 2D opto-mechanical scanners to dynamically control the position of a high-resolution inset over a wide FOV low-resolution background display. Rolland et al. proposed a scheme for a high-resolution inset display which uses a pair of lenslet arrays to optically duplicate an inset image over a background display and uses a liquid crystal shutter to select one copy corresponding to the gaze point and block the other copies instead of using mechanically moving parts for scanning [121]. Hua and Liu investigated how the spatially variant properties of the HVS, including resolution, contrast, and color sensitivities, can be used to maximize the information throughput and reduce data bandwidth requirements of imaging and display systems and demonstrated an actively foveated dual-sensor system which approximates the foveation properties of the HVS [122]. More recently, Wu et al. demonstrated a dual-resolution bench prototype of an HMD system in which two different magnifications are implemented for two microdisplays so that one yields a high-resolution inset and the other produces a low-resolution background [123].

8.5.2 Dynamic Range Enhancement

The dynamic range (DR) of a real scene, which has continuous luminance levels, is commonly defined as the ratio between the brightest and the darkest luminance in the scene, or as a base-10 or base-2 logarithmic value of the ratio. A real-world scene can have a very broad DR as large as 14 orders of magnitude and has infinitely fine luminous levels. The perceivable luminance variation range of the HVS is above five orders of magnitude without adaptation. State-of-the-art color displays, including HMDs, typically render a finite and discrete number of luminous levels, and their DR is commonly characterized by the maximum command level (CL) that a display can produce or the bit depth (BD) for each pixel (or each channel in the case of color displays). Most state-of-the-art color displays, including HMDs, are only capable of rendering images with an 8-BD per color channel or maximally 256 discrete intensity levels. Such low BD is far below the capability of accurately rendering the broad dynamic range of real-world scenes and below the visual capability of the eye.

Enhancing the dynamic range of HMDs becomes critical for VR and AR applications, especially for OST-HMDs. For immersive VR applications, the images rendered by LDR HMDs lack the capability to render scenes with large contrast variations, which may result in loss of fine structural details, high image fidelity, and sense of immersion. For optical see-through AR applications, the virtual images displayed by LDR HMDs may appear to be washed out with highly compromised spatial details when merged with a real scene which likely contains a dynamic range which is wider by several orders of magnitude.

Methods of enhancing the DR of displays can be categorized into two types [124]. The most common method is an algorithmic approach aiming to display a HDR image on a conventional LDR display. It typically adopts a tone-mapping technique that compresses the HDR of the image to fit the dynamic range of an LDR device while maintaining the image integrity. Although such a tone-mapping technique can make HDR images viewable on conventional displays of nominal dynamic range, it comes at the cost of reduced image contrast which is limited by the device dynamic range and it does not prevent the displayed images being washed out in an AR display.

The second method of enhancing display DR is a hardware approach which aims to extend the DR of the display hardware so that HDR images can be faithfully rendered. Most effort has been made toward developing hardware solutions to HDR displays for direct-view desktop applications. Perhaps the most straightforward approach to HDR displays is to increase the maximum displayable luminance level and increase the addressable bit-depth for each of the color channels of the display. However, it requires high-amplitude, high-resolution drive circuits as well as high luminance light sources, which is challenging to implement at affordable cost. An alternative method is to combine two or more layers of SLM to simultaneously control the pixel output values, as schematically illustrated in Figure 8.27. A viewer sees a 2D image located at one of the SLM layers (e.g. the front layer SLM1), from which a cone of light is seen from each pixel of the image, and the other layers (e.g. the back layer SLM2) provide spatially-varying light modulation to enhance the dynamic range. When SLMs are placed with negligible separation, the maximum CL that can be rendered by such a multi-layer architecture can be as high as (CL_1*CL_2*…CL_N), where CL_1, CL_2, and CL_N are the maximum

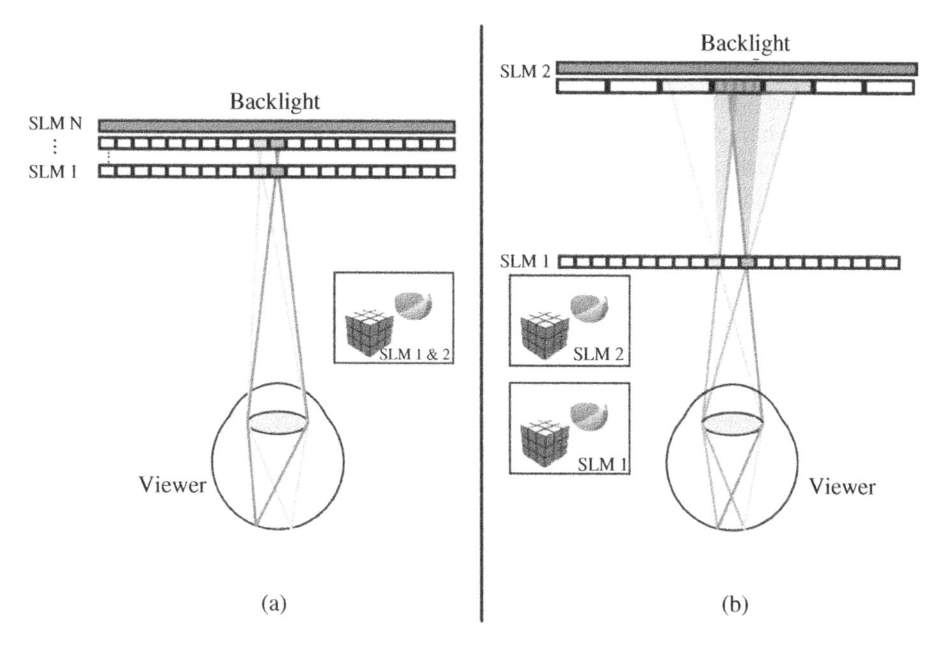

Figure 8.27 Schematic layout of multi-layer modulating scheme of 2D HDR display: (a) SLM layers perfectly overlay and have same spatial resolution and (b) SLM layers are separated with a gap and they may have different spatial resolution [124].

CLs of the N-layers of the SLMs, respectively. A multi-layer modulation architecture gains DR extension at a potential cost of compromised light efficiency because of light loss through the multiple SLMs.

The range and accuracy of the DR modulation based on the multi-layer method greatly depends on the spacing between the SLM layers and the spatial resolution of the layers. When the two SLMs of an HDR display provide the same spatial resolution, as illustrated in Figure 8.27a, making the two SLM layers perfectly overlay without any axial gap is highly desirable, as this offers pixel-by-pixel DR modulation and yields the maximum range and accuracy of DR enhancement. In this case, each pixel on the display layer is only modulated by a corresponding pixel on the modulation layer. When the spatial resolutions of the two SLMs do not match or the gaps between the layers are large, as illustrated in Figure 8.27b, the higher-resolution SLM is typically used as the display layer which displays images including high spatial frequencies while the lower-resolution SLM provides low-frequency, spatially-varying light modulation, similar to the prototype by Seetzen et al. utilizing a low-resolution LED array for spatially-varying modulation [125]. In this case, a reasonable gap between the modulation layer and the display layer is desired so that the pixel structure of the low-resolution modulation layer is not perceptible and does not degrade the overall image quality. With a non-negligible gap, light seen from each pixel on the display layer originates from a finite area of the modulation layer defined by a projection cone which is strongly dependent on the separation of the two SLM layers and the NA of the HDR display to the viewer. When the projected area on the modulation layer is larger than the pixel size of the modulation layer, the perceived illuminance of each pixel of the display is simultaneously modulated by multiple pixels of the modulation layer. Consequently, the maximum range and accuracy of dynamic range enhancement are compromised in comparison to a pixel-by-pixel modulation approach. Furthermore, the low-resolution pixel structure of the modulation layer may have negative effects on the overall quality of the dual-layer modulated image. For instance, the pixel structure of the modulation layer may cast blurred but visible halo or shadow effects near the sharp edges of a displayed image.

Based on the scheme in Figure 8.27, Seetzen et al. proposed an HDR display method for direct-view desktop displays based on a dual-layer SLM scheme [125]. In contrast to conventional LCDs using a uniform backlight, they used a projector to provide a spatially modulated light source for a transmissive LCD to achieve dual-layer modulation and achieved a 16-bit dynamic range with two 8-bit SLMs. They also demonstrated an alternative implementation of the dual-layer modulation scheme, in which an LED array is used to replace the projector unit and provides a spatially-modulated light source for an LCD [125]. This implementation is more similar to the scheme shown in Figure 8.30b where a low-resolution LED array is used as a spatially-modulated DR modulator and a high-resolution LCD with a gap from the LED array provides the display layer. More recently, Wetzstein et al. [126] demonstrated a multi-layer multiplicative modulation and compressive light field factorization method for HDR displays where multiple transmissive LCD layers are stacked together with small designated gaps.

The multi-layer modulation scheme in Figure 8.27 developed for direct-view desktop displays can be adapted to the design of an HDR-HMD system. In principle, directly stacking two transmissive, miniature LCDs along with a backlight source and an eyepiece may be considered the most straightforward and compact realization of a dual-layer modulation scheme for an HDR-HMD system, which would result in a hardware configuration similar to the light field stereoscope approach used by Huang et al. [127]. In practice, however, the direct stacking of multiple SLMs suffers from several critical problems, which make it impractical for HDR-HMD systems. Firstly, transmissive LCDs tend to have LDR and low transmittance. The stacked dual-layer modulation would lead to very low light efficiency and limited dynamic range enhancement. Secondly, transmissive LCDs tend to have low fill factors and the microdisplays used in HMDs typically have pixels as small as a few microns, much smaller than the pixel size of about 100~500 μm for direct-view displays. As a result, light transmitted through a two-layer LCD stack will inevitably suffer from severe diffraction effects and yield poor image resolution following magnification by an eyepiece. Most importantly, due to the physical construction of typical LCD panels, the modulation layers of liquid crystal will inevitably be separated by a gap as large as a few millimeters, depending on the physical thickness of the cover glasses. For direct-view type desktop displays, a gap of a few millimeters between the two SLMs will not have much influence for dynamic range modulation. In an HMD system, even a gap as small as 1 mm in the SLM stack will be elongated tens of times due to the HMD eyepiece magnification, resulting in a large separation in the viewing space. The resulting separation in visual space depends on the physical gap between the SLM layers and the focal length of the eyepiece. For example, assuming a focal length of 20 mm for a typical HMD eyepiece and a typical viewing distance of 2 m in the visual space, a 0.1 mm gap in the SLM stack will lead to an axial separation as large as 0.6 m. Unlike a tensor display where a gap between adjacent SLM layers is required for light field rendering, a large gap between adjacent SLM layers makes accurate dynamic range modulation practically impossible. Xu and Hua recently performed a thorough analysis on how the quality of HDR enhancement is influenced by the physical gap between the SLMs [124].

To minimize the gap between the modulation and display using a dual-layer modulation method, Xu and Hua proposed a new HDR-HMD system design, as shown by the schematic layout in Figure 8.28a, consisting of an HDR image generator and viewing optics [124]. The HDR image generator is composed of two F-LCoS

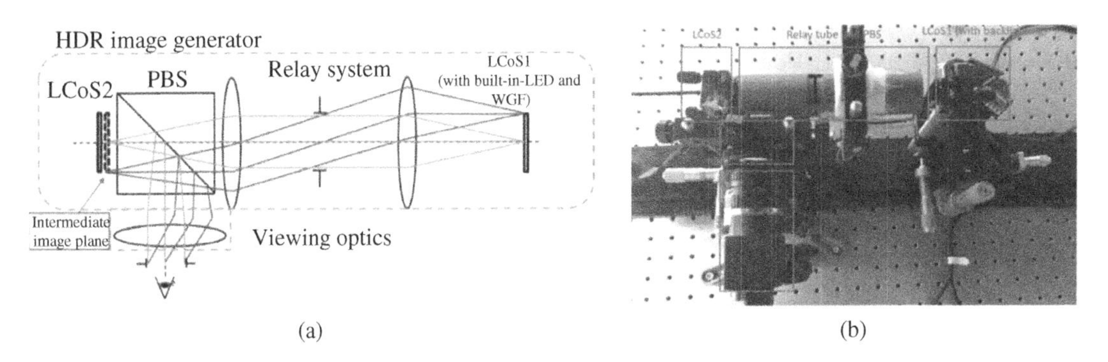

Figure 8.28 (a) Schematic layout and (b) bench prototype of a monocular HDR-HMD based on dual-layer modulation scheme [124].

microdisplays and a custom relay system which was designed to optically minimize the physical separation between the SLMs. Compared to the drawbacks of low fill factor and low light efficiency of transmissive LCDs, reflective type F-LCoS microdisplays offer both high pixel resolution and high fill factor as well as large contrast ratio and optical efficiency, which help to minimize diffraction artifacts and enhance the dynamic range of the system. The relay optics allows the system to optically overlay the SLMs with a minimal gap as small as a few microns and achieve pixel-by-pixel contrast modulation. Figure 8.28b shows a bench prototype built from stock lenses and two 0.4″ F-LCoS devices with a pixel resolution of 1280 by 960 and a 6.35 μm pixel pitch. Given that both of the LCoS displays have an 8-BD for each color channel, the HDR image generator is able to achieve a combined 16-bit modulation with an enhanced dynamic range above 60 000 : 1.

Although the prototype in Figure 8.28 demonstrated significantly enhanced DR in an OST-HMD system, the optics is very bulky and not suitable for wearable systems. More research effort is clearly needed to develop innovative solutions for DR enhancements in HMD systems.

8.5.3 Addressable Focus Cues in HMDs

A wide range of cues are exploited by the HVS for 3D percepts and depth estimation. Examples include linear or aerial perspective, occlusion, shadows and shading, apparent size, texture gradient, binocular disparity, motion parallax, retinal blur, accommodation, and vergence. Among these cues, the vergence cue refers to rotation of the eyes so that their visual axes intersect at a 3D object of interest at near or far distances, while the accommodation cue refers to focusing of the eye by the ciliary muscles which change the refractive power of the crystalline lens to bring the object of interest into sharp focus. Associated with eye accommodation is the retinal image blur effect in which objects away from the eye's accommodation depth appear blurred in the retinal image due to the limited DOF of the eye. The degree of image blur varies with the distance of the object from the focus distance. The accommodation and retinal image blur effects together are known as the focus cues. In natural vision, the actions of eye convergence and accommodation are highly coupled, as schematically illustrated in Figure 8.29a. The vergence depth of the eyes coincides with their accommodation depth and the retinal image blur of objects in the scene increases with their distances from the eye accommodation depth.

In viewing conventional HMD images, however, the stimuli for convergence and accommodation may be decoupled and in conflict, which is known as VAC. As schematically illustrated in Figure 8.29b when viewing conventional stereoscopic displays with a fixed screen depth, the vergence depth of the eyes is cued by the binocular disparity rendered in the pair of stereoscopic images, and is mismatched to the accommodation depth of the eyes which is cued by the depth of the screen. In addition, the retinal blur effects of virtual objects are also mismatched. Virtual objects rendered via stereoscopic images, regardless of their rendered depths, are in focus if the viewer accommodates on the image plane, or are uniformly blurred if the user accommodates at a point separated from the image plane by a distance greater than the eye's DOF. The VAC problem stems from the fact that the image source is normally a 2D flat surface located at a fixed distance from the eye. Conventional HMDs, regardless of their being monocular or binocular, being see-through or immersive, form a virtual display at a fixed depth optically conjugate to the microdisplay and thus lack the ability to render correct focus cues for information which may appear at distances other than that corresponding to the virtual image plane. As a result, conventional HMDs fail to stimulate natural eye accommodation response and retinal blur effects.

The problem of incorrect focus cues in HMDs causes several visual cue conflicts. Marran and Schor [128] and Hua [129] provided thorough reviews on the various types of accommodative cue conflicts in both VR and AR systems. For instance, one of the conflicts in a see-through AR display is the mismatch of the accommodation cues between the 2D virtual image and the real-world scene. Other examples of conflicts include the mismatch of accommodation and convergence between the 2D image plane and the 3D virtual world rendered in a conventional stereoscopic display and the mismatch of the retinal image blur cues between the virtual and real-world scenes, as shown in Figure 8.29b. Though not conclusive, many studies have provided strong supportive evidence that these visual conflicts and incorrectly rendered focus cues in conventional

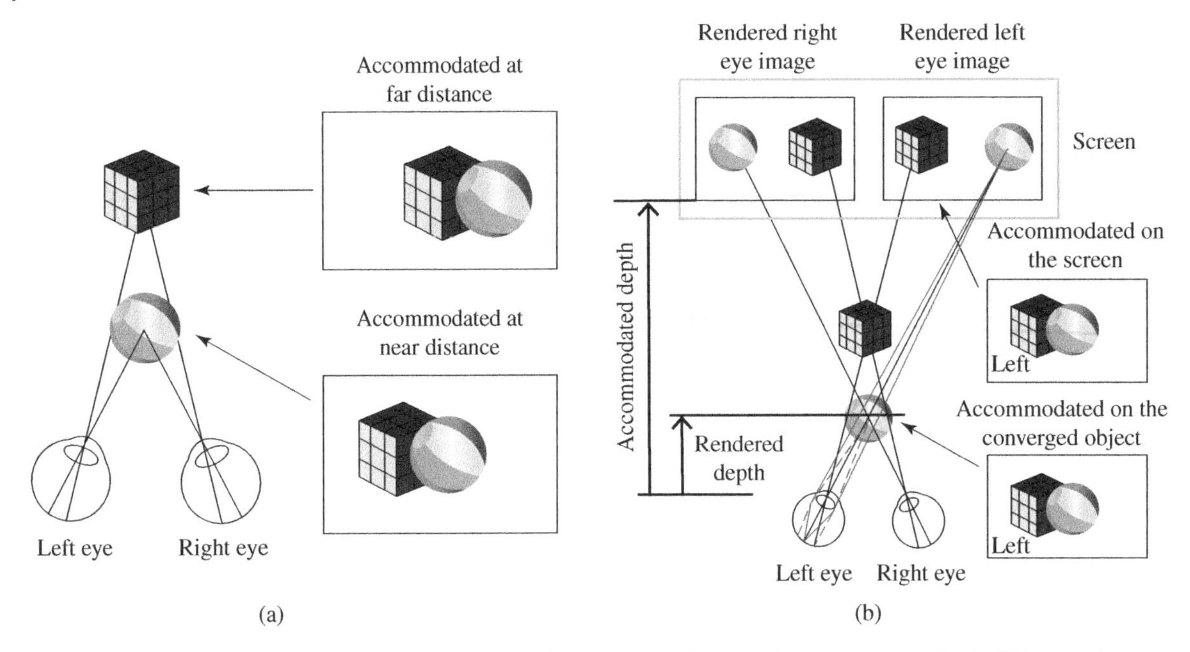

Figure 8.29 Illustration of (a) naturally coupled actions of eye accommodation and convergence and retinal blur cues in real-world view, and (b) mismatch of accommodation and vergence depths of the eye and mismatch of retinal blur cues in viewing conventional stereoscopic displays with a fixed screen depth.

HMDs may contribute to various visual artifacts and degraded visual performance. In a nutshell, incorrect focus cues may contribute to the two commonly recognized issues: distorted depth perception [130, 131] and visual discomfort. Examples of discomfort include diplopic vision [132], visual fatigue [133], and degradation in oculomotor responses [134], especially after viewing such displays for an extended period of time.

In recent years, many attempts have been made to address the VAC problem in HMDs and develop displays that aim to approximate the visual effects created by the focus cues in natural vision. Additionally, over the past decades many efforts have been made to develop direct-view 3D display technologies, including volumetric displays [135] that render all voxels within a volumetric space regardless of the viewer's point-of-interest or viewpoint, holographic video displays [136] and super-multi-view autostereoscopic displays [137, 138]. For the purpose of this chapter, however, we will concentrate on the methods most readily applicable to HMD systems aimed at addressing the VAC problem, which may be categorized into four general types; extended depth of field (EDOF) displays, VFP displays, MFP displays and light field displays, based on different optical mechanisms for controlling focus cues. Historically, holographic video displays have been regarded as the Holy Grail and ultimate goal of true 3D display technologies. Many efforts have been invested in developing direct-view holographic video displays, though full-color, large-size, holographic video displays continue to be a grand challenge. In recent years, some efforts have moved toward developing head-mounted holographic video displays [139, 140]. As there is, as yet, no complete solution to the problems encountered in providing head-mounted holographic displays, and in view of the extensive technical background which would have to be introduced to provide a meaningful account of the few existing efforts in this direction, this chapter does not attempt to cover the topic, but interested readers are encouraged to refer to recent publications in the field [139–141].

Among the four types of methods to be discussed, the EDOF method aims to eliminate the need for eye accommodation change when viewing the display so that the eye can rest at an accommodation state which may be induced by the vergence stimuli. The VFP display method aims to address the VAC problem by dynamically controlling the focal depth of the display hardware to mitigate the conflict. The other methods can be considered true 3D display technologies that are potentially able to faithfully render both aspects of focus cues in the same way as a real-world scene. They aim to create proper cues to stimulate the eye to accommodate and converge at the same depth typically by reconstructing an apparent 3D volume, and to recreate the same retinal blur effects as those observed for a real-world scene.

8.5.3.1 Extended Depth of Field Displays

Extending the DOF of the virtual display is a simple way that may potentially mitigate the VAC problem in HMDs to some degree. When viewing a conventional HMD, the DOF of the virtual display is inversely proportional to the diameter of the ray bundles instantaneously entering the eye pupil, and typically has a limited range. Naturally, the eyes need to accommodate at or near the depth of the virtual display in order to perceive sharp images rendered on the display, while the eyes may converge at other depths due to the stimulation of the rendered disparity.

The methods used in EDOF displays fall into two categories, pinhole optics and PSF engineering. A pinhole optics design is expected to have a very large DOF where a very thin bundle of paraxial rays from each field of the display is focused on the retina through a small area on the eye pupil. When the diameter of the ray projections on the pupil is substantially smaller than that of the eye pupil itself, the DOF of the display is extended and the display is considered as being accommodation free. A Maxwellian view display is one of the well-known architectures for implementing this pinhole optics method [142]. As illustrated in Figure 8.30a, the diverging rays from a point source *S* are collimated by a condenser lens then focused by an eyepiece onto the entrance pupil of the eye instead of being imaged directly on the retina. A target or typically a SLM is placed at a focal distance away from the eyepiece such that an image conjugate to the SLM is formed on the retina. Due to the fact that each field on the SLM is imaged through a narrow aperture on the pupil (pinhole), the image is observed with very large DOF without requiring eye accommodation.

Several efforts have been made to develop this type of display. A Maxwellian view retinal projection display, as shown schematically in Figure 8.31a, was proposed by Ando et al. for accommodation-free HMDs [144]. A diverging laser source is collimated by a lens. The parallel rays pass through or reflect from the pixels of a SLM and are focused by an HOE at the center of the eye pupil. Each parallel ray from each pixel projects directly onto the retina and stimulates a different point. A spatial filter along with additional focusing and collimation lenses is inserted between the HOE and SLM to mitigate the diffraction effects of the small pixel apertures on the coherent laser rays. As a result, the image pattern shown on the SLM is directly imaged on the retina. Based on the Maxwellian view principle, Von Waldkirch et al. demonstrated retinal projection display prototypes in which a partially coherent LED illuminates an LCD through a condenser lens [146, 147].

Besides noticeable image quality degradation, the most significant limitation of these conventional Maxwellian view displays is their small entrance pupil diameter due to the nature of Maxwellian viewing which requires that all the rays converge to a small area on the pupil. Yuuki et al. demonstrated an implementation of a dense Maxwellian view display consisting of a regular base display such as a LCD panel, a light

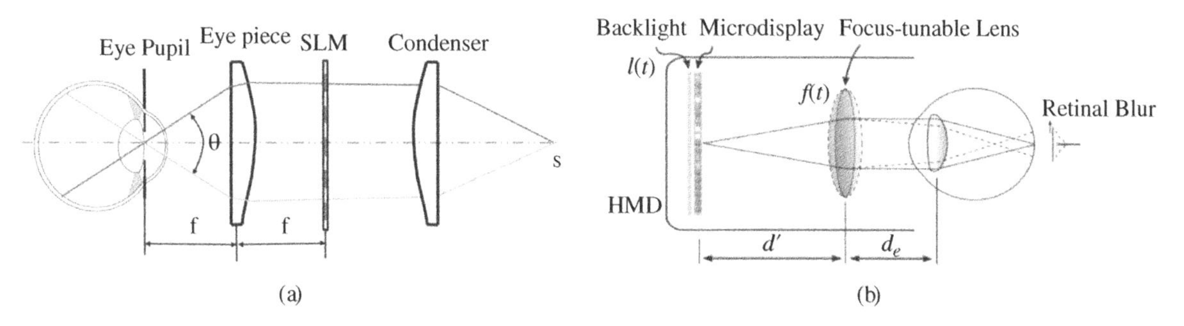

Figure 8.30 Optical methods for EDOF displays: (a) Maxwellian view pinhole display; and (b) accommodation-invariant display via PSF engineering [143].

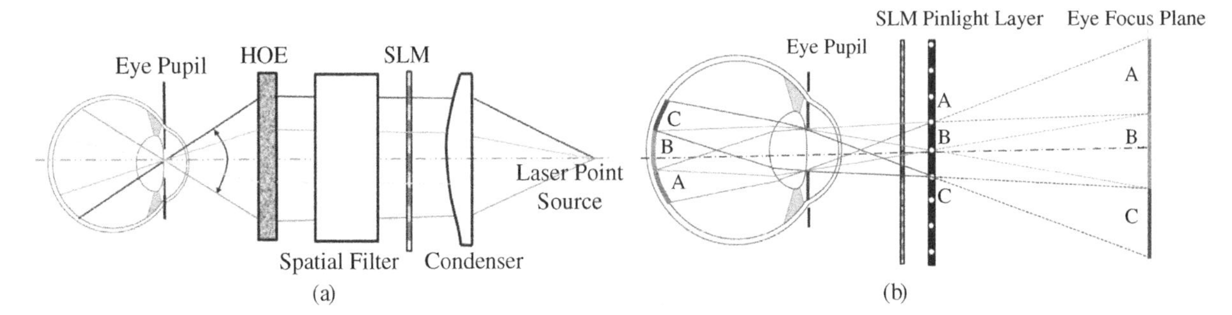

Figure 8.31 Examples of EDOF displays based on the pinhole optics method: (a) schematic illustration of a Maxwellian view retinal projection display [144]; and (b) schematic layout of a pinlight display [145].

absorption layer with pinhole patterns, and a plano-convex fly-eye lens sheet which led to the formation of an array of 7×7 ray intersection points on the eye pupil, allowing an observer to see all pixels on the base display in clear focus by placing their eye at any of the ray intersecting points [148]. More recently, Maimone et al. demonstrated a compact, wide-FOV OST-HMD design, termed a Pinlight display, as schematically shown in Figure 8.31b, mainly consisting of an LCD panel and an array of point light sources placed directly in front of the eye [145]. The array of point sources, ("pinlights"), is formed on an edge-lit acrylic sheet into which small cavities are etched, and LEDs placed around the edge of the plastic sheet light up the cavities and form the point sources. Light rays emitted by each pinlight are modulated by a section of the LCD panel and the modulated light rays are refracted by the eye lens and create a "sharp" copy of the modulated image on the retina. The small subset images projected by all the pinlights are tiled together to create a wide FOV image. The pinlight display can be thought of as a dense Maxwellian view display because a viewer can see the entire FOV of the displayed image by placing their eye at any of the ray intersecting points on the eye pupil. A prototype display in the form factor of large glasses was demonstrated, offering a 110° diagonal FOV.

Besides the use of pinhole optics for extending the DOF, an alternative computational method was proposed by Konrad et al. [143] by adopting a method developed for EDOF microscopy [149]. In Figure 8.30b, a focus-tunable lens is used to continuously sweep the focal depth of the virtual display between far and near points with the microdisplay located at a fixed position. The sweep action of the focus-tunable lens leads to a cumulative retinal image blurring of the virtual display and the level of blurring is generally constant independent of the eye accommodation state. The properties of invariant accommodative system response were rigorously demonstrated for EDOF microscopy in [150], and full-resolution microscopic images can be recovered from the captured blurring image via a digital deconvolution process given the invariant system response. However, due to the fact that the retinal images cannot be post-processed and microdisplays are unable to render negative values of intensity, deconvolution is not a possible approach to recovering the loss of resolution due to focus sweeping. As a result, the display suffers from severe image quality degradation and the level of degradation increases as the sweeping depth range increases.

8.5.3.2 Vari-Focal Plane (VFP) Displays

The vari-focal approach illustrated in Figure 8.32, provides a simple remedy to the VAC problem which generates the proper cues for eye accommodation response by dynamically compensating the focal distance, Z, of a single-plane display to match it with the convergence depth of the eye. It adds the least hardware complexity to a conventional HMD design by directly manipulating the apparent distance of the virtual display.

The focus compensation can be achieved by a mechanical mechanism or by an electronically-driven active optical element. Using a mechanical mechanism, focal depth adjustment may be implemented by mechanically zooming the eyepiece of the display or adjusting the distance between the micro-display and the eyepiece (Δz). For instance, Wann et al. suggested the use of an oscillating lens to adjust the focal depth of the virtual display

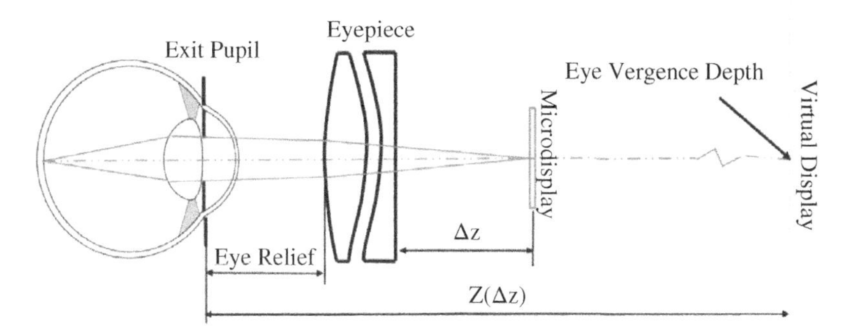

Figure 8.32 Schematic illustration of a vari-focal plane HMD.

[130] and Shiwa et al. demonstrated the first vari-focal display prototype in which a relay lens group inserted between a display device and eyepiece was mechanically moved back and forth along the optical axis by a stepper motor to control the overall optical power of the optics [151]. Shibata et al. demonstrated a bench prototype with a similar vari-focal function for focus control achieved by axially translating the microdisplay which was mounted on a micro-controlled stage [152].

Instead of varying the focal depth by mechanical means, a range of electronically-driven active optical elements can be used, including a liquid lens [153], liquid crystal lens [154], and deformable mirror [155]. Liu and Hua demonstrated the first OST-HMD prototype with focus control using an eyepiece design integrated with a liquid lens [156]. Figure 8.33a–c show the optical layout, a photograph of the prototype, and experimental demonstration respectively, of such an OST-HMD system. The prototype is able to dynamically control the focal depth of the virtual display from 5 to 0 diopters (the near point of the eye to infinity) by driving the liquid lens from 38 to 51 Vrms. Figure 8.33c shows two images captured through the prototype. In each of the images, three bar resolution targets were physically placed 16, 33, and 100 cm, respectively, away from the eye position along the visual axis of the HMD and were used as references to the virtual image with a variable accommodation cue. A virtual torus was rendered at approximately 16 cm away from the eye and the voltage applied to the liquid lens was adjusted to match with the torus depth. The two images show photos captured with the camera focused at 6 and 1 diopters, respectively. Studies on depth perception and eye accommodative response to a vari-focal display [38] suggest an improved depth perception induced by correctly rendered focus cues, in comparison to findings on traditional S3D displays, which suggest distorted and compressed perceptions of depth.

The VFP method requires dynamically tracking the eye convergence distance in real time. It is able to render correct focus cues for the fixated objects if the focal distance of the virtual display is in synchrony with the distance of eye fixation, but objects at other depths would still have incorrect focus cues. The vari-focal method may be further improved by applying a depth-dependent blur filter to simulate retinal image blur and this may improve depth perception to some extent. The prototype design shown in Figure 8.33b was further extended by integrating a binocular eyetracking system which tracked the eye convergence depth of the viewer, addressing the focal distance of the virtual image according to the vergence depth, and rendering gaze-contingent retinal blur cues in real time [109].

8.5.3.3 Multi-Focal Plane (MFP) Displays

Unlike a VFP approach which dynamically adjusts the focal depth of the virtual display to match a single focal plane with the vergence depth of the eye as needed, MFP displays create a stack of discrete focal planes dividing an extended 3D scene volume into multiple zones along the visual axis. Each of the focal planes samples the projection of 3D objects located within a depth range centered on the corresponding focal plane and these discrete focal planes additively reconstruct a large scene volume. A MFP-based HMD can be thought of as a method to sample the light field projections of a 3D scene at different depths along the visual axis, as seen from

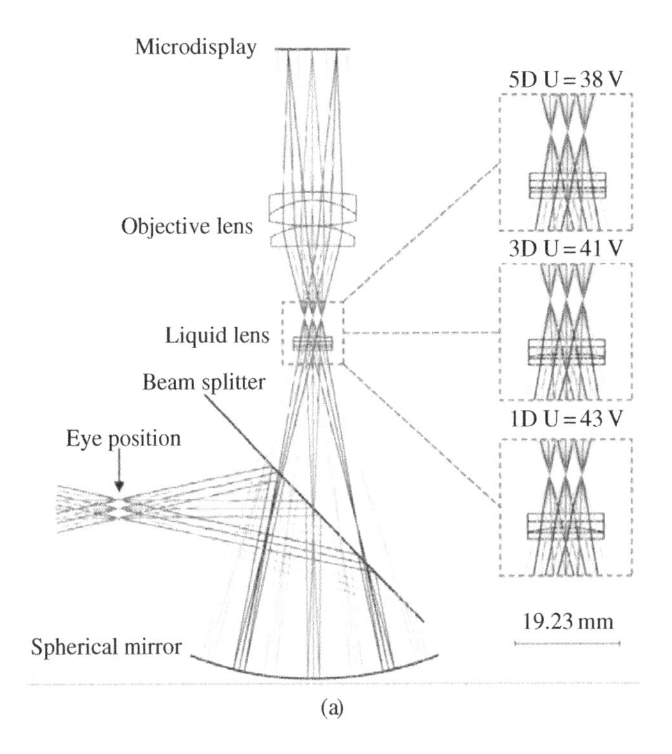

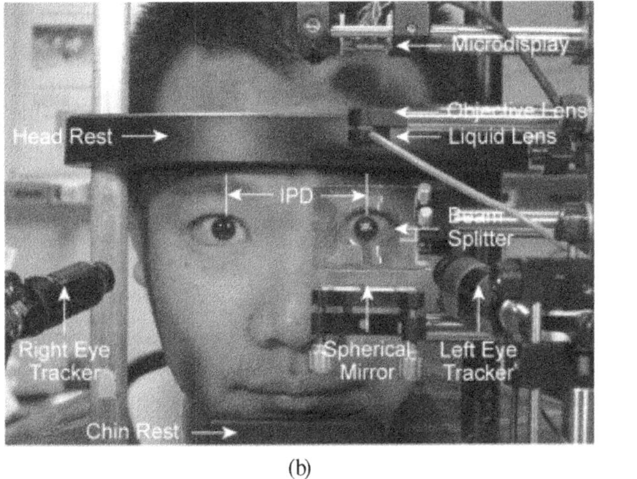

Figure 8.33 Example of a vari-focal OST-HMD prototype using a liquid lens: (a) optical layout; (b) prototype; and (c) experimental results [156].

a fixed viewpoint. Unlike a vari-focal method, the MFP method does not require any mechanism to determine the point of vergence and is capable of rendering correct focus cues through a 3D volume.

Figure 8.34 shows a schematic diagram of a dual-focal plane example in which the projections of a 3D donut and a sphere are rendered by FPI and FPII, respectively. Additionally, the projections of the objects rendered by the front layer can be rendered as occlusion masks on rear plane (e.g. the projection of the donut on FPII) to render correct occlusion relationships among these objects. An MFP-based display may be implemented either by spatially multiplexing a stack of 2D displays or by rapidly switching the focal distance of a single 2D

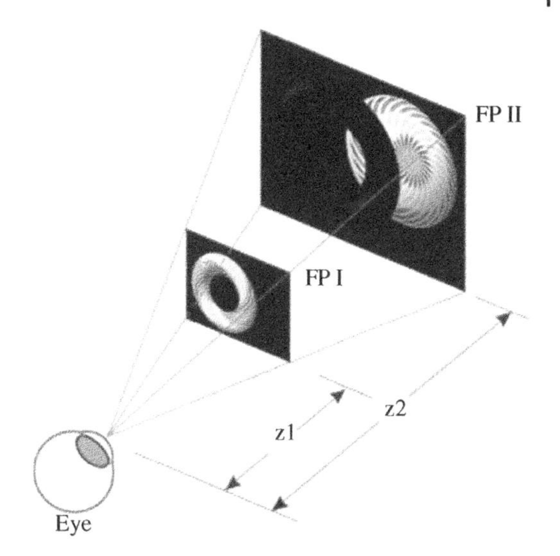

Figure 8.34 Schematic illustration of a multi-focal plane HMD.

display sequentially by a high-speed VFE, synchronized with the frame rendering of multi-focal images (i.e. in a time-multiplexed fashion).

The pursuit of a MFP-based HMD starts with the spatially-multiplexed approach. For instance, the use of bifocal lenses was suggested by Marran and Schor to create different focal depths for different parts of the visual field in HMDs [128]. Koetting suggested a monocular lens addition method, also known as a monovision technique, which adds a lens to one eye for near viewing and the other for distant viewing [157]. More recently Johnson et al. [158] implemented the monovision method in HMDs where lenses with different focal lengths were used for the display paths of the left and right eyes. Crabtree demonstrated an early experimental implementation of a volumetric projection display [159], in which a series of 2D images are projected into a multiplanar optical element (MOE) acting as a projection screen with a variable depth. One of their prototypes projected a total of 13.1 million voxels covering a volume of 39.8 cm × 34.3 cm × 23 cm with 50 planes and 512 × 512 pixels per plane. Rolland et al. theoretically investigated the engineering requirements for adapting the multiplane volumetric display method by Crabtree to HMDs by using a thick stack of planar displays [160]. The analysis concluded that 14 equally-spaced (in dioptric spacing) focal planes would be required to accommodate the focusing range from infinity to 0.5 m for a VA of 1 arc minute and a 4 mm pupil diameter. Akeley et al. demonstrated the first experimental implementation of a spatially-multiplexed three-focal plane display intended for HMD applications with an equal spacing of 0.67 diopters between adjacent focal planes [161]. The prototype covers a fixed depth range from 0.311 to 0.536 m by dividing a flat panel display into three focal planes through three beamsplitters placed at different distances from the viewer. More recently, Schowengerdt et al. suggested a spatially-multiplexed RSD using a fiber array to produce a multi-focal bundle of beams [60]. Cheng et al. designed a spatially multiplexed dual-focal plane system by stacking two freeform prisms [162].

Alternatively, a stack of focal planes can be implemented in a time-multiplexed fashion by incorporating an electrically-controlled high-speed VFE, through which the focal depth of a 2D focal plane can be controlled in synchronization with the depth of the objects being rendered so that all objects in a large 3D volume are rendered sequentially at a flicker-free rate and appear to have correct focus cues simultaneously. Several time-multiplexing MFP prototypes were demonstrated over the last decade. McQuaide et al. [163] demonstrated a dual-focal plane RSD in which the focal depth of a modulated laser beam is dynamically controlled through a deformable membrane mirror device (DMMD) while the beam is scanned to render a 3D volume on a pixel-by-pixel basis. Love et al. demonstrated the use of birefringent lenses as the VFEs and high refresh rate CRTs as the image sources for a prototype with four fixed focal planes [164]. Using a liquid lens as the VFE and an OLED micro-display as the image source, Liu and Hua demonstrated the first prototype of a dual-focal plane OST-HMD [165], where the front and back focal planes were placed at a depth of 5 diopters and 1

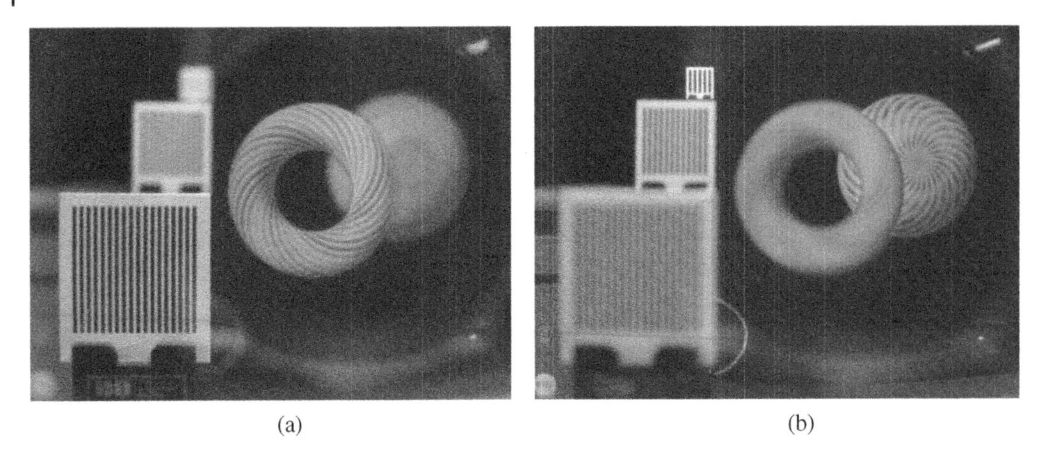

Figure 8.35 Photos captured through a dual-focal plane OST-HMD prototype with camera focused at (a) the front resolution target located at 5 diopters and (b) the back resolution target located at 1 diopter distances [165].

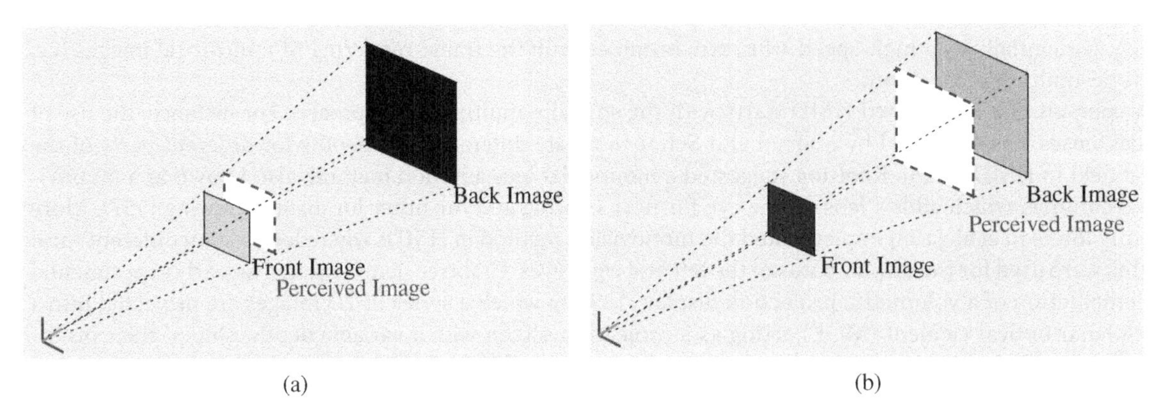

Figure 8.36 Schematic model of a DFD-MFP display: the luminance ratio between the front and back images are modulated to change the perceived depth of the fused image to be near to the front image plane (a) and near the back image (b).

diopter, respectively, from the eye. Figure 8.35 shows two photos captured by a camera from the exit pupil of the display with focus at 5 and 1 diopter, respectively. Three printed resolution targets were placed in the scene at 5, 2, and 1 diopter, respectively, as physical references. A virtual sphere located at a 1 diopter distance was rendered by the back focal plane while a donut located at a 4 diopter distance along with the occlusion mask of the sphere was rendered by the front plane.

In both spatially- and time-multiplexed methods, a large number of focal planes and small dioptric spacing are desirable for creating a large scene volume with accurate focus cues and high image quality. To achieve good image quality with a significantly reduced number of focal planes, the multi-focal plane approach may be further improved by incorporating depth-fused 3D (DFD) perception [166]. Figure 8.36 illustrates the depth fusion concept of two focal planes separated by a dioptric distance [26, 27]. The two overlapped images displayed at two different depths may be perceived as a single-depth fused image. The perceived depth of the fused image can be manipulated by modulating the luminance ratio between the two images through a depth-weighted fusion function which optimizes the contrast magnitude and gradient of the fused image to create cues which stimulate and stabilize the eye accommodation response. A MFP display incorporating the DFD effect (DFD-MFP) can render correct or nearly-correct focus cues for 3D objects spanning a large depth range to the viewer while significantly reducing the number of necessary focal planes to an affordable level.

In the design of a DFD-MFP display, the optimal spacing between adjacent focal planes depends on several factors. The most significant factor is the desired image resolution and contrast of the fused image which vary as a function of the dioptric spacing. Prior studies suggest that a spacing of ~0.6 diopter between two adjacent focal planes is appropriate to render non-conflicting focus cues for MFP displays with a Nyquist frequency of ~18 cpd, and a spacing of 0.4 diopters or smaller would be desired for displays affording a resolution of 30 cpd (i.e. 1 arc minute) [26, 27, 167].

The appropriate fusion functions applied to adjacent focal planes depend on many factors, such as the tolerance to focus cue error, focal plane spacing, display characteristics, and eye model parameters. They can have a significant impact on the effectiveness of rendering correct or nearly correct focus cues and minimizing conflicting cues. For instance, a pair of linear depth-blending functions on adjacent focal planes was suggested in Akeley et al. [161] and more recently by Ravikumar et al. [168] to mitigate image artifacts caused by discrete sampling of focal planes. A linear blending function was demonstrated to be effective in creating a continuous 3D scene through discrete focal planes and can be fairly effective in creating nearly-correct focus, but it does not necessarily minimize conflicting cues. An optimization framework for DFD-MFP system design was developed to optimize the retinal image fused by the two images overlapping along the visual axis. This allowed selection of focal plane spacing and fusion functions that maximize the contrast magnitude and gradient to avoid conflicting focusing cues and create a smooth contrast gradient that helps to stimulate and stabilize the eye accommodation response [26, 27, 167]. A set of non-linear fusion functions was developed and applied to a system of five focal planes spaced by 0.6 diopters to ensure less than ±0.1 diopters difference between the simulated depth and the depth where the maximum retinal image contrast is obtained [167]. These non-linear functions require that the pixel intensities distributed on two adjacent focal-planes are biased toward the focal plane which is further away from the simulated depth.

The depth-weighted MFP blending method in [27, 167] concentrates on optimizing retinal image quality and minimizing accommodation cue conflicts when rendering a continuous 3D scene volume through a few discrete focal planes with uniform focal-plane spacing. Narain et al. recently proposed an alternative depth-blending weighting scheme by computationally optimizing luminance weight to accurately reproduce the defocus behavior of occlusions, reflections, and other non-local effects as a function of accommodation [169]. Wu et al. explored optimization of the dynamic configuration of a finite number of focal planes based on the characteristics of the 3D content to be rendered. They optimized an objective function which characterized the overall perceptual quality of the rendered 3D scene, and their results suggest that the perceived visual quality of the content can be improved by explicitly reducing the overall loss of contrast for aggregated accommodation states within the rendered image volume [170].

Several DFD-MFP based prototypes have been designed and built. Figure 8.37a shows the monocular optical layout of one of the prototypes by Hu and Hua, consisting of a composite optical see-though eyepiece and an image generation subsystem (IGS) [39]. The composite eyepiece was composed of a wedge-shaped freeform eyepiece, a freeform see-through compensator lens, and a cylindrical lens and enabled the successful integration of MFP display and OST-HMD techniques. The IGS, consisting of a high-speed DMD microdisplay, a DMMD, and relay lens groups, achieves the core function of generating the multi-focal-plane contents. By changing the optical power of the DMMD, the intermediate image of the DMD display shifts axially with respect to the eyepiece without magnification change, resulting in the depth of the virtual display changing from far to close. Synchronized operation of the DMMD and image rendering at a high speed such as 360 Hz enables the creation and multiplexing of six focal planes at a flicker-free rate of 60 Hz. Figure 8.37b shows the actual setup built on an optical bench, while Figure 8.37c and d show photos of a 3D scene rendered by the virtual display and captured with a camera placed at the exit pupil position. Six focal planes were dynamically formed at 3.0, 2.4, 1.8, 1.2, 0.6, and 0.0 diopters respectively, at an overall refresh rate of 60 Hz. The 3D scene consists of a green floor grid extending from 3.0 to 0.6 diopters, a green wall grid and UoA logo at 0.6 diopters, a grating target extending from 3.0 to 0.6 diopters, as well as an OSC logo placed at the 3.0 diopter limit. Each focal plane displays a different part of the 3D scene. By incorporating the depth fusion functions in [27, 39], this 3D scene was rendered continuously by five of the focal planes. Figure 8.37c and d show images from the

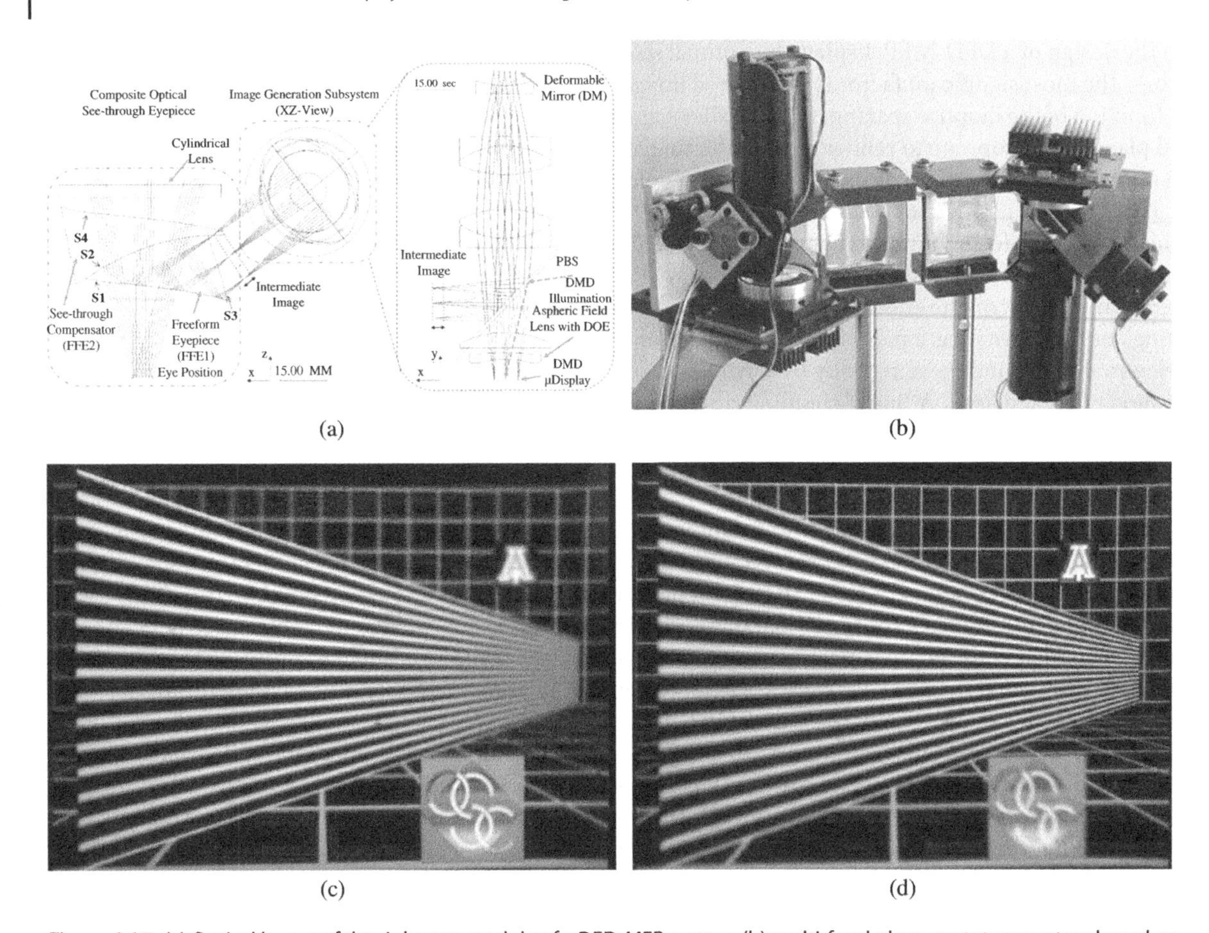

Figure 8.37 (a) Optical layout of the right-eye module of a DFD-MFP system, (b) multi-focal plane prototype system based on the layout in (a), (c–d) photos captured with a f/4.8 camera focused at 3 diopters and 0.6 diopters, respectively [39].

camera when focused at 3 diopters (near) and 0.6 diopters (far), respectively. Natural focus cues are clearly demonstrated and high-contrast continuous targets were correctly fused across five spatially separated focal planes, which visually validates the depth-fusion display method.

More recently, Rathinavel et al. proposed an alternative method for implementing the rendering pipeline of a full-color DFD-MFP display [171]. Rather than driving a high-speed VFE into discrete focal states as in the example in Figure 8.37, they proposed to oscillate the VFE in continuously so it does not settle at specific focal depths, and a stack of binary images are rendered at high speed such that the displayed stack of images is perceived as slices of a continuous full-color volume. The binary images decompose the rendered 3D scene volume locally on a per-voxel basis and the decomposition is distributed based on the location of the voxel. They demonstrated a prototype system based on low-level hardware access to a high-speed DMD, similar to the one used in Figure 8.37 and a high-speed RGB LED light source synchronized with the bit-depth of the DMD binary frame, and the system demonstrated impressive near-accurate per-pixel focus across a depth volume from 6.7 to 0.25 diopters through a sequence of 280 synchronized binary images.

In general, the spatially-multiplexed MFP approach allows rendering of multiple focal planes in parallel and reduces the speed requirements for the display technology. On the other hand, its practical implementation is challenged by the lack of stack display technologies with high transmittance and by the demand for computational power to simultaneously render a stack of 2D images of a 3D scene. It often becomes impractical for implementing a large number of focal planes due to optical path complexity and low optical efficiency through a stack of SLMs. Another drawback associated with the spatially-multiplexed MFP approach is the lack of flexibility to vary the positions and spacing of the focal planes. By adopting a depth-weighted blending method, the time-multiplexed method becomes technically capable of rendering nearly correct focus cues for addressing the VAC problem with a small number of focal planes. In the meantime, even rendering a small number of time-multiplexed focal planes at a flicker-free rate demands high response speed for the active optical element, the display device, and the graphics rendering engine and the response speed is proportional to the number of flicker-free focal planes to be portrayed. This approach continues to address several critical technical obstacles to it becoming a viable solution to truly wearable light field AR displays.

8.5.3.4 Head-Mounted Light Field (LF) Displays

A head-mounted light field display, also referred to as LF-3D HMD, is considered one of the most promising 3D display techniques to address the VAC problem. As illustrated in Figure 8.38, a LF-3D HMD renders the perception of a 3D object (e.g. the cube) by reproducing directional samples of the light rays apparently emitted by each point on the object. The light intensities corresponding to the angular samples of the light rays are anisotropically modulated to reconstruct a 3D scene that approximates the visual effect of viewing a natural 3D scene. Each directional sample represents the subtle difference of the object when viewed from slightly different positions and thus is regarded as an elemental view of the object.

To enable the eye to accommodate at the depth of a 3D object rather than that of the sources from which the rays originate, a true LF-3D display requires that multiple different samples of ray directions, commonly referred to as elemental views, are seen through each of the eye pupils and that they integrally sum together to form the perception of the object. When the eye is accommodated at a given depth (e.g. the blue corner of the cube in Figure 8.38), the directional ray samples apparently emitted by a point at the same depth naturally form a sharply focused image on the retina, while the rays for points at other depths (e.g. the orange and green corners in Figure 8.38) show retinal blur effects varying with their corresponding depths. Therefore, a LF-3D display has the potential to render correct or nearly correct focus cues for 3D scenes and resolve the VAC problem. Furthermore, it can also potentially render correct motion parallax, which is the change in retinal image caused by a change of eye pupil position, and further improve 3D perception. Conventional two-view stereoscopic displays lack the ability to render motion parallax because the retinal images remain unchanged when the eyes move within each of their own viewing windows. More details about LF-3D HMD will be discussed in the Section 8.5.4.

Among the different methods summarized above for enabling focus cues in HMDs, the Maxwellian view display attempts to extend the DOF of the virtual display to the extent such that eye accommodation is no longer required to view the virtual scene. This allows the eye to accommodate and converge freely to view real-world scenes, without compromising the image sharpness of the virtual display, which partially resolves the VAC in OST-HMDs. However, it is unable to produce natural retinal blur cues in virtual scenes. The vari-focal approach attempts to adapt the focus of the display in real-time to the vergence depth corresponding to the region of interest. It can be a simple and effective modification to conventional HMDs and is able to address the VAC problems to a great extent, but it is unable to produce natural retinal blur cues. The MFP and LF-3D methods render a true 3D scene by sampling either the projections of the scene at different depths or the directions of the light rays apparently emitted by the scene and viewed from different eye positions. The MFP method is an extension to the bifocal lens method described earlier in [5] but provides a larger number

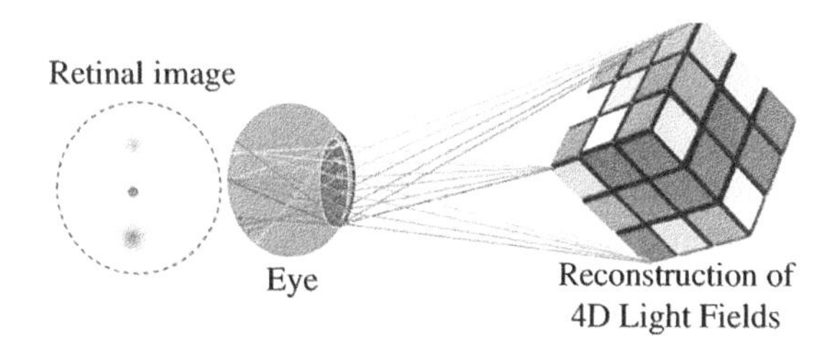

Figure 8.38 Schematic illustration of a light field based 3D display which reproduces the directional light rays apparently emitted by an object at a given position.

of focal plane samples and improved 3D continuity and the LF-3D method shares similarities with some of the pinhole optics solutions.

8.5.4 Head-Mounted Light Field Displays

The light field of a 3D scene can be represented by the well-known 4D light field function which characterizes the radiance of a light ray as a function of a 2D ray position and 2D direction [172]. As illustrated in Figure 8.38, a LF-3D display aims to reconstruct the 4D light field of a 3D scene by sampling the angular directions of the light rays apparently emitted by the 3D scene. Rendering of a 3D point in space is accomplished by multiple bundles of light rays, each of which represents a different perspective of a 3D scene. These ray bundles are typically created by spatially separated pixels defining the positional information of the 4D light field function and an array of optical elements defines the directional information of the light field function. The lights rays from these spatially separate pixels integrally create the 3D point that appears to emit light in different directions and appears to be located at a different field angle and depth. As noted earlier, however, to stimulate the eye to accommodate at the depth of the 3D point being rendered and address the VAC problem, two or more samples having different ray directions are required to enter each of the eye pupils.

Several different methods have been explored to implement LF-3D displays, including super multi-view (SMV) displays [137, 138], integral-imaging (InI) based displays [173, 174], and computational multi-layer light field displays [175, 176]. The SMV display is analogous to the camera array method for light field capture, and is schematically illustrated in Figure 8.39a. It generally employs an array of 2D displays or a single rapidly scanned 2D display to create an array of projection directions, each of which renders an elemental view of a 3D scene from a given viewpoint, to produce dense samples of the light rays apparently emitted by the scene. For instance, Lee et al. demonstrated the construction of a 100 in., 300-Mpixel, and horizontal-parallax only light field display system using 304 projectors arranged in a 19 by 16 array with about a 0.17° horizontal parallax interval [177]. Instead of using an array of displays, Jones et al. demonstrated a 360° horizontal-parallax light field display system by projecting light-field patterns onto a rapidly spinning anisotropic reflective surface through a single high-speed projector [138]. Most of the existing systems based on the SMV-like method, however, only render the horizontal parallax of the light field due to the enormously increased complexity if vertical parallax were to be considered.

As schematically illustrated in Figure 8.39b, an InI-based display, using the same principle as the integral photography technique invented by Lipmann in 1908 [178], typically consists of a display panel and a 2D optics array which can be a micro-lens array (MLA) [173, 174] or aperture array [179]. The 2D optics array provides angular sampling of the directional light rays of a 3D scene, while the display renders a set of 2D elemental images, each of which presents a different perspective of the 3D scene through each microlens or aperture. The conical ray bundles emitted by the corresponding pixels in the elemental images intersect and integrally create the perception of a 3D scene that appears to emit light and occupy the 3D space. An InI-based display using 2D arrays allows the reconstruction of a 3D shape with full-parallax information in both horizontal and vertical directions, which is its main difference from the conventional autostereoscopic displays with only horizontal parallax using one-dimensional parallax barriers or cylindrical lenticular lenses [180].

As schematically illustrated in Figure 8.39c, a computational multi-layer display is a relatively new, emerging class of LF-3D display methods that samples the directional rays through multi-layers of pixel arrays. It typically consists of a stack of light-attenuating layers of SLMs illuminated by either a uniform or directional backlight [175, 176]. The light field of a 3D scene is computationally decomposed into a number of masks representing the transmittance of each layer of the light attenuators. The intensity of each light ray entering the eye from the backlight depends on the product of the pixel values of the attenuation layers at the points which the ray intersects. The computational multi-layer displays operate in a multiplicative fashion and approximate the light fields of the scene by angularly sampling the directions of the light rays apparently emitted by a 3D scene in a fashion similar to the InI-based display. The pinlight display discussed in Section 8.5.3 may also be

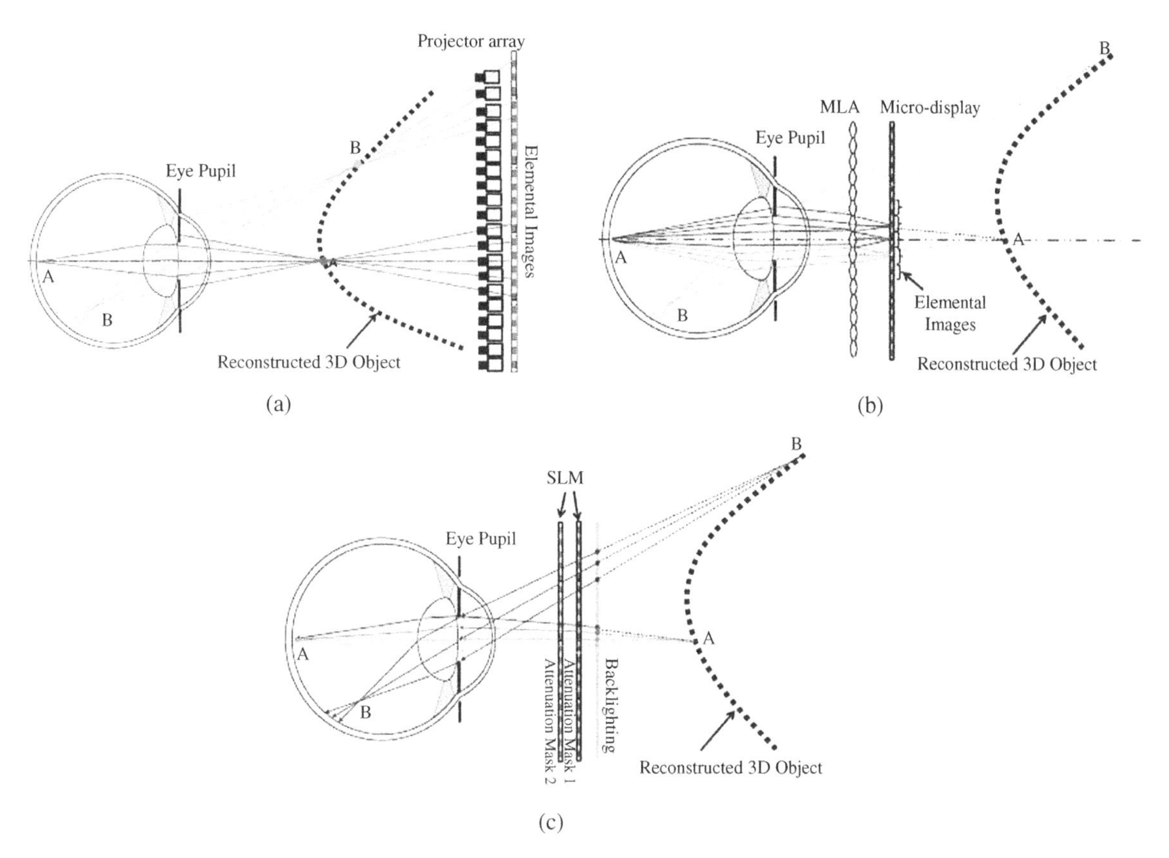

Figure 8.39 Schematic illustration of LF-3D display methods: (a) super multi-view displays implemented with an array of 2D displays, (b) integral imaging based display; and (c) computational multi-layer LF display.

considered a computational multi-layer display where the back layer of light attenuators next to the backlight is replaced by a pinhole array or an array of point sources [145].

Among these three methods of implementing LF-3D displays, the construction of SMV-based displays typically is very bulky and not suitable for adaption to a HMD format. On the other hand, both the InI-based method and the computational multi-layer method can be adapted to a HMD system. Thus, the rest of the section will focus only on recent advances in these two methods.

8.5.4.1 InI-Based Head-Mounted Light Field Displays

Since its invention by Lippmann in 1908, the InI-based technique has been widely explored both for capturing the light fields of real scenes [181, 182] and for use in eyewear-free autostereoscopic displays [183]. It has been known for its limitations in low lateral and longitudinal resolutions, narrow DOF, and narrow viewing angle. However, the simple optical architecture of an InI technique makes it attractive to integrate with an HMD optical system and create a wearable light field display.

One way of implementing an InI-based LF-3D HMD is a direct-view configuration which directly places a microdisplay and array optics in front of the eyes, similar to that illustrated in Figure 8.39b. Rays from the pixels corresponding to the same scene point are imaged by multiple lenslets toward the eye pupil and virtually intersect to form a reconstruction point at the rendered depth. Based on such a direct-view configuration, Lanman et al. demonstrated a prototype of an immersive LF-3D HMD design for VR applications [184]. The prototype system has an FOV of about 29° × 16° and a spatial resolution of 146 × 78 pixels. The main advantage of the direct-view configuration is its compactness due to the thin profile of the microdisplay-MLA assembly,

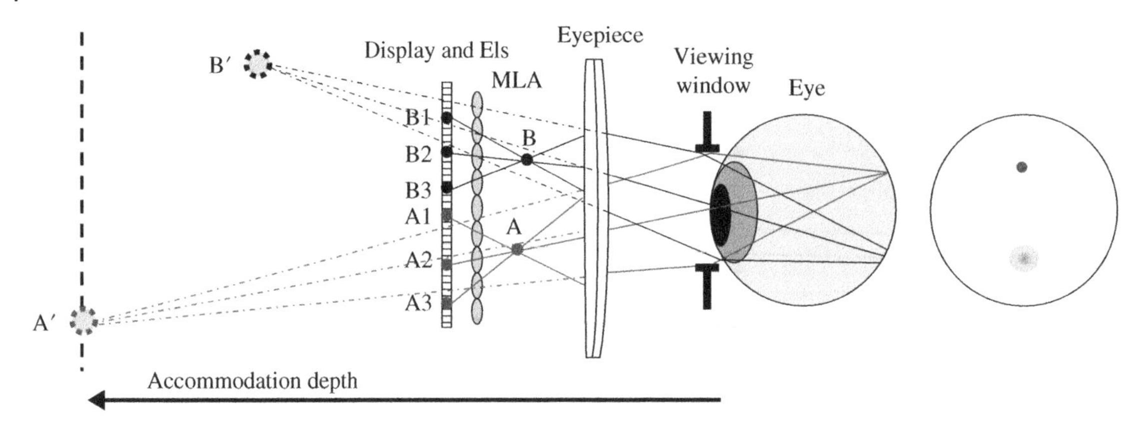

Figure 8.40 Schematic illustration of a magnified-view configuration.

but its obvious limitation is the lack of a see-through capability. Recently, based on the same direct-view configuration, see-through prototypes were demonstrated where transparent gaps are created between adjacent micro lenses and a transparent microdisplay is used [185]. In general, the main limitations of the direct-view approach include low lateral and longitudinal resolutions as well as shallow DOF, especially when a relatively large FOV is required. The depth of the reconstructed scene tends to be very shallow, appearing to be only a few inches away from the eye, due to the limited imaging capability of a stand-alone MLA with a long object-image conjugate.

Alternatively Hua and Javidi proposed a magnified-view architecture, as schematically illustrated in Figure 8.40, where a microscopic InI (micro-InI) unit is combined with a magnifying eyepiece to improve the overall depth of reconstruction and image quality [174]. The micro-InI unit, consisting of a microdisplay and array optics such as an MLA or pinhole array, reproduces the full-parallax light fields of a 3D scene and forms a 3D miniature scene (virtual or real). The 3D reconstructed scene takes the place of a conventional 2D microdisplay as the image source viewed by the eyepiece which forms a magnified, virtual 3D scene occupying a large depth volume for the eye to view.

Hua and Javidi further demonstrated the first practical implementation of an OST-HMD design by integrating a micro-InI unit for full-parallax 3D scene visualization with a freeform eyepiece [174]. Figure 8.41a and b show the optical layout and a photograph of a proof-of-concept monocular prototype. The prototype used a 0.8″ OLED, offering 1920 × 1200 color pixels with a pixel size of 9.6 μm, an MLA of a 3.3 mm focal length and 0.985 mm pitch, and a wedge-shaped freeform eyepiece with an EFL of 28 mm, along with a see-through compensator. An array of 12 × 11 elemental images were simulated, each of which consists of 102 × 102 color pixels. The FOV of the reconstructed 3D scene by the eyepiece was about 33.4° diagonally. The viewing angle of the micro-InI unit was about 14°, which yields a crosstalk-free EPD of 5 mm in which the 3D InI view can be observed. The system demonstrates an angular resolution of 2.7 arc minutes on the virtual reference plane in the visual space. Figure 8.41c shows a photograph captured by a camera focusing at 0.25 diopters. The captured scene consists of two physical references, a Snellen eye chart and a resolution grating, placed at the depths of 0.25 diopters and at 3 diopters, respectively, and three columns of letter "E" rendered by 12 × 11 elemental images at the depths of 0.25, 1, and 3 diopters, respectively. Song et al. demonstrated another OST InI-HMD design using a pinhole array together with a similar freeform eyepiece [179].

Although the prototype examples above demonstrated that an InI-based HMD method can potentially produce correct focus cues and true 3D viewing, like other InI-based display and imaging technologies [182, 183], conventional InI-based displays suffer from several major limitations when applied to HMD systems [174, 179, 184] such as a narrow DOF when providing good spatial resolution of the 3D scene, or a constant but low spatial resolution over a long DOF, and a small viewing window due to crosstalk between the neighboring elemental images on the display panel. For instance, the prototype system in Figure 8.45 demonstrated

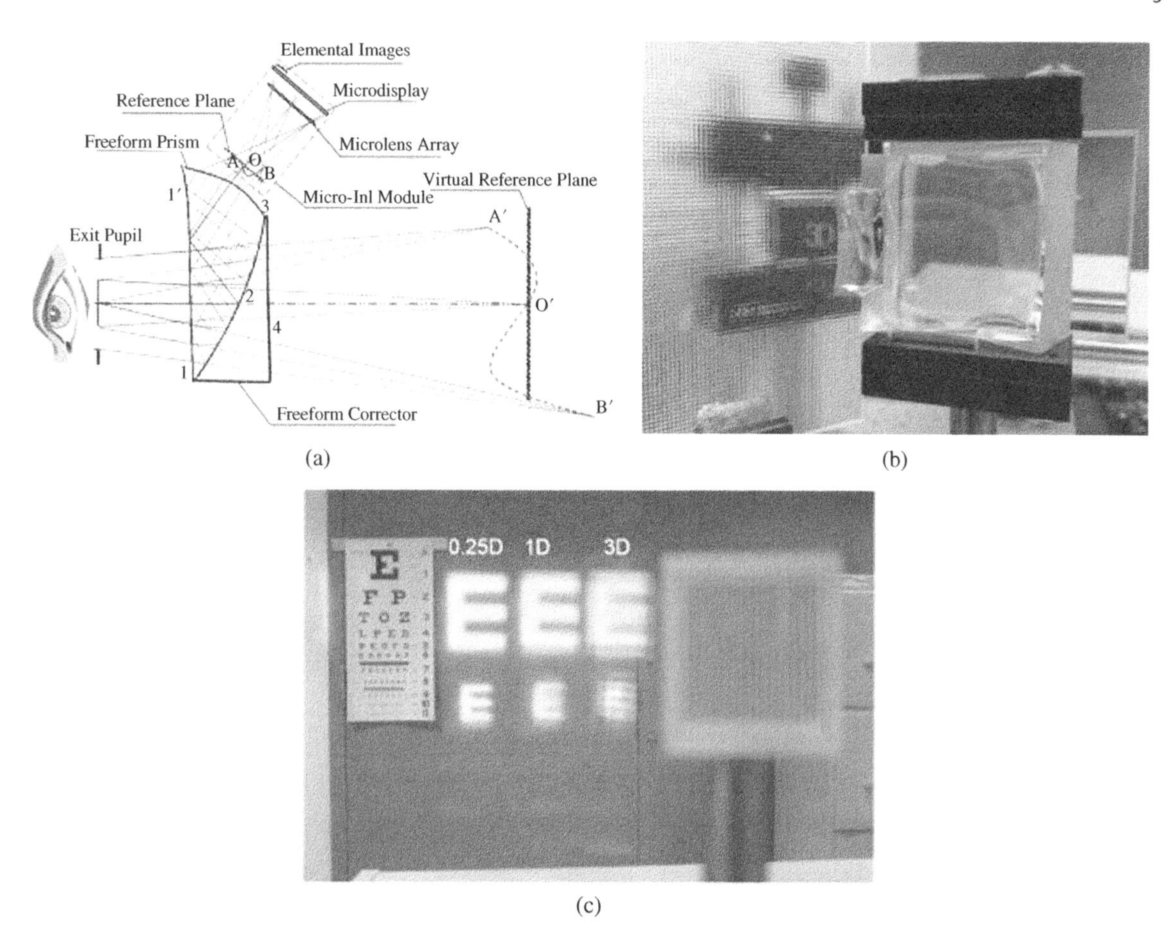

Figure 8.41 Demonstration of a 3D integral imaging optical see-through HMD using freeform optical technology: (a) optical design layout; (b) prototype of an InI-OST-HMD; (c) photograph captured with a camera focusing at 0.25 diopters [174].

a low spatial resolution of about 10 arc minutes per pixel in the visual space, a low longitudinal resolution of about 0.5 diopters, a narrow DOF of about 1 diopter for a 10-arc minute resolution criterion, noticeable crosstalk within a 4 mm view window, and low view density which barely affords two different views to fill a 3 mm eye pupil [174].

To address these limitations, Huang and Hua proposed a new optical architecture that improves the performance of an InI-based light field HMD by incorporating a tunable lens to extend the DOF without sacrificing the spatial resolution and an aperture array to reduce crosstalk, or equivalently expand the viewing window [76, 186]. Figure 8.42a shows the optical layout for one of the prototypes based on this new architecture [186]. The optical system consists of three key sub-units, a micro-InI unit integrated with a custom aspherical MLA and a custom aperture array, a tunable relay group that is able to dynamically tune the axial position of the reconstructed light field of a 3D miniature scene, and a custom freeform eyepiece. Two new rendering methods were also proposed that use the capability of the tunable relay optics to render a light field with a large depth volume from very near to very far from the viewer without compromising the spatial resolution. Combining fully-optimized custom optics with the new light field rendering methods significantly improves the performance of InI-based 3D light field displays in terms of spatial resolution, DOF, FOV, and crosstalk. Figure 8.42b shows a photograph of a binocular, wearable prototype system based on the proposed optical design architecture. Figure 8.42c and \d show two photographs captured with a camera placed at the exit pupil of the display.

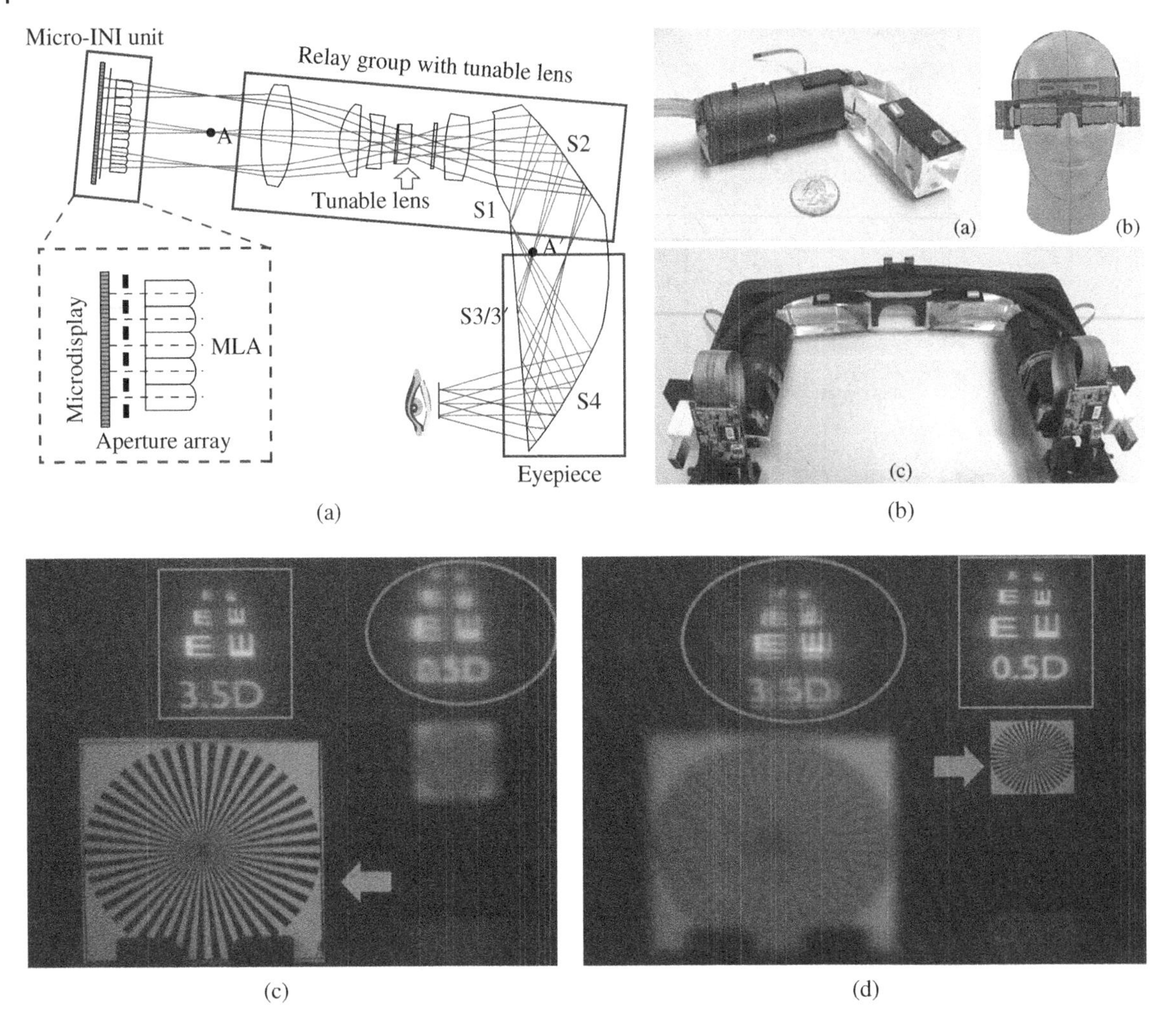

Figure 8.42 Example of a high-performance InI-based LF-3D OST-HMD: (a) the optical layout, (b) prototype, (c) and (d) images captured through the prototype with the camera focused at the depths of 3.5 and 0.5 diopters, respectively [76].

The scene was composed of two virtual targets rendered at depths of 3.5 and 0.5 diopters respectively, through their elemental images along with two physical references placed at the same depths as their corresponding virtual targets. The gaps and strokes of the smallest letters on the virtual target subtend a visual angle of 3 arc minutes. The prototype, with a total weight of about 450 g and with a volume of about 210 mm (width) by 80 mm (depth) by 40 mm (height), is capable of rendering a true 3D light field at a constant spatial resolution of 3 arc minutes across a very large depth range of over 3 diopters (from optical infinity to as close as 30 cm to the viewer) and removed crosstalk within an eye box about 6 mm by 6 mm. It not only maintains high quality imagery across a depth range of over 3 diopters for the virtual display path but also achieves high visual resolution and wide FOV for the see-through view.

A key challenge in developing LF-3D HMD technology is to establish methods for quantifying the accuracy of the focus cues rendered by a LF-3D display, establish the threshold requirements to render correct or nearly correct focus cues; and understand the relationship between light field sampling strategies in terms of ray directions, ray positions, accommodative cue accuracy and image quality. Several pioneering works have attempted to address some of these issues. Takaki suggested that the angular separation of the directional ray bundles is required to be around 0.2°–0.4° to allow more than two view samples for each eye and to stimulate an accommodation response so that the eyes focus at the rendered depth instead of at the 2D screens [187]. It remains unclear how accurate the rendered focus cues will be if such a view density is satisfied. Kim et al. experimentally measured the accommodative responses in viewing a real object and a digital 3D object rendered through integral imaging (InI) and suggested that over 73% of the 71 participants were able to accommodate at the depth of the rendered object instead of a display screen [188]. Stern et al. attempted to combine major perceptual and HVS requirements with analytical tools to build an analytical framework for establishing perceivable light fields and determining display device specifications [189]. Overall, however, a critical gap is the lack of a systematic method to quantify the relationships between the accuracy of focus cue rendering and the number of samples per pupil area and a systematic investigation on the tradeoff relationship between number of views and retinal image quality. More recently, Huang and Hua presented a systematic approach to fully address the three aforementioned fundamental issues in designing a LF-3D display. By extracting the common characteristics of existing LF-3D methods, a generalized framework was proposed to model their image formation processes, based on which systematic method was developed to simulate and characterize the retinal image quality and the accommodative response perceived from a LF-3D display by accounting for both the ocular and display factors [28, 29]. Using this framework and simulation approach, optimal view sampling strategies were investigated in terms of ray direction and position for LF-3D display designs offering a balance between accommodation cue accuracy and retinal image quality.

8.5.4.2 Computational Multi-Layer Head-Mounted Light Field Displays

Instead of using the InI-based architecture for angular sampling of directional light rays through a pinhole array or MLA, a computational multi-layer light field display samples directional rays through multi-layers of pixel arrays. It typically employs multiple layers of SLM illuminated by either a uniform or directional backlight. As illustrated in Figure 8.39c, the light field of a 3D scene is computationally decomposed into a number of masks representing the transmittance of each layer of the light attenuators. The intensity value of each light ray entering the eye from the backlight is the product of the pixel values of the attenuation layers which the ray intersects. Unlike the additive nature of the MFP approach discussed in Section 8.5.3 which renders the volume of a 3D scene by summing up the projected intensity values of a 3D scene sampled by focal planes of different depths, the computational multi-layer displays render the luminance values of the light rays apparently emitted from a 3D scene by multiplying the values of the different SLM layers with the luminance of the rays from a backlight. The rendered light rays integrally generate the effects of light fields of the 3D scene. Owing to this difference, a MFP display can achieve high spatial resolution and a large DOF with a small number of focal planes as demonstrated by some of the prototype examples. On the other hand, it is regarded as a volume display with a fixed viewing point due to the light addition nature of the projections on the different focal planes. A computational multi-layer display offers the ability to view the content from different viewing points, although often within a limited viewing window, due to its nature of ray rendering. It, however, often offers limited spatial resolution and DOF mostly due to diffraction effects and high cost of sampling light rays.

A computational multi-layer light field display is essentially an adaptation of the well-known parallax-barrier autostereoscopic displays which introduces content-adaptive optimization to compute the proper attenuation values for the mask layers. For instance, Wetzstein et al. demonstrated a new tensor display comprised of a stack of time-multiplexed light-attenuating layers illuminated by a directional backlight for autostereoscopic displays [175]. Unlike a conventional parallax barrier display which uses a fixed barrier to create multiple views, here the modulation pattern of each attenuation layer is optimized to produce images for a desired viewing zone. In this sense, the barrier layer is dynamically controlled based on the 3D scene.

Maimone and Fuchs pioneered work to apply the multilayer computational light field display technique for use in HMDs and demonstrated the first computational multilayer AR display [176]. A stack of transparent SLMs, a thin, transparent backlight, and a high-speed shutter were sandwiched together with a small spacing between the SLM layers. The sandwiched stack was placed directly in front of the eye, without any focusing optics between the display stack and the eye. The device operates in two modes: augmented image rendering mode (shutter off) and the occluded real-world image formation mode (shutter on). In the augmented view mode, the real-world view is blocked and a set of optimized patterns are rendered on the SLM layers to attenuate the light rays from the backlight and produce the final color of the rays entering the eye, which is the product of the attenuation values assigned to each of the intersected pixels across the layers. The multiple SLM layers reproduce a set of light rays with adequate angular resolution over the eye pupil that appear to be emitted from a virtual object at an apparent location far from the device stack. The eye perceives the sum of these rays integrally which synthesizes a virtual object at a desired depth. Therefore, the image formation process of the multi-layer display is similar to that of integral imaging and able to reconstruct the light field of a 3D virtual object with correct focus cues. In real-world image formation mode, the backlight is turned off and the shutter is turned on. Occlusion masks can be displayed on the SLM layers to allow selective transmission of the real-world rays, enabling mutual occlusion between virtual and real-world scenes. Due to the close proximity of the SLM stack to the eye, this method could potentially achieve a compact OST-HMD with a wide FOV. Owing to the light field rendering nature, it can also potentially render correct focus cues and mutual occlusion capabilities. Their early prototype demonstrated these capabilities to some extent. On the other hand, this approach is subject to major limitations. For instance, both the augmented image rendering mode and the occlusion-enabled see-through mode, although recognizable, suffer dramatic resolution loss due to diffraction effects through the SLM stack. The see-through view of the real world is blurred and low contrast due to the diffraction effects in the SLMs as well as their limited transparency.

More recently, Wetzstein and coworkers extended their multi-layer factored light field autostereoscopic display method and demonstrated a light field stereoscope for immersive VR applications [127]. The schematic layout and the prototype implementation are shown in Figure 8.43a and b, respectively. The prototype consists of two stacked liquid crystal display panels with a small gap in between and a pair of simple magnifier lenses with a focal length of 5 cm for each eye. Modulation patterns are computed using a rank-1 factorization process to synthesize and render the light field of a 3D scene. Although their preliminary demonstration is promising, this approach is subject to a diffraction limit due to the fact that the virtual image pattern on the rear display panel is observed through the front panel.

Overall, the computational multi-layer approach to head-worn light field display is subject to a number of obvious limitations and requires significant innovations to enable improvements. First of all, it suffers from significantly reduced spatial resolution due to diffraction artifacts caused by the small pixel apertures of high-resolution attenuation layers. The diffraction artifacts can be reduced by using reflective SLMs with high fill factors at the cost of increasing optic complexity and volume. Secondly, the technique is computationally intensive, requiring significant reduction of the optimization time to enable low-latency applications for AR displays. Finally, the prototype also suffers from high light loss due to the low transmittance of the SLM stack.

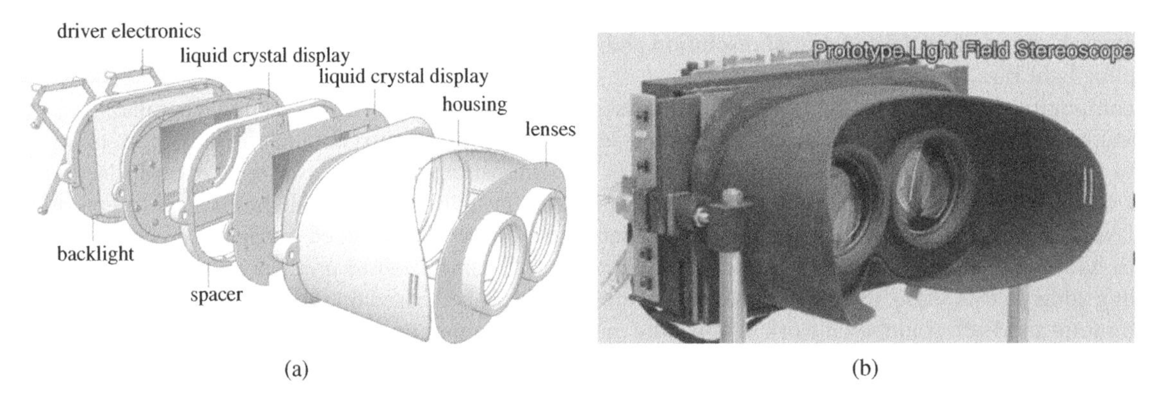

Figure 8.43 A computational multi-layer light field stereoscope: (a) schematic layout (b) prototype implementation [127].

8.5.5 Mutual Occlusion Capability

Occlusion is a natural light blocking behavior between overlapping objects. When an opaque object fully or partially blocks the view of another object, it is perceived to be closer to the viewer than the occluded object. Occlusion, as a relative depth ranking cue, is probably one of the most pervasive and reliable visual cues for depth perception. When virtual and real objects are mixed through AR or MR displays, an opaque virtual object should appear to be fully opaque and block the view of real objects that are located behind it; and a real object should naturally occlude the view of virtual objects located behind the real one. Therefore, mutual occlusion has two different aspects, the occlusion of real-scene objects by virtual ones and the occlusion of virtual objects by real scenes. The occlusion of a virtual object by a real object can be achieved straightforwardly, by simply not rendering the part of the virtual object that overlaps with the occluding real object, when the location of the real object relative to the virtual scene is known. The occlusion of a real object by a virtual one may be conveniently implemented in video see-through HMDs where the real-world view is digitally captured and thus can be digitally processed for proper occlusion with virtual objects. In an optical see-through HMD, however, implementing correct occlusion of real object by a virtual one presents a much more complicated problem because it requires the ability to selectively block the light of the real scene from reaching the eye in real time.

State-of-the-art OST-HMDs typically rely on an optical combiner to blend light from the real scene with virtual objects, and lack the ability to selectively block the light of the real world from reaching the eye. As a result, digitally rendered virtual objects viewed through OST-HMDs typically appear "ghost-like," floating "in front of" the real world. Figure 8.44 shows an un-edited AR view captured by a camera through a typical OST-HMD lacking occlusion capability where the virtual airplane appears not only washed out and non-opaque but also low-contrast.

Creating a mutual occlusion-capable optical see-through HMD (OCOST-HMD) poses a complex challenge. In the last decade, few OCOST-HMD concepts have been proposed, with even fewer designs being prototyped. Existing methods for implementing OCOST-HMDs fall into two types: direct ray blocking and per-pixel modulation. The direct ray blocking method, as schematically illustrated in Figure 8.45, selectively blocks the rays from the real-world scene without focusing them. It can be implemented by selectively

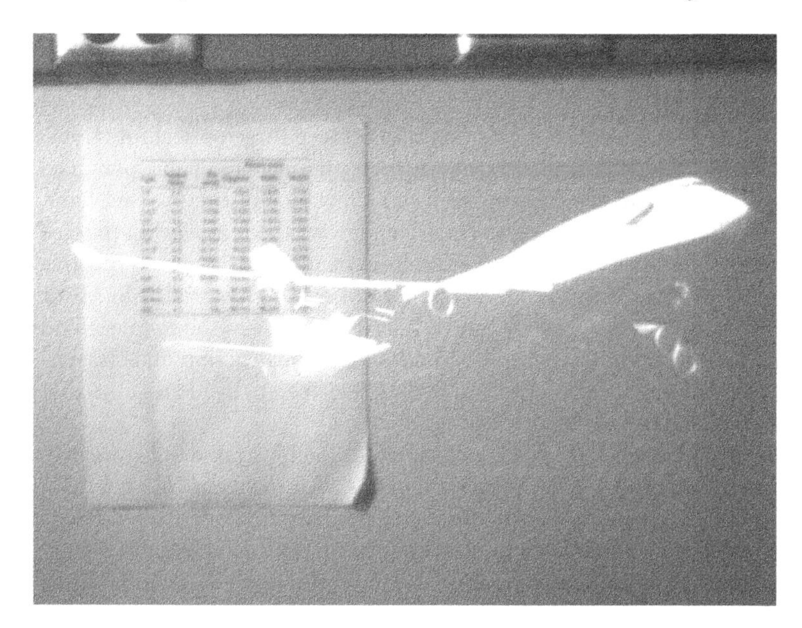

Figure 8.44 Superimposing a virtual airplane in a well-lit real world environment: AR view captured through a typical OST-HMD without occlusion capability.

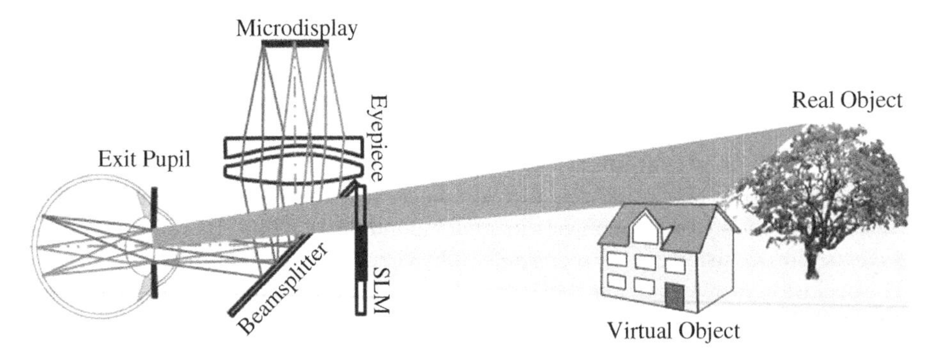

Figure 8.45 Schematic illustration of direct-ray blocking method for occlusion in OST-HMDs.

modifying the reflective properties of physical objects or by passing the light from the real scene through single or multiple layers of SLM placed directly near the eye. For instance, Hua et al. investigated the idea of creating natural occlusion of virtual objects by physical ones via a HMPD device, which involved the use of retroreflective screens onto physical objects and thus can only be used in limited setups [45]. Tatham demonstrated the occlusion function through a transmissive SLM directly placed near the eye with no imaging optics [190]. The direct ray blocking method via an SLM would be a straightforward and adequate solution if the eye were a pinhole aperture allowing a single ray from each real-world point to reach the retina. Instead, the eye has an area aperture, which makes it practically impossible to block all the rays seen by the eye from an object without blocking the rays from other surrounding objects using a single-layer SLM. Recently, Maimone and Fuchs proposed a lensless computational multi-layer OST-HMD design which consists of a pair of stacked transmissive SLMs, a thin and transparent backlight, and a high-speed optical shutter [176]. Multiple occlusion patterns can be generated using a multi-layer computational light field method so that the occlusion light field of the see-through view can be rendered properly. Although the multi-layer light field rendering method can in theory overcome some of the limitations of a single-layer ray blocking method, it is subject to several major limitations such as the significantly degraded see-through view, limited accuracy of the occlusion mask, and the low transmission. The unfavorable results can be attributed to the lack of imaging optics, low light efficiency of the SLMs, and most importantly the diffraction artifacts caused by the fine pixels of the SLMs located close to the eye pupil.

The per-pixel occlusion method requires a telescope-like optical system to achieve three optical functions: (i) capture the light from the real-world scene and create an intermediate focused image to enable per-pixel opacity control of the real-world scene; (ii) modulate the transmission or reflectance of the focused light, and (iii) relay and collimate the modulated light to optically reproduce the view of the real-world scene. A first-order optical representation of these functions is illustrated in Figure 8.46, mainly consisting of objective optics for creating an intermediate focus of the real-world scene where a SLM can be inserted, a SLM for light modulation, an eyepiece for collimating the modulated light, and a relay lens between the eyepiece and objective to provide an erect see-through view. This optical setup is similar to the construction of a classical telescope system except that here the optical powers of the eyepiece and objective need to be identical so that there is no artificial angular magnification applied to the see-through view.

Although it offers an accurate occlusion capability, the per-pixel occlusion method illustrated in Figure 8.46 adds significant complexity to the HMD optics and has been very challenging to implement in a compact and lightweight system. Based on this principle, perhaps the most complete demonstration of OCOST-HMD systems is the ELMO series displays with occlusion capabilities by Kiyokawa et al. [191, 192]. All the ELMO series prototypes were implemented using conventional lenses, prisms, and mirrors. Figure 8.47a and b show the schematic optical layout and the prototype of the latest ELMO-4 system. The ELMO-4 prototype contains four lenses, two prisms, and three mirrors arranged in a ring structure that requires a very bulky package blocking most of the user's face. Limited by the microdisplay and SLM technologies at that time, the ELMO

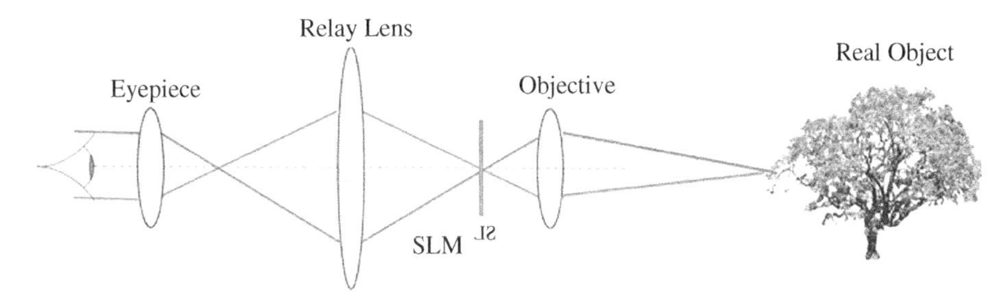

Figure 8.46 Schematic illustration of per-pixel method for occlusion in OST-HMDs.

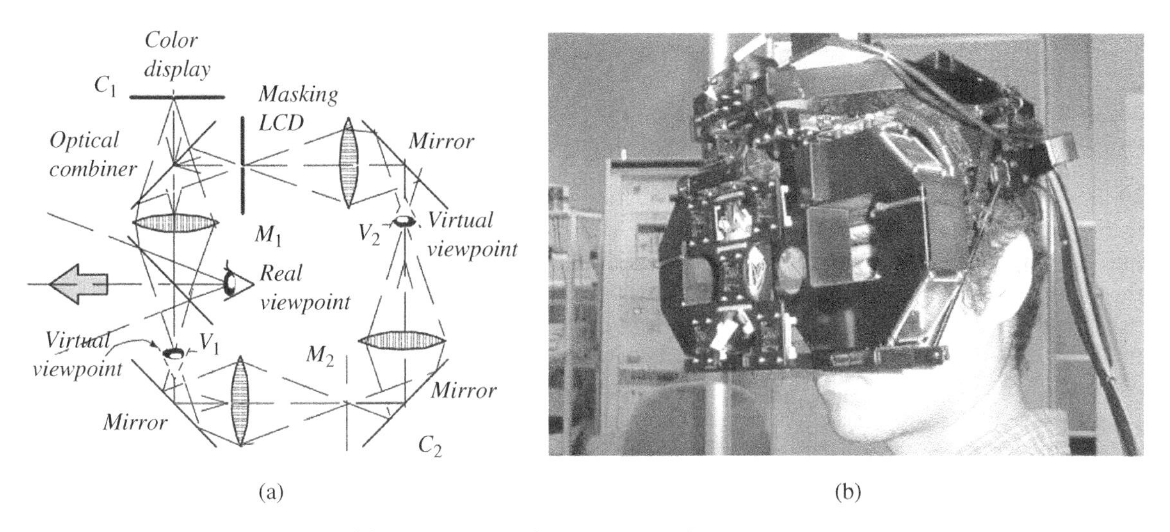

Figure 8.47 ELMO-4 occlusion-capable OST-HMD: (a) schematic layout; (b) prototype [192].

prototypes have fairly low resolutions for both the see-through and virtual display paths, both of which used a 1.5 in. QVGA (320 × 240) transmissive LCD module. Using a transmissive LCD as a SLM becomes problematic because when coupled with a polarizing beamsplitter (PBS), it allows limited light (<20%) from the real scene to pass through to the user, causing the device to become ineffective in dim environments. Not only because of the number of elements being used, but also more importantly due to the rotationally symmetric nature of the optical systems, the ELMO series of OCST-HMDs inevitably have a helmet-like, bulky form factor. Cakmakci et al. attempted to improve the compactness of the overall system by using polarization-based optics and a reflective SLM [193]. They used a reflective LCoS in conjunction with an OLED display to give an extended contrast ratio of 1 : 200. An x-cube prism was proposed for coupling the two optical paths to achieve a more compact form factor. However, the design failed to erect the see-through view correctly.

Recently, Gao et al. proposed to use freeform optics, a two-layer folded optical architecture, along with a reflective SLM to create a compact high resolution, low distortion OCOST-HMD [194]. Figure 8.48a shows the basic concept of the design. The green arrows represent the light propagation path of the real-world view and the red arrows represent the light path of the virtual view. The design mainly consists of two freeform prisms both of which are made of multiple freeform optical surfaces. The eyepiece prism is the sole optical element required for virtual display, while the objective prism and the eyepiece prism together act as an afocal optic for the real-world view. The objective prism collects incoming light from the physical environment and forms an intermediate image at its focal plane and the SLM placed at the focal plane of the objective prism controls the opaqueness of the real view. Through a beamsplitter the modulated light is folded toward the eyepiece prism and effectively merges with the light path of the virtual display for viewing. The focal planes of the two

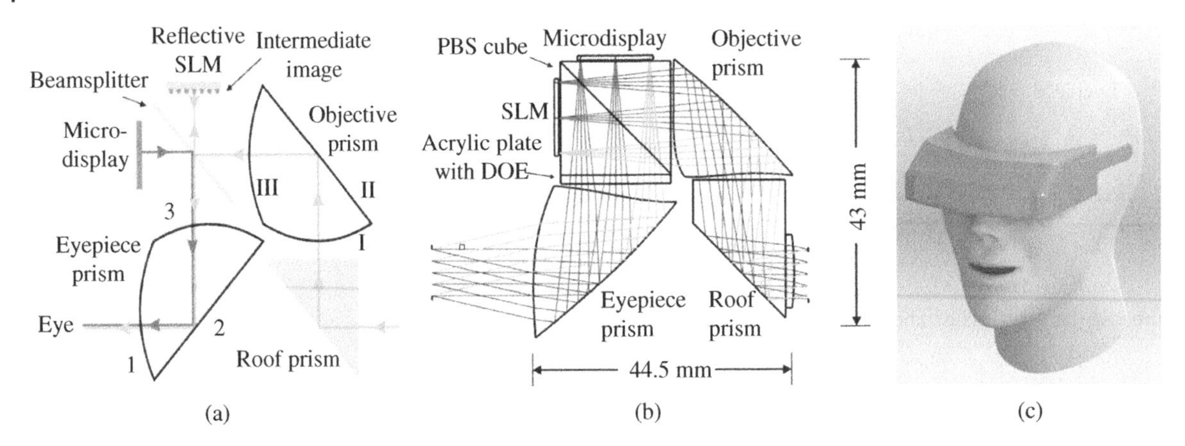

Figure 8.48 Example of a two-layer folded occlusion-capable OST-HMD design: (a) schematic layout; (b) optical layout of an optimized design with freeform prisms; and (c) mechanical package of a binocular system based on the design in (b) [194].

prisms are optically conjugate with each other through the beamsplitter, which makes the pixel-by-pixel level occlusion manipulation possible. See-through view erection is achieved using a right-angle roof prism which not only serves the purpose of folding the optical path of the real view for compactness but also erects the see-through view. Figure 8.48b shows the optical layout of a fully optimized design based on a 0.8″ microdisplay with a 5 : 4 aspect ratio, a 1280 × 1024 pixel resolution and a pixel size of 12 μm. The design uses a SLM of the same size and resolution as the microdisplay and achieves a diagonal FOV of 40°, that is 31.7° horizontally and 25.6° vertically, an EPD of 8 mm (non-vignetted), and an eye clearance of 18 mm. In this design, a DOE plate is employed to correct chromatic aberrations. Figure 8.48c shows a true-size mechanical package of a binocular system. With use of a reflective LCoS device as the SLM, the system allowed for a high luminance throughput and high optical resolution for both virtual and see-through paths. The optical design and preliminary experiments demonstrated great potential for a very compelling form factor and high optical performances, but the design was dependent on the use of expensive freeform lenses and, regrettably, was not fully prototyped.

More recently, based on the two-layer folding layout by Gao et al. [194], Wilson and Hua designed and built a prototype system using available stock lenses, as shown by the optical layout in Figure 8.49a [195]. The light path for the virtual display (eyepiece) is denoted by the blue rays, while the light path for the see-through view is shown by red rays. The red rays for the see-through view overlap the blue rays of the eyepiece after the PBS and thus only the blue rays are traced to the eye pupil in Figure 8.49a. The design consists of 11 glass lenses, 2 folding mirrors, 1 PBS, and 1 roof prism, all of which are stock components except for the meniscus which is made of flint glass with an aperture diameter greater than 40 mm. The design was based on the choice of a 0.7″ Sony color OLED microdisplay with a pixel size of 12 μm and a native resolution of 1280 × 720 pixels and a 0.7″ gray-scale LCoS as the SLM for the see-through path. The LCoS offers a native resolution of 1400 × 1050 pixels, a pixel pitch of 10.7 μm, and an aspect ratio of 4 : 3. The design achieves a prototype, shown in Figure 8.49b, with a diagonal FOV of 30°, or 26.5° horizontally and 15° vertically, an angular resolution of 1.24 arc minutes per pixel for the virtual display, and highly accurate opaqueness control (pixel by pixel) on the real-world scene at an angular resolution of 0.85 arc minutes per pixel. For the purpose of qualitative demonstration of the occlusion capability of the OCOST-HMD prototype, we created a real-world scene composed of a mixture of laboratory objects with a well-illuminated white background wall (~300–500 cd/m^2) while the virtual 3D scene was a simple image of a teapot. Figure 8.50a–c show a set of images captured with a digital camera placed at the exit pupil of the eyepiece. The camera lens has a focal length of 16 mm with its aperture set at about 3 mm to match the f/# setting equivalent to that of human eyes under typical lighting conditions. Figure 8.50a is the view of the natural background scene only captured through the occlusion module when the SLM is turned on for light transmission without modulation and with the OLED microdisplay turned off.

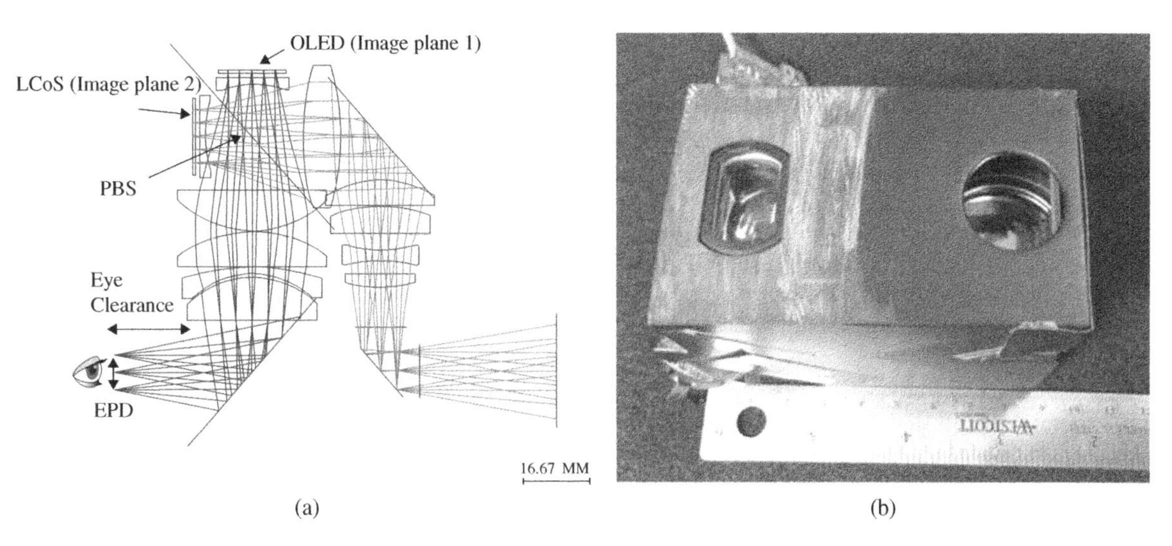

Figure 8.49 Example of a two-layer folded occlusion-capable OST-HMD using stock lenses: (a) optical layout; and (b) prototype assembly [195].

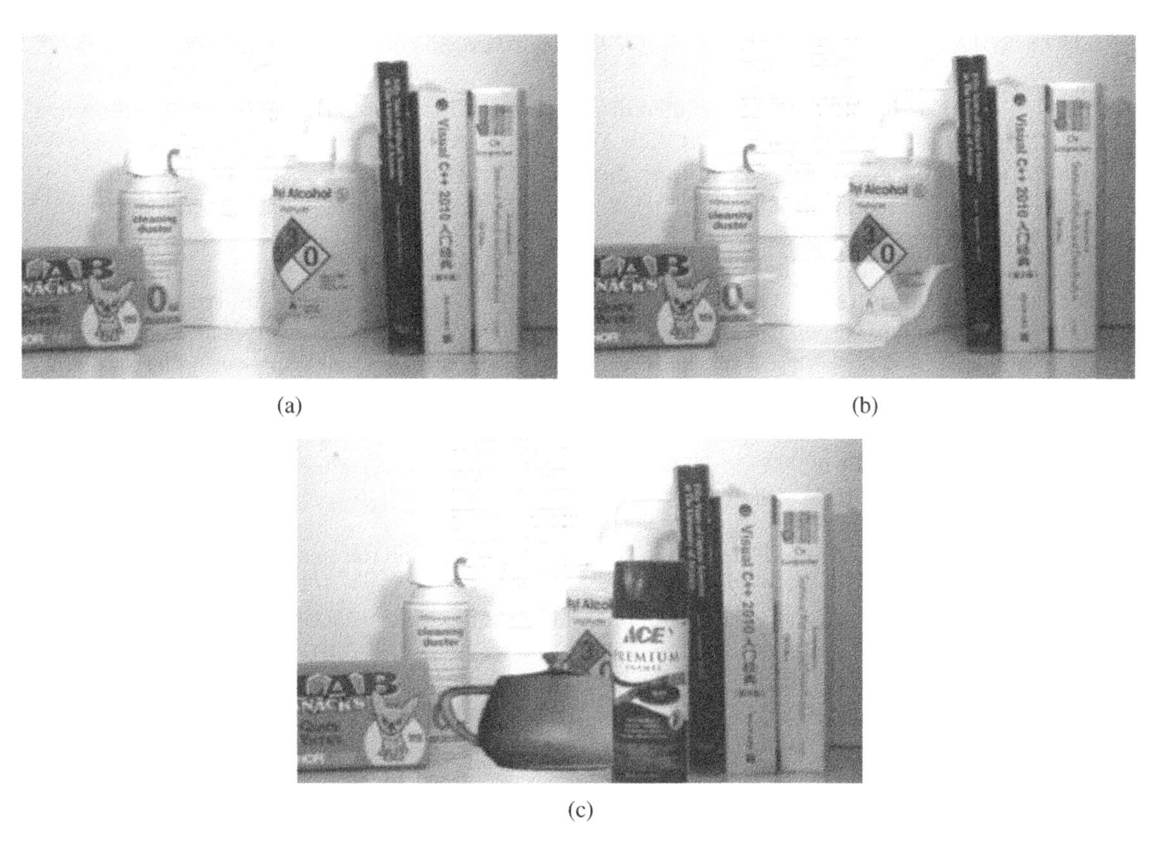

Figure 8.50 Demonstration of occlusion capability in our OCOST-HMD prototype [195].

Figure 8.50b is the view of an augmented scene captured through the eyepiece module when no occlusion mask was rendered; and Figure 8.50c is the augmented view when the occlusion mask was enabled.

All of the per-pixel occlusion examples based on the scheme in Figure 8.46 create a fixed occlusion plane at a depth optically conjugate to the SLM through an objective lens. As a result, the occlusion mask is rendered at a single, fixed focal depth. With the emergence of HMDs that offer the ability to render correct focus cues as discussed in Section 8.5.3, the focal depth of the occlusion mask may mismatch the apparent focal depth of the occluding object, leading to conflicting focus cues. Hamasaki and Itoh recently demonstrate a proof-of-concept varifocal-occlusion system in which an SLM is mechanically adjusted along the optical axis of the occlusion path so that the apparent depth of the occlusion mask can be adjusted and match with the depth of the occluding virtual object [196].

In summary, HMD technology has come a long way since its inception and will play a critical role in the era of spatial computing and in the rapidly growing revenues in VR and AR industries and beyond. In the meantime, it is worth pointing out that a wide range of engineering challenges continue to exist and new forms of optical materials, components, and architecture are required for evolutional development of HMD technology. It is also worth noting that there have not been adequate psychophysical studies to reveal the impacts of various HMD technologies to the perception, health, and social behavior of the human users.

References

1 Melzer, J.E. and Moffitt, K. (1997). *Head Mounted Displays: Designing for the User*. New York: McGraw Hill.
2 Rolland, J.P. and Hua, H. (2005). Head-mounted display systems. In: *Encyclopedia of Optical Engineering* (eds. R.B. Johnson and R.G. Driggers), 1–13. New York, NY: Marcel Dekker.
3 Cakmakci, O. and Rolland, J. (2006). Head-work displays: a review. *J. Disp. Technol.* 2 (3): 199–216.
4 Azuma, R., Baillot, Y., Behringer, R. et al. (2001). Recent advances in augmented reality. *IEEE Comput. Graphics Appl.* 21 (6): 34–47.
5 Feiner, S. (2002). Augmented reality: a new way of seeing. *Sci. Am.* 54: 2002.
6 Rolland, J.P. and Fuchs, H. (2001). Optical versus video see-through head-mounted displays. In: *Fundamentals of Wearable Computers and Augmented Reality* (eds. T. Caudell and W. Barfield). Erlbaum.
7 Zhang, R. and Hua, H. (2008). Design of a polarized head-mounted projection display using ferroelectric liquid-crystal-on-silicon microdisplays. *Appl. Opt.* 47 (15): 2888–2896.
8 Wheatstone, S.C. (1838). Contributions to the physiology of vision.—Part the first. On some remarkable, and hitherto unobserved, phenomena of binocular vision. *Philos. Trans. R. Soc. London* 128: 371–394.
9 Gruber, G.J. (2015). *The Biography of William B. Gruber*. Mill City Press, Inc.
10 M. Heilig, Sensorama simulator, US Patent #3,050,870, 1962.
11 Fisher, S., Wenzel, E., Coler, C., and Mcgreevy, M.W. (1988). Virtual interface environment workstations. *Proc. Hum. Factors Soc. Annu. Meet.* 32 (10).
12 Sutherland, I.E. (1968). A head-mounted three-dimensional display. *Proc. Fall Joint Comput. Conf. AFIPS* 33: 757–764.
13 Caudell, T.P. and Mizell, D.W. (1992). Augmented reality: an application of heads-up display technology to manual manufacturing processes. In: *IEEE Proceedings of the Twenty-Fifth Hawaii International Conference on System Sciences*, 659–669. IEEE.
14 Feiner, S., MacIntyre, B., and Seligmann, D. (1993). Knowledge-based augmented reality. *Commun. ACM* 36 (7): 53–62.
15 L. B. Rosenberg, "The Use of Virtual Fixtures As Perceptual Overlays to Enhance Operator Performance in Remote Environments," Technical Report AL-TR-0089, USAF Armstrong Laboratory, Wright-Patterson AFB OH, 1992.

16 Feiner, S., MacIntyre, B., Höllerer, T. et al. (1997). A touring machine: prototyping 3D mobile augmented reality systems for exploring the urban environment. *Personal Technologies* 1 (4): 208–217.
17 Mann, S. (1997). Wearable computing: a first step toward personal imaging. *IEEE Computer* 30 (2): 25–32.
18 Billinghurst, M., Weghorst, S., and Furness, T.A. III, (1998). Shared space: an augmented reality approach for computer supported collaborative work. *Virtual Real.* 3: 25–36.
19 Billinghurst, M., Kato, H., and Poupyrev, I. (2001). The MagicBook – moving seamlessly between reality and virtuality. *IEEE Comput. Graphics Appl.* 21 (3): 2–4.
20 H. Kato, M. Billinghurst, "Marker tracking and HMD calibration for a video-based augmented reality conferencing system. "*In Proceedings of the 2nd IEEE and ACM International Workshop on Augmented Reality (IWAR 99),* October 1999.
21 Albert B. Pratt, Weapon, US patent 1, 183,492, filed in 1915.
22 Rash, C.E. (ed.) (2001). *Helmet-Mounted Displays: Design Issues for Rotary-Wing Aircraft.* Bellingham: SPIE Press PM.
23 Schwiegerling, J. (2004). *Field Guide to Visual and Ophthalmic Optics.* Washington, DC, USA: SPIE Press.
24 Schwiegerling, J. (2018). The human eye and its aberrations. In: *Advanced Optical Instruments and Techniques,* vol. 2 (eds. D. Malacara-Hernández and B.J. Thompson). Boca Raton, FL: Taylor & Francis.
25 Navaro, R. (2009). The optical design of the human eye: a critical review. *J. Optom.* 2 (1): 3–18.
26 Liu, S. and Hua, H. (2010). A systematic method for designing depth-fused multi-focal plane three-dimensional displays. *Opt. Express* 18 (11): 11562–11573.
27 Hu, X. and Hua, H. (2014). Design and assessment of a depth-fused multi-focal-plane display prototype. *IEEE/OSA J. Disp. Technol.* 10 (4): 308–316.
28 Huang, H. and Hua, H. (2017). Systematic characterization and optimization of 3D light field displays. *Opt. Express* 25 (16): 18508–18525.
29 Huang, H. and Hua, H. (2019). Effects of ray position sampling on the visual responses of 3D light field displays. *Opt. Express* 2 (7): 9343–9360.
30 Ogle, K.N. and Schwartz, J.T. (1959). Depth of focus of the human eye. *J. Opt. Soc. Am.* 49: 273–280.
31 Barten, P.G.J. (1999). *Contrast Sensitivity of the Human Eye and Its Effects on Image Quality.* SPIE Press Book.
32 H. T. E. Hertzberg, G. S. Daniels, and E. Churchill, "Antropometry of Flying Personnel--1950 (Wright Air Development Center, Wright-Patterson Air Force Base, OH, WADC TR-52-321, 1954.
33 S.S. Grigsby, B.H. Tsou. "Visual factors in the design of partial overlap binocular helmet-mounted displays." Society for Information Displays Inernational Symposium Digest of Technical Papers, Vol. XXVI, 1993.
34 Melzer, J.E. (1998). Overcoming the field of view: resolution invariant in head mounted displays. In: *Proceedings of SPIE,* Helmet- and Head-Mounted Displays III, vol. 3362 (eds. R.J. Lewandowski, L.A. Haworth and H.J. Girolamo), 284–293. SPIE.
35 Klymenko, V., Verona, R.W., Beasley, H.H., and Martin, J.S. (1994). Convergent and divergent viewing a ffect luning, visual thresholds, and field-of-view fragmentation in partial binocular overlap helmet-mounted displays. In: *Proceedings SPIE,* Helmet- and Head-Mounted Displays and Symbology Design Requirements, vol. 2218, 82–96. SPIE.
36 Greivenkamp, J.E. (2004). *Field Guide to Geometrical Optics.* Washington, DC: SPIE Press.
37 Y. Iba, "Image observation device," US Patent 5,384,654, Jan 1995.
38 Liu, S., Hua, H., and Cheng, D. (2010). A novel prototype for an optical see-through head-mounted display with addressable focus cues. *IEEE Trans. Visual Comput. Graphics* 16: 381–393.
39 Hu, X. and Hua, H. (2014). High-resolution optical see-through multi-focal-plane head-mounted display using freeform optics. *Opt. Express* 22 (11): 13896–13903.

40 Kijima, R. and Hirose, M. (1995). A compound virtual environment using the projective head-mounted display. In: *Proceedings of ACM International Conference on Artificial Reality and Tele-Existence/ACM Conference on Virtual Reality Software and Technology (ICAT/VRST)*, 111–121. ACM.
41 R. Fisher, "Head-mounted projection display system featuring beam splitter and method of making same," U.S. Patent 5,572,229, 1996.
42 J. Fergason, "Optical system for head mounted display using a retro-reflector and method of displaying an image," U.S. patent 5,621,572, 1997.
43 Hua, H., Girardot, A., Gao, C., and Rolland, J.P. (2000). Engineering of head-mounted projective displays. *Appl. Opt.* 39 (22): 3814–3824.
44 Hua, H., Gao, C., Biocca, F., and Rolland, J.P. (2001). An ultra-light and compact design and implementation of head-mounted projective displays. In: *Proceedings of IEEE Virtual Reality Annual International Symposium*, 175–182. IEEE.
45 Hua, H., Gao, C., and Brown, L.D. (2002). A Testbed for precise registration, natural occlusion, and interaction in an augmented environment using head-mounted projective display. In: *Proceedings of IEEE Virtual Reality*, 81–89. IEEE.
46 Hua, H., Brown, L.D., and Gao, C. (2004). SCAPE: supporting stereoscopic collaboration in augmented and projective environments. *IEEE Comput. Graphics Appl.* 24 (1): 66–75.
47 Hua, H., Ha, Y., and Rolland, J.P. (2003). Design of an ultralight and compact projection lens. *Appl. Opt.* 42: 97–107.
48 Hua, H. and Gao, C. (2007). Design of a bright polarized head-mounted projection display. *Appl. Opt.* 46: 2600–2610.
49 Martins, R., Shaoulov, V., Ha, Y., and Rolland, J.P. (2007). A mobile head-worn projection display. *Opt. Express* 15 (22): 14530–14538.
50 Rolland, J.P., Biocca, F., Hamza-Lup, F. et al. (2005). Development of head-mounted projection displays for distributed, collaborative augmented reality applications. *Presence Teleop. Virt. Environ.* 14 (5): 528–549.
51 Brown, L.D. and Hua, H. (2006). Magic lenses for augmented virtual environments. *IEEE Comput. Graphics Appl.* 26 (4): 64–73.
52 Hua, H., Brown, L., and Zhang, R. (2011). Head-mounted projection display technology and application. In: *Handbook of Augmented Reality* (ed. B. Furht), 123–156. Springer Science+Business Media.
53 Zhang, R. and Hua, H. (2009). Imaging quality of a retroreflective screen in head-mounted projection displays. *J. Opt. Soc. Am. A* 26 (5): 1240–1249.
54 A. Becker, "Miniature video display system," U.S. Patent 4 934 773, Jun. 19, 1990.
55 T. Furness, Virtual Retinal Display," U.S. Patent 5 467 104, Nov. 14, 1995.
56 Urey, H. (2005). Retinal scanning displays. In: *Encyclopedia of Optical Engineering*, vol. 3 (ed. R. Driggers), 2445–2457. New York: Marcel Dekker.
57 Urey, H. (2005). Vibration mode frequency formulae for micromechanical scanners. *J. Micromech. Microeng.* 15: 1713–1721.
58 Holmgren, D. and Robinett, W. (1993). Scanned laser displays for virtual reality: a feasibility study. *Presence Teleop. Virt. Environ.* 2: 171–184.
59 Urey, H. (2001). Diffractive exit-pupil expander for display applications. *Appl. Opt.* 40 (32): 5840–5851.
60 B. T. Schowengerdt, M. Murari, E. J. Seibel, "Volumetric display using scanned fiber array," SID Symposium Digest of Technical Papers, 2010.
61 T. J. Edwards, M. Rud, "Dual combiner eyepiece," US Patent 5,506,728, 1996.
62 M. Spitzer, Eyeglass display lens system employing off-axis optical design, US Patent 6,353,503.
63 Cheng, D., Wang, Y., Hua, H., and Talha, M.M. (2009). Design of an optical see-through head-mounted display with a low f-number and large field of view using a freeform prism. *Appl. Opt.* 48 (14): 2655–2668.
64 J. Legerton, R. Sprague, Method and apparatus for constructing a contact lens with optics, US patent 8,142,016 B2, March 2012.

65 T. Levola, "Method and system for beam expansion in a display device," US Patent US US 7,206,107 B2, 2007.
66 Y. Amitai, "Light guide optical device," US Patent 7,457,040 B2, 2008.
67 B.S. Fritz, Head mounted display using mangin mirror combiner, US Patent 5,838,490.
68 J. D. Robinson, C. M. Schor, P.H. Muller, W.A. Yankee, Video headset, US Patent 5,696,521, December 1997.
69 D. Kessler and M. Bablani, Head-mounted optical apparatus using an OLED display, US Patent 8,094,377 B2, Jan 2012.
70 M. B. Spitzer, X. Miao, B. Amirpartiz, Method and apparatus for a near-to-eye display, US Patent 8,767,305 B2, July 2014.
71 Cakmakci, O. and Rolland, J.P. (2007). Design and fabrication of a dual-element off-axis near-eye optical magnifier. *Opt. Lett.* 32 (11): 1363–1365.
72 Zheng, Z., Liu, X., Li, H., and Xu, L. (2010). Design and fabrication of an off-axis see-through head-mounted display with an x–y polynomial surface. *Appl. Opt.* 49 (19): 3661–3668.
73 S. Yamazaki, K. Inoguchi, Image display apparatus, US Patent 6,384,983, May 2002.
74 Cheng, D., Wang, Y., Hua, H., and Sasian, J. (2011). Design of a wide-angle, lightweight head-mounted display using free-form optics tiling. *Opt. Lett.* 36 (11): 2098–2100.
75 Hua, H., Hu, X., and Gao, C. (2013). A high-resolution optical see-through head-mounted display with eyetracking capability. *Opt. Express* 21 (25): 30993–30998.
76 Huang, H. and Hua, H. (2018). High-performance integral-imaging-based light field augmented reality display using freeform optics. *Opt. Express* 26 (13): 17578–17590.
77 Levola, T. (2006). Diffractive optics for virtual reality displays. *J. Soc. Inf. Disp.* 14 (5): 467–475.
78 Mukawa, H., Akutsu, K., Matsumura, I. et al. (2009). A full-color eyewear display using planar waveguides with reflection volume holograms. *J. Soc. Inf. Disp.* 17 (3): 185–193.
79 Piao, M.L. and Kim, N. (2014). Achieving high levels of color uniformity and optical efficiency for a wedge-shaped waveguide head-mounted display using a photopolymer. *Appl. Opt.* 53 (10): 2180–2186.
80 Han, J., Liu, J., Yao, X., and Wang, Y. (2015). Portable waveguide display system with a large field of view by integrating freeform elements and volume holograms. *Opt. Express* 23 (3): 3534–3549.
81 Kress, B.C. and Cummings, W.J. (2017). 11-1: Invited Paper: Towards the ultimate mixed reality experience: holoLens display architecture choices. *SID Digest of Technical Papers* 48 (1): 127–131.
82 Waldern, J.D., Grant, A.J., and Popovich, M.M. (2018). DigiLens switchable Bragg grating waveguide optics for augmented reality applications. In: *Proceedings of SPIE*, Digital Optics for Immersive Displays, 106760G:1-16,, vol. 10676. SPIE.
83 Weng, Y., Xu, D., Zhang, Y. et al. (2016). Polarization volume grating with high efficiency and large diffraction angle. *Opt. Express* 24 (16): 17746–17759.
84 Lee, Y., Yin, K., and Wu, S. (2017). Reflective polarization volume gratings for high efficiency waveguide-coupling augmented reality displays. *Opt. Express* 25 (22): 27008–27014.
85 Xu, M. and Hua, H. (2019). Methods of optimizing and evaluating geometrical lightguides with microstructure mirrors for augmented reality displays. *Opt. Express* 27 (4): 5523–5543.
86 Cheng, D., Wang, Y., Xu, C. et al. (2014). Design of an ultra-thin near-eye display with geometrical waveguide and freeform optics. *Opt. Express* 22 (17): 20705–20719.
87 Wang, Q., Cheng, D., Hou, Q. et al. (2015). Stray light and tolerance analysis of an ultrathin waveguide display. *Appl. Opt.* 54 (28): 8354–8362.
88 C. J. Wang and B. Amirparviz, "Image waveguide with mirror arrays," U.S. Patent No. 8,189,263 (2012).
89 B. Pascal, D. Guilhem, and S. Khaled, "Optical guide and ocular vision optical system," U.S. Patent, No. 8,433,172 (2013).
90 H. Erfle, Ocular. US Patent 1,478,704, Dec 25 1923.
91 E. M. Howlett, Wide angle photography method and systems, U.S. Patent 4,406,532 (1983).

92 Geng, Y., Gollier, J., Wheelwright, B. et al. (2018). Viewing optics for immersive near-eye displays: pupil swim/size and weight/stray light. In: *Proceedings of SPIE*, Digital Optics for Immersive Displays, 1067606:1-17,, vol. 10676. SPIE.

93 M. D. Missig, G. M. Morris, "Wide-angle eyepiece optical system employing refractive and diffractive optical elements," US Patent 5,446,588 (1995).

94 K. Takahashi, "Head or face mounted image display apparatus," US Patent 5,701,202, 1997.

95 Wong, T.L., Yun, Z., Ambur, G., and Jo, E. (2017). Folded optics with birefringent reflective polarizers. In: *Proceedings of SPIE*, Digital Optical Technologies, 103350E:1-7, vol. 10335. SPIE.

96 Huxford, R.B. (2004). Wide FOV head-mounted display using hybrid optics. In: *Proceedingsof SPIE*, Optical Design and Engineering, vol. 5249 (eds. L. Mazuray, P.J. Rogers and R. Wartmann), 230–237. SPIE.

97 J. A. LaRussa, Image forming apparatus. US Patent 3,940,203, 1976.

98 A. L. Berman, J. E. Melzer, Optical collimating apparatus. US Patent 4,859,031, 1989.

99 Gu, L., Cheng, D., and Wang, Y.T. (2018). Design of an immersive head mounted display with coaxial catadioptric optics. In: *Proceedings of SPIE*, Digital Optics for Immersive Displays, 106761F: 1-6, vol. 10676. SPIE.

100 Droessler, J.G. and Rotier, D.J. (1995). Tilted cat helmet-mounted display. *Opt. Eng.* 29 (8): 24–49.

101 B. Chen, "Helmet visor display employing reflective, refractive, and diffractive optical elements," US Patent 5,526,183, 1996.

102 Chunyu Gao, "Ergonomic head mounted display device and optical system," US Patent 9,740,006 B2, August 2017.

103 T. M. Lippert and C. T. Tegreene, "Scanned display with plurality of scanning assemblies," US Patent 6,762,867 B2.

104 Shenker, M. (1994). Image quality considerations for head-mounted displays. In: *OSA Proceedings of the International Optical Design Conference*, vol. 22 (ed. G.W. Forbes), 334–338. OSA.

105 Ha, Y. and Rolland, J.P. (2002). Optical assessment of head-mounted displays in visual space. *Appl. Opt.* 41 (25): 5282–5289.

106 H. C. Self, Optical tolerances for alignment and image differences for binocular helmet-mounted displays Armstrong Aerospace Medical Research Lab., Dayton, OH, AAMRL-TR-86-019, 1986

107 Thomas, M.L., Siegmund, W.P., Antos, S.E., and Robinson, R.M. (1989). Fiber optic development for use on the fiber optic helmet-mounted display. In: *Proceedings of SPIE*, Helmet-Mounted Displays, vol. 116 (ed. J.T. Carollo), 90–101. SPIE.

108 Iwamoto, K., Katsumata, S., and Tanie, K. (1994). An eye movement tracking type head mounted display for virtual reality system: -evaluation experiments of a prototype system. In: *Proceedings of 1994 IEEE International Conference on Systems, Man, and Cybernetics*, vol. 1, 13–18. Humans, Information and Technology (Cat. No.94CH3571–5).

109 S. Liu, Methods for generating addressable focus cues in stereoscopic displays, Ph.D. Dissertation, The University of Arizona, 2010.

110 Duchowski, A.T. (1998). Incorporating the viewer's Point-Of-Regard (POR) in gaze-contingent virtual environments. In: *Proceedings of SPIE – the International Society for Optical Engineering*, vol. 3295, 332–343. SPIE-International Society for Optical Engineering.

111 L. Vaissie and J. P. Rolland, "Eyetracking integration in head-mounted displays," U.S. Patent 6,433,760B1, August 13, 2002.

112 Hua, H. (2001). Integration of eye tracking capability into optical see-through head-mounted displays. *Proc. SPIE* 4792: 496–503.

113 Hua, H., Pansing, C., and Rolland, J.P. (2007). Modeling of an eye-imaging system for optimizing illumination schemes in an eye-tracked head-mounted display. *Appl. Opt.* 46 (31): 7757–7770.

114 Hua, H., Krishnaswamy, P., and Rolland, J.P. (2006). Video-based eyetracking methods and algorithms in head-mounted displays. *Opt. Express* 14 (10): 4328–4350.

115 C. Curatu, Hong Hua, and J. P. Rolland, "Projection-based head-mounted display with eye-tracking capabilities," Proceedings of the SPIE International Society for Optical Engineering, Vol. 5875, 2005.

116 Baumgarten, J., Schuchert, T., Voth, S. et al. (2011). Aspects of a head-mounted eye-tracker based on a bidirectional OLED microdisplay. *J. Inf. Disp.* 13 (2): 1–5.

117 Loschky, L.C. and Wolverton, G.S. (2007). How late can you update gaze-contingent multiresolutional displays without detection? *ACM Trans. Mult. Comp. Commun. App.* 3 (4): 1–10.

118 Reingold, E.M., Loschky, L.C., McConkie, G.W., and Stampe, D.M. (2003). Gaze-contingent multiresolutional displays: an integrative review. *Hum. Factors* 45: 307–328.

119 Duchowski, A.T. and Çöltekin, A. (2007). Foveated gaze-contingent displays for peripheral LOD management, 3D visualization, and stereo imaging. *ACM Trans. Mult. Comp. Commun. App.* 3: 1–21.

120 Patney, A. et al. (2016). Towards foveated rendering for gaze-tracked virtual reality. *ACM Trans. Graphics* 35 (6): 179.

121 Rolland, J.P., Yoshida, A., Davis, L.D., and Reif, J.H. (1998). High-resolution inset head-mounted display. *Appl. Opt.* 37: 4183–4193.

122 Hua, H. and Liu, S. (2008). A dual-sensor foveated imaging system. *Appl. Opt.* 47 (3): 317–327.

123 Tan, G., Lee, Y., Zhan, T. et al. (2018). Foveated imaging for near-eye displays. *Opt. Express* 26 (19): 25076–25085.

124 Xu, M. and Hua, H. (2017). High dynamic range head mounted display based on dual-layer spatial modulation. *Opt. Express* 25 (19): 23320–23333.

125 Seetzen, H., Heidrich, W., Stuerzlinger, W. et al. (2004). High dynamic range display systems. *ACM Trans. Graphics* 23 (3): 760–768.

126 Wetzstein, G., Lanman, D., Heidrich, W., and Raskar, R. (2011). Layered 3D: tomographic image synthesis for attenuation-based light field and high dynamic range displays. *ACM Trans. Graphics* 30 (4): 95.

127 Huang, F., Chen, K., and Wetzstein, G. (2015). The light field stereoscope: immersive computer graphics via factored near-eye light field displays with focus cues", ACM SIGGRAPH. *ACM Trans. Graphics* 33 (5): 36.

128 Marran, L. and Schor, C. (1997). Multiaccommodative stimuli in VR systems: problems & solutions. *Hum. Factors* 39 (3): 382–388.

129 Hua, H. (2017). Enabling focus cues in head-mounted displays. *Proc. IEEE* 105 (5): 805–824.

130 Wann, J.P., Rushton, S., and Mon-Williams, M. (1995). Natural problems for stereoscopic depth perception in virtual environments. *Vision Res.* 35 (19): 2731–2736.

131 Watt, S.J., Akeley, K., Ernst, M.O., and Banks, M.S. (2005). Focus cues affect perceived depth. *J. Vision* 5 (10): 834–862.

132 Howarth, P. (2011). Potential hazards of viewing 3-D stereoscopic television, cinema, and computer games: a review. *Ophthalmic Physiol. Opt.* 31: 111–122.

133 Hoffman, D.M., Girshick, A.R., Akeley, K., and Banks, M.S. (2008). Vergence-accommodation conflicts hinder visual performance and cause visual fatigue. *J. Vision* 8 (3): 1–30.

134 Vienne, C., Sorin, L., Blonde, L. et al. (2014). Effect of the accommodation-vergence conflict on vergence eye movements. *Vision Res.* 100: 124–133.

135 Blundell, B.G. and Schwarz, A.J. (2002). The classification of volumetric display systems: characteristics and predictability of the image space. *IEEE Trans. Visual. Comput. Graphics* 8 (1): 66–75.

136 Onural, L., Yaraş, F., and Kang, H. (2011). Digital holographic three-dimensional video displays. *Proc. IEEE* 99 (4): 576–589.

137 Takaki, Y. and Nago, N. (2010). Multi-projection of lenticular displays to construct a 256-view super multi-view display. *Opt. Express* 18 (9): 8824–8835.

138 Jones, A., McDowall, I., Yamada, H. et al. (2007). Rendering for an interactive 360° light field display. *ACM Trans. Graphics* (Proc. of SIGGRAPH 2007), 26 (3) 40): 1–10.

139 Moon, E., Kim, M., Roh, J. et al. (2014). Holographic head-mounted display with RGB light emitting diode light source. *Opt. Express* 22 (6): 6526034.

140 Yeom, H.J., Kim, H.J., Kim, S.B. et al. (2015). 3D holographic head mounted display using holographic optical elements with astigmatism aberration compensation. *Opt. Express* 23 (25): 32025–32034.

141 Mainone, A., Georgiou, A., and Kollin, J.S. (2017). Holographic near-eye displays for virtual and augmented reality. *ACM Trans. Graphics* 36 (4): 85.

142 Westheimer, G. (1966). Maxwellian viewing system. *Vision Res.* 6: 669–682.

143 Konrad, R., Padmanaban, N., Molner, K. et al. (2017). Accommodation-invariant computational near-eye displays. *ACM Trans. Graphics* 36 (4): 88.

144 Ando, T., Yamasaki, K., Okamoto, M. et al. (2000). Retinal projection display using holographic optical element. *Proc. SPIE* 3956: 211–216.

145 Maimone, A., Lanman, D., Rathinavel, K. et al. (2014). Pinlight displays: wide field of view augmented reality eyeglasses using defocused point light sources. *ACM Trans. Graphics* 33 (4) 89:): 1–11.

146 von Waldkirch, M., Lukowicz, P., and Tröster, G. (2003). LCD-based coherent wearable projection display for quasi accommodation-free imaging. *Opt. Commun.* 217: 133–140.

147 von Waldkirch, M., Lukowicz, P., and Troster, G. (2005). Oscillating fluid lens in coherent retinal projection displays for extending depth of focus. *Opt. Commun.* 253: 407–418.

148 Yuuki, A., Itoga, K., and Satake, T. (2012). A new Maxwellian view display for trouble-free accommodation. *J. SID* 20 (10): 581–588.

149 Liu, S. and Hua, H. (2011). Extended depth-of-field microscopic imaging with a variable focus microscope objective. *Opt. Express* 19 (1): 353–362.

150 Sheng-huei, L. and Hua, H. (2015). Imaging properties of extended depth of field microscopy through single-shot focus scanning. *Opt. Express* 23 (8): 10714–10731.

151 Shiwa, S., Omura, K., and Kishino, F. (1996). Proposal for a 3-D display with accommodative compensation: 3DDAC. *J. SID* 4 (4): 255–261.

152 Shibata, T., Kawai, T., Ohta, K. et al. (2005). Stereoscopic 3-D display with optical correction for the reduction of the discrepancy between accommodation and convergence. *J. SID* 13 (8): 665–671.

153 Kuiper, S. and Hendriks, B.H.W. (2004). Variable-focus liquid lens for miniature cameras. *Appl. Phys. Lett.* 85 (7): 1128–1130.

154 Ren, H., Fox, D., Wu, B., and Wu, S.T. (2007). Liquid crystal lens with large focal length tunability and low operating voltage. *Opt. Express* 15 (18): 11328–11335.

155 Fernandez, E.J. and Artal, P. (2003). Membrane deformable mirror for adaptive optics: performance limits in visual optics. *Opt. Express* 11 (9): 1056–1069.

156 S. Liu, D. Cheng, and H. Hua, "An optical see-through head-mounted display with addressable focal planes," in Proc. of IEEE and ACM International Symposium on Mixed and Augmented Reality (ISMAR 2008), 2008.

157 Koetting, R.A. (1970). Stereopsis in presbyopes fitted with single vision contact lenses. *Am. J. Optom. Arch. Am. Acad. Optom.* 47: 557–561.

158 Johnson, P.V., Parnell, J., Kim, J. et al. (2016). Dynamic lens and monovision 3D displays to improve viewer comfort. *Opt. Express* 24 (11): 11808–11827.

159 A. F. Crabtree, "Method and apparatus for manipulating, projecting, and displaying light in a volumetric format," U.S. patent 5,572,375, November 1996.

160 Rolland, J.P., Kureger, M., and Goon, A. (2000). Multifocal planes head-mounted displays. *Appl. Opt.* 39 (19): 3209–3214.

161 Akeley, K., Watt, S.J., Girshick, A.R., and Banks, M.S. (2004). A stereo display prototype with multiple focal distances. *ACM Trans. Graphics* 23: 804–813.

162 Cheng, D., Wang, Q., Wang, Y., and Jin, G. (2013). Lightweight spatial-multiplexed dual focal-plane head-mounted display using two freeform prisms. *Chin. Opt. Lett.* 11 (3): 031201.

163 McQuaide, S.C., Seibel, E.J., Kelly, J.P. et al. (2003). A retinal scanning display system that produces multiple focal planes with a deformable membrane mirror. *Displays* 24 (2): 65–72.

164 Love, G.D., Hoffman, D.M., Hands, P.J.W. et al. (2009). High-speed switchable lens enables the development of a volumetric stereoscopic display. *Opt. Express* 17 (18): 15716–15725.

165 Liu, S. and Hua, H. (2009). Time-multiplexed dual-focal plane head-mounted display with a fast liquid lens. *Opt. Lett.* 34 (11): 1642–1644.

166 Suyama, S., Ohtsuka, S., Takada, H. et al. (2004). Apparent 3-D image perceived from luminance-modulated two 2-D images displayed at different depths. *Vision Res.* 44 (8): 785–793.

167 X. Hu, Development of the Depth-Fused Multi-Focal Plane Display Technology, Ph.D. Dissertation, College of Optical Sciences, University of Arizona, 2014.

168 Ravikumar, S., Akeley, K., and Banks, M.S. (2011). Creating effective focus cues in multi-plane 3D displays. *Opt. Express* 19: 20940–20952.

169 Narain, R., Albert, R.A., Bulbul, A. et al. (2015). Optimal presentation of imagery with focus cues on multi-plane displays. *ACM Trans. Graphics* 34 (4): 59.1–59.12.

170 W. Wu, P. Llull, I. Tosic, N. Bedard, K. Berkner, N. Balram, "Content-adaptive focus configuration for near-eye multi-focal display," 2016 IEEE International Conference on Multimedia and Expo (ICME), 2016.

171 Rathinavel, K., Wang, H., Blate, A., and Fuchs, H. (2018). An extended depth-of-field volumetric near-eye augmented reality display. *IEEE Trans. Vis. Comput. Graphics* 24 (11): 2857–2866.

172 Levoy, M. and Hanrahan, P. (1996). Light field rendering. In: *Proceedings of the 23rd Annual Conference on Computer Graphics and Interactive Techniques*, 31–36. ACM.

173 Arimoto, H. and Javidi, B. (2001). Integral three-dimensional imaging with digital reconstruction. *Opt. Lett.* 26: 157–159.

174 Hua, H. and Javidi, B. (2014). A 3D integral imaging optical see-through head-mounted display. *Opt. Express* 22: 13484–13491.

175 Wetzstein, G., Lanman, D., Hirsch, M., and Raskar, R. (2012). Tensor displays: compressive light field synthesis using multilayer displays with directional backlighting. *ACM Trans. Graphics* 31 (4): 80.

176 Malmone, A. and Fuchs, H. (2013). Computational augmented reality eyeglasses. *Proc. ISMAR*: 29–38.

177 Lee, J., Park, J., Nam, D. et al. (2013). Optimal projector configuration design for 300-Mpixel multi-projection 3D display. *Opt. Express* 21: 26820–26835.

178 Lippmann, G. (1908). Epreuves reversibles donnant la sensation du relief. *J. Phys. (Paris)* 7: 821–825.

179 Song, W., Wang, Y., Cheng, D., and Liu, Y. (2014). Light field head-mounted display with correct focus cue using micro structure array. *Chin. Opt. Lett.* 12: 060010.

180 Urey, H., Chellappan, K., Erden, E., and Surman, P. (2011). State of the art in stereoscopic and autostereoscopic displays. *Proc. IEEE* 99 (4): 540–555.

181 Levoy, M., Ng, R., Adams, A. et al. (2006). Light field microscopy. *ACM Trans. Graphics* 25 (3).

182 Javidi, B., Sola-Pikabea, J., and Martinez-Corral, M. (2015). Breakthroughs in photonics 2014: recent advances in 3-D integral imaging sensing and display. *IEEE Photonics J.* 7 (3): 0700907.

183 Kim, Y., Hong, K., and Lee, B. (2010). Recent researches based on integral imaging display method. In: *3D Research*, vol. 01, 17–27. Springer.

184 Lanman, D. and Luebke, D. (2013). Near-eye light field displays, "Proc. ACM SIGGRAPH. *ACM Trans. Graphics* 32 (6): 1–10.

185 Yao, C., Cheng, D., and Wang, Y. (2018). Design and stray light analysis of a lenslet-array-based see-through light-field near-eye display. In: *Proceedings of SPIE*, Digital Optics for Immersive Displays, 106761A,, vol. 10676. SPIE.

186 Huang, H. and Hua, H. (2017). An integral-imaging-based head-mounted light field display using a tunable lens and aperture array. *J. Soc. Inf. Disp.* 25 (3): 200–207.

187 Takaki, Y. (2006). High-density directional display for generating natural three-dimensional images. *Proc. IEEE* 94: 654–663.

188 Kim, Y., Kim, J., Hong, K. et al. (2012). Accommodative response of integral imaging in near distance. *J. Disp. Technol.* 8: 70–78.

189 Stern, A., Yitzhaky, Y., and Javidi, B. (2014). Perceivable light fields: matching the requirements between the human visual system and autostereoscopic 3D displays. *Proc. IEEE* 102: 1571–1587.

190 Tatham, E. (1999). Getting the best of both real and virtual worlds. *Commun. ACM* 42 (9): 96–98.

191 K. Kiyokawa Y. Kurata and H. Ohno "An Optical See-through Display for Mutual Occlusion with a Real-time Stereo Vision System " Elsevier Computer & Graphics Special Issue on "Mixed Realities - Beyond Conventions " 25(5):2765-779, 2001.

192 Kiyokawa, K., Billinghurst, M., Campbell, B., and Woods, E. (2003). An occlusion capable optical see-through head mount display for supporting co-located collaboration. In: *Proceedings of the Second IEEE and ACM International Symposium on Mixed and Augmented Reality*, 1–9. IEEE.

193 Cakmakci, O., Ha, Y., and Rolland, J.P. (2004). A compact optical see-through head-worn display with occlusion support. In: *Proceedings of the Third IEEE and ACM International Symposium on Mixed and Augmented Reality (ISMAR)*, 16–25. IEEE.

194 Gao, C., Lin, Y., and Hua, H. (2012). Occlusion capable optical see-through head-mounted display using freeform optics. In: *IEEE International Symposium on Mixed and Augmented Reality (ISMAR)*, 281–282. IEEE.

195 Wilson, A. and Hua, H. (2017). Design and prototype of an augmented reality display with per-pixel mutual occlusion capability. *Opt. Express* 25 (24): 30539–30550.

196 Hamasaki, T. and Itoh, Y. (2019). Varifocal occlusion for optical see-through head-mounted displays using a slide occlusion mask. *IEEE Trans. Visual. Comput. Graphics* 25 (5): 1961–1969.

9

Touch Panel Technology

9.1 INTRODUCTION

Human beings have five senses with which to interact with the world: sight, hearing, taste, smell, and touch. As discussed in Chapter 1, displays are used to stimulate human vision, which is one of the dominant senses. Machines (such as mobile phones, TV, and laptop computers) can output not only images but also sounds which stimulate human hearing. On the other hand, when considering communication between humans and machines, we need to provide inputs to the machine [1]. For example, a remote control is an input device for a TV. A keyboard and mouse are used to control a computer. A touch panel is typically integrated onto a display panel to use as an input device for machines, such as mobile phones, tablets, and some car navigation systems. Obviously, as the touch panel is typically put on the top side of the display, it must be optically transparent. By integrating the senses of human sight and touch, a touch panel display can provide an intuitive user-experience to communicate with the machine. For example, by touching images on the display, items can be selected. A "dragging" function results in the movement of the images. Gestures can be recognized by the machine, providing an immersive user experience. Compared to the display, the resolution of the touch panel is typically much lower; the "touch" is typically provided by a human finger so mm resolution is quite sufficient for typical applications.

To feel "touch" (from a finger or any other object), a sensor device must be embedded in the machine. A "resistive" touch panel consists of two conductive layers separated by a small gap [2]. When the panel is touched, the top conductor deforms and contacts the bottom one. Electric current flows between two electrodes, and the resistance value changes according to the touch position. When a human finger touches the screen, it provides a channel for electric charge to ground potential. This is the basic principle of the "capacitive" touch panel [3]. Note that unlike the resistive sensor, mechanical deformation is not needed for the capacitive touch panel. In a resistive touch panel a gap is needed to separate the two conductor layers which results in lower transmittance, compared to the capacitive type. Also, the two substrates used in resistive touch panels increase the module thickness. To reduce the module thickness and fabrication cost, it is possible to integrate the touch sensor and the display panel together. A LCD consists of two substrates sandwiching the LC materials. If the touch panel is fabricated on the top substrate, it is called an on-cell configuration. An organic light-emitting diode (OLED) can be fabricated on a single substrate. However, a second substrate is sometimes used to protect the OLED display and the touch panel function can be fabricated on it. In an active-matrix LCD display, thin-film transistors (TFTs) are used to drive the LC device. By suitable design of the layout of the TFT and its associated elements (such as capacitor and conductor), it is possible to embed the touch panel function with the TFT substrate of the LCD, which is then called an in-cell configuration.

Capacitive touch panels have been widely applied to small and medium displays (<20 in.). However, the limited conductivity of the transparent conductor layer (usually indium tin oxide [ITO]) required for both resistive and capacitive touch panels results in signal distortion and degrades performance of larger diagonal touch panels. Thin metal mesh or other transparent conductors can be used to replace ITO. Other sensing techniques

Introduction to Flat Panel Display, Second Edition. Jiun-Haw Lee, I-Chun Cheng, Hong Hua, and Shin-Tson Wu.

can also be used for large-size applications. Optical sensing is one of the most promising technologies. In such a system, one or more transmitters emit light (typically in the infrared region) which is received by arrays of detectors. When the panel is touched, some light is blocked or reflected and the corresponding position can be detected. Such an optical sensor represents a form of machine vision which can be extended beyond simple touch sensing. Elaborate systems can not merely sense touch in a 2D domain, but can detect depth information which means that 3D sensing is possible. In other words, when a user looks at a display, the machine looks at and responds to the user at the same time, which begins to imitate face-to-face communication. (Of course, for natural human-to-human interaction, voice exchange is also required. This is possible for machines but is not included in this chapter.) For some applications, it is necessary to distinguish which objects are touching the panel. For example, we may want to use a pen or stylus to write or draw on the display, as we do on paper. The hand holding the pen may also rest on the panel but the touch panel should respond to the pen only and ignore the hand contact. A touch pen which transmits an electromagnetic wave with a receiver inside the display can fulfill this requirement.

In the following sections, we will first introduce the resistive touch panel, followed by the capacitive touch panel. Then, integration of touch and display panels will be discussed including on-cell and in-cell configurations. Optical sensing is favorable for large-sized panels due to the limited conductivity of transparent conductors. Hence 1D, 2D, and 3D optical sensing techniques will be illustrated.

9.2 RESISTIVE TOUCH PANEL

Figure 9.1a shows a schematic diagram of a resistive touch panel, which consists of two substrates, each coated with a transparent conductor on the inner surface. The top substrate must be deformable. The substrates are separated by dot spacers so without any touch input the top and bottom conductors are electrically isolated. When the panel is touched, as shown in Figure 9.1b, the upper membrane deforms and the two conductors make contact. Different contact points result in different resistivity changes which can be detected. Here, the top-substrate must be deformable, such as a PET or polyimide (PI) foil. The bottom-substrate, should be rigid, or laminated onto an inflexible substrate. Otherwise, flexing of the whole panel can easily produce false contacts.

Figure 9.2 shows the structure and the operation principle of the 4-wire resistive touch panel. Two parallel electrodes on the top-substrate provide the electric field along the x-direction. On the other hand, electrodes on the bottom conductive layer give a potential gradient in the y-direction. To perform the touch sensing function, a voltage difference is applied to the two electrodes (such as $X_1 = 5$ V and $X_2 = 0$ V) of the top-substrate (which is typically called the scan signal). When a pen touches a certain position on the panel, the top- and bottom-substrates make contact, and hence the voltage of the touch point can be sensed at the bottom substrate (which is typically called the sensing signal), and encodes the position of the contact point along the x-axis, according to the ratio of the effective resistances R_2 and $(R_1 + R_2)$. Then, by applying the voltage difference on the bottom electrodes and sensing on the top conductive membrane, y-axis information can be

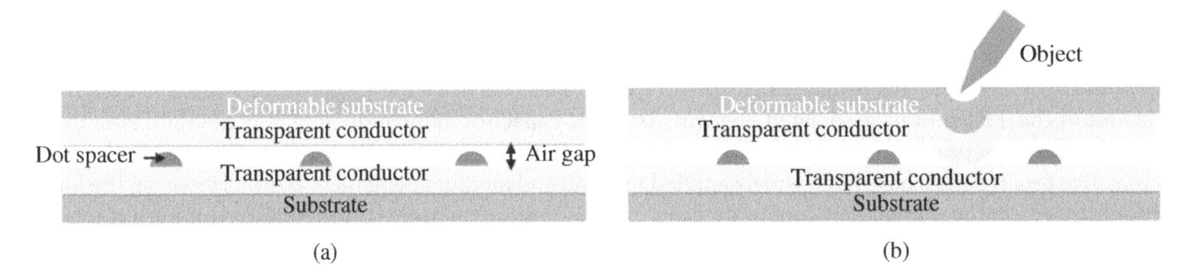

Figure 9.1 (a) Structure of resistive touch panel, and (b) the resistive touch panel upon touch.

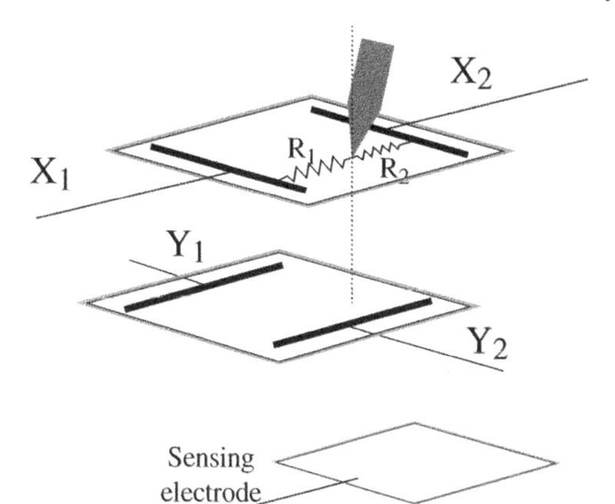

Figure 9.2 Electrode structure and operation principle of 4-wire resistive touch panel.

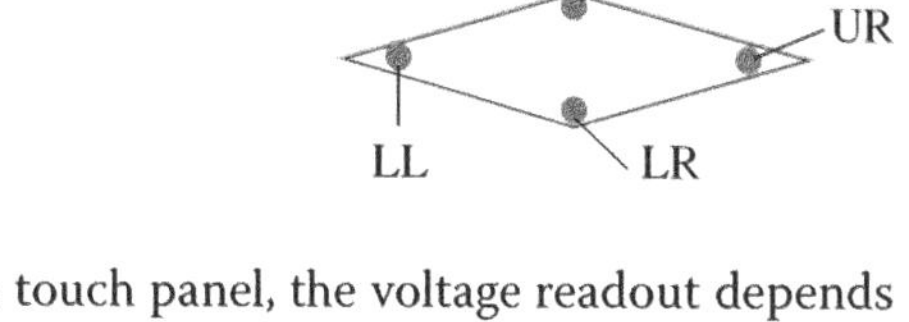

Figure 9.3 Structure of 5-wire resistive touch panel.

independently established from the measured potential. In such a touch panel, the voltage readout depends on the resistance values in the x- and y-directions, hence it is termed a resistive touch panel. There are four connections to this device, so it is also called a 4-wire touch panel.

ITO is a common material for the transparent conductor. Typically, high-quality ITO thin films are deposited onto the substrate by a sputtering process at an elevated temperature of ~300 °C, which will destroy PET film and some PIs. For the flexible top-substrate, the ITO sputtering temperature must be decreased, which results in a higher resistivity. In addition, the continuous deformation of the top-substrate caused by touch sensing, may crack the ITO film after long-term use. Therefore, ITO is poorly suited to providing the scanning signal on the upper substrate. A 5-wire resistive touch panel has been proposed to solve this problem. As shown in Figure 9.3, four contacts are placed at the four corners of the bottom substrate. The entire top-conductor is used as the sensing electrode. By setting UR and LR (upper right; lower right, etc.) at 5 V and UL and LL at 0 V, the x-position of the touch point can be obtained from the potential of the sensing electrode (i.e. the top-conductor). Similarly, by setting UR and UL at 5 V and grounding LR and LL, the y-axis of the touching position can be measured.

Note that typically the 4- and 5-wire resistive touch panels shown in Figures 9.2 and 9.3 do not support a multi-touch function. By patterning the ITO on the top- and bottom-electrodes, the structure is equivalent to a matrix of many 4-wire resistive touch panels and a multi-touch function can be achieved, as shown in Figure 9.4a,b [4].

One of the disadvantages of the resistive touch panel is its low optical transmittance. At an interface between layers of different refractive index, the reflection which occurs increases with an increasing refractive index difference. As shown in Figure 9.5, the refractive index of ITO (~2.0), results in 10% reflection at an interface with air. There are two ITO/air interfaces in the resistive touch panel and hence the total transmittance is limited to ~80%.

9.3 CAPACITIVE TOUCH PANEL

Figure 9.6a shows the schematic layer structure of a capacitive touch panel. An upper transmitter electrode is supplied with an alternating current (AC) signal, and forms a parallel plate capacitor with a grounded reference electrode. In the absence of any touch input, the capacitance value is C_s as shown in Figure 9.6a. When a finger touches the outer surface of the capacitor, the human body provides an additional path to ground, denoted as

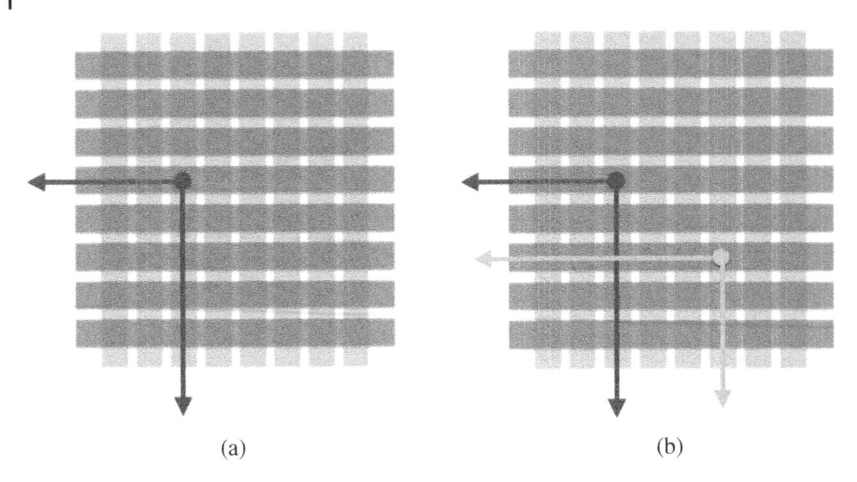

Figure 9.4 Structure of multi-touch resistive touch panel upon (a) single, and (b) multiple touched points.

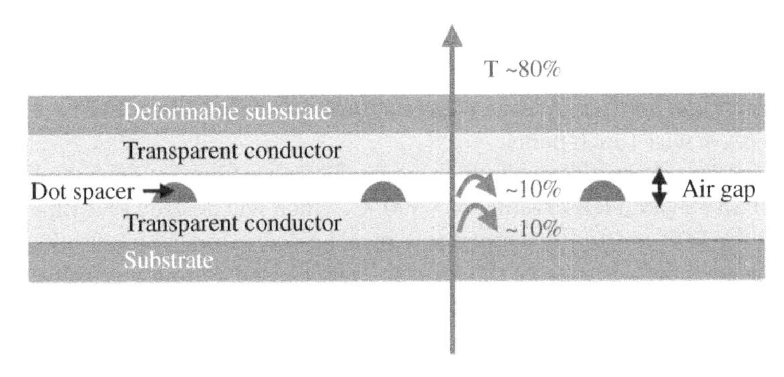

Figure 9.5 Light transmission from display through touch panel to the human eye.

C_f in Figure 9.6b, and hence the capacitance value increases according to the usual law for parallel connection of capacitors ($C = C_s + C_f$), as shown in Figure 9.6b. In this situation, the capacitance which is measured is that between the sensing electrode and the ground, referred to as the self-capacitance of the electrode. Figure 9.6c shows the situation in an entire panel when a point is touched. An AC signal is sent in sequence from each of the four corners of the transmitter electrode. The equivalent circuit of the panel in the presence of a touch input is represented in Figure 9.6d. The touch position is obtained by analysis of the potentials generated by this RC network; the whole assembly is called a "surface" capacitive touch panel. Note that no mechanical deformation is required in the capacitive touch panel. Obviously, such a sensor needs to be touched by a conductor connected to the ground. Capacitive sensors readily detect touch by a finger, but insulating objects like a wooden pencil provide no response.

As in the resistive touch panel, a multi-touch function can be obtained in a capacitive touch panel by patterning the transparent conductor as shown in Figure 9.7. Here, two layers of transparent conductor are arranged in perpendicular stripes, separated by an insulator layer. This device is called a projected capacitive touch panel. Each diamond shaped pattern along the x- and y-directions can be regarded as a small surface capacitance touch panel. By supplying voltage pulses to the x-electrodes sequentially and connecting all the y-electrodes into a common sensor electrode, the x-positions of touched points can be identified from the capacitance values (called "self-capacitance"). The y-coordinates can be obtained by using the y- and x-electrodes for scanning and sensing, respectively. For example, in Figure 9.7a, a finger is shown overlapping two columns and three rows, to differing degrees. The position of the touch is determined by reading signals at connections made at the panel edges.

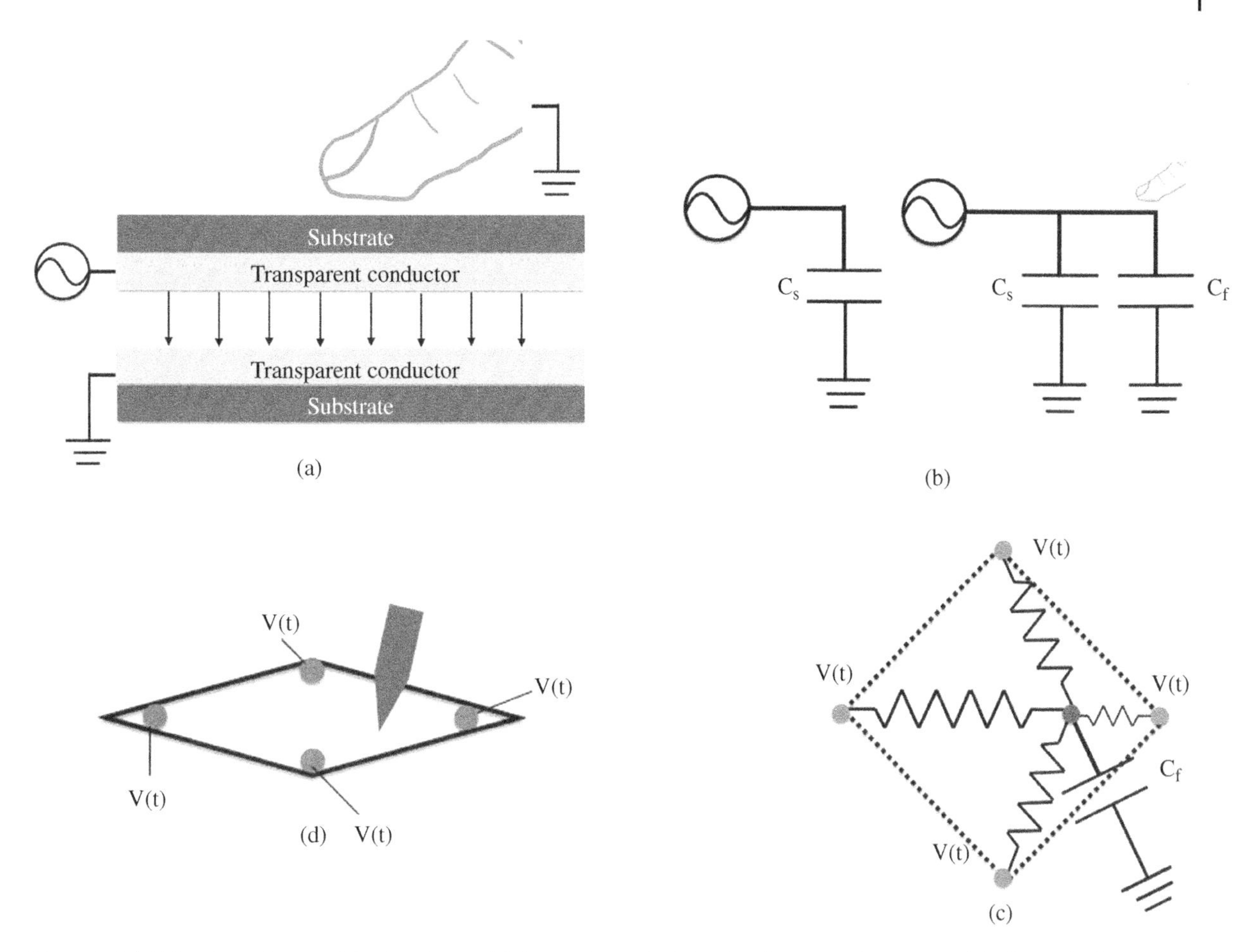

Figure 9.6 (a) Structure of a capacitive touch panel, (b) operation principle of surface capacitive sensor, (c) schematic diagram of a touched point on the capacitive touch panel, and its (d) equivalent circuit.

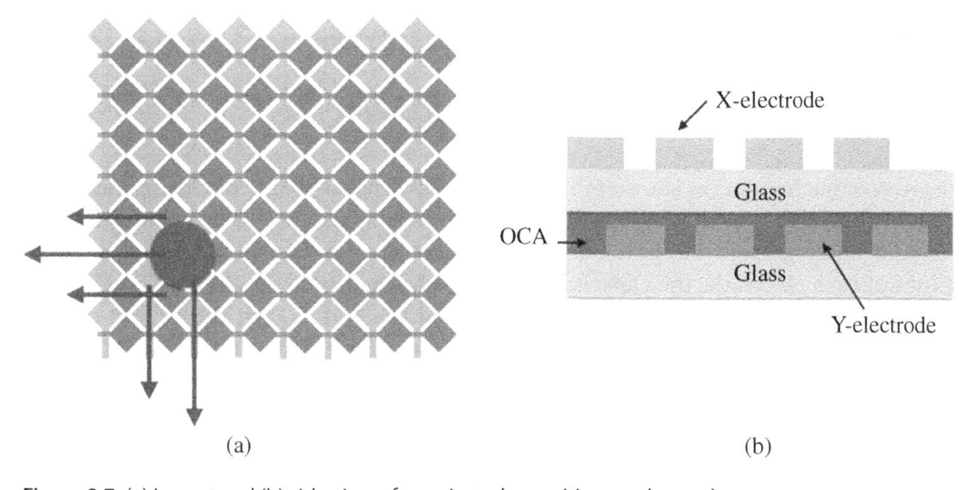

Figure 9.7 (a) Layout and (b) side view of a projected capacitive touch panel.

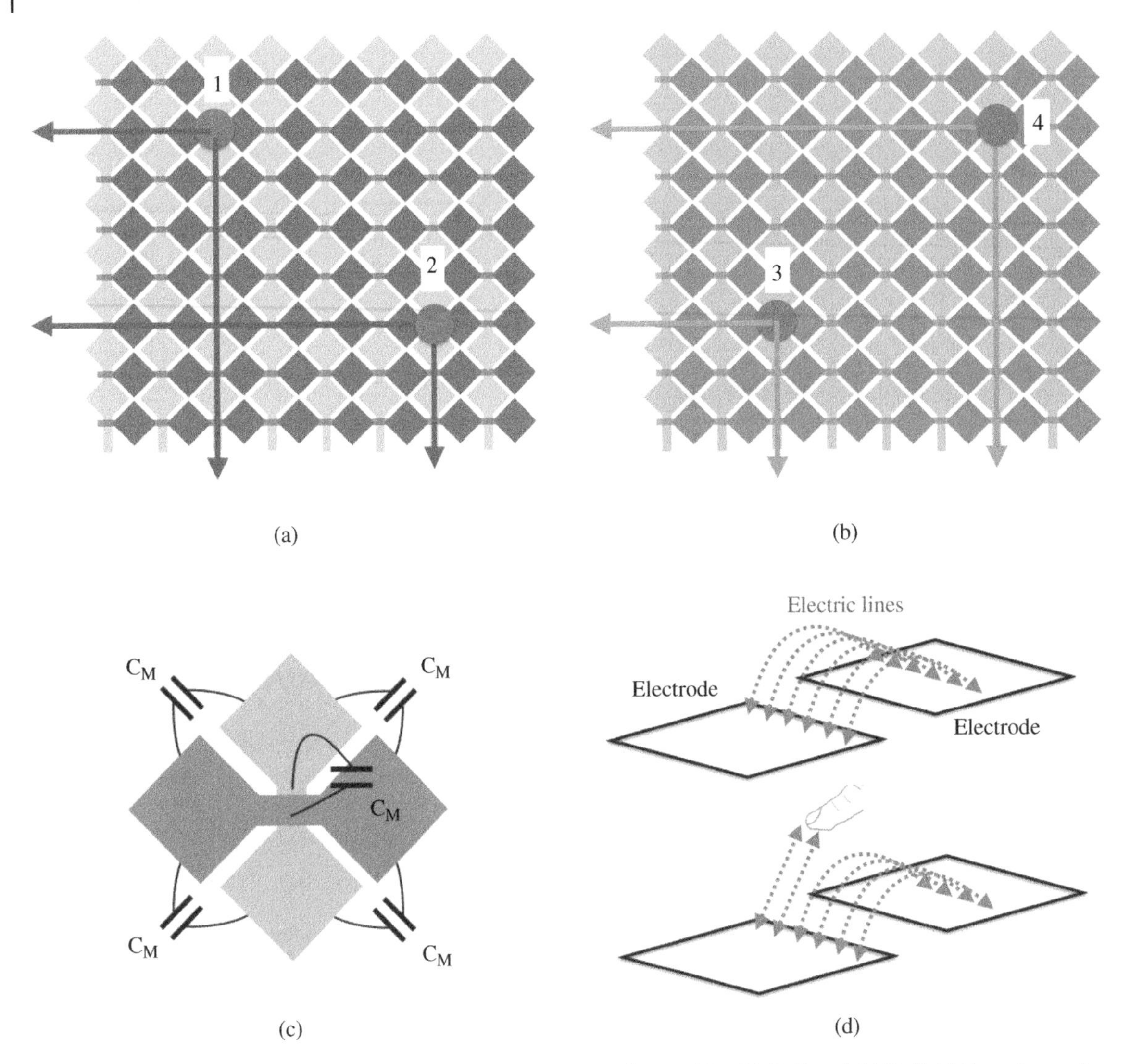

Figure 9.8 The same responses from self-capacitance measurement by touching (a) (1, 2), and (b) (3, 4). (c) Mutual capacitances between *x*- and *y*-electrodes. (d) Electric field shielding in mutual capacitance measurement.

However, when there are two touch points as shown in Figure 9.8a,b, it is impossible to distinguish whether the touch is occurring at positions [1, 2] or at [3, 4], by analyzing the self-capacitance values obtained from the *x*- and *y*-electrodes. Considering again the unit cell of the projected capacitive touch panel, in addition to the self-capacitance of each electrode, the capacitance can also be measured between pairs of *x*- and *y*-conductors, which is called their "mutual-capacitance," and denoted C_M and $C_{M'}$, in Figure 9.8c. When the panel is touched, the mutual-capacitance values change and can be used to determine the exact input positions for multi-touch applications. Considering a touch screen with an electrode matrix of $m \times n$ stripes, $(m + n)$ measurements are needed to detect the self-capacitance values, by scanning the sensing signal across the rows and then the columns. However, it requires $(m \times n)$ scans to measure the mutual-capacitance values. The transmitter signal is scanned across all of the columns, and for each scanned column the capacitance is measured at each row in turn. As mentioned in Figure 9.6b, the self-capacitance value increases when the electrode is touched, because the finger provides another route to the ground. The parallel connection of capacitors results in a higher capacitance value. On the other hand, the mutual capacitance value decreases when the panel is touched, because the finger shields electric field lines and diverts some charge from the transmitter away from the receiver, as shown in Figure 9.8d.

To have perpendicular arranged conductor lines separated by the insulator layer, it is possible to stack two substrates with transparent conductors patterned along the x- and y-directions. Figure 9.7b shows an example of the construction of a projected capacitive touch panel. First, the diamond patterns are fabricated on each substrate. Then, they are stacked together, and bonded with an optically clear adhesive (OCA). Typically, a cover glass is needed outside the touch panel, whose surface is strengthened (by chemical or thermal treatment), so that the hardness of the glass is higher. Sharp edges and corners of the cover glass should be removed for safety reasons. Some surface grinding and hole drilling may be needed to meet overall product design needs. The surface reflectivity of the cover glass should be as low as possible to optimize the ambient contrast ratio of the display. Typically, this is achieved by thin-film coating. To reduce visible marks left by water and oil left behind after repeated touching, a suitable surface treatment is desirable, to provide hydrophobic and oil-repellent properties.

The structure in Figures 9.7b and 9.10a is called the "GGG" configuration because there are three glass substrates. An OCA is needed between the substrates to maximize the optical transmittance. Obviously, it is physically thick, especially for use in mobile devices. A "GG" configuration is possible by: (i) fabricating the ITO patterns on both sides of a single substrate, or (ii) depositing both x- and y-electrodes on one side of a substrate, with insulator bridges to separate the x- and y-conductors at their crossover points. These arrangements are called double-sided indium tin oxide patterning (DITO) and single-sided indium tin oxide patterning (SITO), respectively, as shown in Figures 9.9a,b and 9.10b,c. For the DITO structure, ITO must be deposited and patterned on both sides of a common substrate. For the SITO structure, an insulator layer is needed to separate the crossovers in the x- and y-directions, as shown in Figure 9.9b. If the mutual capacitance is employed for a SITO touch panel, note that the overlap region of the x- and y-conductors results in larger parasitic capacitance, which reduces the signal-to-noise ratio. In the DITO configuration the spacing between the conductors (i.e. the glass substrate thickness, ~550 μm) is much larger than in the SITO device (~3 μm for the insulator thickness), so a touch panel with the DITO configuration exhibits a better performance. As a final step, it is possible to integrate the touch function onto the cover glass, providing a "one glass solution (OGS)," as shown in Figure 9.10d. On the other hand, it is also possible to use a polymer film as the touch substrate to replace glass which exhibits the advantages of being thinner, lighter, and has a lower price. Corresponding to the GGG and DITO structures, GFF and DITO-film constructions have been proposed, as shown in Figure 9.10e,f. GFF means that two films were used to replace two glass substrates, compared to the GGG case. Although GFF has one more film than DITO-film, the fabrication process is easier. The cost of the patterned ITO film is not high, and the thickness and weight does not increase too much. Hence, GFF is popular as it achieves a good combination of cost and performance in the touch panel, compared to other out-cell structures. It is relatively difficult to carry out a multi-layer process on one side of a film substrate. Hence, the SITO-film is not popular.

In summary, capacitive touch panels have had great success due to their many advantages, such as their operation with a soft touch which does not cause mechanical deformation, high optical transmittance, and their

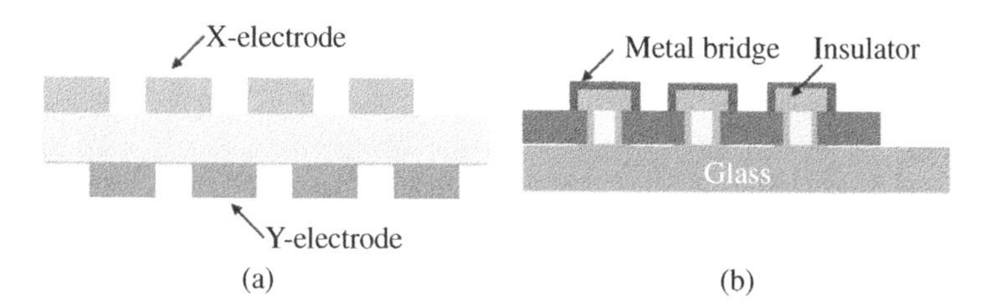

Figure 9.9 Side views of (a) DITO, (b) SITO configurations.

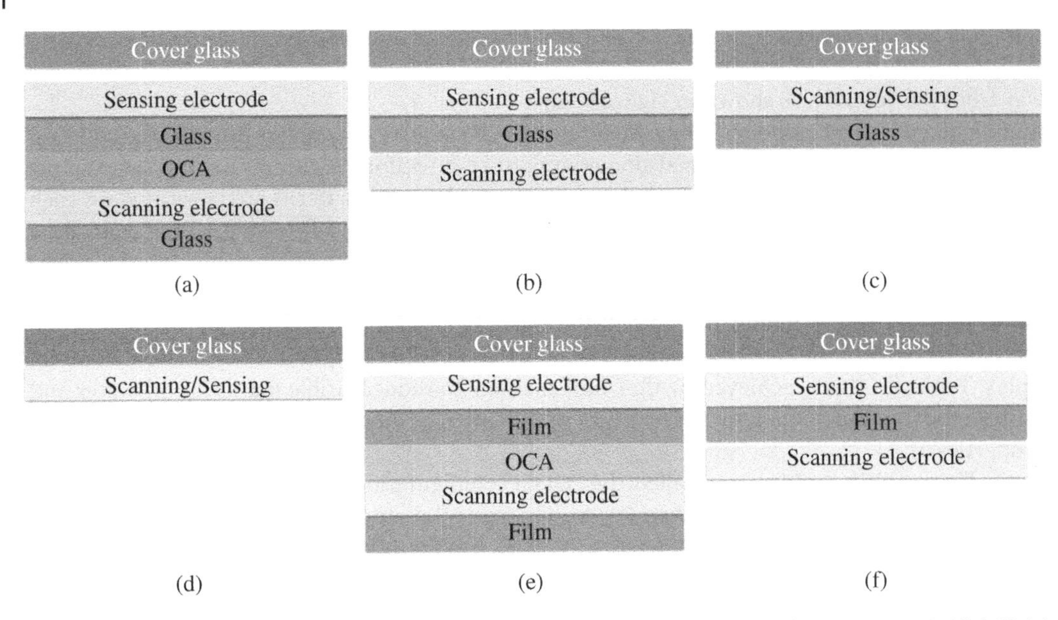

Figure 9.10 Projected capacitive touch panel with different configurations: (a) GGG, (b) DITO, (c) SITO, (d) OGS, (e) GFF, and (f) DITO-film.

resistance to dirt and scratches on the cover glass. However, there are still some restrictions to the capacitive touch panel. For example, the touching object must be a conductor connected to the ground. There is no response when the capacitive touch panel is touched by an insulator. Also, it cannot sense the amount of pressure applied to the panel.

9.4 ON-CELL AND IN-CELL TOUCH PANEL

In LCDs, an LC material is contained between two substrates, with the TFT array fabricated on the bottom substrate. Apart from displays using the in-plane switching (IPS) or fringe field switching (FFS) modes, there are typically common electrodes in the inside surface of the top substrate. Hence, there are already transparent conductors on the panel. With a suitable layout of these conductors (perpendicularly arranged transmitter and receivers), a touch panel function can be embedded into the display panel. The on-cell/in-cell touch panel means the conductors for the touch function is on the top-/bottom-substrate, respectively. Figure 9.11a shows an example of the on-cell touch panel embedded in a LCD display panel. A SITO electrode structure is first

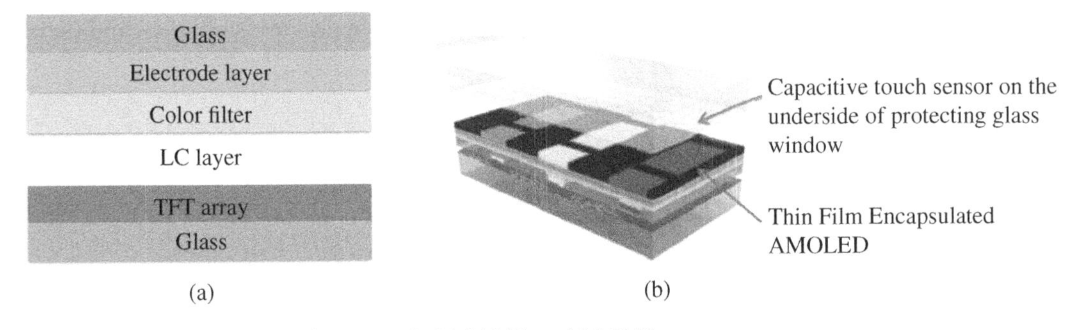

Figure 9.11 On-cell touch panel integrated with (a) LCD, and (b) OLED.

fabricated on the substrate, followed by fabrication of the CF. The structure is also similar to the OGS structure, but with the conductors moved from the inner side of the cover glass into the inner (or outer) side of the CF glass. In a top-emission OLED display, only one substrate (carrying the TFTs and OLED layers) is needed for display operation. The other glass on the top of the display is only used for device encapsulation and to provide mechanical protection [5]. Hence a touch panel can easily be integrated onto the cover glass, as shown in Figure 9.11b.

Compared to the TFT circuits used for display driving, the circuit needed for the touch panel is relatively simple. Hence, it is possible to integrate the two functions together in the TFT circuits. As shown in Figure 9.12, self-capacitance and mutual-capacitance sensing are both possible by adding an electrode layer onto the TFT structure [6]. Arrows in Figure 9.12 show the electric field lines for each of these two modes. Separation of the signals in the time domain is needed to drive this hybrid display/touch panel. For the touch panel, the resolution is lower and the acquisition interval for the electrical signals can be relatively long. Therefore, the proportion of time required for the touch function is quite short, compared to the display addressing. However, as the resolution, size, and frame rate are increased, there is still a trade-off in the time allocated to the display addressing and touch functions.

Amorphous-Si and low-temperature polycrystalline silicon can be used not only for the TFTs which control the LCD or OLED, but also in optical sensors due to their good light-absorbing capability. It then becomes possible to achieve a touch function using arrays of optical sensors embedded in the TFT array [7]. Figure 9.13a shows the layout of the a-Si TFT array. An active pixel sensor (APS) which acts as a light detector is located at the center of the pixel. The detailed layout of the APS is shown in Figures 9.13b, c shows the equivalent circuit of the pixels. An extra voltage line (V_{DD}) is needed to provide bias for the optical sensor. Because the resolution of the touch panel is not as high as the display panel, it is not necessary to put sensors in every pixel. Of course, such a photodetector receives light from the backlight and ambient light, as well as the signal light. Hence, suitable driving circuits are needed which filter out noise to avoid false activation.

By integrating two photodetectors and a suitable optical shield in one pixel, depth information can be obtained which provides 3D sensing, beyond the function of a simple touch panel. Figure 9.14a shows the idea of this configuration [8]. Light emitted from the display panel is reflected and scattered by objects close to the panel. If the backward emission can be measured by the panel itself, the position of the object can be identified. Note that no touch is needed in this configuration. Besides, when arranging two sensors in one pixel with different "viewing" directions, it is possible to obtain 3D information about the location of the object. Figure 9.14b shows the device structure of the optical sensor. With careful design of the upper light shield, the photodetector only accepts the photons from a limited range of solid angles. Figure 9.14c shows the schematic diagram of the measurement. When the object is within a certain distance (typically ~2 cm) of the panel, the 3D position can be calculated from the signals at different pixels.

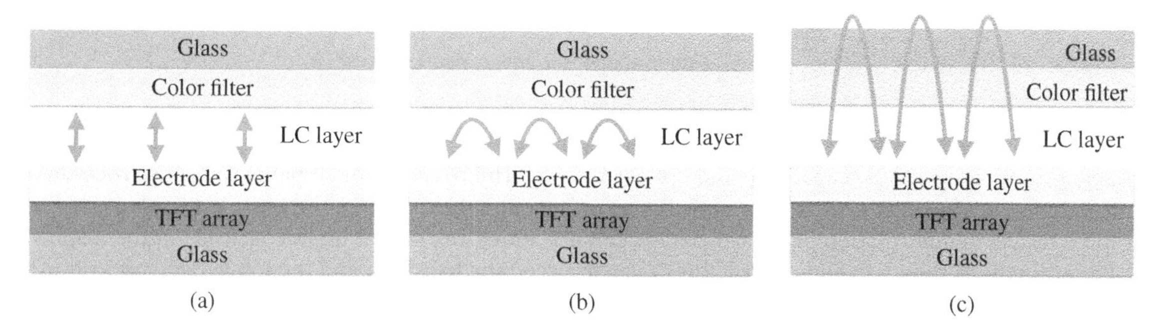

Figure 9.12 Configuration of in-cell touch panel with (a) self- and (b) mutual-capacitance mode. (The detailed arrangement and interconnection of the electrodes allowing both display addressing and touch sensing, are typically kept as proprietary information by panel manufacturers.)

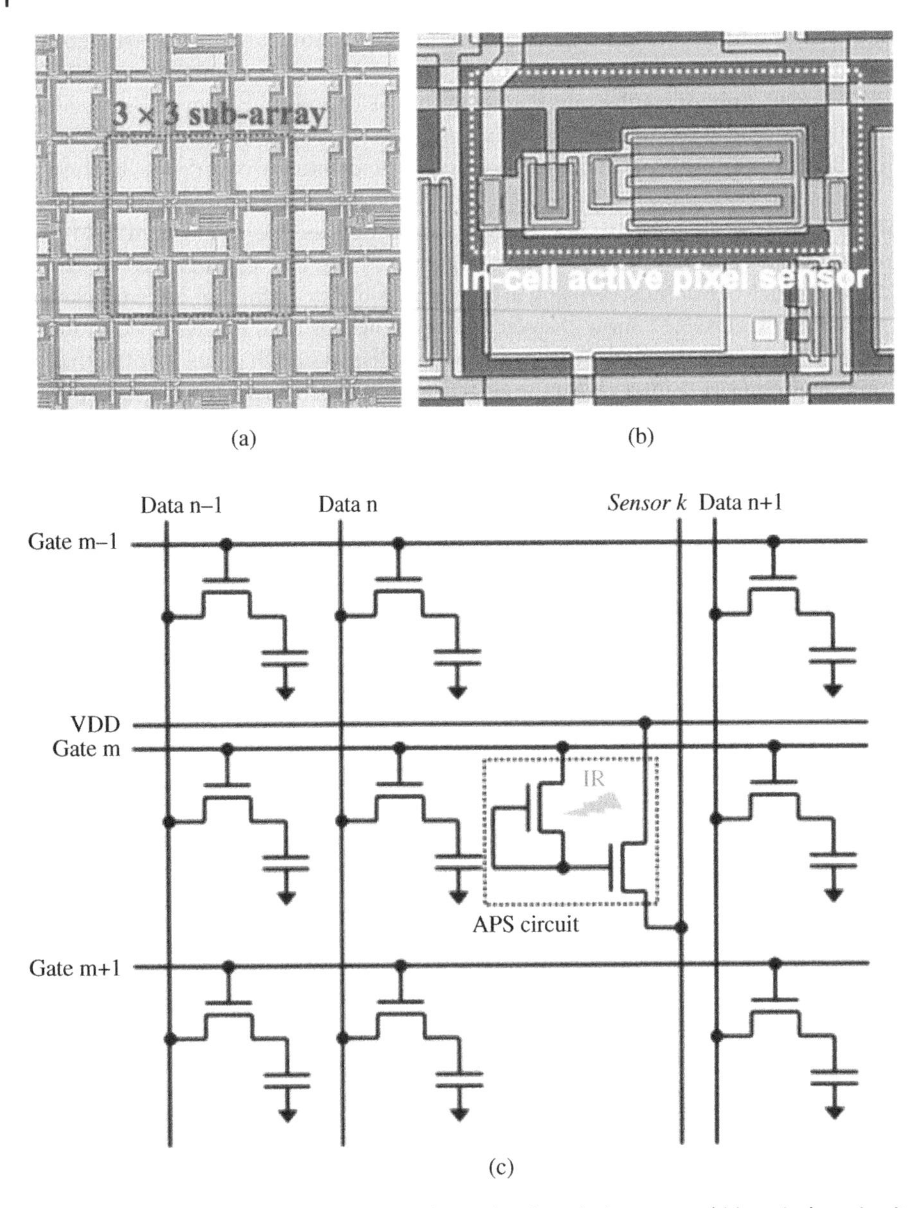

Figure 9.13 (a) and (b) Layout of in-cell touch panel with optical sensor, and (c) equivalent circuit.

As well as the position of a touch on a display, one may want to know the pressure on the panel, which cannot be obtained from conventional capacitive and optical sensing. The resistive touch panel can sense pressure, but it has many disadvantages. In an alternative approach to pressure sensing, the basic principle is that a larger force results in deformation of the panel. One can use a pressure sensor on the backside of the display panel. For example, as shown in Figure 9.15, the air gap between the backlight and the frame decreases under pressure and can be used as a transducer for pressure sensing [9].

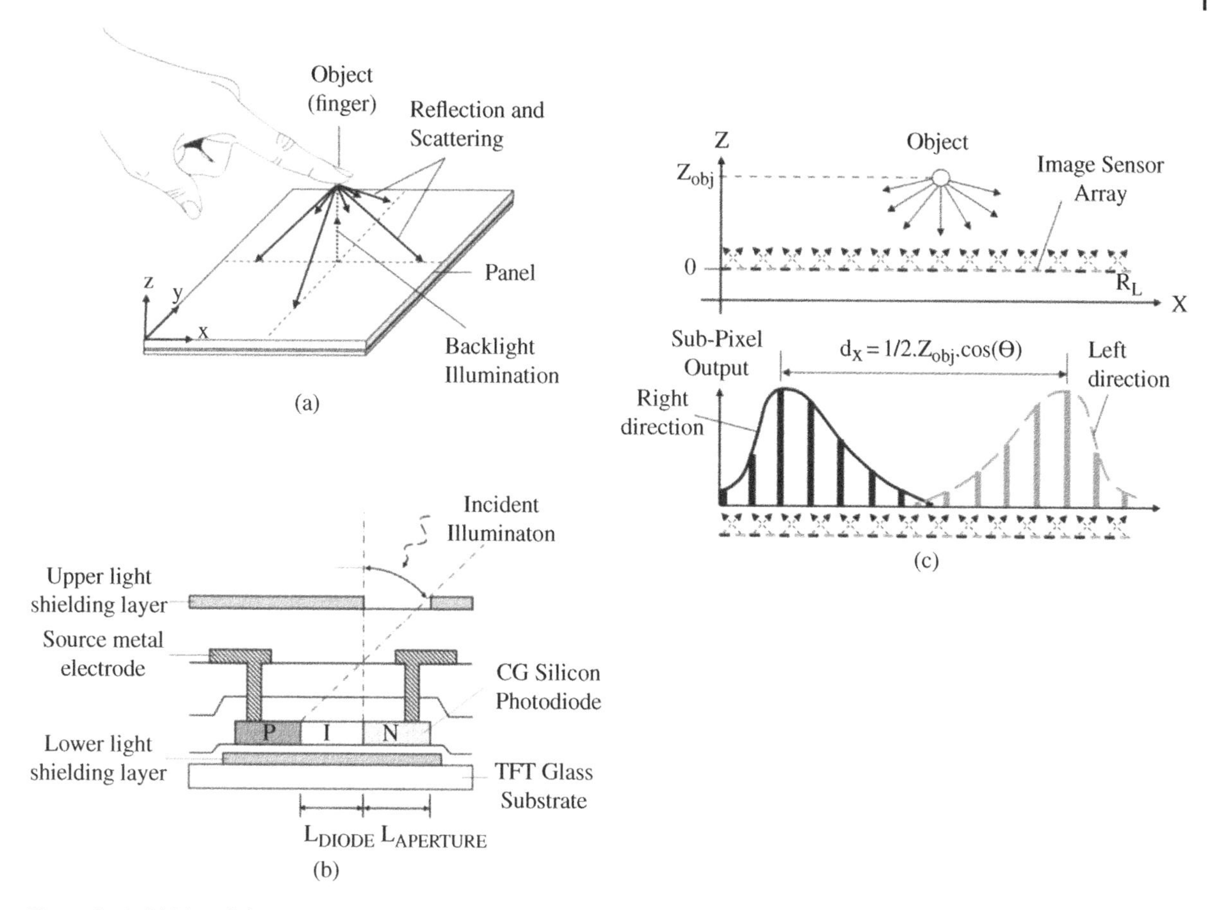

Figure 9.14 (a) Idea of the 3D optical sensing, (b) photo-sensor structure, and (c) operation principle of this 3D optical sensing.

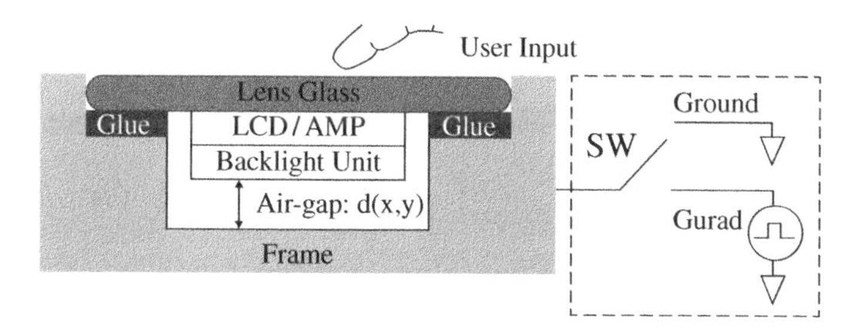

Figure 9.15 A configuration for sensing the pressure.

9.5 OPTICAL SENSING FOR LARGE PANELS

With increasing display size, the low conductivity of the transparent conductor (such as ITO) results in a high RC value, which significantly distorts the electrical signals on the touch panel. Metal mesh is a possible solution but brings the penalty of low optical transmittance. On the other hand, optical sensing is a solution well suited

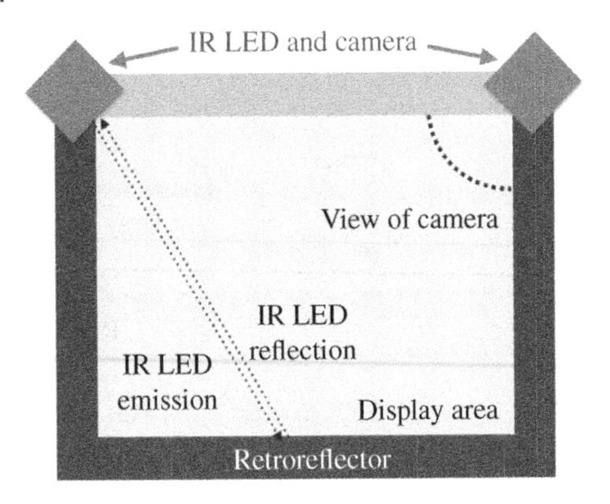

Figure 9.16 Optical touch panel for large display.

to this application. Typically, infrared (IR) light is used for optical sensing. It is invisible and therefore does not interfere with the display of information. As shown in Figure 9.16, two IR LEDs and linear detector arrays are placed at the top corners of the display. Retroreflectors are put at the edges of the display to reflect the light rays back to the camera. When the panel is touched, light is blocked which can be seen from two sensor arrays (like the two eyes of the human). Then, the touched position can be calculated. Because a reflector is needed to confine the IR light for detection, the sidewall of the touch panel is about 1 cm in height, which is suitable for wall-mounted displays. To reduce the height of the frame at the edge of the display, a waveguiding structure has been proposed [10]. The side-emitting LED couples light into the waveguide. When it is touched, the condition for total internal reflection condition is broken; IR light is scattered out of the waveguide and can be detected by the camera underneath which determines the touch position. This sensor arrangement is called a frustrated total internal reflection (FTIR) panel. A more versatile capability can be provided using a 3D camera, which can sense the 3D location of an object. This type of sensor does not use touch but can control equipment based on motion and gestures of the operator. The working distance for such camera-based input can be up to several meters [11].

Homework Problems

9.1 What properties and functions should be offered by the perfect man–machine interface system?

9.2 What will happen if a water-drop lands on capacitive touch panels using (a) self- and (b) mutual-capacitance measurement?

9.3 Discuss the effects of the cover glass thickness for the self- and mutual-capacitive touch panels.

9.4 Describe the technical problems and application possibilities which arise from considering flexible touch panels using different kinds of sensors?

References

1 Bhowmik, A.K. (ed.) (2015). *Interactive Displays: Natural Human-Interface Technologies*. Wiley.
2 Westinghouse Electric, *Interface Device and Display System*. US Patent 3,522,664 (1970).
3 Johnson, E.A. (1965). Touch display – a novel input/output device for computers. *Electronics Letters* 1: 219.
4 Wang, W.C., Chang, T.Y., Su, K.C. et al. (2010). The structure and driving method of multi-touch resistive touch panel. *SID 10 Digest* 41 (1): 541–543.
5 Shim, H., Kim, S., Chun, Y. et al. (2011). Mutual capacitance touch screen integrated into thin film encapsulated active-matrix OLED. *SID 11 Digest* 42 (1): 6211–6624.

6 Chiang, C.H., Wu, Y.E., Ho, K.T., and Chan, P.Y. (2016). Mutual-capacitance in-cell touch panel. *SID 16 Digest* 47 (1): 498–501.
7 Chiang, W.J., Kung, C.P., Chen, S.W. et al. (2012). Flexible in-cell infrared a-Si sensor. *SID 12 Digest* 43 (1): 338–341.
8 Brown, C., Montgomery, D., Castagner, J.L. et al. (2010). A system LCD with integrated 3-dimensional input device. *SID 10 Digest* 41 (1): 453–456.
9 Reynolds, K., Shepelev, P., and Graf, A. (2016). Touch and display integration with force. *SID 16 Digest*: 617.
10 J. Y. Han, "Low Cost Multi-Touch Sensing through Frustrated Total Internal Reflection." *Symposium on User Interface Software and Technology: Proceedings of the 18th annual ACM symposium on User interface software and technology. Seattle, WA, USA,* 115 (2005).
11 Shum, H.P.H., Ho, E.S.L., Jiang, Y., and Takagi, S. (2013). Real-time posture reconstruction for Microsoft Kinect. *IEEE Trans. Cybern.* 43: 1357.

Index

Introduction to Flat Panel Display, Second Edition. Jiun-Haw Lee, I-Chun Cheng, Hong Hua, and Shin-Tson Wu.

Printed and bound by CPI Group (UK) Ltd, Croydon, CR0 4YY
16/04/2025
14658554-0009